VETERINARY ETHICS

Navigating Tough Cases

VETERINARY ETHICS

Navigating Tough Cases

Siobhan Mullan and Anne Fawcett

5m Publishing

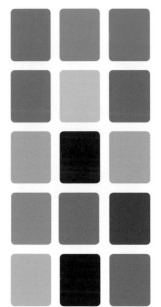

First published 2017
Reprinted 2021

Published by
5M Publishing Ltd,
Benchmark House,
8 Smithy Wood Drive,
Sheffield, S35 1QN, UK
Tel: +44 (0) 1234 81 81 80
www.5mpublishing.com

A Catalogue record for this book is available from the British Library

ISBN 978-1-910455-68-5

Book design and layout by Alex Lazarou

Printed by Replika Press Pvt. Ltd, India

|||||||||||||||||||||||||||||||||||||||||||||||||||||||||||||||||||||||||||||||||||||||||||||||

SIOBHAN

For Chris, Flora and Amber

ANNE

For Michael and Debbie

– the most patient, accepting and encouraging parents anyone could hope for. I hope this one's going straight to the pool room.

CONTENTS

||||||||||||||||||||||||||||||||||||

||||||||||||||||||||||||||||||||||||

WHAT IS VETERINARY ETHICS AND WHY DOES IT MATTER? /1

MAKING ETHICAL DECISIONS /37

IIIIIIIIIIIIIIIIIIIIIIIIIIIIIIIIIIIIII
CONTENTS
continued

CONTRIBUTORS' BIOGRAPHIES

ARMITAGE-CHAN, Elizabeth
Vet MB DipACVAA FHEA

Elizabeth Armitage-Chan is a Lecturer in Veterinary Education, and Departmental Teaching Director, at the Royal Veterinary College, and is currently enrolled in a PhD looking at professional identity formation in veterinary graduates. Liz leads the Professional Studies teaching at the RVC, and is particularly interested in teaching strategies that support the development of an ethical and autonomous veterinary professional.

ASHALL, Vanessa
BVSc MA CertWEL DipECAWBM(AWSEL)MRCVS

Vanessa Ashall worked as a veterinary surgeon in mixed and then companion animal practice before discovering ethics through her role as Named Veterinary Surgeon at an animal research facility. She then joined the newly formed Pet Blood Bank UK as welfare and ethics supervisor and was awarded the RCVS certificate in Welfare Science, Ethics and Law. Vanessa now holds a Master's Degree in Medical Ethics and Law and is Wellcome Trust Senior Research Fellow in ethics and society at the University of Nottingham, studying the social and ethical implications of extending the medical concept of donation to animals. She is a European Specialist in Animal Welfare Science, Ethics and Law.

BAGULEY, John
BVSc MBA PhD GradCert (HigherEd) MANZCVS

John Baguley is the Registrar of the Veterinary Practitioners Board of NSW. He graduated from the University of Sydney and worked mostly in small animal practice in Sydney for just over 10 years. John joined the University of Sydney Faculty of Veterinary Science in 2003. He coordinated their Professional Practice Program and taught in a variety of areas including practice management, ethics and law. He developed the final year mentor program, was academic coordinator of the final year Veterinary Student Intern Program and Honours Program, and served as Sub Dean for Learning and Teaching. John's research interests in veterinary practice management and education continue despite his move to the Veterinary Practitioners Board in 2012.

BULLER, Henry
BA(Hons) PhD

Henry Buller is professor of (non)-Human Geography at the University of Exeter. A social scientist with a longstanding interest in farm animals (including farmed fish) and farm animal welfare, he has many journal articles and book chapters on human/animal relations and a forthcoming book on farm animals and the human food chain. He is a member of the Farm Animal Welfare Committee and chair of the Defra/FAWC Welfare at Killing Committee and in Autumn 2016 was Visiting Professor at the Swedish Centre of Excellence in Animal Welfare Science in Uppsala.

CAMPBELL, Madeleine
BVetMed (Hons) MA (Oxon) MA (Keele) PhD DipECAR DipECAWBM(AWSEL) MRCVS

Madeleine Campbell is a RCVS, a European Recognised Specialist in Equine Reproduction, and a member of the Board of the European College of Animal Reproduction. In addition to her veterinary qualifications, Madeleine holds a MA in Medical Ethics and Madeleine recently completed a term as Chair of the BVA's Ethics and Welfare Group. Between 2011 and 2015, Madeleine was the Wellcome Trust Clinical Research Fellow in Veterinary Ethics at the Royal Veterinary College, researching bioethical decision-making in the use of assisted reproductive techniques in non-human mammals. Madeleine is the sole partner at Hobgoblins Equine Reproduction Centre, and the Director of Equine Ethics Consultancy.

CHAPURIN, Nikita
MHSc

Nikita Chapurin is a third year medical student at Duke University. He was born and spent his childhood in the small European country of Moldova until 2001, when his family moved to the United States. Interestingly, as a college student at Cornell University, taking a number of really engaging courses offered through the veterinary school strongly influenced his interest in medicine and surgery. He is currently completing his research year under the mentorship of Dr Walter Lee. Nikita spends his spare time playing sports, enjoying the outdoors, and traveling.

COE, Jason B
B DVM, PhD

Dr Jason Coe is an Associate Professor at the Ontario Veterinary College in Guelph, Ontario, Canada. After graduating from the College in 2001, he returned from mixed-animal practice to complete a PhD in veterinary communications in 2008. Jason is now a faculty member at the College where he has established an internationally recognised research programme in veterinary communications. Jason is also an active researcher in the areas of veterinary education and the relationships that exist between people and animals. In his current role at the College, he coordinates the clinical-communication curriculum across all four years of the veterinary programme.

COGHLAN, Simon
BVSc, PhD GradCert (HigherEd)

Simon Coghlan is a veterinarian, teacher and moral philosopher. After obtaining a veterinary degree from the University of Melbourne, he completed a Masters in bioethics at Monash University and a PhD in moral philosophy at the Australian Catholic University. Currently he does some lecturing in philosophy at the Australian Catholic University, and also works as a private veterinarian at the Epsom Road Veterinary Clinic in Melbourne.

CORR, Sandra

Sandra Corr is Professor of Small Animal Orthopaedic Surgery at the University of Glasgow. She is a European and RCVS Recognised Specialist in Small Animal Surgery, who spends her time teaching veterinary students and working in the University Small Animal Hospital. She has published widely in veterinary and comparative journals on surgical topics, and more recently, has developed an interest in veterinary ethics. She has co-authored a book on *Companion Animal Ethics* with Peter Sandøe and Clare Palmer (published by Wiley/Blackwell 2016).

COUSQUER, Glen

BSc(Hons) BVM&S CertZooMed PGDOE MSc (Outdoor Education) MSc (Education Research) MRCVS

Glen Cousquer is both a vet and an International Mountain Leader and is currently President of the British Association of International Mountain Leaders. He qualified from the Royal (Dick) School of Veterinary Studies in 1997. After ten years in veterinary practice pursuing special interests in avian and wildlife medicine, he returned to the University of Edinburgh to pursue a postgraduate diploma and MSc in Outdoor Education. A further MSc in educational research followed. Glen is now a researcher based in the Institute of Geography, where he has developed a research programme focusing on pack animal welfare in mountain tourism. Glen's immersion in experiential education and educational philosophy, coupled with his interests in professional ethics and professionalism, have helped shape his unconventional approach to animal ethics. His work challenges the terms, categories, boundaries and other constructs that have been, and continue to be, relied upon by humans who use and abuse the non-human.

DEGELING, Chris

BVSc PhD Grad Dip Sci MRCVS

Chris Degeling is a veterinarian, health social scientist and philosopher whose research interests include the social and cultural dimensions and ethics of human–animal interactions. He is a Research Fellow at the Centre for Values, Ethics and the Law in Medicine at the School of Public Health, University of Sydney.

DIXON, Greg

MA VetMB CertWel PhD MRCVS

Greg Dixon holds the RCVS certificate in Animal Welfare Science, Ethics and Law, and received a PhD from the University of Bristol for a thesis on the effects of dietary change on feather pecking in layer chickens. He is currently a practising veterinary surgeon in south Wales, where he remains interested in the issues of animal ethics.

FAWCETT, Anne

BA(Hons)BSc(Vet)(Hons)BVSc(Hons) MVetStud GradCertEduStud (HigherEd) MANZCVS (Animal Welfare) DipECAWBM(AWSEL)

Anne Fawcett completed an honours degree in philosophy, writing a thesis on the metaphysics and ethics of Spinoza, before studying veterinary science. In addition to working in companion animal practice, she lectures in veterinary ethics and professional practice at the University of Sydney and the University of Queensland. She has written over 30 peer-reviewed articles on veterinary ethics and aspects of primary veterinary care. Anne continues to work in, and enjoy the challenges of, companion animal practice. She has also worked as a journalist, writing for *The Sydney Morning Herald* and *The Veterinarian* magazine. She currently writes a column for News Local and blogs about animal welfare.

GARDINER, Andrew

BVM&S CertSAS MSc PhD DipECAWBM(AWSEL) MRCVS

Andrew Gardiner qualified from Edinburgh in 1992 and spent 14 years in general practice, gaining the RCVS certificate in Small Animal Surgery in 2000. He then received a Wellcome Trust Medical Humanities scholarship for a PhD in History of Medicine and researched the historical development of the British veterinary profession. He returned to the Royal (Dick) School of Veterinary Studies in 2008 and is currently Senior Lecturer. His research interests are in primary care, community animal health and welfare, and interdisciplinary animal studies. He is a European Veterinary Specialist in Animal Welfare Science, Ethics and Law.

GIUFFRE, Emmanuel

BA/LLB (Hons)

Emmanuel Giuffre is Legal Counsel of Voiceless, the Animal Protection Institute. Voiceless is a not-for-profit think tank, focused on raising awareness and alleviating the suffering of animals in factory farming and the commercial kangaroo industry in Australia. Emmanuel's expertise is in the legal frameworks governing factory farming and the commercial kangaroo industry in Australia, and advises government on the development and implementation of animal welfare policy and law reform in these areas. Emmanuel also focuses on community education and awareness building, and is a regular guest lecturer on the ethics of animal use and various aspects of animal law.

GODDARD, Pete
BVetMed PhD DipECSRHM DipECAWBM(AWSEL) MRCVS

Pete Goddard is a veterinary surgeon whose focus is on animal welfare. He has authored over 70 scientific papers and a number of book chapters on animal behaviour, physiology, health and welfare. He is a member of the Editorial Advisory Board of *Applied Animal Behaviour Science*. He is a former chair of the Animal Welfare Science, Ethics and Law Veterinary Association (AWSELVA) and was for six years a trustee of the Animal Welfare Foundation. A diplomat of the European Colleges of Small Ruminant Health Management and Animal Welfare and Behavioural Medicine, he currently chairs the Wild Animal Welfare Committee.

GOODFELLOW, Jed
BA/LLB (Hons) GDipLegalPrac PhD

Jed Goodfellow is an Australian animal welfare lawyer. He works as a Senior Policy Officer for RSPCA Australia with a focus on legislative and regulatory issues affecting animal welfare and teaches Animal Law at Macquarie Law School in Sydney. Jed recently completed a PhD in animal welfare law examining the democratic legitimacy of Australia's farm animal welfare regulatory framework. Prior to this, Jed practised as a prosecutor for RSPCA South Australia and worked as an inspector for RSPCA Queensland.

GREEN, Richard
BVSc CertWEL MRCVS

Richard Green qualified as a vet in 1989, then worked for a variety of practices, both in the UK and abroad, including the SPCA in Hong Kong. After returning to the UK, he spent some years working in television production, before taking on the role of Chief Veterinary Surgeon for the Blue Cross in Hammersmith, where he gained his RCVS certificate in Animal Welfare Science, Ethics and Law. He is currently a Claims Advisor for the Veterinary Defence Society, part-time clinician at the Blue Cross, and maintains a keen interest in veterinary ethics.

HANLON, Alison
BSc MSc PhD

Alison Hanlon is an Associate Professor at the School of Veterinary Medicine, University College Dublin, specialising in animal welfare science and ethics. Her research interests include working with stakeholders to improve animal welfare outcomes and to support the development of national policy. In 2014 she embarked on a collaborative study to explore the current and future ethical challenges facing the veterinary profession in Ireland. Dr Hanlon is a member of the Farm Animal Welfare Advisory Council and Scientific Advisory Committee on Animal Health and Welfare (Ireland) and an associate member of the European College for Animal Welfare and Behavioural Medicine.

HEPPLE, Sophia
BVSc PhD DWEL DipECAWBM(AWSEL) MRCVS

Sophia Hepple graduated from Bristol University in Veterinary Science (BVSc) in 1995. She is a Government Veterinary Adviser for animal welfare (farm, transport and killing on farm) supporting delivery functions of the Animal & Plant Health Agency, as well as providing expertise on animal welfare science, ethics and law to the UK Government's agency partners. Sophia is a recognised RCVS Specialist and European Diplomat in Animal Welfare Science, Ethics and Law, acts as an expert witness for court prosecutions and subsidy appeals panels and is part of the European TAIEX teaching group, responsible for knowledge transfer to pre-accession and EU countries.

HORSEMAN, Sue
BSc(Hons) MSc PhD

Sue Horseman completed her PhD within the Animal Welfare and Behaviour group at the University of Bristol. Her research looked at stakeholder perceptions of welfare challenges facing horses in England and Wales with an emphasis on human perceptions of these problems. Sue now lectures at Duchy College on their Equitation Science undergraduate and postgraduate courses and has a particular interest in training methods that support positive horse welfare.

JOHNSON, Jane
BSc Arch (Hons) Dip Arts (Hons) PhD

Jane Johnson is a philosopher who works in applied ethics. Her current research concerns the ethics of animal experimentation and involves developing new ways of construing animals in research to improve their situation as well as the quality of scientific research. She divides her time between the Centre for Values, Ethics and the Law in Medicine at the University of Sydney where she is a Research Fellow, and Macquarie University, where she is a Scholarly Teaching Fellow in the Department of Philosophy.

JOHNSON, Robert
BVSc MANZCVS (Feline) CertZooMed BA

Robert Johnson is a graduate of the University of Sydney (1977) and runs a small animal, reptile, zoo and wildlife practice with his veterinarian wife Jane in Sydney, NSW. He has extensive experience in clinical zoo and wildlife practice and has published and presented widely on reptile, wildlife and zoo animal topics. As president of the Australian Veterinary Association and past chair of the AVA Policy Advisory Council he is well briefed on current ethical and animal welfare issues. For relaxation Robert enjoys cartooning, listening to music, reading, learning languages and hanging out with his family.

KNIGHT, Andrew
BSc (Vet Biol), BVMS, CertAW, MANZCVS (Animal Welfare), DipECAWBM (AWSEL), DACAW, PhD, MRCVS, SFHEA

Andrew is a European, American and RCVS-recognised Veterinary Specialist in animal welfare, a Professor of Animal Welfare and Ethics, and Director of the Centre for Animal Welfare, at the University of Winchester. He has over 80 academic publications and a series of YouTube videos on animal issues. These include an extensive series examining the contributions to human healthcare, veterinary and other education, of invasive procedures on animals, which formed the basis for his 2010 PhD and his subsequent book, *The Costs and Benefits of Animal Experiments*.

LEE, Walter T
MD MHSc, FACS

Walter Lee is an Associate Professor of Otolaryngology-Head and Neck Surgery at Duke University Medical Center, Durham, North Carolina, USA. He is Co-Director of the Duke Cancer Institute Head and Neck Program as well as Staff Surgeon at the Durham VA Medical Center. He has an undergraduate degree in Philosophy with a concentration in ethics from George Washington University in Washington DC. His educational research focus is on developing leaders in medicine through a virtue-based approach to medical education and professionalism.

MAGALHÃES-SANT'ANA, Manuel
DVM MSc PhD DipECAWBM(AWSEL)

Manuel Magalhães-Sant'Ana is a European Veterinary Specialist in Animal Welfare Science, Ethics and Law. From the University of Porto, Portugal, he received his DVM in 2001, his MSc in bioethics and his PhD in veterinary ethics education. He was part of the FVE/EAEVE working group to develop a core curriculum in animal welfare in European veterinary education (2012–13), and was Veterinary Council Educational Trust Newman Fellow in Veterinary Ethics at the School of Veterinary Medicine, University College Dublin, Ireland (2014–16). He is currently Vice-President of the Disciplinary Committee of the Portuguese Veterinary Order and teaches at Escola Universitária Vasco da Gama, Coimbra, Portugal. He is external expert for the World Animal Protection Global Animal Network.

MAIN, David
BVetMed PhD CertVR DWEL DipECAWBM(AWSEL) MRCVS

David Main is a veterinary surgeon and Professor of Animal Welfare at the University of Bristol Veterinary School. His research interests include welfare assessment, intervention strategies to improve welfare and animal welfare education. He has been a member of the Farm Animal Welfare Council, Soil Association Council and Farm Animal Welfare Forum. He has been involved in several animal welfare initiatives, including as coordinator of the AssureWel project, a collaboration with RSPCA and Soil Association, that aimed to embed welfare outcomes into certification schemes, and the

Healthy Feet project that aimed to reduce lameness in UK dairy cattle.

McCULLOCH, Steven P
BVSc BA PhD DipECAWBM(AWSEL) MRCVS

Steven qualified as a veterinary surgeon in 2002 from Bristol University and holds a BA in Philosophy from Birkbeck College, London University. He has a PhD from the RVC, London for his thesis "The British animal health and welfare policy process: accounting for the interests of sentient species". Steven is a Diplomat of ECAWBM and a recognised specialist in Animal Welfare Science, Ethics and Law. Steven is Treasurer and sub-editor (ethics) for the *Journal of the AWSELVA*. He was the AWSELVA representative organising the ECAWBM/AWSELVA/ESVCE/PsiAnimal Congress in Portugal 2016. Steven coordinates the monthly "Everyday Ethics" column in the professional veterinary journal *In Practice*.

MELLOR, David J
BSc (Hons), PhD, HonAssocRCVS, ONZM

David Mellor's academic interests include foetal and neonatal physiology, stress physiology, pain assessment and management, livestock slaughter, conceptual frameworks in animal welfare science, and bioethics applied to animal welfare. He has ~528 publications, 300 being significant works of scholarship, including 6 books. His publications in foetal and neonatal physiology span 46 years and in animal welfare science and bioethics 28 years. Currently he is Professor of Animal Welfare Science, Professor of Applied Physiology and Bioethics, and Foundation Director of the Animal Welfare Science and Bioethics Centre at Massey University in New Zealand where he has occupied various professorial positions since 1988.

MORTON, David
CBE BVSc PhD FSB Cert Biol DipECLAM(ret) DipECAWBM(AWSEL) MRCVS

David Morton is Professor Emeritus of Biomedical Science and Ethics at the University of Birmingham, UK. He is a veterinary surgeon whose research interests concern the recognition, assessment and alleviation of pain and distress in animals. He is a founding member of the Animal Welfare Science, Ethics and Law Veterinary Association, and a lecturer in human healthcare and veterinary ethics. He has contributed to several books on veterinary ethical issues, particularly those issues that analyse human–animal interactions. He is a scientific and ethics reviewer for EU grant applications for human and animal research, and the author/co-author of over 250 scientific papers.

MULLAN, Siobhan
BVMS PhD DWEL DipECAWBM(AWSEL) MRCVS

Siobhan Mullan developed her interest in veterinary ethics as an undergraduate and continued to question how we should treat animals, and people, through her career in practice and now as a researcher. Her research interests at the University of Bristol are focused on welfare assessment and routes to welfare improvement for a range of farmed and companion animals. She teaches ethical decision-making to veterinary undergraduates. She initiated and coordinated the "Everyday Ethics" column in the *In Practice* veterinary journal for 10 years. She is a founder member of the European College for veterinary specialists in Animal Welfare Science, Ethics and Law and an RCVS Recognised Specialist in Animal Welfare Science, Ethics and Law.

PHILLIPS, Clive
BSc MA PhD

Clive Phillips obtained a PhD in dairy cow nutrition and behaviour from the University of Glasgow. He lectured in and researched livestock production and welfare at the Universities of Cambridge and Wales. In 2003 he joined the University of Queensland as the inaugural Chair in Animal Welfare, where he established the Centre for Animal Welfare and Ethics. Since that time he has been largely involved in research and policy development in animal welfare. He edits a journal in the field, *Animals*, and a series of books on animal welfare for Springer. Recent books include *The Animal Trade*, published by CABI in 2015, and the *Welfare of Animals, the Silent Majority*, published by Springer in 2009.

PRESTMO, Pia
DVM, MRCVS

Pia is a resident in veterinary public health at the University of Bristol School of Veterinary Sciences. She has previously worked as an Official Veterinarian in slaughterhouses in the UK and in Norway. She is currently involved in research focusing on the use of slaughterhouse inspection data to improve animal health and welfare on farm, in addition to teaching students in the university abattoir.

REISS, Michael J
MA PhD MBA FRSB FRSA FAcSS

Michael Reiss is Professor of Science Education at UCL Institute of Education, University College London, Visiting Professor at the Universities of Leeds and York and the Royal Veterinary College, Docent at the University of Helsinki, a Fellow of the Academy of Social Sciences and a Priest in the Church of England. He was a member of the Farm Animal Welfare Council/Committee (2004–12), Director of Education at the Royal Society (2006–08), a member of the GM Science Review Panel (2002–04), Specialist Advisor to the House of Lords Select Committee on Animals in Scientific Procedures (2001–02) and Chair of EuropaBio's External Advisory Group on Ethics (2000–01). For further information see www.reiss.tc

SANDØE, Peter

Peter Sandøe has been Professor of Bioethics at the University of Copenhagen since 1997. He was educated in philosophy at the University of Copenhagen and the University of Oxford. Since 1990 the major part of his research has been within bioethics with particular emphasis on ethical issues related to animals, biotechnology and food production. He is committed to interdisciplinary work combining perspectives from natural science, social sciences and philosophy. His books include *Ethics of Animal Use* (co-authored with Stine B. Christiansen), published by Blackwell 2008, and *Companion Animal Ethics* (co-authored with Sandra Corr and Clare Palmer), published by Wiley/Blackwell 2016.

TAMMEN, Imke
Dr. med. vet., MEd (Higher Education), MBEth

Imke Tammen completed undergraduate training in veterinary medicine at the Hanover School of Veterinary Medicine in Germany. Her postgraduate research doctorate degree (Dr. med. vet) and a postdoctoral fellowship at the same institution focused on molecular genetics of inherited disease in livestock. She joined the University of Sydney in 1997, initially in a research position and since 2002 continuously in a teaching and research role. Imke completed a Master in Education (Higher Education) in 2006 and a Master in Bioethics in 2011. Imke has research and teaching interests in veterinary science, animal genetics and bioethics, and aims to embed generic skills and leadership training in her disciplinary teaching.

TURNER, Patricia V
MS DVM DVSc DACLAM DABT DipECAWBM(AWSEL)

Patricia Turner is a Professor and Program Leader, Laboratory Animal Science in the Department of Pathobiology at the Ontario Veterinary College, University of Guelph, Canada. She has animal welfare and infectious disease research interests, and teaches comparative medicine and pathology, animal welfare and toxicology to undergraduate, veterinary and graduate students. Following graduation from veterinary school, Pat worked in mixed, mostly food animal practice in rural Ontario, before returning to the OVC for post-graduate specialisation. She is past Chair of the Canadian Veterinary Medical Association's Animal Welfare Committee and she continues to work with various Canadian farm animal producers and marketing groups to enhance on-farm animal welfare practices.

VAN KLINK, Ed
DVM, PhD, Dipl. ECVPH, MRCVS

Ed van Klink is Senior Lecturer Veterinary Public Health at the University of Bristol School of Veterinary Sciences. He has had a long career in animal health, public health and animal welfare policy. He has worked as a government vet in Zambia and has carried out a number of consultancy missions on animal health, public health and animal welfare issues around the world, including in eastern European countries, South America and the Far East.

VERRINDER, Joy
BA DipT MBA MA PhD

Joy Verrinder worked on a doctoral thesis in animal ethics education particularly for veterinarians and other animal-related professions, through the Centre for Animal Welfare and Ethics at the University of Queensland, Australia from 2012 to 2016. Prior to this, she was Strategic Development Officer for the Animal Welfare League of Queensland, working with governments, veterinarians and other stakeholders at local, state and national levels to develop and implement legislation and other strategies to prevent the abandonment and killing of cats and dogs in pounds and shelters. She has also worked in education as a teacher and Deputy Principal.

WEBSTER, John
MA VetMB PhD MRCVS

John Webster graduated from Cambridge as a veterinary surgeon in 1963, half inclined to be a cow vet and half inclined to work for the good of all farm animals at an international level. After a varied career in Scotland and Canada he was appointed Professor of Animal Husbandry at Bristol University in 1977 where he established the unit for the study of animal behaviour and welfare, which is now over 60 strong. He was a founder member of the Farm Animal Welfare Council and first propounded the "Five Freedoms" as standards for defining the elements of good welfare in domestic animals. He is now Professor Emeritus. He has written several books. If the word Eden is in the title, they deal with animal welfare.

WHITING, Martin
BSc BVetMed MA PG Cert VetEd PhD
DipECAWBM(AWSEL) MAcadMEd MRCVS FRSA FHEA

Martin Whiting is the Lecturer in Veterinary Ethics and Law at the Royal Veterinary College, London. He qualified from the RVC as a veterinarian and has a degree in Philosophy and a masters in Medical Ethics and Law from King's College London. Martin is a Diplomat of the European College of Animal Welfare and Behavioural Medicine. His PhD is on the public interest in the veterinary profession. Martin is a chair of the Social Science Ethical Review Board at the RVC and member of the Zoological Society of London Ethics Committee. He is a member of the Disciplinary Committee of the Royal College of Veterinary Surgeons.

YEATES, James
BVSC BSc(Hons) DWEL DipECAWBM(AWSEL) PhD MRCVS

James graduated from Bristol University in 2004 as a veterinarian, where he also completed a bioethics degree, certificate, diploma and PhD. He worked in Gloucestershire in private practice, then at an RSPCA branch. He became head of the RSPCA's companion animals department in 2011 and their Chief Veterinary Officer in 2012. He is based at the RSPCA's South Support Centre in Sussex. He is also the author of *Animal Welfare in Veterinary Practice*.

ZAKI, Sanaa
BVSc MACVSc GradCertEdStud (Higher Ed)

Sanaa Zaki is Senior Lecturer in Professional Practice and Sub Dean Post Graduate Studies (Coursework) in the Faculty of Veterinary Science, at the University of Sydney. Her teaching interests include animal welfare, ethical reasoning and resilience and self-care. Sanaa is also a Veterinary Anaesthetist, combining teaching, clinical service and clinical research activities in the area of anaesthesia and pain management. Sanaa completed a Graduate Certificate in Educational Studies (Higher Education) in 2005, and was appointed Director of the University Veterinary Teaching Hospital-Sydney in 2009. Sanaa is currently completing a PhD investigating the mechanisms of osteoarthritis pain. She is a staunch supporter of the veterinary profession and a regular invited speaker at seminars, workshops and conferences.

CONTRIBUTORS
TO CHAPTER SCENARIOS

Chapter 3 ANIMAL DEATH
Expert contributors: Simon Coghlan; Greg Dixon and David Main; Ed van Klink and Pia Prestmo; Andrew Gardiner; Patricia V Turner; Steven P McCulloch; Richard Green.

Chapter 4 ANIMAL USE
Expert contributors: Emmanuel Giuffre; John Webster; David Mellor; Steven P McCulloch; Simon Coghlan; Richard Green; Glen Cousquer; David Morton; Clive Phillips and Joy Verrinder.

Chapter 5 VETERINARY TREATMENT
Expert contributors: Andrew Gardiner; James Yeates; Chris Degeling; David Morton; Vanessa Ashall; Peter Sandøe and Sandra Corr; Imke Tammen; Martin Whiting; Patricia V Turner.

Chapter 6 MONEY
Expert contributors: Andrew Knight; Peter Sandøe and Sandra Corr; John Baguley; James Yeates; Richard Green.

Chapter 7 PROFESSIONALISM
Expert contributors: Jason B Coe; Glen Cousquer; Jane Johnson; Andrew Knight; Patricia V Turner; Martin Whiting and Elizabeth Armitage-Chan; Pete Goddard; Peter Sandøe and Sandra Corr; Chris Degeling; Steven P McCulloch.

Chapter 8 ERRORS AND COMPLICATIONS
Expert contributors: Manuel Magalhães-Sant'Ana; Alison Hanlon; Sanaa Zaki; Walter Lee and Nikita Chapurin; Martin Whiting.

Chapter 9 CONSENT
Expert contributors: Andrew Gardiner; Andrew
Knight; Jane Johnson; Steven P McCulloch;
Jane Johnson.

Chapter 10 EDUCATION AND TRAINING
Expert contributors: Manuel Magalhães-Sant'Ana;
Glen Cousquer; Walter Lee and Nikita Chapurin;
Martin Whiting; Andrew Gardiner; John Baguley.

Chapter 11 TEAM RELATIONS
Expert contributors: Jason B Coe; Alison Hanlon;
John Baguley; Sanaa Zaki.

Chapter 12 WORKING WITH THE LAW
Expert contributors: Pete Goddard; Ed van Klink and Pia
Prestmo; Sophia Hepple; Patricia V Turner; Jed Good-
fellow; Martin Whiting; Manuel Magalhães-Sant'Ana.

Chapter 13 ONE HEALTH
Expert contributors: Alison Hanlon; Henry Buller;
Andrew Knight; Manuel Magalhães-Sant'Ana;
Sanaa Zaki.

Chapter 14 WILDLIFE
Expert contributors: Pete Goddard; Steven P
McCulloch and Michael J Reiss; Richard Green;
James Yeates.

Chapter 15 CHANGING AND CLONING ANIMALS
Expert contributors: David Morton; Imke Tammen;
Madeleine Campbell; John Webster; Sue Horseman.

FOREWORD

The central importance of ethics to veterinary surgeons and veterinary nurses is apparent to anyone who considers the role and future of the veterinary profession. During my year as President of the British Veterinary Association (BVA), at least two activities have made this clear.

Firstly, we developed, through consultation, a BVA animal welfare strategy: "Vets speaking up for animal welfare". The strategy notes that, as in society at large, there are differences between veterinary professionals in their interpretation and attitudes towards animal welfare. Such differences can play out in veterinary practice – where different team members may draw different value-based conclusions on ethical dilemmas – and in veterinary policy – where different members, with different perspectives, can make different and opposing policy recommendations. In seeking to support members in their professional lives and to develop national policy based on transparent and democratic debate, "ethics" – including developing member guidance and applying ethical frameworks to policy formulation – has been identified as one of the strategy's priority areas.

Secondly, BVA has partnered with our UK regulator, the Royal College of Veterinary Surgeons, on the *Vet Futures* project, seeking to prepare for, and shape, the profession's future. Following consultation, "leadership in animal health and welfare" emerged as one of the report's key ambitions, with a specific recommendation to "enhance moral reasoning and ethical decision-making in education, policy-making, practice-based research and everyday veterinary work". It is clear that vets and vet nurses view leadership in animal health and welfare as central to their professional identity, but that they want support to help navigate complex ethical issues. Further, reducing the significant moral stress that can accompany ethical dilemmas will contribute to another of the *Vet Futures* ambitions, to improve veterinary wellbeing.

Veterinary ethics is in its relative infancy compared to other medical professions and leadership is essential. Two of those leaders are Siobhan Mullan and Anne Fawcett. Both hold advanced postgraduate qualifications and teaching responsibilities in Animal Welfare Science, Ethics and Law (AWSEL) and have been instrumental in advancing and promoting this area of veterinary specialism. In this book, they have gathered the talents of many other leaders in the field, including several who have challenged and informed my own views during my career.

During the animal welfare strategy consultation, I asked a recently graduated doctor about her undergraduate ethics training. She referenced ethical frameworks in her comprehensive answer and spoke of their relevance in a recent case where a 15-year-old girl was refusing treatment, against her distressed mother's wishes, following a deliberate paracetamol overdose. The anecdote brought into sharp focus the imperative for medical training in ethics and law, where human lives and interests are at stake.

It is worth reflecting on why lesser importance may have historically been attached to veterinary ethics, perhaps linked to the moral status afforded to animals. Regardless, as animal welfare becomes a prominent social ethic, and the status of professions changes, the veterinary profession's ethical reasoning abilities are assuming centre stage. As BVA President I have been challenged in the national UK media on both perceived overtreatment of animals by vets, and whether high spending on pets can be justified alongside other worthy causes. Meanwhile, vets everywhere are being challenged on their fees, on how to manage client interactions on social media, on the extent they should influence clients to improve animal welfare, on how to charge for repeat surgery following unavoidable complications, and much more. All of these issues are covered in this book.

In fact, Siobhan and Anne have quashed any suggestion that ethics is a distraction from the proper business of becoming or being a vet or vet nurse. The sheer volume and variety of everyday scenarios gathered and discussed here serve to remind us all that veterinary professionals are dealing with ethical dilemmas throughout their professional lives; so much so, that many of us cease to realise the considered judgements we are constantly having to make. The cases made me proud that our profession takes so many thorny dilemmas in its stride. Yet, they *are* stressful and "because I'm a professional" is no longer acceptable as a justification for our actions. Vets and vet nurses have clearly indicated that they want support and this book, with its tools and compendium of everyday scenarios, will cement the relevance of ethics to the profession. Importantly, it will help ensure that individual veterinary professionals go home at night not worrying about whether they made the "right" decisions, but content that they made the most justifiable and animal welfare-focused decisions possible.

Sean Wensley
BVA President
2015–2016

ACKNOWLEDGEMENTS

Writing a book is a team effort. The authors and formal contributors get the kudos, but colleagues, friends and family provide endless advice, proof-reading, real-life ethical scenarios, counterexamples, links to relevant journal articles and robust debate. Those in the inner circle provide nutritional support, and bear the burden of living with an increasingly preoccupied author as the manuscript develops. They are the wind beneath our wings.

The team at 5M, in particular Sarah Hulbert and Denise Power, were very easy to work with and respected our evolving approach to this project.

The following people reviewed the manuscript or sections of it, and made helpful suggestions: Dr Sean Wensley, Dr Bruce Kaplan, Dr Elaine Cheong, Christopher Croese, Dr Julie Strous, Dr Andrea Harvey, Dr Louise Rabbitt, Dr Bob Doneley, Dr Steve McCulloch, Kristina Vesk OAM, Joy Verrinder and Dr Jo Hockenhull. We are particularly grateful to Dr Wensley for taking the time out of his very busy schedule to write the foreword.

Anne's acknowledgements

Dr Robert Johnson not only provided several of the wonderful cartoons in this book, he is a generous mentor. He and his wife Dr Jane Roffey maintain one of the friendliest, calmest veterinary practices and their compassion for and gentle manner with the less familiar species is something we should all aspire to.

Many colleagues and friends provided thoughtful and challenging discussions, moral support and revelations about ethical dilemmas encountered in a range of contexts. These include Dr Stephen Cutter, Associate Professor Tom Gottlieb, Dr Emma Jane, Dr Nicole Vincent, Professor Graeme Allan, Professor James Serpell, Professor Ben

Mepham, Dr Stephen Page, Dr Robert Stabler, Dr Jane Johnson, Dr Andrea Harvey, Dr Kim Frost, Dr John Culvenor, Professor Richard Malik, Professor Marc Bekoff, Professor David Gunkel and Professor Mark Coeckelbergh.

Colleagues at the University of Sydney supported me in allowing me to undertake research, explore new ideas in teaching and discuss their own views and values. In particular, Drs Jacqui Norris, Susan Matthew, Sanaa Zaki, Glenn Shea, Jeni Hood and Professor Paul McGreevy acted as sounding boards and provided encouragement.

Drs John Baguley and Christine Hawke encouraged me to teach veterinary ethics, and workshop many of the ideas that sections of this book are based on. Dr Hawke, for me, has been an exceptional role model and taught me as much about compassionate, ethical practice as she has about veterinary dentistry.

I work with a brilliant team of veterinarians and nurses – Angela Phillips, May Chin Oh, Esther Tarszisz, Felicity Spicer, Belinda Dowcra, Jenna Moss Davis, Aileen Devine, Asti May, Vivian Lang, Kim Howatt, Annie Shen, Katia Oliveira, Maddie Giuliano, Amanda Duran and Elizabeth Williams – who throughout the writing of this book acted as a "brains trust" – reminding me of hairy ethical situations, suggesting scenarios, reading raw chapters and going on long dog-walks to provide a much-needed break. As the project progressed, Dr Tarszisz and I conducted remote "shut up and write sessions" online while she was stationed in Antarctica. These provided a necessary haven from the white noise and information overload of modern existence.

My family, human and non-human, didn't complain when I isolated myself in the study, bath or library. To Mick, Deb, Bosca, Betty, Chris, Kat and Genevieve, I could not have wished for a more supportive, loving family. Draga Dubaich took over the running of the household and became part of that family. My husband Jamie has been incredibly supportive, allowing me the freedom to explore ideas, accepting my often antisocial working hours and making endless cups of tea.

Gloria Morales, Michael Neil and Shannon Dooley reminded me that ordinary people can do seemingly impossible things, and that being means doing.

My days were topped, tailed and punctuated spending time with animals that I consider family members. Michael and Hero (cats) sat on my keyboard and occasionally tore up papers on my desk. At one point Hero alerted me to the fact that he had calcium oxalate uroliths by voiding on a landmark paper about evidence-based medicine. Phil (dog) slept under the desk and reminded me to go for walks. Cornflake, Osler, and the late Randy and Cushing (cavies) enriched our lives and (slowly) mowed the grass.

Finally I would like to thank my patients, clients and students from whom I continue to learn.

Siobhan's acknowledgements

Thanks to my family, friends and colleagues for the many stimulating conversations and experiences that help me to think about how we should treat animals. I feel enriched by them.

HOW TO USE THIS BOOK

||||||||||||||||||||||||||||||||||||||||||||||||||||||||||||||||||||||||||||||||

"Hard thinking is humbling. Probably no one who has attempted to clear a path through thickets of difficult ideas has emerged brimming with confidence that every turn was the right one, made for the right reasons."
Tom Regan, 1981

The aim of this book is not to provide an answer to every ethical dilemma one might confront in practice. We don't believe that is possible nor desirable. Rather, we want to provide tools for veterinarians, technicians, nurses and other staff to use to explore and justify ethical propositions, increase their ethical sensitivity or awareness and provide a basis upon which sound ethical decisions can be made.

The first part of the book (chapters 1–2) provides the toolkit, the second (chapters 3–15) demonstrates its use, contextualises some of the ethical issues that veterinarians, nurses and technicians face, and provides a range of examples of ethical reasoning. We have chosen scenarios that represent fundamental issues of the moral status of animals, common veterinary dilemmas and some less common situations. Such are the plethora of animal ethical issues that we had many more that did not make the final cut to be sent to our contributors. We deliberately present a range of voices in the responses, with different ethical orientations, allowing an opportunity to test the robustness of one's own views through challenge by other perspectives.

We hope that this book facilitates open discussion of ethical issues, within and outside of the veterinary practice and academic settings. As David Main notes:

"Open discussion among colleagues and clients is a healthy activity that increases the transparency of the veterinary profession and can reduce conflict arising from ethical dilemmas in veterinary practice."

(Main 2006)

Importantly, veterinarians and allied professionals work with animals daily and while their decision-making is critical it is their actions that are important. As John Webster said: "Ethics is terribly important, but it matters not to the animal what we think, but what we do" (personal communication, 2015).

Dr Anne Fawcett and Dr Siobhan Mullan
anne.fawcett@sydney.edu.au
Siobhan.Mullan@bristol.ac.uk
|||||||||||||||||||||||||||||||||||||||||||||||||||||||||

WHAT IS VETERINARY ETHICS AND WHY DOES IT MATTER?

Introduction

This chapter provides a brief overview of what ethics is – just what do we mean when we talk about ethics and morality? What makes an ethical decision different from any other and why should veterinarians and associated professionals be concerned with ethics? We will explore why many of us find ethics challenging, and the place for teaching and developing ethical reasoning skills.

When it comes to animal health professions, ethics and animal welfare are inextricably linked, so we have provided a brief discussion of key concepts in animal welfare and how our values impact welfare assessment. We discuss the contentious issue of which animals are worthy of ethical consideration.

In this chapter we also look at common ethical dilemmas and sources of ethical conflict in veterinary practice, as well as the impact of those on veterinarians and associated professionals. Finally, we briefly explore policy, based on ethical reasoning, as an aid to decision-making in veterinary practice.

◁

1.1 Cartoon

REPRODUCED WITH PERMISSION OF MATTHEW BOYD AND IAN MCCONVILLE

1.1
What is ethics?

Ethics is a branch of philosophy. Philosophy, which loosely translated means love of wisdom, is the study of general concepts such as principles of reasoning, the nature of knowledge and truth, reality, perception, and so forth. It asks the big questions, such as "Do I really exist?"

Unlike other areas of philosophy (for example, metaphysics), ethics generally presupposes reality – and in fact in the broadest sense considers the question "What should I do?" It assumes that there are, if not right and wrong answers to that question, better and worse answers. The word "ethics" is an umbrella term for beliefs, principles and rules determining what is right and wrong.

Let's say that you are granted the power of invisibility. The ethics of invisibility concerns itself with how you use that power – whether to use that power for your personal gain by walking into a bank and stealing money in broad daylight, or for achieving a greater good, such as helping to expose the actions of those who perpetrate injustice.

Similarly, after years of study, veterinary, nursing and animal science students develop special skills and knowledge, as well as entitlement to registration ("powers") that come with these roles – for example, the ability to diagnose and treat health problems in animals, to perform acts of veterinary science. It is possible to use these powers to help or harm others. Veterinary ethics concerns itself with how you use these powers.

Of course, poor ethical decisions may not simply result from deliberate abuse of power, but may also arise out of ignorance and laziness. One of the aims of ethics teaching is to generate awareness about decision-making so we don't fall into these traps.

1.1.1 What do we mean when we use the word "ethics"?

Bernard Rollin, who taught the first veterinary ethics course at Colorado State University in the 1970s, makes a helpful distinction between Ethics$_1$ and Ethics$_2$ (Rollin 2006). Ethics$_1$ is our set of beliefs – what is right, what is wrong, what is just, what is unjust, good and bad and so on. Rollin argues that these beliefs are acquired from multiple sources: parents, school friends, teachers, authority figures and the mass media. Ethics$_1$ comprises our personal, social and even professional ethics.

Ethics$_2$ is the systematic study of Ethics$_1$. It examines Ethics$_1$ propositions (for example, "it is wrong to kill animals for sport") and looks for consistency, contradictions and wider implications, as well as examining the way Ethics$_1$ propositions are justified.

According to Rollin, examples of non-debatable, consensus, socio-ethical principles include prohibitions against murder and other forms of violence (Rollin 2000). He argues that "personal ethics begin where social ethics are silent", for example on matters such as whether we give to charity, how many offspring we have, what we eat or whether we adhere to religious tradition (Rollin 2000). Social media is full of examples where there is apparent crossover.

Anthrozoologist and author Hal Herzog writes that ethics is similar to journalism. Just as journalists investigate who, what and why (as well as when and where) something happened, ethicists look at who, what and why questions: "who is entitled to moral concern, what obligations we have to them, and why one course of action is better than another" (Herzog 2010).

There are other terms that are used frequently in the ethics literature.

ETHICS$_1$	ETHICS$_2$
Right vs wrong	The study of Ethics$_1$
Good vs bad	Analysis of ethical propositions
Fair vs unfair	How are Ethics$_1$ propositions justified?
Propositions from variable sources	Are Ethics$_1$ propositions held by this person/group/ organisation consistently?
Contradictory or conflicting beliefs	
Personal ethics	How can we address conflict between Ethics$_1$ propositions?
Professional ethics	
Social ethics	

△

Table 1.1 Ethics$_1$ and Ethics$_2$

Normative ethics addresses the question "what should I do and why?" Ethical theories or frameworks attempt to generate and justify these norms or ethical propositions. These propositions are used to judge whether an action is right or wrong. An example of such a principle may be the widely accepted belief that it is wrong to harm another human being.

Descriptive ethics refers to the factual investigation of moral beliefs and conduct – the psychology, neurobiology, sociology and anthropology of beliefs. It describes moral and ethical reasoning and behaviour without judging or *prescribing* beliefs and conduct.

Metaethics is the study of ethical reasoning, moral knowledge and ethical "truth". It is concerned with questions such as "what is good?", "what are right and wrong?" and "how can we tell the difference between good and bad or right and wrong?"

△

1.2 It helps to become familiar with ethical terminology.

1.2

What makes a decision ethical?

For philosopher William Shaw, ethical or moral decisions differ from non-ethical decisions in three ways. Firstly, they are concerned with actions that can seriously impact the welfare and in some cases survival of others. Whether or not to wear a particular coloured dress is not an ethical decision in most cases. Deciding whether to euthanase or treat an animal is an ethical decision.

Secondly, because of their importance, moral standards take priority over other standards. For example, if a veterinary nurse believes strongly that it is wrong to kill a healthy companion animal,

most would argue that he or she should not participate in this just because it is a service a client has demanded.

Thirdly, the soundness of moral standards depends on the reasons used to justify them. Thus it is unsound to justify an ethical action (for example, treating a wildlife casualty) on the grounds that it was done "because I could" (Shaw 2010). We would expect a justification referring to broader ethical principles – for example, "I treated this animal because the suffering involved will be short-lived and is outweighed by the likely successful rehabilitation" (a utilitarian justification) or "I treated this animal because I was acting in accordance with my duty to respect the value of the life of all animals" (a deontological justification).

In the clinical setting, ethical reasoning has been proposed to consist of four key components:

• knowledge and understanding of ethical theory or frameworks;
• awareness of different stakeholders and their interests in a given scenario;

—
1.3

WHY SHOULD VETERINARIANS AND
ASSOCIATED PROFESSIONALS BE
CONCERNED WITH ETHICS?

- awareness of one's own morals as a practitioner;
- the ability to incorporate all of the above in the clinical reasoning process (Edwards & Delaney 2008).

Veterinary medical ethics, along with medical ethics, has been criticised previously for focusing too much on intraprofessional etiquette (Magalhães-Sant'Ana 2015, Rollin 2000). To date, many codes of conduct focus on issues such as advertising, referral, client relations and not denigrating colleagues or bringing the profession into disrepute. It is clear that veterinary ethics is much broader than that; however, there is a large amount of overlap between ethics and etiquette.

1.3

Why should veterinarians and associated professionals be concerned with ethics?

While we might all want to live a "good" life, doing the right thing, some scholars argue that the nature of the work of veterinary and associated professionals gives rise to a greater responsibility to develop ethical sensitivity and reasoning skills:

> "Veterinary professionals have the same general responsibilities to animals as other people but are **more accountable** because we have **more opportunities to cause greater harms** and **fewer excuses because of our greater knowledge.**"
>
> (Yeates 2013; emphasis added)

According to veterinarian and educator Liz Mossop, "decision-making is the cornerstone of the veterinarian's role, and an expectation of all healthcare professionals" (Mossop 2015).

Vets are in a powerful position. They are widely trusted, as evidenced by opinion polls (e.g. Royal College of Veterinary Surgeons 2015), and can influence both life-and-death decisions for individuals and the welfare of many. Part of that trust may stem from the meaning of the word "profession", encompassing both an occupation and a promise or a vow. The promise concerns clients as well as wider society. Thus Allister asks, "What is the promise that the veterinary profession makes to wider society? Do we achieve what we set out to? And how does that translate on a personal level, to vets making sense of and enacting in our working environments?" (Allister 2016).

As well as the capacity to cause harm veterinary professionals can, and should, use their influence to improve animal welfare and encourage more ethically acceptable practices for animals (British Veterinary Association 2016). Poor decision-making, on the other hand, can lead to a negative impact on the welfare of animals, unhappy clients, and undesirable effects on the wellbeing of the veterinarian (for example, reflected by lack of job satisfaction or stress) (Mossop 2015). For example, research has shown that veterinary practitioners are inconsistent in making decisions regarding patient care, and give preferential care to clients they assess positively. This has a direct impact on animal care and will lead to some animals receiving better care than others [Morgan 2009 unpublished cited in (Batchelor, et al. 2015)].

Veterinarians and associated professionals are expected to understand ethical concepts – at least enough to be able to make and justify their decisions. Indeed, awareness and understanding of ethical responsibilities is, in many countries, an expected Day One Competency for veterinary graduates. For example, in its "RCVS Day One

CHAPTER 1 VETERINARY ETHICS AND WHY IT MATTERS

—
1.3
WHY SHOULD VETERINARIANS AND
ASSOCIATED PROFESSIONALS BE
CONCERNED WITH ETHICS?

Competencies", the UK's Royal College of Veterinary Surgeons makes explicit reference to ethics, stating that the new veterinary graduate should be able to "Understand the **ethical** and legal responsibilities of the veterinary surgeon in relation to patients, clients, society and the environment"; and have "a breath of underpinning knowledge and understanding" about "the **ethical framework** within which veterinary surgeons should work, including important **ethical theories** that inform decision-making in professional and animal **welfare-related ethics**" (Royal College of Veterinary Surgeons 2014; emphasis added).

Other listed competencies assume the ability to make sound ethical decisions, for example, the new veterinary graduate should be able to undertake the following:

> "34: **Recognise when euthanasia is appropriate** and perform it humanely, using an appropriate method, whilst showing sensitivity to the feelings of owners and others, with due regard to the safety of those present; advise on disposal of the carcase."
>
> (Royal College of Veterinary Surgeons 2014;
> emphasis added)

It is likely that Day One Competencies pertaining to ethics and animal welfare will be expanded. In a joint report the Federation of Veterinarians of Europe (FVE) and the European Association of Establishments for Veterinary Education (EAEVE) found that animal welfare and ethics are inextricably linked, and called for more uniform, comprehensive teaching of animal welfare, ethics and the law across veterinary schools. The report states that "One cannot be a good clinician without being aware of the ethical issues in decision-making in practice" (Morton, et al. 2013).

The FVE/EAEVE working group developed an animal welfare curriculum, and recommended a number of ethics-related day-one learning outcomes for veterinary graduates, some directly pertaining to ethics while others require a solid foundation of ethical knowledge:

"Graduates should have the ability to
1. Appraise different concepts as well as analytical frameworks of animal welfare and how they relate to practice and the context in which they are set.
2. Apply sound principles to objectively evaluate the welfare status of animals and to recognise good and poor welfare.
3. Participate in animal welfare assessment, monitoring and auditing with the aims of improving the physical and mental health of animals.
4. Formulate an informed, science-based, view on animal welfare matters and communicate effectively with those involved in keeping animals.
5. Appraise the social context and participate in societal debates about animal welfare and ethics.
6. Retrieve up-to-date and reliable information regarding local, national and international animal welfare regulations/standards in order to describe humane methods for animal keeping, transport and killing (including slaughter)."

(Morton, et al. 2013)

Aside from regulatory and professional requirements there is a societal expectation that veterinarians and related professionals have an understanding of ethical issues and the ability to navigate ethical conflict.

One study found that veterinarians and nurses have a very strong sense of career identity, of which one's ethical and moral approach was a big part (Page-Jones & Abbey 2015). Ethical and moral mismatch between the individual and employer was a source of tension which the authors predicted would rise with increasing corporatisation of the industry.

The teaching of veterinary ethics is in its infancy when compared to subjects such as surgery and anatomy. Future questions for universities include whether prerequisites related to student ethical standards should be considered at admission; how ethical education and training are coordinated with the wider university, veterinary professional organisations and registration bodies; and how ethical reasoning can be integrated in other subjects.

1.4

Why is ethics challenging?

Ethical practice can be challenging for veterinarians for a number of reasons. Cultural, legal and economic factors may lead to animals being kept or treated in a manner that is not conducive to their welfare (for example, hens in battery cages) (Verrinder & Phillips 2014). In addition, animal care is frequently inconsistent, within and across species, for example, different husbandry standards for rabbits and rodents depending on their use by humans (Verrinder & Phillips 2014).

In a study of Australian veterinary students, the third most important motivator (after a desire to work with animals and the wish to help sick or injured animals) was the desire to improve the way animals are treated (Verrinder & Phillips 2014). The majority of students were concerned about animal ethics issues (Verrinder & Phillips 2014).

Translating ethical principles into action can be challenging. In a study of 258 clinical psychology students, only 37 per cent of those who identified

△

1.3 A desire to work with animals, the desire to help sick or injured animals and the desire to improve the way animals are treated were the most important motivators for veterinary students.
PHOTO ANNE FAWCETT

what they felt to be an appropriate response to an ethical dilemma said they would act on it (cited in Verrinder & Phillips 2014). In another study, 54 per cent of veterinary students who said they were concerned about ethical issues admitted to doing little or nothing to resolve these (Verrinder & Phillips 2014). While over 90 per cent of students believed that the veterinary profession should be involved in addressing animal ethics issues in the wider community, only one third agreed that it was "sufficiently involved" (Verrinder & Phillips 2014).

Why might this be so? Well, there is a vast difference between believing something and acting on it. Action requires conviction, effort, time, and often the belief that the benefits of acting will outweigh the costs.

In veterinary practice, there may be conflicts between the interests of animals and those of clients, industry or society as a whole. For example, in most contexts veterinarians charge owners for their services, a well-recognised potential source of conflict as seen in the following quotes:

"The problem faced by advocates of the model of veterinary medicine as a business is fitting their sincere care and concern into a model that also wants to view professional life in terms of selling and buying. This task is like mixing oil and water."

(Tannenbaum 1995)

"Every time we recommend a course of treatment we face a potential conflict of interest between benefit to the patient, the cost to the owner and the benefit of the practice from the financial profit generated."

(Viner 2010)

Indeed, this dual involvement in providing care and making a profit may be viewed with cynical distrust by clients and even some veterinarians. This situation may be further clouded by the veterinarian's relationship to their employer. For example, a veterinarian may be employed by a practice which has policies such as recommending particular products over others due to a commercial deal with a wholesaler. The employee may feel conflicted in situations where use of that product is not in the animal's or client's best interests.

What if we don't actively consider ethical issues? The risk is that we may make poor decisions – or even wrong decisions. Our default moral cognition is at the mercy of complex psychological mechanisms that we may not be aware of. Thus, we may be subconsciously geared toward avoiding punishment rather than actually doing good – an approach that doesn't always withstand analysis. As such, many people will avoid committing an act of commission (actively bringing something about), but may willingly perform an act of omission (neglecting to do something) even if this leads to the same or a worse outcome than the act of commission (DeScioli, et al. 2011).

For example, in India many people are unwilling to kill cows (an act of commission), but will allow cows to die by starvation (an act of omission).

△
1.4a–b Cows are revered in India but may also be allowed to die from starvation rather than be euthanased.
PHOTO ISTOCK

Similarly, euthanasia of human patients is illegal in many countries, while withdrawal of treatment is not. In a study involving a simulated economic game in which one person could take money from another person by omission or commission, participants were more likely to choose omission even when this destroyed welfare, and resulted in poorer outcomes for themselves and others (DeScioli, et al. 2011). Omission was a strategy, the authors concluded, designed to avoid the condemnation of others and subsequent punishment.

Guilt is an emotional state that occurs when one believes, rightly or wrongly, that one has violated a moral code (Fordyce 2011). In a meta-analysis, guilt – and the desire to avoid it – was one of the most influential emotions in decision-making, over and above the motivation to "do the right thing" (Angie, et al. 2011).

Ethical reasoning is important because our intuitions and feelings about what is the right or wrong thing to do can be unreliable. For example, most people would argue there is a moral distinction between a person committing a murder, and one failing to prevent a murder, even if both result in the same amount of suffering. Many laws reflect such a distinction, which raises the question: if we cannot appeal to our intuition as the basis of sound ethical judgement, what about appealing to the law?

Unfortunately the law reflects the predominant social values of the time, many of which are subsequently recognised as outdated and in some cases unethical. For example, in the not-so-distant past the law has prohibited women from voting, limited the rights (including voting rights) of indigenous peoples and even permitted slavery. The application of legal principles is not straightforward either, hence the need for legal experts and specialists and lengthy court proceedings. Similarly, appealing to religion is challenging when many religious teachings are ambiguous and open to different readings.

> "The difference of opinions on a range of issues within the veterinary profession demonstrates a need for ethical analysis."

The difference of opinions on a range of issues within the veterinary profession demonstrates a need for ethical analysis. Being able to articulate our concerns with reference to ethical frameworks at least gives us a common base to facilitate communication around ethical issues.

Whether our motivations matter depends on the approach you take to ethics. As we will discuss, motivation or intent may not matter if you judge an action as ethical based on the consequences, but it certainly matters according to non-consequentialist ethical models (discussed in chapter 2).

1.5
Why should we study veterinary ethics?

There are a variety of reasons we should learn about ethics but fundamentally we are aiming to make more "right" decisions in the variety of settings that vets find themselves in.

> "Vets are not just at the front line, they are also on an ethical highwire, constantly balancing their concern with animal welfare against the demands of the industries, clients and practices they work for, without necessarily having been given any training in how to do this."
> (Rawles 2000)

BOX 1.1 MOTIVATIONS FOR BEHAVING ETHICALLY

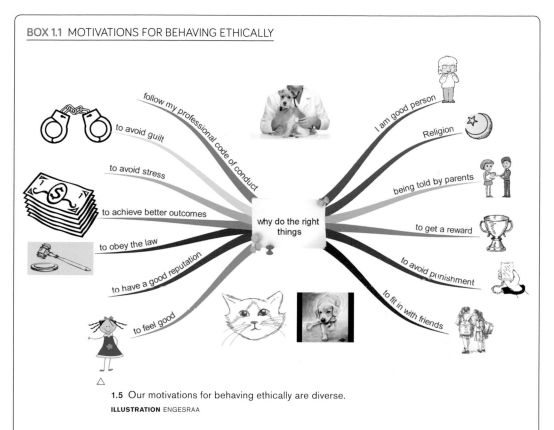

1.5 Our motivations for behaving ethically are diverse.
ILLUSTRATION ENGESRAA

Our motivations for behaving ethically are diverse. When asked to give a reason for behaving ethically, or engaging in ethical decision-making, a group of 120 Australian second-year veterinary students gave a range of responses:

- Because my religion dictates it
- Because my parents told me
- Because I am a role model
- Because it will increase the likeliness of me being rewarded
- Because it will decrease the likelihood of me being punished
- To fit in with my peers

- To get warm, fuzzy or positive feelings
- To maintain a good reputation
- Because it is the law
- Because it leads to better outcomes for others
- Because ethical behaviour is dictated by my professional code of conduct
- To avoid stress/follow the path of least resistance
- To increase the likelihood of the best outcome
- To avoid feeling guilty
- Because a good person would behave according to ethical principles and I am a good person (class survey, Fawcett 2012, unpublished).

△
1.6 Vets are on an ethical highwire.
CARTOON RAFAEL GALLARDO ARJONILLA

Many qualified veterinarians have had no formal ethics training and have found their own path through the ethical minefield in which they operate. Unfortunately, increased experience in practice does not reduce the stress of ethical dilemmas (Batchelor & McKeegan 2012). Furthermore, moral reasoning is not simply learned from repeated exposure to ethical dilemmas (Batchelor, et al. 2015). A preliminary investigation into the moral reasoning of UK veterinarians found that, despite having a professional degree, the moral reasoning skills of practising veterinarians were highly variable, and were often no better than those of members of the public (Batchelor, et al. 2015). The authors concluded that the moral reasoning skills of veterinarians may be insufficient to meet the demands of such an ethically challenging job. In this study veterinarians working in an academic setting fared better – perhaps because they are in a working environment that promotes critical thinking and discussion (Batchelor, et al. 2015).

Concerningly, a study found that the moral reasoning of veterinary students was not improved during veterinary school (Self, et al. 1996). In one study, 78 per cent of veterinarians reported that their veterinary degree did not provide them with adequate training to deal with ethical dilemmas (Batchelor & McKeegan 2012). In a survey of veterinary educators who taught ethics, one motivation was to equip students to deal with ethical tension to prevent them "dropping out after a few years in practice" (Magalhães-Sant'Ana, et al. 2014). This involves recognising ethical issues and developing ethical reasoning and decision-making skills (Figure 1.7). What we don't yet know is to what extent ethics teaching alters such outcomes.

Increasingly, veterinary students are being taught ethical theories and reasoning, not least to comply with the expected competencies of veterinary regulators. To determine the underlying reasons behind ethics teaching Manuel Magalhães-Sant'Ana and colleagues surveyed veterinary educators and found four major themes: raising ethical awareness, providing an ethical knowledge base, developing ethical skills and developing individual and professional qualities (Magalhães-Sant'Ana, et al. 2014).

These themes are associated with learning objectives as outlined in Figure 1.7.

Classroom discussions provide a relatively safe environment to discuss personal views and values (Magalhães-Sant'Ana, et al. 2014). There is, however, no accepted gold standard for veterinary ethics education and ethics curricula vary in terms of how the themes are prioritised. In addition, ethics can be taught within different subjects. A qualitative study showed that veterinary ethics teaching is grounded or framed within animal welfare science, laws and regulations, theories and concepts and professionalism.

Batchelor, et al. (2015) suggested that veterinary medicine adopt the successful teaching methods for ethics employed in medicine and nursing. In most cases student-centred, group discussions of ethical dilemmas or scenarios are

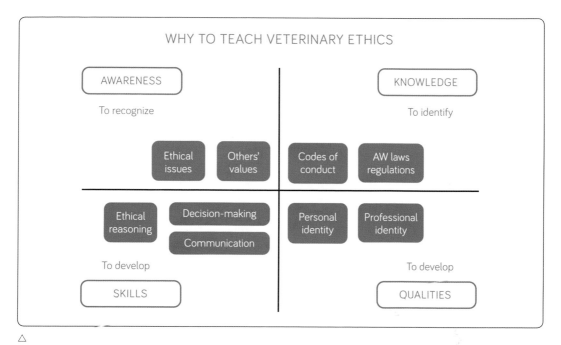

△
1.7 Learning objectives corresponding to themes in ethics teaching.
ADAPTED FROM MAGALHÃES-SANT'ANA, ET AL. (2014)

used, with the focus on developing critical reasoning rather than simply transferring facts and values.

Sant'Ana proposed a common framework for the teaching of human and veterinary medical ethics based on three concepts: professional rules, moral virtues and ethical skills (Magalhães-Sant'Ana 2015). The professional rules approach is based on transmitting norms or deontological principles (see chapter 2) in legislation, professional codes, oaths and principles. The values-based approach is focused on development of moral attitudes and behaviours, and promoting the values and beliefs underpinning the rules. Teachers and senior colleagues act as role models, through example and socialisation (see chapter 2 for a discussion of virtue ethics). The skills-based approach aims to

equip students with the tools for moral reasoning, allowing them to assess ethical dilemmas and conflict from different perspectives and take moral responsibility by using these tools or frameworks to come to their own decisions. We will discuss key ethical frameworks in chapter 2.

Each concept, when employed alone, has weaknesses which may impede ethical decision-making. For example, the rules-based approach fails to recognise differences between the law and morality, and the fact that there cannot be a rule, law or guidelines for conduct in every ethically challenging situation in veterinary practice or medicine. In addition, ethical reflection becomes redundant if one can simply follow the rules (Magalhães-Sant'Ana 2015). The cultivation of virtues cannot be cultivated within the timeframe it takes to teach a single subject, and vices can be role-modelled just as easily as virtues (Magalhães-Sant'Ana 2015).

A common criticism of the skills-based approach is that in providing numerous frameworks there is a risk that students may not have time to become properly acquainted with the merits and limitations of different ethical frameworks, hampering ethical decision-making (Magalhães-Sant'Ana 2015). Ethics courses that pit one theory against another, and require students to critique each theory, may give the impression that all ethical frameworks have flaws and thus all ethical opinions are simply that: opinion (Verrinder, et al. 2016).

Accepting that ethical frameworks are essentially complementary may be a means of avoiding such confusion and "disenchantment of relativism and pluralism often associated with ethics" (Verrinder, et al. 2016).

1.6

Who is worthy of ethical consideration?

"In contrast to other professions, veterinarians must deal with a centrally contested moral claim – the moral status of animals – in their day to day interactions with clients and patients"
(Morgan & McDonald 2007)

Ethical status, also referred to as moral standing, moral status, or ethical standing, refers to the property of an individual of being worthy of at least some degree of ethical consideration. We refer to intrinsic value when a being is worthy of ethical consideration in their own right, independently of or additional to being useful as a means to an end. For example, we feel that people have their own worth that is important to themselves and requires protecting.

But do we see animals in the same light? In Western philosophy, until recently, animals were overwhelmingly considered to be "lesser beings". The philosopher René Descartes, who famously performed live dissections (vivisection) of animals, compared them to machines, like clocks, and argued that they had no reason, no intelligence and no rational soul. Thus, the argument goes, we can treat them as we would machines – we may do with them and dispense with them as we wish.

This is an extreme view. One of the challenges of veterinary ethics is that the ethical status of animals is contested. Some believe that animals have instrumental value – that is, they are valuable because of their use to us. This could apply to farm animals, who provide a source of food, working dogs who assist humans in their tasks, or pets who provide companionship and in some cases protection. Others believe that animals have intrinsic value – that is, these animals have value in and of themselves, as living beings, irrespective of and independent to their value for us.

For example, in companion animal practice some owners of dogs or cats may view these animals as effectively property, considered only in terms of how they are useful to that person (instrumental value). Others view dogs and cats as true companions or family members (intrinsic value).

Of course an animal or human being may have both intrinsic and instrumental value. For example, most of us believe that humans have intrinsic value. A veterinarian or veterinary nurse, as a human being, has intrinsic value, but also instrumental value because of the useful tasks they perform.

If we attribute moral standing to animals, we imply that in addition to their instrumental value (for example, a dairy goat providing milk or a guard dog protecting a family), we have a duty to respect them as ethical subjects.

As stated by Ben Mepham, "The idea that animals have *merely* instrumental value, as we commonly assumed until very recently, now seems totally discredited. Putting it starkly to emphasize a point, if someone destroyed one of their valuable books we might, at worst, think him a fool – but if he destroyed his healthy cat (even painlessly, by poisoning when it was asleep) we should think him depraved" (Mepham 2008; emphasis in original).

Increasingly, legislation recognises the intrinsic value of animals. For example, a court ruling in Argentina found that a captive orangutan was a "non-human person" unlawfully deprived of her freedom. The Association of Officials and Lawyers for Animal Rights filed a habeas corpus petition – typically used to challenge imprisonment of a person – on behalf of the animal. The court found that the animal has sufficient cognitive functions and should not be treated as an object. The finding paved the way for the orangutan to be transferred from the zoo at which she was kept to a sanctuary (Lough 2014). However, in contrast, a New York court did not agree that chimpanzees were "non-human persons" and therefore keeping them captive was not an infringement of their right to liberty (Stern 2015).

It is not surprising that the successful petition was filed on behalf of a primate and not, for example, an otter. This is because society as a whole operates on the assumption that there is a scale of moral standing. When considering differences in moral worth, we need to consider which differences are morally relevant. For example, we no longer consider skin colour a morally relevant difference between people and therefore discrimination on this basis, racism, is widely condemned.

In addition to intrinsic value and instrumental value, animals can also have no value at all, such as pests. One good example is rabbits. Rabbits can be at the same time pets (intrinsic value), experimental animals (instrumental value) or pests (no value). It could be said that in the case of farm

△

1.8 An orangutan in Argentina was declared by the courts to be a "non-human person" who therefore could not be deprived of her liberty.
PHOTO ISTOCK

animals, we can reach the same outcomes in terms of welfare by considering their instrumental value (because we want to improve human needs, e.g. the meat quality) or by taking into account their intrinsic moral value (because we want to provide them with a good life).

If we distinguish between humans and other animals in terms of their moral standing, we need to establish a morally relevant difference between us. Increasingly the morally relevant difference between humans and various animal species considered in such situations is sentience, which encompasses the ability to think, suffer and experience emotions. Sentio-centrism prioritises sentience, and holds that sentient individuals (human or animal) have their own intrinsic moral worth – their welfare matters *to them*, and as such should be considered at the heart of ethical questions.

Despite scientific definitions delineating sentience and affective state, here they are considered functionally equivalent, where to be sentient is to experience positive and negative feelings. However, determining just which individuals or species are likely to have such conscious experiences is difficult as, by their very nature, they are privately experienced. This is where we commonly "argue by analogy", comparing neurophysiology and behaviour of animals to the "gold standard" for sentience: (currently) humans (Low 2012). This approach recognises that animals do not need to possess the same neuro-anatomy or act the same as humans to be sentient, but they must have relevantly similar systems for processing the world around them and responses to that world. Precisely where the line is drawn to define which animals are sentient depends on interpretation of the ever-increasing scientific literature, and the judgements of people who make up society. It's fair to say that more and more animal species appear to be at least knocking on the door of, if not yet fully joining, the reasonably well-established vertebrate "sentience club" as welfare science advances further. Cephalopods (Mather 2008), decapod crustaceans (Elwood 2012), insects (Bateson, et al. 2011) and other invertebrates (Sherwin 2001) have all been shown to exhibit some neuro-anatomy and/or behaviours that are similar to those of sentient humans.

Being outside the sentience club does not mean that an animal should not be considered in any ethical decision, only that, for those people primarily concerned about the feelings of animals, the consideration need not include an evaluation of their welfare, as by definition non-sentient animals cannot experience welfare. However, we

△

1.9 More and more species are being included in the so-called "Sentience Club".
CARTOON DR ROBERT JOHNSON

could have an ethical concern for welfare that encompasses naturalness or physical functioning which could also apply to non-sentient animals where the impact of any decision on an animal's natural behaviour, ability to reproduce or on the survival of the species would be examples of relevant concerns. Likewise, we could have a greater concern for conservation of species, preservation of ecosystems and maintenance of biodiversity than for the welfare of individuals. We must also remember that the absence of evidence does not imply the evidence of absence – it may be that a sentient species is not considered such because science has not proven it so or has not advanced sufficiently to provide any useful evidence.

WHO IS WORTHY OF ETHICAL CONSIDERATION?

What do you think?

△

1.10 When your container ship sinks, which people or animals would you prioritise to take in a lifeboat to a nearby island?
PHOTO ISTOCK

"Lifeboat dilemmas" are thought experiments that allow us to affirm and articulate our values, particularly in discussion with other people. Here is one that allows us to explore the relative moral worth of some people and animals.

A container ship is rapidly sinking with only one remaining lifeboat. Nearby, certainly in range of the lifeboat, is a large forested island with a small human settlement.

There are 10 "units" of people and animals on your container ship (listed below) that you might be able to rescue by bringing into the lifeboat. But, in which order would you start to fill up the lifeboat? Any predators will not prey on any other beings in the lifeboat. Each of the "units" takes up the same space on the lifeboat (from Kawall 1999):

ONE	An intelligent, healthy, morally virtuous human
TWO	An intelligent, healthy, morally evil human
THREE	A healthy moose (there is an indigenous moose population on the island)
FOUR	A collie with a permanently lame leg
FIVE	A severely mentally disabled human
SIX	Ten chickens
SEVEN	A breeding pair of an endangered species of bird, once native to the island
EIGHT	A human in a coma (who will almost certainly never recover)
NINE	A breeding pair of common, but beautiful, indigenous songbirds
TEN	Two breeding pairs of a non-indigenous variety of rapidly breeding wild rabbits (with no known predators on the island, and an extensive food supply).

—

1.6

WHO IS WORTHY OF ETHICAL
CONSIDERATION?

In the veterinary context, there is major scope for conflict between the interests of humans (for example, clients, industry, professional organisations and associations, and veterinarians) and the interests of animals.

The so-called fundamental problem of veterinary ethics is often expressed thus:

"…should the veterinary surgeon give primary consideration to the animal or the client?"
(Batchelor & McKeegan 2012)
or
"…to whom does the veterinary owe primary obligation: owner or animal?"
(Rollin 2006)

The answer rests on how we value animals. If we see animals as having intrinsic value, then we have obligations to that animal. The idea of killing that animal because a person no longer wants it is ethically objectionable. But if we view an animal as property of the client, the decision to destroy the animal due to the client's wishes is not problematic. Rollin draws the analogy of the paediatrician, who prioritises the needs of the child, even though the parent is paying the bills, versus a garage mechanic who repairs or destroys the client's car based entirely on the wishes of the client (Rollin 2006).

If we acknowledge that animals have intrinsic as well as instrumental value, we have to

△

1.11 Veterinary ethicist Bernard Rollin drew the analogy of the paediatrician versus the garage mechanic. Is the veterinarian fundamentally more like the paediatrician or the garage mechanic?
CARTOON MALBON DESIGNS

determine – at some point – how we navigate conflict between the interests of humans and non-human animals.

It is argued repeatedly that, at least in Western countries, we are generally very inconsistent in ascribing moral standing to animals, as illustrated in this example from Coeckelbergh and Gunkel:

"…it is generally believed that animals that are more human-like (i.e. can feel pain as we feel pain, exhibit sentience or even conscious behaviour etc.), have a higher moral status than those who do not exhibit these properties, and ought to be treated accordingly (i.e. better than "lower" animals). On the other hand, our actual practices and treatment of animals do not really fit this framework. We kill and eat animals that are very similar to us and that can feel pain, such as pigs, whereas we treat other animals such as dogs and cats like companions, friends or children for reasons that have little to do with their biological properties."

(Coeckelbergh & Gunkel 2014)

They add that even where philosophers argue that animals are worthy of moral standing, their assessment is derived from our "unexamined anthropocentric privilege" – notably we accept as having moral standing only animals that have individual properties just like us. Our assessment of who is morally significant and what is not may have more to do with our upbringing and cultural contexts (for example, in cultures where pet ownership or meat eating is common) than it does with a truly well-thought-out ethical justification.

For these philosophers the question of how we ascribe moral standing to human and non-human others requires rigorous examination.

WHO IS WORTHY OF ETHICAL CONSIDERATION?

What do you think?

ONE _____ Do animals have value?

TWO _____ If so, how would you describe or characterise this value?

THREE _____ How much of their value comes from belonging to or benefitting a human being?

FOUR _____ How acceptable is it that animals are instruments or tools employed by human users for various purposes (for example, knowledge production, scientific research, companionship, practical employments and so on)?

WHO IS WORTHY OF ETHICAL CONSIDERATION?

What do you think?

ONE _____ In working with or providing treatment to an animal, to whom are veterinarian practitioners responsible or accountable?

TWO _____ Who is it you are benefiting, when things go right?

THREE _____ Who is harmed, if something goes wrong?

FOUR _____ Who, in other words, is the "Other" to whom you owe moral respect and consideration? Do you owe it to the animal? Do you owe it to the owner of the animal? Do you owe it to society at large? And why?

1.7

Ethics and animal welfare

1.7.1 How our values affect welfare assessment

The welfare of an individual or group of animals is usually central to any ethical decision about them. For example, the answer to the question "How much is my dog suffering?" might influence a decision about euthanasia; "What effect will box rest have on the welfare of my horse?" might affect which treatment option is preferable; "How much can mice suffer?" or indeed "How much can primates suffer?" may influence our decision (or the decision of an ethics committee) to approve the use of these animals in experiments. The very concept of "welfare" has been open to ethical examination with relative consensus that welfare science is not "value-free" (Rollin 1996).

There has been an increasing recognition that welfare science and philosophy must come together for an inter-disciplinary discussion on the nature of welfare and how best to assess it (Fraser 1999, Lund, et al. 2006, Sandøe & Simonsen 1992, Thompson 1999). In the main, welfare is considered as a continuum from extremely poor to excellent although not all components that contribute to this overall welfare have a similar continuum. For example, whereas happiness ranges through neutral to unhappiness and hunger passes through neutral to being positively satiated, thirst and pain do not have a positive equivalent, a lack of them being only neutral.

Fraser, et al. (1997) proposed a model of animal welfare that reflects three ethical concerns people might hold:

"(1) that animals should lead natural lives through the development and use of their natural adaptations and capabilities,

△
1.12 The question "How much can mice suffer?" may influence the decision to use these animals in experiments.
PHOTO ANNE FAWCETT

(2) that animals should feel well by being free from prolonged and intense fear, pain, and other negative states, and by experiencing normal pleasures, and

(3) that animals should function well, in the sense of satisfactory health, growth and normal functioning of physiological and behavioural systems."

(Fraser, et al. 1997)

Proponents who value natural living are concerned that animals should live in a natural environment, fulfil a range of natural behaviours or live according to their naturally evolved characteristics, even in human-controlled environments. Many domestic animals have largely retained the core behavioural traits of their wild counterparts despite domestication as farm animals, for example, pigs (Stolba & Woodgush 1989) or pet animals such as rabbits (Stodart & Myers 1964). The concept of respect for "telos" – the essence of an animal, which can be thought of as "the dogness of a dog" or "the cowness of a cow" – is based upon a concern about naturalness of animals (Rollin 1993). However, telos, and indeed naturalness, are not always easily defined, particularly in a world of increasing man-made alterations to natural habitats. Natural living can result in negative feelings, such as fear of a predator or discomfort from cold, or in reduced physical fitness, such as poor breeding success when food is scarce.

The second ethical concern, for the feelings that animals have, both positive and negative, corresponds to ethical frameworks for human quality of life that focus on hedonism (pleasures) or the fulfilment of desires (which may or may not be pleasurable in themselves) (Jensen & Sandøe 1997). If affective state is considered important then welfare is promoted through increasing positive feelings such as contentment or excitement and minimising negative emotions such as fear, and avoiding suffering. Suffering has been variously defined but could be considered as:

"substantial physical discomfort and/or mental distress which affects our whole being and sidelines most (if not all) other considerations normally important to us."

(Aaltola 2012)

Suffering is relevant even if it occurs naturally, for example through heat stress in a wild animal, or without a reduction in capacity to function, for example if the animal were able to reproduce. Animals may be able to fulfil longer-term desires or "life goals", such as nurturing their young or

△
1.13 Many domestic animals have largely retained the core behavioural traits of their wild counterparts despite domestication as farm animals.
PHOTO ANNE FAWCETT

△
1.14 Natural living can result in negative feelings, such as fear of a predator or discomfort from cold, or in reduced physical fitness, such as poor breeding success when food is scarce.
PHOTO ANNE FAWCETT

△

1.15 Red squirrels may demonstrate longer-term "life goals" through caching hazelnuts for use when food is scarce later in the winter.
PHOTO ISTOCK

△

1.16 Racehorses epitomise physical fitness, but this is lessened when injured, with the presence of gastric ulcers or if infertile.
PHOTO ISTOCK

planning for the future through caching food, rather than having a purely hedonistic focus. Here it could be possible to define welfare in terms of feelings, as long as they are not harmful to these overriding "life goals".

The final ethical concern values physical fitness, which can be considered in the widest sense to include not just physical health and function but also behavioural systems. These behavioural systems would not have to be operating at all times, but would have to have the capacity should the need arise. For example, a physically fit prey animal may employ escape mechanisms only in the face of a predator. With focus on physical fitness, welfare can be considered to be reduced even in the absence of any conscious negative feelings. Examples include some types of infertility or early neoplasias, regardless of whether it satisfied naturalness criteria or not.

As Fraser, et al. (1997) point out, these concerns are often overlapping and considered to a lesser or greater degree by individuals and society who, despite favouring one concern, rarely exclude the others. For example, when considering animals that we have responsibility for, most people would be appalled at the idea of keeping animals on euphoria-inducing drugs (promoting positive feelings but against naturalness); they would be shocked if live gazelle were supplied for lions to hunt and kill in zoos (promoting naturalness and maybe fitness); and they would be concerned if the genetic determinants for faster racehorses were coupled with increased anxiety (promoting physical fitness but negative emotional states).

Each of us must consider where our values place these concerns in order to assess welfare holistically and be able to defend it to clients, colleagues and the public. Researchers have investigated the primary ethical concern of certain groups of people when making an overall judgement of animal welfare. For example, conventional farmers tended to favour a view of welfare that

△△
1.17 More natural systems for keeping pigs allow extensive foraging behaviour.
PHOTO ISTOCK

△
1.18 Intensive pig farms only provide some basic elements of physical welfare such as food and water.
PHOTO ISTOCK

focuses on the physical fitness of the animal and can be assessed by health and productivity indicators (Kling-Eveillard, et al. 2007) whereas organic farmers also valued naturalness (Bock & van Huik 2007). A survey of American citizens found that 40 per cent of respondents were primarily concerned that basic elements of physical welfare (food, water) should be provided to farm animals, whereas 46 per cent placed a strong emphasis on naturalness, for example the ability to exercise outdoors. In this study only 14 per cent of citizens were unconcerned about animal welfare, valuing a low product price above all other concerns, including welfare concerns (Prickett, et al. 2010).

1.7

ETHICS AND ANIMAL WELFARE

Vanhonacker, et al. found citizens in Belgium, compared to farmers, were more in favour of natural living for animals (Vanhonacker, et al. 2008). However, the researchers suggest that farmers were not disinterested in naturalness, but have a discord between their values and their interests in farming profitably through more intensive systems.

Historically, welfare scientists have investigated physical welfare and associated risk factors, but they have increasingly focused on the emotional capacities of animals and the effect of husbandry practices on their feelings. These are expressed through behaviour, occurring both unprompted and in response to tests such as for cognitive bias, where pessimistic and optimistic biases may be able to indicate the underlying affective state (Mendl, et al. 2009).

In order to ensure that welfare assessments – especially those aimed to deliver benefits to society such as in farm assurance schemes – reflect the ethical concerns of citizens, welfare scientists have worked in partnership with societal representatives. In the development of the Welfare Quality® farm animal assessment protocols, citizen focus groups and juries highlighted that, in

comparison with the scientists, they more highly valued low-input natural farming systems, positive welfare and a holistic appraisal of welfare. This holistic concept of welfare was inextricably linked to other attributes such as environmental impact or product quality. The protocols were subsequently developed by scientists to take these concerns into account (Miele, et al. 2011).

1.7.2 The science of welfare assessment

Our ethical decisions can only ever be as good as the evidence they are based on. When welfare assessment of an animal is key to a decision it's important to have the most accurate assessment available. But, how exactly can we do that? Ideally we would ask the animal themselves. We are gradually learning to understand what they are telling us. Animal welfare science has developed over recent decades to determine valid indicators of welfare or the preferences of animals.

So, we now have a much better understanding of how restrictive types of housing for farm

◁

1.19 In this T maze, a form of preference test, the hen was released into the centre and, having been previously trained to understand what is on offer in each of the two options, has chosen the one to the left.

PHOTO CHRISTINE NICOL

animals affect their welfare. Using the pig as an example, compared to those kept in group housing, sows in individual confinement stalls have been shown to have higher levels of stereotypies (Chapinal, et al. 2010, Zhou, et al. 2014); inability to express normal behaviours (Weaver & Morris 2004); higher levels of some health problems, such as bursitis, but lower lameness (Diaz, et al. 2014); and fewer aggressive interactions (Jansen, et al. 2007). In preference tests, stalled sows were shown to prefer shorter (30-minute) compared to longer (240-minute) periods of restriction (Spinka, et al. 1998).

However, there are problems inherent in the scientific assessment of animal welfare. Firstly, in terms of the welfare indicators, we must still interpret what such indicators mean. For example, stereotypies – repetitive, non-functional movements – are usually associated with other measures of decreased welfare. However, these may be a coping strategy and it may be that within stereotypy-inducing environments those individual animals not stereotyping actually have the worst welfare (Mason & Latham 2004). Even indicators that we assume to be associated with poor welfare, such as lameness, may be hard to quantify. It may help to assess the impact of lameness on other indicators. For example, lame broilers demonstrated changes in behaviour (Weeks, et al. 2000). Studies demonstrating self-selection of analgesic food give more insight into the mental state of lame broilers (Danbury, et al. 2000).

The second problem is how to extrapolate results from individuals to populations, and from populations to individuals. Individual animals may have different preferences, and thus be differentially affected by the same housing, human interactions and so on. Indeed, in a choice test, individual chickens have been shown to consistently choose their preferred housing environment which varied between birds (Browne, et al. 2010). Inferences about the welfare of individuals, as deduced from

data about the group, are subject to the "ecological fallacy". Thus in measuring a flock of chickens, it is impossible to determine if each individual within that flock experiences the same risk factors. If they aren't the same, then inferences about the association between risk factors and outcomes are biased by the group effect (Siegford, et al. 2016). Trying to determine the best environment for a flock of hens may therefore be problematic. A sensible solution may be to offer choice to animals to allow them to maximise their welfare through exercising autonomy (Edgar, et al. 2013). As technology is refined, monitoring individual animals within large groups may become a viable option, but at present the cost is generally prohibitive (Siegford, et al. 2016).

Thirdly, preference testing – a direct way to "ask the animal" – has some limitations as animals can only choose between what is on offer (which could be the equivalent of a rock and a hard place) and we may not understand what motivates their choice and how well this relates to their feelings. Operant tests, where animals are actively engaged in a process to achieve something, for example pressing a lever to gain access to a resource, may be useful in determining the strength of motivation. However, they cannot indicate the effect on welfare if the animal has never had access to the resource, or the resource is withdrawn (Kirkden & Pajor 2006).

Fourthly, trying to combine the evidence from a range of welfare indicators into a holistic assessment is difficult in both principle and practice. Consider if one's main ethical concern is even for just one element of welfare, such as affective state. For a pony outdoors on a mountain in winter, how do we weigh the relative importance of thermal discomfort, hunger, social interactions and freedom of movement in relation to how it feels? A single welfare index has been developed for cattle, pig and poultry farms by weighting in the Welfare Quality® scheme. In Welfare Quality®, indicators assess the welfare of individuals or groups of animals on the farm (Veissier, et al. 2011). However, further examination has suggested that the weighting may not be adequate to give a reliable assessment of whole farm welfare (Heath, et al. 2014).

Finally, welfare assessments are often performed at a point in time, and even if repeated over time may be inadequate to give a good indication of the welfare of an animal over its whole life. Here again we have the difficulty of weighing up a variety of welfare indicators and an additional problem that we just may not assess an animal enough – for example, we may not observe a single but very stressful experience during an animal's life. This is problematic, since we often want to evaluate the whole life of an individual and rate husbandry systems based on their total living experience. In practice, part of a whole life assessment often relies on an evaluation of the risk of poor welfare or likelihood of good welfare.

1.7.3 How to conduct a welfare assessment

Despite the limitations already discussed, we should make use of appropriate evidence wherever possible. In assessing the effect of an action, disease or husbandry practice we should make full use of the literature and relevant experience of experts, be they scientists, animal keepers or others. When we are assessing the welfare of an individual animal, or group of animals, it is helpful to have a checklist to ensure that what we are assessing accords with our ethical concerns and that we cover all elements of welfare at that time, or over time as necessary. Our checklist should be based on evidence of valid indicators and how to combine them. Where the evidence is lacking we must make our best estimate.

> **BOX 1.2**
>
> STEPS FOR DEVISING A WELFARE ASSESSMENT
>
> Answer the following questions:
>
> - What are the *best* indicators for assessing the positive and negative elements of welfare of this animal/ this group of animals?
> - Who can best perform the welfare assessment?
> - How can these indicators best be interpreted as a holistic welfare assessment?
> - How can this welfare assessment reflect the welfare of all animals in the group?
> - How can this welfare assessment reflect the lifetime welfare of this/ these animal(s)?

It's clear that most welfare science has focussed on indicators of poor welfare but there are good reasons for wanting to also assess positive welfare experiences (Yeates & Main 2008). For example, when discussing a euthanasia decision with a client, it can be very helpful to identify what their dog still enjoys in life. Unfortunately we have fewer validated measures of positive welfare but many owners will feel confident in telling you what their animal enjoys. Here owners have the great advantage when making, or contributing to, the welfare assessment that they know the individual animal(s) very well and will probably be best able to detect even small changes in behaviour. However, they may be limited by their ability to interpret some welfare indicators which their animal is displaying. For example, only 7 per cent of owners of geriatric dogs reported increased thirst when it was subsequently found to present in 56 per cent, and problems that were not recognised by owners were found after veterinary examination in

△

1.20 Owners know their animals best but may not always recognise signs of poor welfare.
PHOTO ISTOCK

80 per cent of the dogs (Davies 2012). Owners may also have difficulty interpreting animal behaviour, particularly if limited by a lack of exposure to the species of animal experiencing a range in welfare states. To illustrate, one author met a UK pig farmer who had kept pigs in barren pens for 40 years but who had never been onto another pig farm, and he had never observed pigs exploring, manipulating, chewing and rooting in a substrate such as straw.

There is also a possibility of conscious or unconscious denial. In the case of companion animals, owners may fail to recognise welfare problems due to the implications of doing so. For example, someone who is very attached to a horse may not wish to recognise the extent to which that animal's quality of life is compromised, if they know that this means the best thing for that animal is euthanasia. Similarly, someone who is concerned about costs of veterinary fees may not wish to acknowledge the seriousness of an animal's clinical signs.

We have discussed "argument by analogy" with regards to sentience, and, in the absence of other information, it is also a reasonable approach for a welfare assessment, where it can form the basis of our best estimate. Caulfield and Cambridge (2008) suggest the following as a starting point when considering the case of sow stalls:

> "'On the basis of human knowledge of the preferences of sentient animals, putting a sow in a sow stall is very likely to be distressing for that animal'. The next question is 'What scientific evidence is there to disprove this?'"
>
> (Caulfield & Cambridge 2008)

They argue that too often a lack of scientific evidence to prove non-harm to animals has impeded good judgement and enabled a system that would seem intuitively to be harmful to continue. Sometimes the original benefits that led to the introduction of such systems become redundant, for example through improved management practices. By appealing to the precautionary principle, and erring on the side of sentience or that some systems and interventions might cause an animal suffering, we are more likely to avoid poor welfare. However, this may come with a monetary cost to some groups, for example the expense associated with providing better conditions for more animals. Ghandi stated that the greatness of a nation can be judged by how it treats animals. Is modern society willing to pay the increased cost to improve the welfare of sentient animals? And what role do veterinarians have in helping to bring about greatness in our respective nations?

—
1.8
VETERINARY ETHICAL DILEMMAS

1.8

Veterinary ethical dilemmas

1.8.1 What is an ethical dilemma?

Technically an ethical or moral dilemma occurs when there is a conflict between responsibilities or obligations of equal moral weight (Morgan & McDonald 2007). Ethical dilemmas arise when we have competing responsibilities with no obvious way to prioritise one responsibility over others (Morgan & McDonald 2007). Put in a different way, moral dilemmas occur where moral obligations "appear to demand that a person adopt each of two (or more) alternative but incompatible actions, such that the person cannot perform all the required actions" (Beauchamp & Childress 2013). In such a case, one may feel torn between two equally appealing (or unappealing) actions.

Here is an example of an ethical dilemma: You have two children, both of whom are suffering from life-threatening medical conditions. With a particular treatment, there is a 98 per cent chance that one child can be saved – but this treatment will require all of your resources, leaving none for the other child. With another treatment, there is a 10 per cent chance that both children will be saved. Which treatment do you choose?

Part of your answer will involve weighing up the interests of each child. But "moral weight" is difficult to define, and matters are complicated further for veterinarians as the moral status of animals is hotly contested.

Thus for example, if the scenario is altered such that it now involves not two children, but a child and a dog – both with medical conditions, only one of which you can treat – the dilemma evaporates for many people. They believe that the interests of the child have greater moral weight

and therefore the child should receive the treatment. Or one might argue that the nature of the relationships – the fact that one relationship is that of a parent to a child – is important here, because certain duties flow from being a parent.

In practice, the strict definition of an ethical dilemma is expanded to include any difficult ethical situations, which may give rise to ethical conflict.

1.8.2 Sources of ethical dilemmas in veterinary practice

Dilemmas can occur due to differences in beliefs about the value or status of animals, differences in beliefs regarding obligations and responsibilities to animals, differences in the assessment and weighting of interests of stakeholders, differences in assessment of outcomes or consequences of actions, or a combination of these.

Examples of potential ethical dilemmas in veterinary practice are outlined in the box overleaf.

While most of the discussion in veterinary ethics examines dilemmas in a companion animal private practice setting, many dilemmas arise outside of this context. For example, veterinarians and associated professionals may be involved in the use of animals in sport. Here, as in human sports medicine, there can be an overriding, economically driven demand to return the athlete to competition (Campbell 2013). There can be a conflict between the desire to give the best treatment to maximise long-term welfare, and treatment that will yield improvement in performance in the short term. Additionally, clinicians may be pressured to treat beyond their expertise, use treatments for which there is little to no evidence base in the hope of a "quick fix", undertake harmful treatment at the client's request or disclose clinical information selectively (Campbell 2013).

Similarly, those working with wildlife face unique challenges. A survey of 60 primatologists

BOX 1.3

EXAMPLES OF ETHICAL DILEMMAS IN
VETERINARY PRACTICE.

- Requests to perform cosmetic procedures
  (tail docking, beak trimming, declawing) on
  animals
- Requests to destroy a healthy animal
- Inability or unwillingness of the client to fund
  treatment that is in the animal's interests
- Client wishing to continue treatment despite
  poor quality of life/welfare
- Breaching client confidentiality to protect an
  animal or herd, or a client
- To what extent is it appropriate to influence
  a client?
- Performing a procedure on a patient for the
  first time
- Whether or not to refer a patient

SOURCES BATCHELOR & MCKEEGAN (2012),
MORGAN & MCDONALD (2007), YEATES & MAIN (2011).

△
1.21 Requests to perform cosmetic
procedures are a source of ethical
conflict in veterinary practice.
CARTOON RAFAEL GALLARDO ARJONILLA

found that ethical dilemmas were common (Fedigan 2010). Even the most seemingly non-invasive approach, such as observing a population of wild animals in the field, can have negative effects which researchers need to weigh up. For example, habituation of primates to the presence of humans may render them vulnerable to harm by reducing their fear of humans and potentially facilitating undesirable behaviours such as crop-raiding (Fedigan 2010). Is research worthwhile if the primate population being studied is then decimated by hunters?

Farm animal practitioners also have to deal with difficult problems, often treating animals to compensate for poor systems or management practices. The individual animals may benefit but future animals may be harmed through propping up unsustainable practices and allowing their continuation.

1.8.3 How common are ethical dilemmas and ethical conflict in practice?

Few studies have examined the incidence of ethical dilemmas and ethical conflict in veterinary settings. One factor that would affect the incidence of ethical dilemmas and ethical conflict would be ethical sensitivity or awareness of ethical dilemmas.

In one study, 58 veterinarians completed a survey reporting how frequently they faced ethical dilemmas (Batchelor & McKeegan 2012). Of these, 91 per cent faced at least one ethical dilemma per week (57 per cent faced 1–2 dilemmas per week and 34 per cent faced 3–5 dilemmas per week). Two respondents reported facing more than 10 ethical dilemmas per week, and three respondents stated they faced none (Batchelor & McKeegan 2012).

Another study examined the number of times veterinarians refused euthanasia or wanted to refuse euthanasia requested by pet-owning clients. Although this was an uncommon issue, the majority of the 58 respondents had experienced this situation at least once a year. Two respondents reported refusing euthanasia "most months" (Yeates & Main 2011). Refusing euthanasia may not be a dilemma, as it may be clear to those veterinarians refusing euthanasia that the animal's continued interest in welfare trumps euthanasia. Furthermore, the veterinarian may be legally entitled to refuse to perform euthanasia. But the request gives rise to ethical conflict.

1.9 Impact of ethical dilemmas and ethical conflict on veterinarians and associated professionals

Veterinarians find ethical dilemmas and ethical conflict stressful. For example, in one study veterinarians rated three ethical scenarios (convenience euthanasia of a healthy animal, financial limitations of clients restricting treatment options and a client wishing to continue treatment despite compromised animal welfare/quality of life) as "highly stressful" (Batchelor & McKeegan 2012). Interestingly, stress ratings were not influenced by the number of years spent in practice, suggesting that coping with ethical dilemmas is not effectively self-taught or improved by repeated exposure (Batchelor & McKeegan 2012).

In a study of Australian veterinary students, 69 per cent reported experiencing moral distress in relation to the treatment of animals (Verrinder & Phillips 2014).

Some scholars argue that "moral stress", or stress associated with ethical dilemmas, may severely impact the wellbeing of veterinarians, even contributing to the high rate of suicide in the profession (Bartram & Baldwin 2008).

Rollin identifies moral stress as stress arising from the situation where people such as veterinarians and nurses whose life work is aimed at promoting the wellbeing of animals are called upon to facilitate the killing of animals when they don't agree that euthanasia is warranted, or "being complicit in creating pain, distress, disease, and other noxious states" required in research (Rollin 2011).

CHAPTER 1 VETERINARY ETHICS AND WHY IT MATTERS

—
1.9
IMPACT OF ETHICAL DILEMMAS
AND ETHICAL CONFLICT ON
VETERINARIANS AND ASSOCIATED
PROFESSIONALS

BOX 1.4

WHAT TO DO IF YOU ARE DISTRESSED ABOUT ETHICAL ISSUES?

△
1.21 Distress and stress over ethical issues is not uncommon.
CARTOON SUHADIYONO94

It is important to recognise that distress and stress are not uncommon. Distress can impact our ability to make sound decisions, therefore it is best to avoid making major decisions while in a distressed state.

Reflecting on and understanding the values, beliefs and motivations that feed into the veterinary professional identity can help us better cope with, and assist colleagues in coping with, stresses and threats to our identities (Allister 2016). These may include but are not limited to ethical dilemmas and ethical conflict.

There are a range of resources for veterinarians, veterinary students, nurses, technicians and associated health professionals suffering from moral distress or indeed other work-related stressors.

Look for guidelines: In some cases, specific legislation, guidelines or codes may apply to some situations which may dictate or suggest the most appropriate response. Unfortunately this is not always the case and legislation can require interpretation. The dilemma of determining the appropriate response may be due to uncertainty arising from unclear, ambiguous or confusing guidelines and rules (Devitt, et al. 2014).

Write it down: Making a list of sources of concern or anxiety (what you are worried about) can provide clarification.

Seek evidence: Is there existing literature on this particular situation or dilemma? A number of textbooks provide scenarios that may be helpful. The British Veterinary Association's *In Practice* journal includes an *Everyday Ethics* column every month.

Seek professional advice: If you are concerned about making a decision, including regarding the impact of that decision on others, talking confidentially with a senior colleague, veterinary board or academic (such as a student advisor) may help.

Phone a friend: Sometimes it can help to clarify the issue by discussing it with a trusted friend.

Counselling: Many professional organisations such as VetLife (supported by the British Veterinary Association), Australian Veterinary Association or American Veterinary Medical Association offer counselling or referral to counselling for members. Alternatively, your family doctor can refer you to an appropriate counsellor. There are also 24-hour confidential telephone and online counselling services, and veterinary-specific resources available.

"This kind of stress grows out of the radical conflict between one's reasons for entering the field of animal work, and what one in fact ends up doing… Imagine the psychological impact of constant demands to kill healthy animals for appalling reasons: 'the dog is too old to run with me anymore; we have redecorated, and the dog no longer matches the colour scheme; it is cheaper to get another dog when I return from vacation than to pay the fees for a boarding kennel", and, most perniciously, "I do not wish to spend the money on the procedure you recommend to treat the animal" or "it is cheaper to get another dog."

(Rollin 2011)

Similarly, Bartram and Baldwin suggest that veterinarians may experience "uncomfortable tension" between the desire to treat the animal and the desire to fulfil the owner's wishes (Bartram & Baldwin 2008). Stark and Dougall argue that the dissonance between personal values or ideals and the reality of "convenience euthanasia" may be a stressor which can lead to veterinary suicide (Stark & Dougall 2012). The association is unproven, but we know that stress increases the risk of suicide and that ethical decisions can be stressful. It makes sense that veterinarians and associated professionals should develop skills in ethical reasoning to equip them to cope with ethical dilemmas and ethical conflict.

1.10

Ethical policies

When we think about applying ethics we usually think about how ethical reasoning can improve decision-making in individual cases. But studying ethics can increase our ethical sensitivity and may inform policy.

Ethical policies are a useful way of ensuring that day-to-day practice reflects the ethical principles that are central to a group or organisation. The benefit of an ethical policy is that it pertains to ethical situations or dilemmas which have been considered in detail, away from time and emotional pressures that might otherwise impact decision-making. For example, a practice may develop a policy on the treatment and euthanasia of wildlife or managing clients who cannot afford the necessary treatment for their animals/stock.

Of course, any ethical policy is by necessity a "one-size-fits-all" (or at least "one-size-fits-many") approach, and even if one broadly agrees with the principles one may be left feeling uneasy when applying it to exceptional cases.

Codes of conduct and professional ethics may have the advantage of establishing acceptable responses to common ethical dilemmas and protecting veterinarians from the pressure of those who may not be acting in an animal's best interests (Campbell 2013). For example, where an equine veterinarian may be pressured by a trainer to undertake a harmful procedure on a horse to gain a competitive advantage, a code banning this technique or approach can give a veterinarian additional authority to refuse.

However, codes of conduct must be broad and it is difficult to develop a detailed code specific enough to instruct a busy practitioner on complex ethical dilemmas. An analysis of European veterinary codes identified eight overarching themes, including definitions and framing concepts, duties to animals, duties to clients, duties to other professionals, duties to competent authorities, duties to society, professionalism and practice-related issues (Magalhães-Sant'Ana, et al. 2015). The emphasis on different themes varied significantly between codes. For example, according to some codes the veterinarian's primary responsibility is animal welfare, while others placed a greater emphasis on professional relationships. Strict adherence to these codes may result in different decision-making for the same scenario.

Conclusion

Ethical decisions are at the heart of veterinary practice, occurring commonly and in all types of work. Our views on the moral status of animals are key to ethical decision-making. Understanding how any decision may affect the welfare of animals is also important and utilising welfare science and being able to make accurate welfare assessments ourselves are useful skills helping to promote better ethical decisions. Ethical decisions are a cause of stress for many in the veterinary team. Support from colleagues and family can be helpful and there are other sources of support that can be employed. Ethical policies, including codes of professional conduct, provide the accepted standards of practice and support veterinarians in their decision-making.

References

Aaltola E 2012 *Animal Suffering: Philosophy and Culture.* Palgrave Macmillan: New York, USA.

Allister R 2016 What does it mean to be a vet? *Veterinary Record* **178**: 316–317.

Angie AD, Connelly S, Waples EP, and Kligyte V 2011 The influence of discrete emotions on judgement and decision-making: a meta-analytic review. *Cogn Emot* **25**: 1393–1422.

Bartram DJ, and Baldwin DS 2008 Veterinary surgeons and suicide: influences, opportunities and research directions. *Veterinary Record* **162**: 36–40.

Batchelor CEM, Creed A, and McKeegan DEF 2015 A preliminary investigation into the moral reasoning abilities of UK veterinarians. *The Veterinary Record* **177**: 124.

Batchelor CEM, and McKeegan DEF 2012 Survey of the frequency and perceived stressfulness of ethical dilemmas encountered in UK veterinary practice. *Veterinary Record* **170**: 19.

Bateson M, Desire S, Gartside SE, and Wright GA 2011 Agitated honeybees exhibit pessimistic cognitive biases. *Current Biology* **21**: 1070–1073.

Beauchamp TL, and Childress JF 2013 *Principles of Biomedical Ethics.* Oxford University Press: New York, Oxford.

Bock BB, and van Huik MM 2007 Pig farmers and animal welfare: a study of beliefs, attitudes and behaviour of pig producers across Europe. In: Kjaernes U, Miele M, and Roex J (cds) *Attitudes of Consumers, Retailers and Producers to Farm Animal Welfare, Welfare Quality Report no. 2*, 73–124. Lelystad, NL.

British Veterinary Association 2016 Vets speaking up for animal welfare: BVA animal welfare strategy. RVA.

Browne WJ, Caplen G, Edgar J, Wilson LR, and Nicol CJ 2010 Consistency, transitivity and inter-relationships between measures of choice in environmental preference tests with chickens. *Behavioural Processes* **83**: 72–78.

Campbell MLH 2013 The role of veterinarians in equestrian sport: a comparative review of ethical issues surrounding human and equine sports medicine. *Veterinary Journal* **197**: 535–540.

Caulfield MP, and Cambridge H 2008 The questionable value of some science-based 'welfare' assessments in intensive animal farming: sow stalls as an illustrative example. *Australian Veterinary Journal* **86:** 446–448.

Chapinal N, de la Torre JLR, Cerisuelo A, Gasa J, Baucells MD, Coma J, Vidal A, and Manteca X 2010 Evaluation of welfare and productivity in pregnant sows kept in stalls or in 2 different group housing systems. *Journal of Veterinary Behavior-Clinical Applications and Research* **5:** 82–93.

Coeckelbergh M, and Gunkel DJ 2014 Facing animals: a relational, other-oriented approach to moral standing. *Journal of Agricultural & Environmental Ethics* **27:** 715–733.

Danbury TC, Weeks CA, Chambers JP, Waterman-Pearson AE, and Kestin SC 2000 Self-selection of the analgesic drug carprofen by lame broiler chickens. *Veterinary Record* **146:** 307–311.

Davies M 2012 Geriatric screening in first opinion practice - results from 45 dogs. *Journal of Small Animal Practice* **53:** 507–513.

DeScioli P, Christner J, and Kurzban R 2011 The omission strategy. *Psychol Sci* **22:** 442–446.

Devitt C, Kelly P, Blake M, Hanlon A, and More SJ 2014 Dilemmas experienced by government veterinarians when responding professionally to farm animal welfare incidents in Ireland. *Veterinary Record Open* **1:e000003.**

Diaz JAC, Fahey AG, and Boyle LA 2014 Effects of gestation housing system and floor type during lactation on locomotory ability; body, limb, and claw lesions; and lying-down behavior of lactating sows. *Journal of Animal Science* **92:** 1673–1683.

Edgar J, Mullan S, Pritchard J, MacFarlane U, and Main DCJ 2013 Towards a 'good life' for farm animals: development of a resource tier framework to achieve positive welfare for laying hens. *Animals* **3:** 584–605.

Edwards L, and Delaney C 2008 Ethical reasoning. In: Higgs J, Jones M, Loftus S, and Christensen N (eds) *Clinical Reasoning in the Health Professions 3rd Edition*, 279–290. Elsevier: St Louis.

Elwood RW 2012 Evidence for pain in decapod crustaceans. *Animal Welfare* **21:** 23–27.

Fedigan LM 2010 Ethical issues faced by field primatologists: asking the relevant questions. *American Journal of Primatology* **72:** 754–771.

Fordyce P 2011 Everyday ethics. *In Practice* **33:** 94–95.

Fraser D 1999 Animal ethics and animal welfare science: bridging the two cultures. *Applied Animal Behaviour Science* **65:** 171–189.

Fraser D, Weary DM, Pajor EA, and Milligan BN 1997 A scientific conception of animal welfare that reflects ethical concerns. *Animal Welfare* **6:** 187–205.

Heath CAE, Browne WJ, Mullan S, and Main DCJ 2014. Navigating the iceberg: reducing the number of parameters within the Welfare Quality® assessment protocol for dairy cows. *Animal* **8:** 1978–1986.

Herzog H 2010 *Some We Love, Some We Hate, Some We Eat: Why It's So Hard to Think Straight About Animals.* Harper: New York.

Jansen J, Kirkwood RN, Zanella AJ, and Tempelman RJ 2007 Influence of gestation housing on sow behavior and fertility. *Journal of Swine Health and Production* **15:** 132–136.

Jensen KK, and Sandøe P 1997 Animal welfare: relative or absolute? *Applied Animal Behaviour Science* **54:** 33–37.

Kawall J 1999 An introductory exercise in articulating values. *American Philosophical Newsletter* **99:** 4–7.

Kirkden RD, and Pajor EA 2006 Using preference, motivation and aversion tests to ask scientific questions about animals' feelings. *Applied Animal Behaviour Science* **100:** 29–47.

Kling-Eveillard F, Dockes AC, and Souquet C 2007 Attitudes of French pig farmers towards animal welfare. *British Food Journal* **109:** 859–869.

Lough R 2014 Captive orangutan has human right to freedom, Argentine court rules. Reuters News Agency.

Low P 2012 The Cambridge Declaration on Consciousness. http://fcmconference.org/img/CambridgeDeclarationOnConsciousness.pdf

Lund V, Coleman G, Gunnarsson S, Appleby MC, and Karkinen K 2006 Animal welfare science - working at the interface between the natural and social sciences. *Applied Animal Behaviour Science* **97:** 37–49.

Magalhães-Sant'Ana M 2015 A theoretical framework for human and veterinary medical ethics education. *Advances in Health Sciences Education.*

Magalhães-Sant'Ana M, Lassen J, Millar KM, Sandøe P, and Olsson IAS 2014 Examining why ethics is taught to veterinary students: a qualitative study of veterinary educators' perspectives. *Journal of Veterinary Medical Education* **41:** 350–357.

Magalhães-Sant'Ana M, More SJ, Morton DB, Osborne M, and Hanlon A 2015 What do European veterinary codes of conduct actually say and mean? A case study approach. *Veterinary Record* **176:** 654.

Mason GJ, and Latham NR 2004 Can't stop, won't stop: is stereotypy a reliable animal welfare indicator? *Animal Welfare* **13:** S57–S69.

Mather JA 2008 Cephalopod consciousness: behavioural evidence. *Consciousness and Cognition* **17:** 37–48.

Mendl M, Burman OHP, Parker RMA, and Paul ES 2009 Cognitive bias as an indicator of animal emotion and welfare: emerging evidence and underlying mechanisms. *Applied Animal Behaviour Science* **118:** 161–181.

Mepham B 2008 *Bioethics: An Introduction for the Biosciences.* Oxford University Press: Oxford.

Miele M, Veissier I, Evans A, and Botreau R 2011 Animal welfare: establishing a dialogue between science and society. *Animal Welfare* **20:** 103–117.

Morgan CA, and McDonald M 2007 Ethical dilemmas in veterinary medicine. *Veterinary Clinics of North America-Small Animal Practice* **37:** 165–179.

Morton DB, Sant'Ana MM, Ohl F, Ilieski V, Simonin D, Keeling L, Wohr AC, Zemljic R, Neuhaus D, Pesle S, and de Briyne N 2013 FVE & EAEVE Report on European Veterinary Education in Animal Welfare Science, Ethics and Law. L'ordre nationale des vétérinaires.

Mossop L 2015 Novice to expert? Teaching and development of moral reasoning skills. *The Veterinary Record* **177:** 122–123.

Page-Jones S, and Abbey G 2015 Career identity in the veterinary profession. *Veterinary Record* **176:** 433.

Prickett RW, Norwood FB, and Lusk JL 2010 Consumer preferences for farm animal welfare: results from a telephone survey of US households. *Animal Welfare* **19:** 335–347.

Rawles K 2000 Why do vets need to know about ethics? In: LeGood G (ed) *Veterinary Ethics: An Introduction*, 3–16. Continuum Publishing: London.

Rollin BE 1993 Animal welfare, science and value. *Journal of Agricultural and Environmental Ethics* **6:** 44–50.

Rollin BE 1996 Ideology, ''value-free science'', and animal welfare. *Acta Agriculturae Scandinavica Section A-Animal Science* **S27:** 5–10.

Rollin BE 2000 Veterinary ethics and animal welfare. *Journal of the American Animal Hospital Association* **36:** 477–479.

Rollin BE 2006 *An Introduction to Veterinary Medical Ethics: Theory and Cases, 2nd Edition.* Blackwell Publishing: Oxford.

Rollin BE 2011 Euthanasia, moral stress, and chronic illness in veterinary medicine. *Veterinary Clinics of North America-Small Animal Practice* **41:** 651–659.

Royal College of Veterinary Surgeons 2014 *RCVS Day One Competencies*, updated March 2014. http://www.rcvs.org.uk/document-library/day-one-competences/

Royal College of Veterinary Surgeons 2015 Vet Futures: 94% of British public says "We trust you, you're a vet". Royal College of Veterinary Surgeons.

Sandøe P, and Simonsen HB 1992 Assessing animal welfare: where does science end and philosophy begin? *Animal Welfare* **1:** 257–267.

Self DJ, Olivarez M, Baldwin DC, and Shadduck JA 1996 Clarifying the relationship of veterinary medical education and moral development. *Journal of the American Veterinary Medical Association* **209:** 2002–2004.

Shaw WH 2010 An introduction to ethics. In: Shaw WH (ed) *Social and Personal Ethics 7th Edition*, 2–13. Cengage Learning, Boston.

Sherwin CM 2001 Can invertebrates suffer? Or, how robust is argument-by-analogy? *Animal Welfare* **10:** S103–S118.

Siegford JM, Berezowski J, Biswas SK, Daigle CL, Gebhardt-Henrich SG, Hernandez CE, Thurner S, and Toscano MJ 2016 Assessing activity and location of individual laying hens in large groups using modern technology. *Animals (Basel)* **6.**

Spinka M, Duncan IJH, and Widowski TM 1998 Do domestic pigs prefer short-term to medium-term confinement? *Applied Animal Behaviour Science* **58:** 221–232.

Stark C, and Dougall N 2012 Effect of attitudes to euthanasia on vets' suicide risk. *Veterinary Record* **171:** 172–173.

Stern M 2015 Judge denies legal personhood to chimps– for now. Slate.

Stodart E, and Myers K 1964 A comparison of behaviour, reproduction, and mortality of wild and domestic rabbits in confined populations. *C S I R O Wildlife Research* **9**: 144–159.

Stolba A, and Woodgush DGM 1989 The behaviour of pigs in a semi-natural environment. *Animal Production* **48**: 419–425.

Tannenbaum J 1995 *Veterinary Ethics: Animal Welfare, Client Relations, Collegiality*. St Mosby-Year Book.

Thompson PB 1999 From a philosopher's perspective, how should animal scientists meet the challenge of contentious issues? *Journal of Animal Science* **77**: 372–377.

Vanhonacker F, Verbeke W, Van Poucke E, and Tuyttens FAM 2008 Do citizens and farmers interpret the concept of farm animal welfare differently? *Livestock Science* **116**: 126–136.

Veissier I, Jensen KK, Botreau R, and Sandøe P 2011 Highlighting ethical decisions underlying the scoring of animal welfare in the Welfare Quality® scheme. *Animal Welfare* **20**: 89–101.

Verrinder JM, McGrath N, and Phillips CJC 2016 Science, animal ethics and the law. In: Cao D and White S (eds) *Animal Law and Welfare – International Perspectives,* 63–85. Springer Publishing: Cham, Switzerland.

Verrinder JM, and Phillips CJC 2014 Identifying veterinary students' capacity for moral behavior concerning animal ethics issues. *Journal of Veterinary Medical Education* **41**: 358–370.

Viner B 2010 *Success in Veterinary Practice: Maximising Clinical Outcomes and Personal Wellbeing*. Wiley-Blackwell: Oxford.

Weaver SA, and Morris MC 2004 Science, pigs, and politics: A New Zealand perspective on the phase-out of sow stalls. *Journal of Agricultural & Environmental Ethics* **17**: 51–66.

Weeks CA, Danbury TD, Davies HC, Hunt P, and Kestin SC 2000 The behaviour of broiler chickens and its modification by lameness. *Applied Animal Behaviour Science* **67**: 111–125.

Yeates J 2013 *Animal Welfare in Veterinary Practice*. Wiley-Blackwell: Oxford.

Yeates JW, and Main DCJ 2008 Assessment of positive welfare: a review. *Veterinary Journal* **175**: 293–300.

Yeates JW, and Main DCJ 2011 Veterinary opinions on refusing euthanasia: justifications and philosophical frameworks. *Veterinary Record* **168**: 263.

Zhou Q, Sun Q, Wang G, Zhou B, Lu M, Marchant-Forde JN, Yang X, and Zhao R 2014 Group housing during gestation affects the behaviour of sows and the physiological indices of offspring at weaning. *Animal* **8**: 1162–1169.

CHAPTER 2
MAKING ETHICAL DECISIONS

2.1
How are ethical decisions made and justified?

In chapter 1 we talked about the need for ethical and moral reasoning in veterinary and animal health settings. Ethical decisions can be intuitive – that is, we just "know" or "feel" what is right without undertaking extensive conscious deliberation. But there can be substantial disagreement about what one person thinks is the right thing to do, and, as discussed in chapter 1, there is an expectation that professionals do more than simply act on their "gut feeling". Furthermore, there is an expectation that such decisions can be justified. Just as veterinarians have to justify their diagnostic and therapeutic considerations in medicine and surgery, you must base ethical decisions on the best-quality evidence.

In this chapter we will introduce a number of ethical tools or frameworks that are used in veterinary and medical ethics to aid ethical decision-making. Many of these will be used, alone or in combination, in the scenarios contained in the following chapters.

2.2
Ethical theories and ethical frameworks

Philosophers and, more recently, those employed in the field of bioethics have proposed ethical theories and frameworks to help systematise ethical decision-making. The aim is to ensure a robust, logical and consistent approach to decision-making so that we act in accordance with our values.

An ethical theory aims to distinguish all morally right from all morally wrong actions (Kaiser, et al. 2007). It might be a set of statements or rules based on values (what is good and what is bad). An ethical framework is a tool to facilitate practical decision-making, usually based on one or more ethical theories (Kaiser, et al. 2007).

2.3
What is applied ethics?

The term "applied ethics" refers to the application of theoretical approaches – as outlined below – to real-life, practical problems. Ethical theorists are often criticised for impractical, "ivory tower" thinking. One might imagine a continuum with pure theory at one end and practice at the other.

△

2.2 The academic has the luxury of time when it comes to ethical dilemmas.

CARTOON DR ROBERT JOHNSON

At the theory end is the academic, who can think deeply, with the luxury of time and without legal and practical constraints, about ethical dilemmas. At the other end are the practising veterinarians and health professionals who must make decisions, often with little time for deliberation, constrained by relevant legislation and codes.

Some scholars argue that while ethical judgements may implicitly commit us to one ethical theory or another, most of the time we navigate problems well enough "by appealing to virtuous habits instilled in us by our parents, or to various rules of thumb that have provided good guidance in the past, or by groping our way analogically from one case to another" (Arras 2010).

However, there is an expectation that professionals such as veterinarians and allied health workers can provide a higher level of justification for their decisions. Nonetheless, not every ethical

decision is the same. In ethical decision-making, the veterinarian is guided by their conscience, their own beliefs about what constitutes ethical behaviour, legislation, professional guidelines and cultural norms (Devitt, et al. 2014). Different interpretations of professional challenges and different approaches may lead to uncertainty about the appropriate response (Devitt, et al. 2014). Increasingly, medical doctors in major hospitals have access to clinical ethicists. While some larger veterinary hospitals may have ethics committees, there are currently few situations we are aware of where veterinarians can call on ethical advice from those in the committee about managing a particular case.

At very least, the ethical theories and frameworks we are about to introduce can provide guidance and structure for decision-making.

2.4

What does ethical decision-making involve?

Essentially, ethical decision-making involves four steps:

(1) Identifying and characterising ethical concerns
(2) Identifying stakeholders
(3) Determining the information and evidence required to make a decision
(4) Using ethical frameworks to make a considered judgement.

We are first required to recognise that there is an ethical issue and identify the relevant stakeholders. Stakeholders in an ethical issue are those parties who may be directly or indirectly affected (negatively or positively) by a decision.

△
2.3 12-year-old downer cow belonging to a tourist farm gives rise to an ethical dilemma.
PHOTO ISTOCK

For example, you are a veterinarian asked to attend to a 12-year-old Jersey cow that went down in the field yesterday (Mullan 2006). She calved four months ago. Following physical examination you conclude that she most probably is suffering from a hind limb injury, and has a poor prognosis for return to function. The cow belongs to a tourist farm and is a local celebrity. You recommend euthanasia but the owners are adamant that the cow has a chance to recover.

There are a number of stakeholders involved in this scenario: at a minimum, these include the cow, the owners, the veterinarian and the local community.

The next step is to determine the information required to make a decision. What do you know already and what do you need to find out? In order to navigate this scenario you need to have some knowledge about the health and welfare consequences of "downer cows" – and the likely prognosis. To what extent is this animal suffering? Is there an acceptable period to wait for improvement prior to euthanasia? Is there legislation, such

as animal welfare legislation, codes or guidelines, which dictates the appropriate action? Does a tourist farm have greater responsibilities to ensure the welfare of its animals? To what extent are these owners capable of providing appropriate care for this animal?

Step 4 is typically the source of most uncertainty, debate and disagreement. Once you have the information, how do you make your decision? How do you weigh up the interests of each stakeholder? For example, is the owners' interest in making a profit from keeping this cow on display worth more, less or is it equal to the cow's interest in not suffering? Do you consider the rights of stakeholders? For example, does the cow have a right to life? Does the owner have a right to refuse the killing of their animal? Does the veterinarian have the right or authority to dictate the way an animal is treated? Do you attempt to predict the consequences of various courses of action, such as treating the cow or killing the cow? Or are the consequences irrelevant?

This is where different ethical theories and frameworks are employed. They aid decision-making by assisting us in weighing up or ranking values or principles.

2.5

Ethical frameworks used in veterinary practice

Philosophers have been developing, honing and debating about the details of ethical theories for millennia. It is impossible to comprehensively cover even a single ethical theory in a book like this. Instead, we have provided a brief summary of key ethical theories and frameworks used in clinical decision-making. For each, we have provided

—
2.5
ETHICAL FRAMEWORKS USED IN
VETERINARY PRACTICE

a general definition, examples, advantages and limitations. Readers are directed to the recommended reading for additional information if they wish to explore further.

> "Whichever theory or framework you rely on more in a given situation is likely to be influenced by your intuition, based on personal and cultural factors, as well as your formal education and training."

As you read about these approaches, you may recognise one or more as approaches you have used when faced with an ethical issue. Whichever theory or framework you rely on more in a given situation is likely to be influenced by your intuition, based on personal and cultural factors, as well as your formal education and training. Having a logical framework to shape and refine your decisions can help avoid inconsistencies that may be produced by a knee-jerk reaction or influenced by a dominating owner or other stakeholder.

2.5.1 Utilitarianism

DEFINITION

Utilitarianism holds that ethical decision-making is sound if it leads to the greatest good for the greatest number of stakeholders. Good is defined as maximal pleasure and minimal pain. Therefore the utilitarian approach weighs up the costs and benefits of different courses of action, with the aim of arriving at an outcome that produces the greatest good for the greatest number. If the circumstances are such that there cannot be a *good* outcome, the aim is to seek

△

2.4 The utilitarian approach involves weighing up costs such as pain and suffering against benefits such as pleasure and the greater good.
CARTOON RAFAEL GALLARDO ARJONILLA

the outcome that yields the least suffering (or in the case of an ethical dilemma, "the lesser of two evils").

BACKGROUND

Utilitarians consider the expected consequences of any choice. It is therefore known as a consequentialist or teleological theory. For the purposes of this discussion we use the word "utilitarianism".

English philosopher Jeremy Bentham (1748–1832) produced the first comprehensive theory of utilitarianism. John Stuart Mill (1806–1873) and Henry Sidgwick (1838–1900) explored and developed this further.

Bentham considered the best outcome that which yielded the most "happiness" – limited to pleasurable experiences and the absence of pain. The concept has been refined and many now measure a good outcome by the overall "well-being" yielded. When considering animals in this light we need to be aware of their capabilities any sentient being, any being able to suffer, has an interest in not suffering and we should take this interest into account when considering the costs and benefits of an ethical decision.

EXAMPLES

If we consider that we can never know for certain whether an animal (or even another human being) is conscious we need to consider the consequences of treating them as though they can or cannot feel pain (in this case, to provide or withhold pain relief). One might then construct a table like Table 2.1.

In looking at this table, we can see that only the consequence of the action of providing analgesia to an animal that is feeling pain would result in "happiness" or wellbeing (positive pleasure or absence of pain, without wastage of analgesia), with the small potential cost of wasted time and expense if analgesia is not required.

Conversely, if we look at the consequences of withholding analgesia from an animal that is consciously experiencing pain, a larger harm results – untreated pain. There may be no significant harm or benefit to not providing analgesia to animals not consciously experiencing pain, presuming that the expense of providing analgesia is affordable. On this analysis we should provide analgesia.

△
2.5 Mammals, birds, reptiles and fish can experience pleasure and pain, so their interests (in experiencing pleasure and not experiencing pain) should be taken into account when making ethical decisions, according to utilitarians.
PHOTO ANNE FAWCETT

– mammals, birds, reptiles and fish can experience pleasure and pain (although there are still a few scientists who claim this is not consciously experienced and thus not morally relevant).

There are various subtypes of utilitarianism, such as rule utilitarianism, according to which the consequences of adopting certain rules are assessed, but these will not be discussed here.

The most well-known modern utilitarian is Australian philosopher Peter Singer, who argues that

POTENTIAL OUTCOME	ANALGESIA GIVEN	ANALGESIA NOT GIVEN
Animal consciously experiences pain	**Yes** Pain alleviated	**No** Pain
Animal unable to consciously feel pain	**Yes** Analgesia is wasted	**No** Analgesia is not wasted

△
Table 2.1 The consequences of providing analgesia to animals.

2.5
ETHICAL FRAMEWORKS USED IN
VETERINARY PRACTICE

△

2.6 The "3Rs", developed by Russell and Burch in the 1950s, are a utilitarian framework designed to minimise suffering – for example, by refining experiments so that analgesia is used – and maximise benefits associated with research.

CARTOON RAFAEL GALLARDO ARJONILLA

Consider the issue of whether or not to experiment on animals to trial a new cancer drug. If the alternative is testing the drug on human subjects, a scientist might argue that the animal trial is "the lesser of two evils". Even so, to ensure that overall harm is minimised, we would expect the experiments to be conducted according to the 3Rs: replacement of live animals with alternatives such as tissue cultures where possible, reduction of the total number of animals used and refinement of procedures and husbandry to ensure that harm is minimised and wellbeing is maximised (Russell & Burch 1959). This utilitarian framework is used by ethics committees in assessing potential experiments, and is focused on minimising suffering.

STRENGTHS

- Utilitarianism promotes welfare, in the sense that an outcome is good because of how much overall happiness or wellbeing it yields.
- Animals are considered stakeholders because of their capacity to suffer. One does not need to be a "moral agent" (that is, capable of making moral or ethical decisions and acting on these) to benefit from utilitarianism.
- Utilitarianism is impartial, that is, we are obliged to assess outcomes according to the overall wellbeing or good that is produced, rather than our own ends. In this sense, every stakeholder is morally equal to every other, and different treatment of one stakeholder can only be justified if it is good for the majority of stakeholders.
- The consequences of an action are taken into account. Thus for a utilitarian, it is acceptable to break rules (including the law) if a better outcome is produced than by not doing so. A utilitarian would hold that it is ethically acceptable to speed on the way to the veterinary hospital if you are transporting a dog suffering from an anaphylactic reaction if (a) the speeding doesn't result in any fatalities or injuries; (b) getting there faster enables you to perform a tracheostomy and save the dog's life; and (c) you don't get caught for speeding or, if you do, the benefits of saving the dog outweigh any negative consequences incurred.

LIMITATIONS

- Utilitarianism can be used to justify immoral means to an end. If the only way to achieve the maximal outcome is to perform an immoral act, the utilitarian approach suggests that such an immoral act is not only permissible but obligatory (Beauchamp & Childress 2013). For some this is objectionable as utilitarians can argue that it is acceptable to do something wrong (for example, speeding, as above, or even taking the life of another to harvest organs) if it yields a good outcome.

- Our ability to predict the consequences relies on us knowing all of the salient information at the time we make the decision, as well as good luck. Research can sometimes improve the accuracy of our predictions – for example, textbooks and journals, the opinion of a more experienced colleague – but this information may not always be available or adequate. Consider that the patient in the analgesia example is a tarantula. We don't know a lot about neurophysiology and behaviour related to perception of pain in this species. The benefits of providing analgesia may not be so clear, and costs may include potential harm caused by untested drugs and unfamiliar techniques of administration.
- What we consider to be the greatest good can change over time. For example, we may weigh up the pain of an animal's suffering against the costs (financial, resources, time, adverse drug reactions) of providing analgesia. If resources are scarce, would it still be justifiable to spend more money alleviating the pain of one animal where this could be used to save the life of another? In such a case, the greatest good might be served by *not* providing analgesia.
- The utilitarian approach can be used to justify gross inequality. For example, as long as the majority of stakeholders are happy, minorities can be neglected or even persecuted. The minority have no rights. In fact, a criticism of utilitarianism is that "might makes right". The majority rules, and problems arise when the "good" for an individual is at odds with the "good" of the majority. For example, if the organs of one individual could be harvested to save five others, a utilitarian would support such a decision because it yields greater overall good to the majority.
- By focusing on the maximisation of benefits, the fairness of their distribution is not addressed. Thus claims of the worst-off groups may not be considered.
- Utilitarianism does not recognise the rights of the individual. The individual has no right to make decisions that go against the interests of the majority, and no protection.
- The utilitarian approach requires that we weigh positive against negative consequences – which can be challenging. For example, how much weight do you give to the suffering of laboratory animals versus the suffering of people with an intractable disease who may benefit from a medical experiment on those animals? Does that change if you are considering the use of laboratory animals in safety testing a new hair shampoo versus a cancer treatment for children?
- Such a system is open to abuse through underestimation of costs and overestimation of benefits.

△

2.7 Consequentialist theories, including utilitarianism, require us to be able to accurately predict the consequences of our actions.

PHOTO ANNE FAWCETT

—
2.5
**ETHICAL FRAMEWORKS USED IN
VETERINARY PRACTICE**

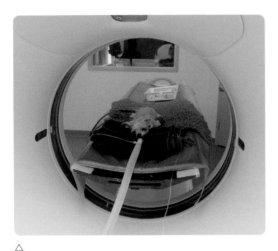

△
2.8 According to a utilitarian analysis, offering a CT scan with every vaccination for the sole purpose of maximising income may be ethically sound if one happens to diagnose a large number of operable tumours.
PHOTO ANNE FAWCETT

- In utilitarian systems it is the outcome, not the intention of the decision-maker, that counts most. This approach can lead to unpalatable outcomes. At best, it can favour the "lucky fool" – someone who just happens to make good decisions by chance. For example, consider a vet who offers Computed Tomography (CT) with every vaccination for the sole purpose of maximising her income. On the first morning she offers advanced imaging to three owners, and all three CTs detect operable, benign brain tumours. All three owners consent to surgery, and two dogs survive and go on to live normal lives. Would we say this veterinarian's actions are inherently good? Probably not. We might say she was very lucky.

Recommended reading

Singer P (1995) *Animal Liberation 3rd Edition*. Pimlico (Random House): London.

Ryan A (ed) (1987) *Utilitarianism and Other Essays.* Penguin Classics: London.

2.5.2 Deontology

DEFINITION
The word deontology is derived from the Greek root "Deon" pertaining to duty or obligation. Deontological ethical theories hold that a decision is correct if it conforms to a moral norm or rule. In other words, we should aim to do the "right" thing, rather than aiming for the most "good". Moral choices should be made following certain duties (or rules), and some choices are morally forbidden. For example, we have a duty to tell the truth (and thus a duty not to lie).

These duties or rules must be followed whatever the situation and in every similar situation regardless of the consequences and regardless of whether or not following the rules will result in the greatest overall good.

△
2.9 Deontologists prioritise conformity to moral norms or rules.
CARTOON RAFAEL GALLARDO ARJONILLA

At the heart of the deontological approach is the premise that rights of individuals place limits on how we may treat them. We cannot justify harms to an individual or group by citing benefits to other individuals or another group.

BACKGROUND

The most well-known deontological philosopher was Immanuel Kant (1724–1804). He developed the "categorical imperative" which states that one should "act only according to that maxim [rule] whereby you can, at the same time, will that it should be universal law".

Or, act as if the principle you are acting by is a law of nature. When it comes to ethical decisions, Kant wants you to base your decision on the belief that everyone else in the world should do the same thing.

American philosopher Tom Regan (1938–) put forward a deontological case for animal rights in his book of the same name (Regan 2004). He argues that humans and animals are "subjects-of-a-life" and thus have similar desires, preferences, beliefs, interests and so on.

According to Kant, only moral agents – those who can apply abstract moral principles in decision-making – share this equal right to be respected and always treated as ends-in-themselves.

Regan differs, arguing that all subjects-of-a-life – even those humans and non-human animals without the capacity to apply abstract moral values – have equal inherent value and therefore are entitled to respect.

Though perhaps less influential on the treatment of animals than the utilitarian school of thought, deontology facilitates the mindset that many animals deserve more respect than they may have been afforded in the past, even if few people admit to subscribing to an "animal rights" view (Coeckelbergh & Gunkel 2014).

△
2.10 You are called upon to treat a horse with colic.
PHOTO ANNE FAWCETT

EXAMPLE

You are called upon to treat a horse with colic. The horse has been showing signs for days, but the owner – who recently lost a family member – had been away for a funeral and left the horse in the care of his neighbours, who are not experienced with horses. By the time you examine the animal it is suffering from endotoxaemic shock, and dies despite intensive care.

The owner approaches you and asks, "Could this have been prevented?"

You know the animal would have suffered less – and likely survived – with early intervention, but you also know that admitting this would cause the owner enormous guilt and strain his relationship with the neighbours. A pure utilitarian may avoid answering the question honestly, possibly deflecting the question or even lying, to spare the feelings of the parties involved.

But this involves telling a lie, which is wrong. A deontologist would not do this under any circumstances.

What if we apply the test of universalisation? If the veterinarian in the scenario does act dishonestly, would we really hope every other vet in the same position acted the same way? The answer has to be no. If everyone took this course of action, there would be no reason for negligent owners or carers to question their actions, and trust in the veterinary profession would be eroded.

The same principle can be used to test rules. For example, one might state that it is wrong to kill a healthy animal. Should that apply in all situations? What if a healthy dog had been badly socialised, and was highly aggressive – to the point that it attacked all persons interacting with it? What about killing animals for meat? If there are too many exceptions, this may not be a suitable rule.

Astute observers will note that this test of universalisation does in fact consider consequences. To this end there is some overlap between utilitarianism and deontology, even if they may be considered at opposite ends of an ethical continuum.

STRENGTHS
- Unlike strict utilitarianism, deontological theories take our intentions into account.
- Deontological theories appeal because they encompass the notion of individual rights, a concept highly valued in liberal democracies. This is an ethical theory that safeguards the interests of the individual.
- The language of deontology reflects that of legislation and codes of professional conduct, which give clear guidelines on how to conduct ourselves.

LIMITATIONS
- To strict deontologists the consequences of an action are irrelevant. By simply ensuring that we do our duty or act according to certain rules, we are "right". The problem is that no consideration is given to anything that results from following

our duty. This can lead to absurd outcomes, where doing our "duty" or the "right thing" can lead to an ethically unsound outcome. One of the most prominent criticisms of the deontological approach is that it fails to take the context of the ethical problem into account. For example, Kant held that it was wrong to lie. As a deontologist, he believed that this was wrong in every circumstance. If it is our duty not to lie, should we help an animal abuser by telling them (if they ask) the whereabouts of their intended victim?
- It is inflexible. One cannot, under any circumstances, do something "bad" to achieve a "good" outcome.
- Another difficulty of this approach is that it doesn't give us guidance as to how we prioritise conflicting rights. What happens when my right to treat my pet as I wish conflicts with their right to avoid suffering – such as the pain and distress associated with neutering? In such cases we need to be able to prioritise our rules. This could be done on the basis of avoiding the worst harms, avoiding harming the worst off or choosing to harm the lowest number of individuals. This problem can become apparent at work – whom do we have a primary duty towards – our patients, our employer or the client? Many veterinarians swear an oath on joining the profession, which states that animal welfare is the primary consideration – but it does not mention the reality of duties towards owners, financial and professional. Animals are also property in law and, although vets may have a degree of moral authority, they have no power to force owners to treat animals in a certain way (Swan 2006). A veterinarian who follows the oath to the letter may find themselves facing untenable conflict with employers and clients.
- Deontological duties are usually phrased as negative constraints on our actions, for example do not kill another; do not lie; do not steal. These tend to represent only the "bottom-line"

of acceptable behaviour and do not promote positive duties. They do not encourage us to act in the "spirit" of the duty. Thus if we can find a loophole – in the way we phrase the duty or our reading of the law – we can get out of it.

Recommended reading

Hill TE (ed) (2009) *The Blackwell Guide to Kant's Ethics.* Wiley Blackwell: London.

Regan T (2004) *The Case for Animal Rights.* The University of California Press: Berkeley.

2.5.3 Virtue ethics

DEFINITION

Virtue ethics assumes that good decisions follow from having a virtuous character. A virtuous person displays virtues, which are traits that are reliably present in an individual, manifested in habitual action. A virtue is good for a person to have, as well as good for the others around that person. Consider the virtue of temperance, or personal restraint. Those who possess this virtue benefit as they do not suffer the consequences of overindulging. But those around benefit as the temperate individual does not deplete resources (such as food) that might be otherwise available to others.

In contrast to the other theories discussed so far, virtue ethics defines good in terms of a person's character rather than the rightness of a decision or action.

BACKGROUND

The Greek philosopher Aristotle (384–322 BC) held that virtuous actions arose from a virtuous character. The "cardinal" virtues were courage, prudence, temperance and justice. For Aristotle, a virtue was defined by corresponding vices, usually those of excess and deficiency. Therefore the virtue courage might lie between foolhardiness at one end of the spectrum, and cowardice on the other. Not only should one possess virtues, but one should also express them in their behaviour.

This approach has been explored and developed by a number of thinkers, including St Thomas Aquinas, and more recently the British philosopher Philippa Foot (1920–2010). It has

△
2.11 Courage is a virtue, but too much (foolhardiness), or too little (cowardice) is problematic.

CARTOON RAFAEL GALLARDO ARJONILLA

2.5

ETHICAL FRAMEWORKS USED IN
VETERINARY PRACTICE

become increasingly popular in the field of bio-medical ethics:

"What matters most in moral life is not adherence to moral rules, but having a reliable character, a good moral sense, and an appropriate emotional responsiveness."

(Beauchamp & Childress 2013)

Proponents of virtue ethics emphasise the importance of personal development:

"We should learn to apply the virtues (courage, honesty and justice) in order to develop our practical wisdom (phronesis) and work

towards the greater good. We should question and scrutinise our motives regularly in order to monitor how our intentions translate into action."

(Cousquer 2011)

As such, mentoring and role-modelling are important methods of conveying virtues to students (Lee 2013).

Beauchamp and Childress identify five focal virtues – compassion, discernment, trustworthiness, integrity, conscientiousness (see Table 2.2) – as well as care, respectfulness, nonmalevolence, benevolence, justice, truthfulness and faithfulness as important virtues for medical professionals

VIRTUE	DEFINITION
Compassion	The virtue of compassion combines an attitude of active regard for another's welfare with an imaginative awareness and emotional response of sympathy, tenderness, and discomfort at another's misfortune or suffering.
Discernment	Discernment involves the ability to make fitting judgements and reach decisions without being unduly influenced by extraneous considerations, fears, personal attachments, and the like.
Trustworthiness	Trust is a confident belief in and reliance on the moral character and competence of another person...Trust entails a confidence that another will reliably act with the right motives and feelings and in accordance with appropriate moral norms. To be trustworthy is to merit confidence in one's character and conduct.
Integrity	In its most general sense, "moral integrity" means soundness, reliability, wholeness and integration of moral character. In a more restricted sense, the term refers to objectivity, impartiality, and fidelity in adherence to moral norms.
Conscientiousness	An individual acts conscientiously if he or she is motivated to do what is right because it is right, has tried with due diligence to determine what is right, intends to do what is right, and exerts appropriate effort to do so. Conscientiousness is the character trait of acting in this way.

△

Table 2.2 Beauchamp and Childress' five focal virtues.
SOURCE BEAUCHAMP & CHILDRESS (2013)

(Beauchamp & Childress 2013). Initiative, self-discipline, responsibility, integrity and accountability were identified as virtues to be taught explicitly in a surgical residency training programme (Lee, et al. 2012).

We consider certain virtues as synonymous with being a "good vet".

EXAMPLE

You are an equine veterinarian called to treat an animal you know will compete in two days. The horse is lame in the left forelimb, and you determine that it requires treatment and rest. The client asks you to provide analgesia to enable the animal to compete.

△

2.12 The question for the virtue ethicist here is, what would the good vet do?

CARTOON RAFAEL GALLARDO ARJONILLA

The question for the virtue ethicist here is, what would the good vet do? The truly virtuous vet of course would not need to deliberate – the appropriate action would follow from their virtuous character.

If we work through the five focal virtues we can appreciate why:

While you may feel sympathetic to the owner's position and sorry for their misfortune, you also have **compassion** for the horse which will suffer if it races. While the owner may be upset if you refuse to treat the animal according to their wishes, your professional **discernment** ensures you are not influenced by these concerns. The **trustworthy** professional would not engage in deceit, even when pressured to do so. Your **integrity** ensures that you act consistently and reliably to ensure the horse's welfare. Ultimately you convince the owner that the horse must be withdrawn from competition because you are a **conscientious** veterinarian.

STRENGTHS

- Virtue ethics recognises that emotions are key in our ethical sensitivity and decision-making.
- The intent or motivation of the individual matters. The virtuous individual may make a decision that yields a bad outcome, but this does not render him or her non-virtuous. The difference is that it is not obedience to a rule that motivates the virtuous, but virtue itself.
- It may lead to more simple or creative approaches to ethical dilemmas than simply following rules or principles.
- The approach is more flexible than the other approaches as virtuous people may behave in different ways despite being in similar circumstances.
- Virtues correspond with societal expectations about professionals. Wherever one sits on the virtue ethics approach, veterinarians and allied health professionals are expected to behave

in ways consistent with the virtues discussed above.

- There is an emphasis on development and reflection. The virtue approach accommodates the fact that individuals can develop and improve.
- The approach allows the virtuous to feel regret rather than focus on the rightness (or wrongness) of a decision or act. In this sense it acknowledges the complexity of ethical decision-making – even in cases where we are convinced we have made the best decision, we can regret that not all stakeholders will be satisfied (Gardiner 2003).

LIMITATIONS

- There is much debate about whether virtue ethics can really offer a stand-alone framework or whether it must supplement an ethical theory that is able to generate rules and norms.
- Philosophers cannot agree about how we should deal with conflict between virtues. For example, if both loyalty and honesty are virtues, it is difficult to determine the right course of action when confronted with the question of whether to intervene when a colleague who has systematically undercharged clients asks you not to tell your employer.
- Similarly, virtues may be misused. Consider industriousness. The productive clinician is widely praised, but the workaholic veterinarian who rejects his family or refuses pleasure would not be considered virtuous.
- Our interpretation of how virtues manifest may vary depending on the role of individual stakeholders. For example, in the case of biomedical ethics, when deciding on whether to treat a child with an incurable condition, doctors make decisions based on their professional ethics and what it means to be a good doctor. But parents make their decisions based on what

it means to be a good parent. There may be substantial conflict. "Like doctors, parents have expertise – their expertise is about family life" (McDougall & Gillam 2014).

Recommended reading
Foot, P (2002) *Virtues and Vices and Other Essays in Moral Philosophy*. Clarendon Press: Oxford.

2.5.4 Justice as fairness, and contractarianism

DEFINITIONS
Justice as fairness is the belief that social institutions such as governments do not confer advantages on some individuals at the expense of others. Accordingly, racial, sexual, religious and class discrimination are condemned, as are many forms of economic inequality.

Contractarianism holds that ethical rules, norms and obligations derive from the idea of a contract or mutual agreement.

△
2.13 Contractarianism holds that ethical rules, norms and obligations flow from a contract or mutual agreement. But can animals enter into a social contract?
PHOTO MICHAEL QUAIN

BACKGROUND

Justice as fairness is attributed to philosopher John Rawls (1921–2002), largely in response to what he argued are major deficiencies of utilitarianism.

In particular, he wrote, "I do not believe that utilitarianism can provide a satisfactory account of the basic rights and liberties of citizens as free and equal persons, a requirement of absolutely first importance for an account of democratic institutions" (Rawls 1999).

Based on the concept of a social contract espoused by the philosophers Locke, Rousseau and to some extent Kant, he developed a theory that would appeal to free, rational people concerned with furthering their own self-interest.

The idea of a social contract is essentially that members of society hypothetically agree on the principles which assign basic rights and duties and determine how social benefits are to be divided. For example, one might imagine that a contract involves individuals giving up some freedom, such as agreeing to obey the laws of society, in exchange for the benefits those same laws confer on them. Individuals may agree to be governed in exchange for the right to elect the government, and so forth.

Of course the social contract doesn't really exist. But if it did, how might it be established?

Rawls used a thought experiment called "the original position". In this position, rational people would develop the principles by which society would operate – but they would make these positions from behind a "veil of ignorance". That is, they would not know their position in society – whether they are rich or poor, working class or upper class – nor would they know about their natural assets such as intelligence and strength. According to Rawls, this veil of ignorance would ensure that principles generated from this position would be fair. After all, those in the original position would not know whether they would be better or worse off, and thus would generate rules and norms impartially.

There is much debate about how Rawls' theory accommodates animals. In fact Rawls excluded animals from his theory because they were not rational, "moral persons" with a sense of justice, and did not contribute to society in a cooperative venture to mutual advantage (Garner 2013, Vickery 2013).

But what if those behind the veil of ignorance didn't know which species they might end up being? Some scholars argue that rational persons would generate norms that are fair to animals, as it would not be in their interests to act according to principles where animals are treated unjustly (Vickery 2013).

Because Rawls didn't include animals in this theory, and there is debate about how animals might be included, it's difficult to apply this scenario. For example, would rational beings behind the veil of ignorance condone any form of animal use, given they might themselves be thrust into the position of an animal? Could animals conceivably agree to "work" just as humans do, and under what conditions might they do so?

Nonetheless, the theory of justice is easily applied when discussing relations between humans.

EXAMPLE

You are employed by a university and involved in a committee to set entry criteria for a veterinary degree. Competition for places is high, and the committee can afford to set the bar high – incoming students are charged costly fees and must achieve almost perfect grades. However, studies have shown that this criterion selects only for students from wealthy families who have attended elite private schools.

Rational persons, behind the veil of ignorance, would recognise that this situation is inherently

△
2.14 If traditional school exams favour some people unfairly should veterinary entry criteria aim to redress this inequality?
PHOTO ISTOCK

unjust. After all, students from poor families didn't choose their circumstances, nor did those from wealthy families. In order to restore justice, one might consider an alternative entry pathway for disadvantaged students. Alternative entry criteria may be established, such as an interview or demonstration of practical skills.

STRENGTHS

- While others had written about a social contract, Rawls had a coherent theory about how such a contract could be generated, why it is necessary and who is party to it.
- Rawls believed that natural attributes, such as age, race, gender, intelligence, beauty and so forth, are not fairly distributed, and therefore the "veil of ignorance" eliminates the risk of people manipulating those attributes to their own advantage at the expense of others.
- Acting ethically (fairly and justly) is argued to be in one's self-interest.

LIMITATIONS

- The concept of justice is open to different interpretations. For example, one person might take justice to emphasise people getting what they deserve in terms of punishment, while another might take justice to emphasise the fair distribution of entitlements (for example, accessible veterinary care).
- The concept of distributive justice (meaning the fair and equal distribution of resources within a society) may also mean very different things to different people. For example, some may argue that resources should be distributed in terms of need, while others may emphasise merit, effort, market forces or equality. The same individual may have different conceptions of justice for humans and non-human animals.
- Contractarianism is often criticised for its anthropocentric (human-centred) perspective. As only humans can act as rational agents in entering a social contract, this approach favours human interests and at best confers animals with indirect rights – subject to human whim.

Recommended reading
Rawls J (1999) *A Theory of Justice (Revised Edition)*. Oxford University Press: Oxford.

2.5.5 Ethics of care

One thing the previous ethical theories have in common is the assumption that ethical decision-making is impartial. But many critics claim that such an approach fails to acknowledge the reality of human relationships. Some argue that the more universal, abstract, impartial and rational ethical decision-making is, the further it is from reality: how can one make an ethical decision that aligns with our values if we cannot take into account our personal relationships? If we must

△

2.15 The ethics of care contends that our relationships with people give rise to obligations to prioritise their interests.

PHOTO ANNE FAWCETT

△

2.16 The ethics of care contends that in the case of a burning building, if we have to choose whom to save, we should first save the person with whom we have an existing relationship over a stranger.

CARTOON RAFAEL GALLARDO ARJONILLA

choose to save someone from a burning building, wouldn't many people choose to save their own child or parent over the life of a stranger? And doesn't our relationship with this person in fact oblige us to do so?

Another issue is that emotion is stripped from ethical decision-making. This is problematic for a number of reasons. Our emotions make us sensitive to particular situations and inform our perceptions. To ignore these is to risk ignoring salient aspects of a scenario. We also need empathy in order to put ourselves in the shoes of another. Finally, while rational, non-sentimental decision-making is valued in our society, persons displaying little or no emotional response are viewed as abnormal or – in the case of psychopaths – dangerous and untrustworthy (Gardiner 2003).

These criticisms have been levelled by feminist ethicists, among others. Psychologist and philosopher Carol Gilligan (1936–) proposed an ethics of care which aimed to recognise our relationships and obligations to those close to us. The care approach to animal ethics is based on understanding the plight of animals in patriarchal and capitalistic institutions. From a care approach, the reason to oppose animal suffering is not to maximise utility or apply rights theory across species, but because we have relationships with animals and care about them (Engster 2006). It recognises that animals share with us many biologic needs, similar capabilities and a desire for survival.

STRENGTHS

- Care ethics recognises that the fact that we make animals dependent on us for their survival, functioning and wellbeing means we have a moral obligation to meet these needs.
- It provides for morally defensible "speciesism". If forced to choose between saving a dog and a mentally impaired human infant, Singer and Regan would prefer the survival of a dog because of the dog's higher level of self-consciousness. However, care theory sidesteps

concerns about complex consciousness, instead arguing that we have special obligations to human beings specifically because we have ourselves claimed care from other human beings, and thus owe other human beings similar moral consideration regardless of their capacities. Speciesism is not then discrimination, akin to sexism and racism, but a moral duty to care for our own kind.

- It demands that animals raised by humans are treated in caring ways, emphasising the quality of life we provide for them.
- It capitalises on our existing moral sentiments about many animals, particularly companion animals, and formalises a duty of care that many already recognise (Engster 2006).

LIMITATIONS

- Care ethics is not a theory in itself.
- The care ethics approach to animal welfare has been criticised for being confusing, vague and underdeveloped.
- There is no consensus about what it actually means to "care" for others and what this entails.
- There remain moral limits to the care we may legitimately expect from others (for example, if caring for us involves danger for the carer, it would compromise their long-term wellbeing or undermine their ability to care for other individuals).

Interested readers are referred to the following resources:

Gilligan C (1993) *In a Different Voice: Psychological Theory and Women's Development, Second Edition.* Harvard University Press, Harvard.

Engster D (2006) Care ethics and animal welfare. *Journal of Social Philosophy* **37**: 521–536.

2.5.6 Common morality

DEFINITION

Common morality is the ethical code shared by members of a society reflected in common values. In other words, all those committed to the objectives of morality share certain common values, rules or norms (see Table 2.3).

When we violate these norms, we feel remorse and attract the judgement of others.

....

COMMON VALUES, RULES AND NORMS
....

Don't kill or harm innocent people

Assist others in distress

Refrain from stealing or damaging property of
others

Keep promises
....

△

Table 2.3 Examples of common values, rules and norms that constitute "common morality".

BACKGROUND

The common morality is seen by many as a combination of all of the above, with an emphasis on utilitarian and deontological approaches.

Utilitarianism and deontology are typically viewed as polar opposites on a continuum of potential ethical positions. To some extent they can lead to similar outcomes. For example, both positions may be used to justify providing first-aid to an injured dog on the side of the road, or promoting an inclusive workplace.

However, as shown in the examples described above, when followed to the letter these approaches can lead to some ethically unpalatable or morally objectionable decisions. For example, in certain circumstances a utilitarian would argue that it is acceptable to torture someone (for example, a terrorist) if it were shown to benefit the wellbeing of others (for example, to find out the location of a bomb that would otherwise detonate and kill hundreds of innocent people). Similarly, a famous objection to deontology is contained within the following example. During the Second World War, in Nazi Germany, a Jewish friend has sought refuge in your home. Nazi soldiers knock on the door and demand to know if your friend is there. Telling the truth would mean certain death for your friend, but a true deontologist would not lie.

The argument is that both theories, when applied alone, can yield decisions which appear to go against the grain of the common morality. While utilitarianism and deontology are often pitched against one another, some argue that a combination of the two is more fitting with our common morality or predominant social ethic.

For example, Bernard Rollin argues that our predominant social ethic is one of modified utilitarianism (Rollin 2011), that is, we try to act according to what will yield the greatest good for the greatest number, but recognise also that moral agents (people) have inviolable rights, for example, the human rights of prisoners of war.

In legislation covering the use of laboratory animals in the UK, both views are encompassed. License applications are judged on a cost:benefit analysis (utilitarianism). The costs to animals involved in the experiment (confinement, pain and so on) are weighed up against the expected benefits (research outcomes). However, the UK Government has also decreed that no licences will be granted for procedures on great apes regardless of potential benefits (deontology) (Select Committee on Animals in Scientific Procedures 2002). Donald Broom has similarly argued that utilitarianism alone is not enough:

△
2.17 In the UK, as in the Netherlands, New Zealand, Sweden, Germany and Austria, experimentation is not allowed on great apes.
PHOTO ISTOCK

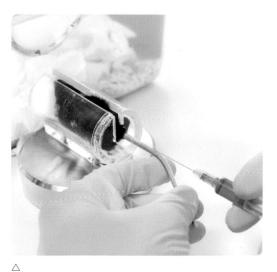

△
2.18 All applications for licences to experiment on non-great ape species are judged on a utilitarian cost:benefit analysis.
PHOTO ISTOCK

"If we use a living animal in a way that gives us some benefit, we have some obligations to that animal. One obligation is to avoid causing poor welfare in the animal except where the action leads to a net benefit to that animal, or to other animals including humans, or to the environment. Such a utilitarian approach is not sufficient to determine all obligations, however: A deontological approach is also needed because there are some degrees of poor welfare that are never justified to benefit others. For example, research scientists should not tear the limbs off a live cat even if they think that key information about curing disease might result from such an action."

(Broom 2016)

This is not an ethical theory so much as a yardstick against which many ethical theories are measured. Thus for example, pure utilitarianism and pure deontology can lead to conclusions that clash with the common morality.

STRENGTHS
- Reflects a "common sense" approach.
- Promotes human flourishing by avoiding or countering situations that cause the quality of people's lives to worsen.
- Recognises that life is more complicated and less predictable than utilitarianism allows, but less utopian than rights-based or deontological approaches suggest.
- Insofar as it can be established, intent does matter. Outcome cannot always be predicted and our social ethic recognises that no matter how good your intentions, the outcome can be bad. Consider two veterinarians, both of whom perform a routine spay, and in both cases the patient dies. Both patients were young, healthy puppies. Vet A took every precaution, proceeded carefully with surgery and dropped a

pedicle causing the patient to bleed to death. Vet B is burnt out. He rushed through the surgery and dropped a pedicle causing a patient to bleed to death. Most of us would like an ethical theory that recognises the differences in care shown by Vet A and Vet B as morally significant.
- Principles can be compromised if there is a good reason to do so.

LIMITATIONS
- There is much discussion about whether there is anthropological or historical evidence that a common morality (one that transcends societies and cultures) exists at all.
- It is difficult to appeal to a common morality when there is conflict across social contexts, cultures, religions and so forth.
- Reconciling deontology and utilitarianism can be difficult, if not impossible. For example, many people hold a utilitarian view around the killing of animals (that is, that it is acceptable because it may benefit others who consume the animal or use the animal for experiments which yield good for many) while simultaneously holding that it is always wrong to kill an innocent human. The problem is that the utilitarian principle as applied to the killing of animals creates a slippery slope – there will be some instances where the same principle permits killing of some human beings (Sandøe & Christiansen 2008).
- The norms or rules of the common morality tell us what not to do, for example do not kill, but they don't specify who is entitled to this protection. Thus it is argued that historically, the common morality has been compatible with slavery or discrimination based on the fact that the status of those enslaved or discriminated against excludes them from moral consideration. Because it tends to be based on

what is intuitively right, rather than a system of values based on reflection, the common morality is often described as the "lowest common denominator" of society's ethical norms, which may invite cynicism (Mepham 2008).

Despite these limitations, many scholars see the value in common morality and have sought to develop ethical approaches and frameworks which articulate and systematise the common morality. The two key examples we will discuss, both very influential in veterinary ethics, are principalism and the ethical matrix.

2.5.7 Principalism

DEFINITION
Principalism is a system of ethics based on four guiding principles of non-maleficence, beneficence, autonomy and justice.

Non-maleficence: this is engendered in the phrase "*Primum non nocere*" or "first, do no harm". Because of this use, the principle of non-maleficence is often invoked as the first principle: whatever we do, we should not make things worse.

Beneficence: this is the principle of promoting good. To this end we should try to improve the welfare of animals under our care in the short term and long term wherever possible. We should also do our best to aid our clients.

Autonomy: this principle pertains to the ability of animals or people to be self-governing. In human medicine this is seen as the patient's right to choose or consent. Conflict can arise between two autonomous people, for example the vet and the client. The bottom line for respecting autonomy usually ends when codes or laws are contravened or broken. Respecting animal autonomy may involve allowing animals to make choices about their life, for example where to eat and sleep.

△
2.19 The most well-known of the four principles is "do no harm", but Beauchamp and Childress argue that all principles are equally important.
CARTOON RAFAEL GALLARDO ARJONILLA

Justice: this principle is about the fair distribution of benefits, risks and costs. According to most interpretations, equals should be treated equally and unequals treated unequally.

Another way of putting the principles is this: "do no harm, only do good, respect the presumed choice(s) of the patients, and be fair and treat similar cases in a similar manner" (Anzuino 2007).

BACKGROUND
The most well-known proponents are Tom Beauchamp and James Childress, authors of *The Principles of Biomedical Ethics*. Their aim was initially to aid those working in human healthcare – doctors, nurses and healthcare professionals – facing ethical dilemmas and issues around patient care.

EXAMPLE
A client makes enquiries about trialling a new chemotherapy agent they have read about on

2.5
ETHICAL FRAMEWORKS USED IN
VETERINARY PRACTICE

their dog. A local referral hospital has been enrolling patients for a clinical trial using this agent.

The principalist approaches the issue by considering each of the guiding principles. In terms of non-maleficence there are a number of considerations: the client's dog may experience predicted and non-predicted adverse effects given this is a trial agent. Adverse effects may be mitigated to some degree through careful monitoring and treatment (for example use of anti-emetic agents in drugs that cause nausea and vomiting). Longer term there is potential harm due to immunosuppression (with risk of infection) as well as extensive follow-up.

The aim of the treatment is to increase the length and quality of the recipient's life, which meets the demands of beneficence if treatment is carried out in a way that minimises risks. There may be a way of achieving this without use of a novel agent, for example use of an existing agent for which the side-effect profile is more established.

Consideration of autonomy when it comes to non-human animals is challenging. The patient

cannot consent to nor refuse treatment, nor can the dog consent to enrol in a clinical trial. All decisions are made by a proxy who does not speak the patient's language nor can they confirm that they have acted in the patient's best interests. A weak argument may be made that the dog with cancer, once successfully palliated or cured, will be able to be autonomous once more.

Justice is also difficult. How can we ensure the patient and other stakeholders, such as the owner, are treated fairly?

When we consider the guiding principles, we need a fairly compelling case to justify enrolling this dog in a trial.

STRENGTHS

- The use of four guiding principles is much easier for non-philosophers than other frameworks (Gardiner 2003).
- The principles may be seen as an attempt to combine utilitarianism and deontology, with the first two principles emphasising utilitarian considerations while the latter two emphasise deontological considerations.
- The principles have been praised for their inclusiveness, being transcultural, transnational, transreligious and transphilosophical (Gillon 1998) – reflecting the common morality.
- In the human health field, the emphasis on autonomy and justice necessitates detailed knowledge sharing and informed decision-making, and ensures that patients are not subjected to controlling constraints.

△

2.20 Consideration of autonomy when it comes to non-human animals is challenging.

CARTOON RAFAEL GALLARDO ARJONILLA

LIMITATIONS

- There is much debate about how to use the principles. For example, the principle of non-maleficence is open to broad interpretation. What is acceptable risk? And how are we to distribute scarce resources when we may expose others to harm in doing so? In reality, any medical or veterinary intervention may inflict short-term harm (for example, invasive surgery) or potential harm (anaesthetic risk) that must be weighed against the longer-term good expected.

- Autonomy has two conditions – freedom (or independence from controlling influences) and agency (a capacity for intentional action). In the human literature, patients are considered competent to consent and make autonomous decisions if they have the capacity to understand information, make a judgement about it in light of their values, intend a certain outcome and communicate their wishes – none of which animals can do.

- The application of the principle of justice to animals is challenging and depends very much on the moral status of animals, which is contested. For example, if we have a finite resource such as food or water we could divide it equally between animals and people. However, some would argue that the needs of animals, having lesser or no moral status, are not of equal moral weight.

- In a particular scenario, the weight of each principle may vary – so good judgement is required to work out which should figure as more important.

Recommended reading

Beauchamp TL and Childress JF (2013) *Principles of Biomedical Ethics, 7th Edition.* Oxford University Press: New York, Oxford.

2.5.8 The ethical matrix

DEFINITION

The ethical matrix is a well-known framework which combines major ethical theories. The matrix is effectively a table (see Table 2.4), in which the first column lists all stakeholders including people, animals and biota (a term which refers to the plant and animal life of a region). The remaining columns correspond to three principles, which align with established ethical theories. "Wellbeing" combines the principles of non-maleficence with beneficence, and broadly corresponds to utilitarianism. "Autonomy" broadly corresponds to a deontological or Kantian theory, and "fairness" broadly corresponds with Rawls' theory of justice as fairness.

BACKGROUND

The matrix was devised by Ben Mepham (1996), formerly Director of the Centre for Applied Bioethics at the University of Nottingham. It has been further refined in collaboration with his colleagues at the Centre and the Food Ethics Council. It has been influential in veterinary ethics teaching around the world.

The design is based on the premise that ethical analysis should be comprehensive, and not exclude issues because they are complex or raise big questions like the concept of the free market. Accordingly, ethical analysis should subject the views of all parties to rigorous, transparent examination in the light of agreed ethical principles. Transparency is important in ensuring broad and inclusive social debate and making decisions that are robust and not easily challenged due to information gaps (Kaiser, et al. 2007).

Mepham (2008) states that, "it is important to appreciate that the aim of the ethical matrix is to facilitate rational decision-making, but not to determine any particular decision." The impacts defined for each separate cell depend on examination of the available evidence, but the matrix itself is ethically neutral – so that it requires weighing and/or ranking of impacts on the different interest groups. Thus the impacts of an action, such as the introduction of a new food product, can be compared with conditions where this action is not performed – so the status quo represents the baseline condition.

It is possible to use the matrix to generate scores. For example, a positive ethical impact might be scored +1 or +2 (indicating "significant" and "very significant" respect for the principle), while a negative score indicates similar degrees of infringement of the principles concerned (and a zero score means an insignificant or neutral impact). But these scores should not be regarded as a means of "calculating" an ethical judgement: they are merely means of recording the extent to which a principle is considered to be respected or infringed.

EXAMPLE

You are on a committee that has been asked to look into the introduction of a hormone to increase dairy cattle milk yield.

2.21 What factors should you consider before introducing a hormone to increase dairy cattle milk yield?
PHOTO ANNE FAWCETT

For any cell there may be conflicting versions of the appropriate "evidence". For example, looking at the autonomy of producers, on the one hand farmers are free to exploit this product. On the other hand, those late adopting this product may suffer financially, and those who don't may go out of business.

Once the cells are completed, users of the matrix can list and spell out the specific concerns of each of the interest groups (see Box 2.1).

RESPECT FOR: TREATED ORGANISM	WELLBEING ANIMAL WELFARE	AUTONOMY BEHAVIOURAL FREEDOM	FAIRNESS TELOS/INTRINSIC VALUE
Producers (e.g. farmers)	Satisfactory income and working conditions	Freedom to adopt or not to adopt	Fair treatment in trade and law
Consumers	Availability of safe food	Respect for consumer choice	Universal affordability of food
Biota	Conservation	Biodiversity	Sustainability

Table 2.4 An ethical matrix applied to the use of a hormone to increase dairy cattle milk yield.
ADAPTED FROM MEPHAM (2000, 2008)

BOX 2.1

DETAILED SPECIFICATION OF PRINCIPLES OF
THE ETHICAL MATRIX FOR USE OF A HORMONE
TO INCREASE DAIRY CATTLE MILK YIELD

Treated organism

WELLBEING: prevention of animal suffering; improving animal health; avoiding/minimising animal welfare risks, e.g. increased disease incidence.

AUTONOMY: ability to express normal behaviour, e.g. grazing, mating.

FAIRNESS: animals treated with respect for their intrinsic value as sentient beings as opposed to being treated just as useful possessions (instrumental value).

Producers (dairy farmers)

WELLBEING: satisfactory incomes and working conditions for farmers and farm workers (the term satisfactory is open to debate).

AUTONOMY: allowing farmers to use their skill and judgement in making managerial decisions, e.g. in choosing a farming system, choosing to opt into/out of using this product.

FAIRNESS: farmers and farm workers receiving a fair price for their work and produce; being treated fairly by trade laws and practices.

Consumers

WELLBEING: access to safe food sources; protection from food-poisoning and food-borne diseases (including adverse effects from drug or pesticide residues); good quality of life due to a productive and profitable farming industry.

AUTONOMY: a good choice of foods, appropriately labelled, together with adequate knowledge to make wise food choices; this principle also encompasses the citizen's democratic choice of how agriculture should be practised.

FAIRNESS: an adequate supply of affordable food for all, ensuring that no one goes hungry because of poverty.

Biota

WELLBEING: protection of wildlife from harm (e.g. by contamination of natural food sources), with remedial measures taken when harm has been caused.

AUTONOMY: protection of biodiversity and preservation of threatened species/rare breeds.

FAIRNESS: ensuring sustainability of life-supporting systems (e.g. soil and water) by responsible use of non-renewable (e.g. fossil fuels) and renewable (e.g. wood) resources; reducing greenhouse gas emissions.

ADAPTED FROM MEPHAM (2008)

STRENGTHS

- At the very least, it is a checklist of stake-holder concerns. This is one reason why it was designed – to ensure those concerns were raised in public policy decision-making discussions, in a transparent manner (Mepham 2000).
- The matrix enables analysis of the ethical impacts of something (for example, application of genetic modification) from the perspective of different stakeholders. Those using the matrix must imagine themselves to be members of each stakeholder group in turn, and consider the impact of the proposed technology on each group (Mepham 2000). This may help overcome bias.
- The matrix requires factual evidence, and as such highlights areas where further evidence is required. In this case we need more infor-mation about the potential welfare impacts of the proposed hormone in cows, incidence of disease and potential effects on consumers and on the environment.
- It aids in identifying areas of agreement and disagreement, and raises potential conflicts of interest rather than giving a false impression of consensus.
- Consideration of the biota is concerned with life on a collective scale – at the level of pop-ulations, breeds and species – and represents fairness in an intergenerational sense. It thus overcomes one of the limitations of the common morality, which collapses when faced with the fact that nature is not "fair" (for example, we can't protect the interests of prey without risk-ing the wellbeing of predators and vice-versa) (Mepham 2008).
- The matrix was designed for scientists, trans-lating abstract principles into concrete issues, which makes it easier for those without a background in ethical theory to use the tool (Mepham 2008).

- The matrix may simplify a problem by bringing all ethically important factors into the frame (Mepham 2008).

LIMITATIONS

- The matrix is not a decision tool – therefore simply completing all of the boxes or even assigning a numerical value to each does not allow a satisfactory decision to be made. In fact the matrix does not show how each cell should be weighed against others. The matrix, like any framework, still requires "competent moral judges".
- Different principles might have different weights for each person using the matrix, which can be a source of disagreement.
- Real-life problems don't slot neatly into boxes. Different applications of the matrix may lead to very different considerations.
- Few, if any, decisions made using the matrix will afford equal value to all ethical principles, such that usually some will need to be overridden by others (Mepham 2008).
- While requiring factual evidence, the nature of the facts – including whether they were obtained reliably and whether they are relevant – raises potential areas for disagreement.
- It is limited by the parameters of the ethical issue raised. For example, in the case of using a hormone to increase milk yield in dairy cattle, the impacts to each stakeholder are relative to a pre-existing condition (in this case, the cur-rent intensive systems of dairy farming) which may not be ethically acceptable to some.
- The matrix is not a tool for addressing all ethi-cal questions. According to its creator, "if farm workers…are paid unfairly low wages, if deceit is entailed in the marketing of a new product, or if an industrial company knowingly and wilfully pollutes the environment to cut costs – we are simply confronted with examples of injustice,

dishonesty and irresponsible behaviour. It is as well to be aware that calls for an ethical analysis in such cases might be cynical moves, calculated to enable the perpetrators to buy time and mount a defence of their unethical practices" (Mepham 2008).

Recommended reading
Mepham B (2008) *Bioethics: An Introduction for the Biosciences, 2nd Edition.* Oxford University Press: Oxford.

Mepham TB (1996) Ethical analysis of food biotechnologies: an evaluative framework. In: TB Mepham (ed) *Food Ethics*, 101–119. Routledge: London.

2.5.9 Narrative ethics

DEFINITION
Narrative ethics uses stories to impart moral values.

BACKGROUND
Do narratives play a role in our moral lives? Ethologists have described the practice of sharing stories which incorporate ethical principles, evaluation of stakeholders and techniques for handling common and unique problems. Thus fairy-tales and folklore are commonly used to impart moral rules to children. For example, despite the fact that the fairy-tale character Cinderella is taunted and bullied by her evil stepmother and stepsisters, she remains gracious and good-hearted, and is rewarded with the love of Prince Charming when he confirms her identity by reuniting her with a lost shoe.

In medical and veterinary settings, the narrative approach substitutes anecdotes for stories. New staff, for example, are introduced to the practice ethos indirectly by hearing stories of previous cases and situations faced by their colleagues and predecessors.

This trend was documented by Sanders:

"The local stories aid regular participants in grounding and justifying the difficult decisions one is forced to make in all medical settings and are presented to newcomers as they are introduced to the practical procedures they can employ and the ethical problems they can expect to encounter."

(Sanders 1994)

Bosk documented the practice of telling "horror stories" in medicine:

"From these stories, we should not infer that hospitals deliver slipshod care; rather, these stories should be seen as moral parables, an element of the oral culture of medicine that remind all that healing is a difficult business that must always be done with care."

(Bosk 2003)

It has also been documented more specifically around euthanasia by Morris:

"Euthanasia-related storytelling falls into two categories, both reflecting the anxieties and concerns of students regarding the real world of veterinary practice: tales of ethically outrageous situations and tales of euthanasia gone wrong. Ethics-based storytelling reminds students of the ambiguity, contradiction, and paradox in the practice of veterinary medicine."

(Morris 2012)

Should the tales we swap with colleagues influence our decision-making?

Readers will note that we have presented cases in subsequent chapters in a narrative form. This is because we believe this structure gives them meaning.

△

2.22 Often the veterinary team shares values and
norms via horror stories.

CARTOON RAFAEL GALLARDO ARJONILLA

EXAMPLE

Soon after you start working in a practice, you
have a conversation with the head nurse over
lunch.

"Adrian [your employer] used to give everyone
massive discounts when they told him a sob story.
I remember one little old lady told him she wanted
everything done for her dog, but couldn't afford
it. He went out of his way to treat the dog, came
in after hours to check the drip was still running,
stayed back to give medications. He gave her a
massive discount, then two weeks later he gets a
written complaint because he didn't offer referral.
It's like they say, no good deed goes unpunished."

STRENGTHS

- Narrative elements are embedded in all forms
  of moral reasoning, and in fact we use examples to demonstrate the adequacy of ethical
  theories and frameworks (Arras 1997).
- Narrative may make it easier for us to empathise with and understand stakeholders (Arras
  1997).
- Stories provide an indirect means of communication, allowing colleagues to obliquely criticise or defend the performance of themselves
  or others by expressing praise or disapproval
  of behaviour in a story (Bosk 2003). In addition, "Horror stories allow participants to communicate in a backhand way their awe at the
  tasks before them, their reverence for sound
  clinical judgement and experience, their appre-
  hensions about the levels of their skills, and the
  secret knowledge that one learns from misadventure" (Bosk 2003).
- Stories and narratives incorporate emotional
  elements, which reflect the reality of practice
  where ethical issues occur "in an emotion-laden
  context" (Gillam, et al. 2014).
- Narratives may facilitate engagement and
  enhance learning (Gillam, et al. 2014).
- Paying close attention to narrative elements in
  a situation – for example the history provided
  by the client – may enable recognition of ethical issues that would go otherwise unnoticed
  (Arras 1997).
- Emotion within a narrative may act as a marker
  or flag for ethically relevant aspects (Gillam, et
  al. 2014).
- Sharing stories about ethically charged situations may help alleviate moral stress, acting as
  both an outlet for the storyteller and a reassurance to the listener that they are not alone
  (Bosk 2003, Morris 2012).
- The study of narratives can provide insight into
  the way that values are constructed and conveyed in the veterinary setting.

LIMITATIONS

- As values are embedded in the narrative, there is little scope for transparent ethical analysis.
- Without another ethical theory there is no way to determine what makes a story morally compelling. According to philosopher John Arras, "the more basic problem for narrative ethics involves the very idea of resorting to a set of abstract criteria for resolving conflicts among more plausible stories. For if we are truly able to pick and choose among competing stories by deploying a set of criteria, then it would appear that the criteria themselves, and not the narratives, are fundamental to the critical function of ethics" (Arras 1997).
- The narrative may presume a value set at its outset (for example, that clients or owners have less of a knowledge base than the veterinary team) which may be controversial or incorrect.
- Narratives tend to be emotive and appeal to a visceral response, which may be misleading.
- They are told from a single narrative point of view and are therefore inherently biased.
- Narratives are open to interpretation. There is no way to ensure that any two listeners derive the same information and draw the same moral conclusions from the same story.

In the biomedical literature there is an interest in ethical approaches that focus on the specific details of situations. Casuistry involves reference to so-called "paradigm cases", cases that strongly reflect our moral intuitions. This is a more formal process than narrative ethics.

For example, we might consider a situation involving a veterinarian failing to report a notifiable disease to authorities because he did not want an adverse outcome for his client. If we are confronted by a situation that approximates or matches this paradigm case, we can use the paradigm case to guide our actions in the current situation. The closer the current case corresponds

to the paradigm case, the greater moral certainty we have. In some sense this is similar to case law, in which judges refer to previous cases to justify their decision-making. Laws and rules may be generalised from specific cases.

Recommended reading

Lindemann Nelson H (ed) (1997) *Stories and Their Limits: Narrative Approaches to Bioethics (Reflective Bioethics)*. Routledge: New York.

Morris P (2012) *Blue Juice: Euthanasia in Veterinary Medicine*. Temple University Press: Philadelphia.

Jonsen AR and Toulmin S (1988) *The Abuse of Casuistry: A History of Moral Reasoning*. University of California Press: Berkeley.

2.5.10 How do these theories relate?

Ethical frameworks are often taught as if they are in competition, but in practice they are complementary.

First, through professional codes of conduct and societal expectations, there is the hope that animal health professionals are instilled with a good character. This aligns most with virtue ethics.

Second, there is an expectation that animal health professionals will be in the best position to BEST predict the consequences, i.e. based on their knowledge, experience (phronesis) and the available evidence. The ethical veterinarian does research, seeks out evidence and knows when they do not know. This aligns best with utilitarianism/consequentialism.

Third, a veterinarian should be aware of all the rules, and we would expect them to act in a way that they would expect others – or at least other ethical vets – to emulate. They would respect the rights of others. This aligns most with deontology.

Finally, we expect that veterinarians should be acting in such a way as to minimise harm whilst promoting good. This means using the best techniques, skills, drugs and equipment they can and performing "best practice" – again, we would expect this to be evidence-informed.

We expect them to be fair to patients and respect the autonomy of patients and clients – aligning with principalism.

Finally, we need to consider our patients and their interests. Because they cannot speak, we need to constantly work to understand their needs and interests – whether it's a pampered lapdog, an orphaned wild bird, a herd of animals or a lab rat.

2.5.11 Limitations of ethical theories and approaches

It is often said in veterinary medicine that if there are many ways to treat a condition, there is little consensus on the perfect treatment. This can also be said with regard to ethical theories. There are many ethical theories precisely because no single theory is universally successful.

Even in a group of people who apply the same ethical theory or approach, disagreements are inevitable. According to Beauchamp and Childress, ethical disagreements can still emerge because of:

(1) Factual disagreements (e.g. about the level of suffering that an action will cause),

(2) Disagreements resulting from insufficient information or evidence (e.g. insufficient information about the benefit of a particular medication or surgical intervention),

(3) Disagreement about which norms or rules are applicable or relevant in the circumstances,

(4) Disagreement about the relative weights or rankings of the relevant norms,

(5) Disagreement about appropriate forms of specification or balancing,

(6) The presence of a genuine moral dilemma,

(7) Scope disagreements about who should be protected by a moral norm (e.g. whether embryos, foetuses, and sentient animals are protected),

(8) Conceptual disagreements about a crucial moral norm (for example, whether refusal to treat an animal with a life-threatening condition constitutes killing). (Beauchamp & Childress 2013)

Nonetheless, having some knowledge of ethical frameworks means that the sources of disagreement can be articulated and documented – as can sources of agreement. These can be revisited as more experience is gained or more information/knowledge becomes available.

ETHICAL FRAMEWORKS USED IN VETERINARY PRACTICE

What do you think?

ONE List three ethical decisions you have made. For each of these, identify the theory or approach that best supports your decision-making.

TWO Take one of these examples and apply a different ethical theory to justify your decision.

THREE Which theory or approach do you find least helpful? Why?

FOUR Philosophers often employ a "counter-example" to criticise an ethical theory. For one of the ethical theories or approaches, can you think of a counter-example?

FIVE Can you think of a story imparted by teachers and colleagues that contains a moral lesson? What do you think that lesson is?

Conclusion

In this chapter we have introduced a number of key ethical theories, frameworks and concepts including consequentialism and utilitarianism, deontology, justice as fairness and contractarianism, ethics of care, virtue ethics, common morality, principalism, the ethical matrix and narrative ethics.

While one can view these theories as competitive, they can also be used in ways that complement one another. In the following chapters we will apply these tools and concepts to scenarios. This will enable a better understanding of the advantages and disadvantages of these concepts and improve ethical reasoning.

References

Anzuino K 2007 Everyday ethics. *In Practice* **29**: 234.

Arras J 2010 Theory and bioethics. In: *Stanford Encyclopaedia of Philosophy*.

Arras JD 1997 Nice story, but so what? In: Lindemann Nelson H (ed) *Stories and Their Limits: Narrative Approaches to Bioethics (Reflective Bioethics). 1st Edition*. Routledge: New York.

Beauchamp TL, and Childress JF 2013 *Principles of Biomedical Ethics*. Oxford University Press: New York, Oxford.

Bosk CL 2003 *Forgive and Remember: Managing Medical Failure*. University of Chicago Press: Chicago.

Broom D 2016 Considering animals' feelings. Précis of Sentience and animal welfare (Broom 2014). *Sentience* **005**: 1–11.

Coeckelbergh M, and Gunkel DJ 2014 Facing animals: a relational, other-oriented approach to moral standing. *Journal of Agricultural & Environmental Ethics* **27**: 715–733.

Cousquer G 2011 Everyday ethics. *In Practice* **33**: 142–143.

Devitt C, Kelly P, Blake M, Hanlon A, and More SJ 2014 Dilemmas experienced by government veterinarians when responding professionally to farm animal welfare incidents in Ireland. *Veterinary Record Open* **1:e000003**.

Engster D 2006 Care ethics and animal welfare. *Journal of Social Philosophy* **37**: 521–536.

Gardiner P 2003 A virtue ethics approach to moral dilemmas in medicine. *Journal of Medical Ethics* **29**: 297–302.

Garner R 2013 *A Theory of Justice for Animals: Animal Rights in a Nonideal World* Oxford University Press: Oxford.

Gillam L, Delany C, Guillemin M, and Warmington S 2014 The role of emotions in health professional ethics teaching. *Journal of Medical Ethics* **40**: 331–335.

Gillon R 1998 Bioethics, overview. In: Chadwick R (ed) *Encyclopedia of Applied Ethics*, 305–317. Academic Press: San Diego.

Kaiser M, Millar K, Thorstensen E, and Tomkins S 2007 Developing the ethical matrix as a decision support framework: GM fish as a case study. *Journal of Agricultural and Environmental Ethics* **20:** 65–80.

Lee WT 2013 Measuring the immeasurable core competency of professionalism. *Journal of the American Medical Association Otolaryngology – Head and Neck Surgery* **139:** 12–13.

Lee WT, Schulz K, Witsell D, and Esclamado R 2012 Channelling Aristotle: virtue-based professionalism training during residency. *Medical Education* **46:** 1129–1130.

McDougall RJ, and Gillam L 2014 Doctors' "judgements" and parents' "wishes": ethical implications in conflict situations. *Medical Journal of Australia* **200:** 372.

Mepham B 2000 A framework for the ethical analysis of novel foods: the ethical matrix. *Journal of Agricultural & Environmental Ethics* **12:** 165–176.

Mepham B 2008 *Bioethics: An Introduction for the Biosciences*. Oxford University Press: Oxford.

Mepham TB 1996 Ethical analysis of food biotechnologies: an evaluative framework. In: TB Mepham (ed) *Food Ethics*, 101–119. Routledge: London.

Morris P 2012 *Blue Juice: Euthanasia in Veterinary Medicine*. Temple University Press: Philadelphia.

Mullan S 2006 Everyday ethics. *In Practice* **28:** 52.

Rawls J 1999 *A Theory of Justice*. Oxford University Press: Oxford.

Regan T 2004 *The Case for Animal Rights*. The University of California Press: Berkeley.

Rollin B 2011 *Putting the Horse Before Descartes: My Life's Work on Behalf of Animals*. Temple University Press: Philadelphia.

Russell WMS, and Burch RL 1959 *The Principles of Humane Experimental Technique*. Methuen: London.

Sanders CR 1994 Annoying owners: routine interactions with problematic clients in a general veterinary practice. *Qualitative Sociology* **17:** 159–170.

Sandøe P, and Christiansen SB 2008 *Ethics of Animal Use 1st Edition*. Blackwell Publishing: Oxford.

Select Committee on Animals in Scientific Procedures 2002 Select Committee on Animals in Scientific Procedures Report. Home Office: London, UK.

Swan P 2006 Everyday ethics. *In Practice* **28:** 104.

Vickery T 2013 Where the wild things are (or should be): Rawls' contractarian theory of justice and non-human rights. *Macquarie Law Journal* **23:** 23–38.

CHAPTER 3
ANIMAL DEATH

Introduction

Animal death at the hands of humans is all around us. Every day, as we view or use meat, leather and other products of animal death, we can't ignore it. There are also the "hidden" deaths – those animals used in medical and veterinary research, day-old male layer chicks and bobby calves killed because they are deemed surplus to industry requirements, or wild animals whose habitats are destroyed, divided or disrupted to name but some.

Veterinary professionals may be involved in overseeing the mass killing of animals for food or disease control, or they may personally euthanase a much-loved companion in a very intimate scenario. Even if not directly responsible for animal deaths, veterinary practitioners will have to face and care for dying animals. The overarching role of veterinary teams in any of these circumstances is to safeguard the welfare of the animals. In this chapter we will consider how veterinarians and others approach some of the challenges of dealing with the deaths of animals.

◁

3.1
Day-old male layer chicks are killed because they can't lay eggs, and are unsuitable for meat production.

PHOTO ANNE FAWCETT

3.1
Does animal death matter?

SCENARIO
DOES ANIMAL DEATH MATTER?

▶ Millions of animals are killed daily by human beings. Veterinarians are often expected to be part of that process, including performing euthanasia to alleviate suffering or overseeing welfare at slaughter. But, does animal death really matter?

RESPONSE
SIMON COGHLAN

▶ Whether animal death matters is one of the most important, and potentially urgent, ethical questions in veterinary medicine. The ethics of animal death has consequences for the role of veterinarians on farms, in laboratories, and in companion and shelter animal medicine. We might even suggest that the question of the significance of animal death relates to the very *meaning* of veterinary life, of what it is to be a veterinarian. In what follows, my focus will be on the value of life to animals, especially on whether animals can be harmed by death.

The obvious way in which death might be morally significant is that it can be bad for that animal. If death itself sometimes constitutes a real

—
3.1
DOES ANIMAL DEATH MATTER?
RESPONSE
SIMON COGHLAN

"We might even suggest that the question of the significance of animal death relates to the very *meaning* of veterinary life, of what it is to be a veterinarian."

harm for the individual who dies, then ethical reflection on killing animals should presumably take that harm into account. But if animals are not harmed by death, then killing animals will likely be a less serious moral issue. Of course, animal death could matter in other ways. Killing an animal could have important negative effects on other animals (or humans) who witness the act or who are part of its social group (Gruen 2014). Furthermore, a dying animal might experience pain and fear. Most, if not all, veterinarians would agree that they have a moral role to play in minimising these negative effects. But the question of whether veterinarians ought to regard killing as morally significant because of what *death itself means for the animal subject* is more controversial.

Veterinarians' moral perspectives on killing animals have partly been shaped by the profession's history, including its long involvement with animal agriculture, production and slaughter. Of course, the profession has also been influenced by growing social concern about animals, and by clients who deeply value their companion animals' lives. Nonetheless, some of the language the profession uses reflects its historical concern with health and suffering rather than with death. For example, *euthanasia* in human medical contexts means medical killing performed in the patient's interests, and debates about voluntary euthanasia for human beings typically revolve around severe and terminal suffering. "Euthanasia" in a veterinary context, by contrast, *can* simply mean *humane* killing, and

its ultimate purpose can even be human convenience rather than the needs of the animal patient. Such language shows that a concern with animal death has taken second place to a concern with animal suffering.

Animal welfare science is another influence on the veterinary profession. Some welfare scientists, like John Webster, claim that "death is not a welfare issue" (Webster 1994). Although some of these scientists also claim that facts about animal welfare are entirely distinct from ethical questions, including moral judgements about killing animals, it is nonetheless natural for veterinarians to draw from assertions like Webster's the conclusion that killing animals is not a significant moral problem and even that it does not matter as long as it is "humane".

The claim that animal welfare science and ethics are quite distinct and separate is misleading. Often the idea here is that welfare specialists, as *scientists*, deliver only the *empirical facts*, and that it is then up to society or moral philosophy to determine what we should do with that purely scientific information. Now it is true that facts about an animal's welfare *underdetermine* the morality of that individual's treatment. After all, many factors can enter real-life moral decision-making. Even so, the concept of animal welfare, despite its links with science, is very different to scientific concepts in physics or chemistry.

Veterinarians, animal welfare scientists, and the general public often give voice to concerns about animal welfare. The term "welfare", as used in these contexts, is intimately connected to judgements about value and about ethics. Welfare helps determine what benefits, or is good for, an animal, and what harms it, and it does so in a way that is necessarily (if often tacitly) linked to *evaluations* about what it means for the animal's life to go well or badly. Philosophers would say that these are questions for *value theory*. An animal's life goes better if its welfare is good; it suffers a misfortune when its welfare is seriously damaged. These statements involve judgements about value. Further, these judgements are intended to be relevant to important *ethical* obligations to animals.

As veterinarians and others use it, the concept of welfare is a significant part of the idea of animal *wellbeing* – or is even equivalent to it (Kasperbauer & Sandøe 2015). Hence the claim that death in itself does not affect welfare appears to imply that animals are not harmed by death, which seems then to entail that killing an animal is, all things being equal, not a significant ethical issue or is morally neutral. These implications might be avoided by holding that death can be a misfortune for reasons other than welfare reasons. But in the absence of any discussion of such reasons, it does appear that the denial that death is a welfare issue or a welfare problem is also the denial that death can be a harm or misfortune for animals plus the denial that animal death is ever morally significant in that way.

The conceptual connection between welfare on the one hand, and harm, misfortune and good lives on the other, helps explain the evolution of the concept of "welfare" within science. Many scientists now believe welfare encompasses not only *negative* states like pain, fear and distress, but also *positive* states like pleasure, happiness and the ability and freedom to perform natural behaviours (Fraser & Duncan 1998). This is fortunate for animal welfare science, since if it held that only negative welfare states are ethically relevant and that they cannot be offset by positive states, it would seem to deliver us a prima facie moral reason to render all animals insensible or dead. But that a concept of welfare should have such an implausible result casts serious doubt upon that understanding. Life for an animal can be worth living, most of us think, despite some negative welfare features, as long as there are sufficient positive features in that life to outweigh the harms and to make the life go well overall.

△

3.2 Life for an animal can be worth living, despite some negative welfare features, as long as there are sufficient positive features in that life to outweigh the harms and to make the life go well overall.
PHOTO SIOBHAN MULLAN

Suppose we agree that positive and negative welfare features, at least of certain kinds, are important constituents of animal wellbeing and are relevant to the ethical treatment of animals. But still, is death an intrinsic harm? Some think that death per se does not harm animals. Consider the argument that death is no harm because, very simply, there are no welfare states in death. In death, feelings of, say, pain or happiness are necessarily and completely *absent*. Therefore, death does not affect or run contrary to welfare. In fact, death is neutral for welfare.

This argument is flawed. We do in fact usually regard the *absence* of some states like pain as being good for the welfare of living beings and the *absence* of other feelings like enjoyment as bad for them. Furthermore, the argument that removing good states or conditions by killing is neutral for that individual's welfare is difficult to reconcile with the common view that the removal of bad states like suffering by way of death can sometimes be good for an animal and its welfare (Yeates 2010). Indeed, euthanasia of animals undergoing prolonged, terminal suffering is usually considered morally obligatory, for the obvious reason that concern for their welfare demands it.

A more promising, and philosophically exciting, argument derives from the ancient Greek philosopher Epicurus. Death, wrote Epicurus (1964), although *apparently*

"the most awful of evils, is nothing to us, seeing that, when we are, death is not come, and, when death is come, we are not. It is nothing, then, either to the living or to the dead, for with the living it is not and the dead exist no longer."

Epicurus observes that death is the cessation of existence, the complete annihilation of the individual. The strange implication of this realisation, he argues, is that after death there is no individual to whom any harm – such as a loss of welfare – could be assigned. In death the subject is no more. Thus, our universal fear of death is actually an intellectual mistake. Indeed – as the Epicurean poet Lucretius argued – calling death a harm or misfortune is as senseless as saying that a creature which never existed is thereby the victim of

the terrible misfortune of never having existed. But just as it cannot be bad for an animal that it never existed, so it is not a misfortune to cease to exist, even if the subject's rosy prospects of a good life are annihilated along with it. In either case, there is simply no identifiable subject of loss or misfortune.

The Epicurean argument is strangely compelling. Still, many philosophers reject it. Philosopher Jeff McMahan (1988) writes:

"In most instances it is a necessary truth that a person must exist to be the subject of some misfortune: I cannot, for example, suffer the pain of a toothache unless I exist. But death is obviously a special case. To insist that it cannot be an evil because it does not meet a condition that most if not all other evils satisfy is tantamount to ruling it out as an evil simply because it has special features."

△
3.3 Dixie the dog is an identifiable subject.
PHOTO ANNE FAWCETT

Furthermore, the cases of the "never existing" animal and the dead animal are not straightforwardly analogous. That comparison appeared to give support to the Epicurean position. But on reflection it seems that in the case of the dead animal there *is* an identifiable subject: it is Dixie the dog, or Daisy the cow, or pig number 3204 on the hook emerging from the killing room. Those individuals may exist no longer, but to say that death necessarily does not harm them, because they are metaphysically equivalent to the "never existing" animal, seems dubious. In any case, the Epicurus/Lucretius position would need further defence.

Philosophers sometimes call death a *privative evil*. Death can be a great harm, on this common view, if it *deprives* the subject of a future that intelligibly would or may have contained significant goods and if the life would have been worth living. Ceasing to exist might be one of the worst misfortunes a creature could suffer, because the loss to the creature is very large. Still, the significance of the harm brought about by death could vary. Many of us feel that a young individual suffers a greater misfortune in death than does an elderly individual. For while the latter may suffer a terrible loss, the young individual who dies loses those opportunities the older individual had the good fortune to benefit from. Some think this affects the ethics of killing – it being prima facie morally worse to kill the very young than the very old. Of course, we may also judge that death under certain conditions, such as intractable suffering, is a blessing or benefit. And that judgement may support moral decisions about the euthanasia of animals who have a bleak future.

Suppose we now say that although animal death may matter, it matters much less than does human death. One view is that the harm of death is partly proportional to the degree of investment an individual has in his or her own future. Normal human beings of a certain age typically have many plans for the future, from falling in love to being a good veterinarian. Death for these individuals is a tragedy because it destroys their dreams, hopes and expectations. In contrast, while animals have desires, they lack this type of *strongly future-oriented* desire. Therefore, death is a less significant event for an animal than for a human being.

This argument has a good deal going for it. However, we might be cautious about its scope. Firstly, many ethicists would point out that some humans, such as the very young and severely intellectually disabled, do not have many, or any, long-term life plans. Secondly, perhaps some animals – like great apes, cetaceans and even domestic species such as pigs and dogs – have some significant future-oriented desires that would be thwarted by death (Singer 2011). Moreover, while the presence of longer-term desires may be ethically important, they alone do not account for our sense of the terrible misfortune that death

△
3.4 Perhaps great apes and other animals such as cetaceans (like this dolphin), and even domestic species such as pigs and dogs, have some significant future-oriented desires that would be thwarted by death.
PHOTO ISTOCK

—
3.1
DOES ANIMAL DEATH MATTER?
RESPONSE
SIMON COGHLAN

sometimes is for human beings. If they did account for death's horror, we would be forced to conclude that the death of a very young child, and the loss of his or her entire future, is no tragedy for that child.

In thinking about why animal death matters, questions of human death are never far away. My discussion has alluded to human death at various points. Why is this? Both humans and animals have interests and can suffer harms and enjoy goods. Their misfortunes and fortunes are connected to judgements about value and about ethics. Furthermore, both humans and animals can have their opportunities for good lives ended by death. This includes not only piglets, calves and lambs – and other species in production systems which are killed when they are still very young – but also young children who lack the complex awareness, hopes and dreams of human adults. When humans die, death appears ordinarily to be a serious misfortune or loss. It is therefore hard to see why, if human death matters in this way, animal death does not matter in this way too. Or to put it the other way around, if we deny that animal death matters, do we thereby threaten some of our deepest views about the terribleness of death for human beings? Philosophers differ on this question (Visak & Garner 2015). Even supposing that animal death does matter, this would not tell us precisely when killing animals is ethical or unethical. Moreover, we could not conclude that killing animals is morally the same as killing humans. It may be significantly less morally significant. But if animal death does matter, then the question of killing animals is indeed an urgent ethical question for the veterinary profession.

DOES ANIMAL DEATH MATTER?

What do you think?

ONE _____ Does animal death matter?

TWO _____ Does death matter in the same way to animals as it does to humans?

THREE _____ How might you explain your answer to someone else?

△
3.5 Why might a veterinarian refuse a request from an owner to kill a healthy goat?
PHOTO ANNE FAWCETT

||||||||||||||||||||||||||||||||||||||||||||||||||||||||||||||||

WE continue our exploration of the significance of animal death with an interview of Greg Dixon by David Main. For many years now these two veterinarians have debated the topic of animal death in front of an audience of first-year veterinary students in the UK. David Main takes a stance broadly reflecting current practices of animal death because human interests trump any interests in living that an animal may have. Dr

Dixon, on the other hand, argues that contrary to the cultural norm there are many things wrong with the mass killing of animals by humans. Both participants, it should be acknowledged, value highly the quality of life of sentient animals and advocate excellent welfare for animals during their lives.

DAVID MAIN ▶ At our annual debate in front of veterinary students we aim to discuss the value of keeping animals alive. This is a complex issue that is fundamental to veterinary science and animal ethics and students find the debate challenging to their existing viewpoint. To maximise student involvement we invite students to join the debate and then vote on which position they preferred. Greg, were you surprised how variable the votes were from one year to the next?

GREG DIXON ▶ I was surprised that I ever won the vote! I know that in the early years, veterinary students are at their most compassionate (Pollard-Williams, et al. 2016) and as we held the debate with first-year students that may be an important factor.

However, I think the views I present are reasonably far from the culturally accepted norms and might even be considered "radical". They therefore take a bit of explaining: maybe some years I did that better than others. I might have been helped by the demographic: nowadays there is a majority of women over men at vet school. Women veterinary students have been shown to display greater concern for animal suffering than male students (Pollard-Williams, et al. 2016, Serpell 2005).

DAVID MAIN ▶ You certainly aimed to be radical but were you really so far from the norm? You use an example of a veterinarian refusing to euthanase a healthy goat that had become surplus to the owner's requirements which seemed to be controversial. The story makes the point that

vets are allowed to have a conscience. Hopefully a veterinary surgeon having a conscience is not itself radical! However, it does seem radical when the conscience differs from conventional etiquette. Some might think that a goat is a farm animal. Most farm animals are killed when it is suitable for us so conventional etiquette might say, why not kill a goat whenever the owner asked? But that draws attention to an inconsistency in the common view: many would object to convenience euthanasia of a dog just because it did not fit an owner's lifestyle. Perhaps you are just radical in trying to be ethically consistent?

GREG DIXON ▶ Yes: the story of the goat is intended to make the point that vets are not just service providers, like car mechanics. In the relationship between the mechanic and the client, the car's viewpoint holds no sway. The car is not the "subject of a life" and can be treated as having only extrinsic value: the use it provides to the owner. The goat, on the other hand, can be argued to have an intrinsic value, being the subject of a life and possessing sentience. Whether or not it ought to be killed in certain situations and whether this represents some moral harm is another question.

In an attempt to explore the issue of killing animals, I introduced the "hermit" thought experiment. A hermit who lives in social isolation can be killed painlessly and without their knowledge by a careful psychopath who derives pleasure from killing. The social isolation of the hermit ensures that other humans are not harmed by feelings of grief, or fear that the same fate awaits them. In the utilitarian calculus, no suffering has been caused to anyone (and some pleasure derived) and so after the murder, is the world a better place? When asked for a show of hands, most students confirm they thought this hermiticide was morally dubious. By exploring why this might be the case and then asking if the moral arguments against killing could apply to moral patients of a different species, I try to tempt the students along the path of consistent thinking on the issue of the comparative morality of killing.

DAVID MAIN ▶ You are in pretty popular territory urging students to think about animals as subject of a life, after all "saving" animals is what veterinarians do, isn't it? What I try to do in the debate is introduce the messy concept of "descriptive ethics", documenting how society does actually value animals in practice. Society is certainly ethically inconsistent with respect to farm animals. During the foot and mouth outbreak in 2001 in the UK there was great gnashing of teeth about the waste of animals' lives needed to control spread of the virus. This was despite the obvious fact that the animals would have been slaughtered for meat anyway.

△
3.6 Foot and mouth disease became widespread in 2001 in the UK, resulting in the direct on-farm slaughter and disposal of at least 6.5 million animals (National Audit Office 2002).
PHOTO ISTOCK

Descriptive ethics around the value of human life is, however, far from inconsistent. Most people would instinctively say that to kill the hermit in your example is wrong and always will be. Why? Because he is human. This may be a simple biological motivation, in line with Richard Dawkins' selfish gene (Dawkins 2006), or a higher moral principle. Either way it is a reflection of how society is.

So a crucial argument I propose is that it is OK to treat animals differently to humans. The philosophers Tom Regan and Peter Singer would critique this as blatant "speciesism". Whilst it may be laudable to try to consider the interests of different animal species equally, does that mean we have to provide the same obligations to animals as we do to humans?

GREG DIXON ▶ The way our societies actually treat animals is pretty consistent. Some say that killing is by far the most common form of human–animal interaction. In 2002, the number of poultry slaughtered in the UK was 850,082,000 and the figure for sheep and lambs was 13,094,000 (The Animal Studies Group 2006). The Animal Studies Group (2006) go on to state: "…it is not just the statistics that are staggering but the fact that almost all areas of human life are at some point or other involved in or directly dependent on the killing of animals", which is a "…structural feature of all human-animal relations. It reflects human power over animals at its most extreme and yet also at its most commonplace."

The authors go on to make the point that all this goes on despite widespread humanitarian sensibilities and professed love of animals. Here is your ethical inconsistency. Perhaps this reflects some deep unease which we feel at what Jim Mason coined as "misothery", the hatred of animals (Kalof & Mason 2015), which seems to be ingrained in our culture, alongside other systems of domination and hierarchy. Interestingly, I would say that society does take an inconsistent view towards human life: an examination of global politics does not proceed very far before coming up against the notion that some humans seem to be more equal than others.

So I do not dispute your whole description of society and its ethics. The question then becomes: is the description the prescription? It seems there is a danger in adopting a violation of David Hume's law, you cannot logically derive what *ought* to be from what *is*. There are many institutions and practices in modern societies which cannot be ethically defended. That we can probably agree on. I would argue that the widespread killing of farm animals is amongst them. I suspect here we part company. I suspect here also I stray into unpopular territory.

Now, most people would argue that killing the hermit is wrong. To simply state this is "because he is human", merely begs the question. The purpose of the thought experiment is to investigate what it is that is wrong about killing the human hermit and to consider if any of these reasons might apply to creatures of other species.

It might well be OK to treat animals and humans differently. I have not proposed equality of treatment. It makes no sense to ensure my terrier has a right to vote nor buy my cat a ticket to the theatre. I simply adhere to what Peter Singer refers to as "equal consideration of interests" (Singer 2011). Unless one adopts an absolute dismissal of the moral status of animals, then one must concede they have at least some interests. One might still concede some minimal interests to animals, but then claim that human interests always have priority. This is the "speciesism" to which you allude. The particular question here is the thorny one of killing: do animals have an interest in not being killed, in the way (in most situations) we would afford humans that interest? I would say, very often, yes!

DOES ANIMAL DEATH MATTER?

What do you think?

ONE_____**Where is the harm in killing any of the following:**

- **bee**
- **pet cat with chronic renal disease**
- **tuna fish**
- **ancient sequoia tree**
- **genetically modified mouse**
- **non-pathogenic bacteria**
- **endangered frog**
- **day-old male chick**
- **sheep farmed for food**
- **dog hookworm**
- **convicted murderer?**

TWO_____**Which of these can you justify being killed? What arguments would you use for your justification?**

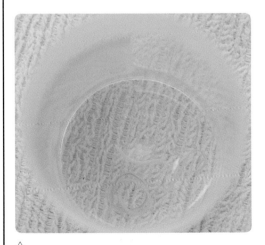

△
3.7 Is there a harm in killing parasites?
PHOTO ANNE FAWCETT

3.2

Ensuring welfare at slaughter

The huge numbers of animals killed and processed in some abattoirs mean that ensuring the highest level of welfare possible within the system actually occurs is no longer a case of spending time with each and every animal as it dies. Instead, it is a matter of implementing a whole system approach, along with associated safeguards and checks, to promote maximal welfare within the system. In this next scenario we consider how to manage such systems, especially when they are not delivering the welfare standards expected.

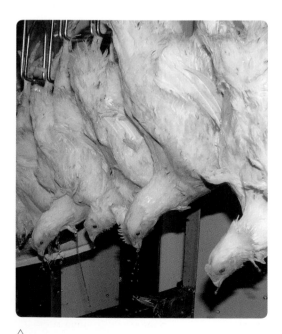

△

3.8 Sub-standard poultry stunning has negative welfare impacts.
PHOTO STEVE WOTTON

SCENARIO
SUB-STANDARD POULTRY STUNNING

▶   You work at a poultry slaughterhouse as the designated Animal Welfare Officer, employed by the slaughterhouse as required under European Union legislation. You notice that the number of birds that have been ineffectively stunned has increased recently, and that some birds have not only missed the neck-cutting machine, but have also been missed by the back-up slaughterperson. You ask to see the CCTV monitoring footage of the area but are told by the line supervisor that it would be "difficult to obtain" and that they are "satisfied with the level of uncut birds and meat quality" so you don't need to worry too much.

How should you proceed?

RESPONSE
ED VAN KLINK AND PIA PRESTMO

▶   It is not uncommon to have to deal with the problem of uncut birds. It happens regularly in abattoirs. There are basically two issues here:

• This is a welfare issue for which the Animal Welfare Officer is responsible because the Animal Welfare Officer must be shown to work on improving the welfare of the animals. If the welfare is compromised or worsening this falls under the immediate responsibility of the Animal Welfare Officer.
• It is a serious welfare problem as the animals would enter the scalding tank (~52°C) without being properly stunned, so fully conscious.

In order to show that the Animal Welfare Officer is improving welfare he/she must be able to monitor the animals and if the plant management refuses to surrender the CCTV footage this is not possible.

Within the European Union, the Regulation 1099/2009 (European Council 2009), on the protection of animals at the time of killing, requires Business Operators in article 17 to employ designated Animal Welfare Officers in slaughterhouses. The task of the Animal Welfare Officer is to assist the Business Operators to ensure compliance with the rules around the welfare of animals at the time of killing.

The Animal Welfare Officer reports directly to the Business Operator, and must be in a position to require that the staff carry out any remedial action needed to ensure compliance with the requirements of the regulation. The Animal Welfare Officer has to keep a record of all actions taken to improve animal welfare. This record has to be kept for at least one year and must be available for audits.

In order to keep a record of performance and improvement, as is required under this regulation, the Animal Welfare Officer has to be able to carry out any monitoring necessary to collect the relevant information. So the real issue here is that the Animal Welfare Officer is hampered in their ability to carry out their duty if they are refused access to the CCTV footage that would enable them to assess the situation.

The Official Veterinarian and the periodic auditor do normally not have to concern themselves with the animal welfare issue at hand, if the Animal Welfare Officer is able to do what is expected. No place is perfect, and there will always be birds that are not of the standard size, resulting in them missing the stunner or the bleeding blade. As long as the Animal Welfare Officer can show that everything is put in place to make sure this is happening as little as possible, there is no legal cause for intervention.

So really the position of the Animal Welfare Officer is at stake here. He/she needs to be fully supported by the management of the plant. Even though there may be a tension between the welfare of the birds and the profit of the company, the consequences of getting an animal welfare case instigated against the company can have serious implications as well.

It is very important to try to identify the causes of the increase in "missed" birds. Is the backup slaughterperson the right person for the task? Has there been an increase in line speed resulting in a higher number of misses? Is the machinery correctly adjusted to fit the size and variability of the birds? How uniform are the birds being brought in? All of this requires the Animal Welfare Officer to have access to all necessary information.

So what should the Animal Welfare Officer do? He/she needs to collect as much information as possible on the performance of the line, and the performance of the slaughterperson, looking back as far as possible in time. If the line supervisor refuses to cooperate, for example on the basis that it influences their performance targets negatively, the Animal Welfare Officer should contact the Business Operator as soon as possible, preferably the same day, making clear what the consequences might be of non-compliance.

There are gas-based stunning systems available already that would preclude most if not all of the risks to animal welfare as a result of incorrect stunning.

||||||||||||||||||||||||||||||||||||||||||||||||||||||||||||||||||||||||||||||

THE welfare of animals at slaughter can be, at times quite literally, hanging in the balance. Systems and procedures to promote welfare may not be the best, may not be maintained or operated well and may be deliberately ridden roughshod over if, for example, the abattoir owner increases the line speed for financial gain or to meet customer orders. The recognition of these powerful motivations in action which may affect many lives is the reason that regulatory oversight is required. In the US the Department of Agriculture recently

—
3.2
ENSURING WELFARE AT SLAUGHTER
RESPONSE
ED VAN KLINK & PIA PRESTMO

declined to allow an increase in poultry line speed from 140 to 175 birds per minute, although food safety and worker protection featured heavily in their reasoning (USDA 2014).

In situations involving compromised welfare at slaughter, the presence of welfare problems – as long as they are below a regulated threshold – may not legally require intervention, but there may be moral reasons to do so. For example, if the aim is to minimise suffering and an alternative method of slaughter involves less suffering, there is a moral obligation to implement this alternative method.

Other strong motivations affecting welfare at slaughter are cultural and religious practices. Many

△

3.9 By the time the animals are processed ready for the consumer, the only way of knowing whether an animal has been stunned prior to slaughter is by labelling.
PHOTO ISTOCK

Jews and Muslims feel required to eat meat only from animals that have undergone religion-specific slaughter practices. Sometimes religious slaughter precludes any stunning and sometimes animals may be stunned immediately after having their throat cut. Notwithstanding welfare problems within stunned systems, the aim of stunning prior to slaughter is to render animals insensible to the pain they would experience from the slaughter process, as acknowledged by several multi-agency reviews (BVA 2015).

In countries where Jews and Muslims are in a minority, religious slaughter is frequently exempt from legislation requiring stunning to protect welfare although some, including Switzerland, Sweden, Norway, Denmark and Iceland at present, do require stunning. In the UK approximately 5 per cent of the population is Jewish or Muslim and 2 per cent of cattle (44,000 animals), 15 per cent of sheep and goats (2.4 million animals) and 3 per cent of poultry (31 million birds) were estimated to be slaughtered without stunning in 2013 (Food Standards Authority 2015).

The main argument for allowing continued non-stun slaughter for religious reasons is the respect for the right to religious practice. However, when challenged under EU Human Rights legislation the European Court of Human Rights ruled that, as long as trade rules allowed the import of suitable meat, the right to buy and consume such meat was not infringed (Whiting in press). An additional ethical element in this issue is that of labelling non-stun meat so all consumers can make an informed choice about what to buy. At present, 72 per cent of consumers in the EU indicated that they would support labelling of non-stun meat (FCEC 2015). This could potentially increase the price of non-stunned meat if it were avoided by the wider communities.

ENSURING WELFARE AT SLAUGHTER

What would you do?

It is not yet clear from research whether banning non-stun slaughter increases demand for meat from countries with lower welfare standards than yours, or speeds up acceptance of pre- or post-slaughter stunning amongst religious communities. Your national veterinary association has asked for volunteers to:

ONE   Sign a petition calling for non-stun slaughter to be banned in your country.

TWO   Demonstrate outside Parliament on the day of a Parliamentary debate on the topic.

What do you do?

△

3.10a–b Would you consider signing an online petition or demonstrating in front of Parliament on an issue?

PHOTO ISTOCK

3.3

Killing some to save others

Sometimes the values of animals' lives end up being pitted against each other. In the case of livestock some individuals are killed to improve the health of the others in the group. For example, control strategies for Johne's disease in cattle (Lu, et al. 2008), Classical Swine Fever in pigs (Duerr, et al. 2013) and Brucellosis in goats (Montiel, et al. 2013) that have included culling individual infected animals from the herd have been shown to be effective. However, each disease has unique features and modelling of Q Fever transmission in dairy goats suggests culling to be the least effective form of control (Bontje, et al. 2016).

The role that research can play in such ethical decisions can't be underestimated. For all but those with an animal rights view (some animal rights people would oppose culling outright), the success of any culling strategy would be an important consideration in determining acceptability. Some may additionally feel that there is a relevant difference between culling animals from within the same domestic herd and killing wild animals to protect domestic animals. To control *Mycobacterium bovis* (Bovine tuberculosis – bTB) a variety of wild animals have been killed, from white-tailed deer in the USA, to wild boar in Spain, possum in New Zealand and buffalo in South Africa, with varying degrees of success (Gortazar, et al. 2015). In this next scenario we consider the case of culling badgers, an iconic species, to control bTB in cattle.

3.3

KILLING SOME TO SAVE OTHERS
SCENARIO
CULLING TO CONTROL INFECTIOUS DISEASE

SCENARIO

CULLING TO CONTROL INFECTIOUS DISEASE

▶ Bovine TB (bTB, *Mycobacterium bovis* infection) is a cattle disease of economic significance and potential zoonotic risk. Many countries in Europe have eliminated bTB using a combination of cattle movement restrictions, test and slaughter policy and abattoir inspection. In some countries wildlife reservoirs of infection are thought to contribute to poor control of the disease in cattle. In the south-west of England and parts of Wales and Ireland, where the disease is endemic, the Eurasian badger (*Meles meles*) has been shown to be a maintenance host of infection.

Is it acceptable to cull badgers as part of a strategy to try and reduce bTB infection in cattle?

△

3.11 Badgers have been culled with the aim of reducing the levels of tuberculosis, a zoonotic disease, in cattle.

PHOTO SUPPLIED, ANONYMOUS

RESPONSE

ANDREW GARDINER

▶ Culling to control an infectious disease is necessarily unique to veterinary medicine. It was first employed on a very large scale during the European cattle plague outbreak of 1865. Although it brought a devastating pandemic under control and helped prove the germ theory of disease, which had enormous implications for animal and human health (Worboys 2000), it was criticised by some as being unethical. Since that time, culling has been used regularly within veterinary medicine during livestock pandemics, most recently and controversially during the 2001 foot and mouth outbreak (Woods 2004). In the case of badgers and TB, culling is directed at an animal group or population thought to be harbouring the disease (badgers) and an animal exposed to the disease but not yet showing clinical signs (the cow, who has reacted positively to an intra-dermal tuberculin test). With badgers, there is no way of easily determining who is infected and who is not, so a contiguous culling technique is used in identified hot spots of the disease (shown by the numbers of cows reacting positively to the TB test) in order to reduce the number of infected badgers in the population in that place and thus the threat to cattle. The approach has some similarities to the contiguous cull carried out during the 2001 UK foot and mouth outbreak, when healthy domestic animals in close proximity to outbreaks were culled to try to create a "fire break" to stop onward spread of the disease. Consideration was also given to culling of wildlife at that time.

The rightness or wrongness of culling badgers is an issue of great public interest as shown by petitions and opinion polls. It has all the features of a public controversy: it is a long-running dispute with a variety of vocal interest groups; there are economic and political pressures at various levels (Carrington 2015); the facts are often disputed;

and there is intense media interest. When a group of independent scientists carried out a trial and concluded that badger culling could make no meaningful contribution to controlling the spread of TB (DEFRA 2007), the debate did not resolve: it polarised even more.

This issue poses particular difficulty for veterinary surgeons – a science-based profession. The ethic of beneficence, central to medical practice, states "First, do no harm". This fundamental ethic guides the patient–doctor interaction. It also holds for the patient–veterinarian interaction in many (but not all) instances. In an ethical sense, the badger could be considered a patient [and is certainly a "moral patient" in the sense used by Regan (2004)], so a fundamental medical principle would seem to be broken if vets promoted the killing of badgers. However, breaking this rule is in itself not too unusual in veterinary medicine. Routine exceptions to "do no harm" could include the neutering of companion animals (Palmer, et al. 2012), euthanasia and very common interventions carried out on livestock, for example dehorning. In these instances, utilitarian considerations take precedence: on balance, more good is thought to be done than harm. This suggests that a utilitarian approach might be used to justify badger culling. However,

this topic generated one of the largest response rates recorded in a public petition against culling – something that would weigh heavily on the side of any utilitarian assessment. The "layered" nature of the concerns in this issue, the fact that they often operate at different and not necessarily comparable qualitative levels (see the list in Table 3.1), might make solving a utilitarian cost–benefit equation difficult. Whose benefit or happiness takes precedence?

When culling of domestic farm animals has previously taken place, veterinarians in official capacities, whether employed full-time or part-time by the state, have in the main acted deontologically, by following various legal control orders which demand culling to stop disease spread. This is an example of one form of deontology ("do no harm") being superseded by another ("follow the state protocol for the control of infectious disease in livestock"). In following the control orders to cull even healthy animals, vets are deemed to be acting for the general good. However, when this rule-based approach was carried out with foot and mouth disease in 2001, a kind of ethical thought experiment was acted out in real time. The cull policy was taken to its extreme point, and at least 6.5 million animals were slaughtered (National Audit Office 2002).

Individual badger in his/her set (Cousquer 2013)	Woodland ecosystem as a whole
Public image of the badger (Cassidy 2012)	Wild reality of the badger
Country life and rural economy	Urban life and rural recreation
Individual farmer's infected prize herd	Health and export status of the national herd
Evidence based on a controlled trial	Evidence based on empirical findings
Opinion of the general public	Opinion of the National Farmers' Union

Table 3.1 Different levels of concerns relating to badger culling.

—

3.3

KILLING SOME TO SAVE OTHERS
RESPONSE
ANDREW GARDINER

RESPECT FOR:	WELLBEING	AUTONOMY	FAIRNESS
Badgers	Individual badgers suffering from TB will be spared further distress if culled Wellbeing of badgers as a whole will be improved if TB is controlled Badger wellbeing will be reduced if culling causes suffering or non-lethal injury in either infected or healthy badgers; culling also disrupts badger social structures and causes dispersal and fighting, which will affect wellbeing	Healthy badgers will have autonomy (and wellbeing) removed or reduced if they are killed or injured Badger autonomy is protected by certain UK laws – healthy badgers being killed may run counter to this	We do not normally cull wildlife to try to control their diseases – badgers are being treated "asymmetrically" It is unfair that healthy badgers are culled rather than just sick ones; badgers, unlike cows, are not tested for the disease before being culled
Farmed cattle	Cows testing positive for TB have wellbeing reduced by being culled, as they are not suffering any distress at this time Wellbeing of the national herd could be reduced if TB takes hold and many cows start to show symptoms (not the case currently)	Cows testing positive have less autonomy than their healthy counterparts – although this has to be seen against norms of autonomy for farm animals (these are used practically in welfare assessments)	There is fairness in the sense that both infected badgers and reactor cows are being culled in this process, i.e. neither is singled out It is fair that only reactor cows are culled (rather than all cows on the farm)
Farmers returning TB test positives on their farms	Farmers' economic and emotional wellbeing is reduced as they will lose their cows (or herd) However, many farmers may support the policy of cow culling if it protects their longer-term future	Famers have no autonomy as once positive tests are shown, there are no other options but culling reactor cows	The process is fair in that all farmers are treated in the same way Compensation is paid, which is fair, although the amount may be disputed
Farmers unaffected by TB	Farmers' wellbeing could be (a) reduced by stress and worry regarding worsening TB and possibly becoming affected; (b) improved by realising that action is being taken to control TB	Culling preserves national farmers' autonomy to continue to farm normally, but there is no autonomy over whether or not to accept culling in reactor cows	As above However, by reducing "competition", unaffected farmers could be given an advantage over poorly compensated affected farmers

△
Table 3.2 Ethical matrix exploring some aspects of culling.

RESPECT FOR:	WELLBEING	AUTONOMY	FAIRNESS
General public	Rural livelihoods and jobs could be affected by the effect of TB on the rural economy There is great public affection for the badger and culling causes distress and anger in both urban and rural dwellers TB could prove a threat to public health (not the case currently, and is preventable by vaccination)	Economic effects of TB in cows could reduce autonomy of employees in various rural industries	The rural community is potentially affected more directly by TB than urban dwellers (although the qualitative differences are hard to compare here), so it could be unfair that numerically larger urban dwellers have more say on what is done
Scientists	Scientists' wellbeing and funding are increased as a result of funding for projects exploring TB and its spread	Scientists retain autonomy and impartiality; however, their reputation and identity are reduced when their evidence is not acted upon	Scientists may argue that it is fair and just that their findings should be privileged over claims made by others (e.g. pro-cull farmers; badger pressure groups) when such claims concern impartial evidence Fairness is generally advanced when all stakeholders have a say but disputes as to "the facts" can occur
Politicians	Politicians' wellbeing and job security may be affected by an on-going crisis in the countryside	Politicians' high autonomy allows them to make policy decisions that do not necessarily relate to impartial evidence offered by scientists and by various stakeholders/ interest groups	Politicians have to fairly represent all their constituents but should also base their policy on available impartial evidence
The biota	The wellbeing of the woodland ecosystem could be either depleted or improved by culling badgers, depending on your view	Badger autonomy as an individual is a key factor in shaping responses to any cull in terms of how this is viewed (Cousquer 2013) Just as there are calls for buildings to have "rights", environmental ethics approaches would promote the autonomy of the woodland as an entity – however, this could translate into very different positions on badger culling Others may argue that there are no purely natural autonomous landscapes in the UK and that this concept is a cultural construct	A solution should be fair to both the farming landscape and the natural landscape, and there are many examples of policies, laws and strategies which aim to ensure this balance. The role of culling could work on either side here depending on your view and the weight that current scientific evidence is given

The ethics of culling on this scale caused deep unease across wide sections of British society, including farmers, some of whom refused state officials access to their land. At this point, political, economic, legal and separate rights-based concerns began to be invoked to challenge the policy and its ethical underpinnings. The science behind the methodology was also questioned. A narrative ethics account of this outbreak, written by an official veterinary surgeon after the event, appears in James Drew's novel, *Following Orders* (Drew 2006). At one level, the book seems to be both a description of deontology and a criticism of it in relation to foot and mouth disease. The book illustrates a common problem with deontology – what do you do when the order you are told to follow is immoral? This well-recognised problem with deontology was most famously discussed during the Nuremberg Trials after World War II.

Individual ethical theories often seem to run into conflict or to operate at different levels in matters such as these, which can make it difficult to arrive at a decision. The ethical matrix approach was developed to try to help resolve such complex issues in relation to public policy. My attempt at a matrix for badger culling is shown (see previous pages). The matrix incorporates mainly deontological and utilitarian approaches operating at different levels (individual, social, ecological/global). For a problem such as this, relative weighting of criteria will be difficult as this will be perceived differently by competing interest groups.

To be comprehensive, a matrix like this would need to be iterative over a lengthy period to account for all stakeholders. Themes could then be extracted from it. The matrix shown here is necessarily incomplete and the reader will be able to see where additional entries could be added.

The matrix cannot provide a solution to this question, but it provides a framework to try to ensure that all the different dimensions of the problem are featured. As mentioned above, problems of relative weighting occur, for example when different deontologies collide (e.g. "We should always act on best scientific evidence" vs "Farmers have the right to kill badgers on their land"; "Do no harm" vs "Follow TB control orders"). Individuals may rank these in different ways. With recent trends towards developing vaccination strategies (Anon 2015), the suggestion is that culling in the form originally intended is being rejected, and that a multi-modal approach to this problem is evolving. An individual veterinarian's ethical position is likely to consist of a mix of elements of different ethical approaches which come together into the overall view, rather than being completely aligned to any one specific ethical theory. A combination of deontology and utilitarianism often seems to apply to veterinary decision-making (Rollin 2011a).

Acknowledgements

Thanks to Dr Elizabeth Mullineaux BVM&S, Cert SHP, DVM&S, MRCVS for her comments on this scenario.

|||||||||||||||||||||||||||||||||||||||||||||||||||||||||||||||

WITH such a complex and emotive subject affecting a large number of stakeholders the author has made it clear that a lengthy dialogue is required to reach a consensus suitable for making policy recommendations. However, as individuals we can identify our preferred course of action more quickly as long as we are able to articulate our values to ourselves and therefore the weight we place on each item in a matrix.

KILLING SOME TO SAVE OTHERS

What do you think?

Brucellosis is a zoonotic disease that can cause infertility, abortions and milk reduction in cattle and bison, and long-term, difficult-to-treat fevers in people. The disease now only remains in the United States within the bison and elk population in and around Yellowstone National Park, with 50 per cent of bison testing positive. Culling some bison to reduce the population to numbers that can be sustained through the winter within the park, reducing ranging beyond the national park boundary, has been undertaken for some years with equivocal results. Cattle herds around the park continue to test positive on occasion, possibly infected by elk (Rhyan, et al. 2013). Campaign groups both for and against this cull are very vocal.

ONE_____What features of such a culling programme would be relevant to you to decide whether to support it?

TWO_____Under what circumstances would you support this cull?

△
3.12 Bison have been culled annually in Yellowstone National Park.
PHOTO ISTOCK

3.4

Convenience euthanasia

"Convenience euthanasia" is a term used to describe killing an animal in the interests of the owner – not necessarily the animal. Unexpected requests for euthanasia can be stressful, the more so when the veterinarian disagrees with the client's request. In one small UK survey, euthanasia refusal occurred yearly for the majority but for a few this was monthly (Yeates & Main 2011). These veterinarians spontaneously reported most commonly that reasons they refuse euthanasia relate to a lack of a legitimate reason, for example the dog was healthy or had only mild health problems. Nearly a quarter of vets reported euthanasia was requested for "convenience" reasons (Yeates & Main 2011). In this scenario we explore a case where a euthanasia request comes as a surprise.

_____
SCENARIO
EUTHANASIA REQUEST DUE TO
NON-TERMINAL ILLNESS

▶   You are working in companion animal practice. Mrs S, a new client, presents you with Toby, a 12-year-old Jack Russell terrier cross, for euthanasia. Mrs S is in tears. Without performing a hands-on examination the dog appears reasonably well, save for a bit of a pot-belly. You ask why the client is electing euthanasia.

"He's been diagnosed with Cushing's disease," Mrs S sobs. "He will only get worse from here. I've had a sick dog before and I don't want to go through that again."

What do you do?

3.4
CONVENIENCE EUTHANASIA
RESPONSE
PATRICIA V TURNER

△

3.13 Toby is a 12-year-old Jack Russell terrier presented for euthanasia.

PHOTO ANNE FAWCETT

RESPONSE

PATRICIA V TURNER

▸   A client request for euthanasia of a companion animal with a non-terminal illness is not an uncommon situation in clinical practice. These can sometimes be frustrating cases for practitioners to deal with, knowing that they have the skills and tools to treat the condition and to improve the quality of life of the patient, but lacking permission to proceed with the recommended course of care.

In veterinary medicine, almost every lifesaving pharmaceutical agent or technique that is available for use in human medicine can be applied to our patients. This leads to other ethical quandaries for the veterinarian, as one can ask whether, because we can do something medically to improve an animal's life or prognosis, does that mean that

we must do it? For example, should every aging companion cat with chronic renal failure be recommended to undergo renal transplantation?

What if it was instead a 10-month-old stray cat with congenital kidney disease that could be permanently cured with renal transplantation? Should a seven-year-old dog that has had three presumptive seizures in the past five years undergo barbiturate dosimetry and be placed on a permanent course of antiepileptic medication for the rest of its life? Where do we draw a line in making value judgements regarding what treatments veterinary patients should receive and which patients are worthy of what type of care?

These are often very personal decisions made using individual ethical frameworks and there is little consensus even within the veterinary profession as to how these types of cases should be managed. Financial cost of treatment is often a limiting consideration in the client's decision, but as in this scenario, it is not always the only consideration.

Coming back to our canine patient with presumptive Cushing's disease, it is important to confirm the state of wellbeing and the diagnosis of the animal before proceeding further with any plans or recommendations. It may be that the animal's condition was diagnosed by a friend or from information extrapolated from an internet site and it may or may not be correct. Assuming the condition is confirmed, an open and unbiased discussion of the actual condition should be conducted, including the animal's prognosis and quality of life with and without treatment, time course of the condition compared with the expected natural life of the animal, the level of care required, and the cost of treatment.

While this may be more information than some clients need or desire, the client should not be asked to make a decision in a state of ignorance. The clinic might recommend that a family member or close friend attend a consultation with the client to help ensure retention of the material discussed. Printed handouts for common medical conditions and their treatment are a useful means of providing information to the client to assist with decision-making and to allow them to process complex material on their own time. Given that the animal appears stable at this point, the client should not be rushed to make a decision that they may regret. At the same time, it would be helpful to have a conversation with the client as to their fears, abilities and financial resources regarding their companion animal.

Their decision to euthanase without delay may be influenced by having experienced a recent and protracted terminal illness or death of a friend or family member, by concern about their ability to administer a treatment over the long term or by limited financial resources to pay for treatment. The latter two concerns can sometimes be dealt with through training in a medication technique or offering clinic payment plans, if available. The first concern could be dealt with by developing an end-of-life plan for the animal, which might give the client confidence to proceed with treatment, knowing that there would be signs to watch for that would be indicative of declining condition in their pet as well as knowing that they would have support from their veterinarian to help them make a difficult decision at a later time.

It is always unwise for a practitioner to make promises, bargain with or bully the client over the care of an animal. Doing this will only lead to client mistrust and, possibly, to malpractice claims, as veterinarians can never know the outcome of a condition in a particular animal with certainty. While many cases of Cushing's disease in dogs are uncomplicated and respond well to medical treatment, some do not. Similarly, while it may be possible to consider taking in and rehoming a three-year-old German Shepherd dog with hip dysplasia that the owner is unwilling to treat, it is not realistic to consider taking in and rehoming every

companion animal that a client is unable to or does not want to treat. Some shelters will accept older pets that are relinquished with chronic medical conditions but there is little hope that all of these animals will be adopted out to permanent homes and these animals may spend the rest of their days housed in a generic shelter pen or enclosure, leading one to question whether any life is better than no life. One may even ask if it is ethical to permanently rehome a 12-year-old dog that is otherwise well cared for (that is, will the animal grieve the previous owner thereby reducing its quality of life, will the stress of rehoming accelerate the course of its disease or other conditions that were undetected in an environment in which the animal was habituated and conditioned, such as partial vision or hearing loss or cognitive dysfunction)?

While it is perfectly appropriate for a veterinarian to offer a recommendation for treating a specific veterinary condition based on current literature and past experience, it is ultimately the client's decision as to how best to proceed, since they will be the one burdened with the care and cost of treatment. In this case, the client could be encouraged to wait and make a long-term decision for their dog based on response to treatment over a month or two. However, ability of the client to comply with the treatment over the long term must be considered. For example, it would be unethical to admit a diabetic cat for insulin dosimetry only to release the animal several days later to a client who is unfit and unable to do daily insulin injections. The client must be comfortable with the treatment and care of the animal before the patient can be discharged. Follow-up telephone communications by the veterinary clinic can help to increase client confidence and ease their concerns about animal care. In a welfare-focused practice, the care of the client and patient does not end once the client exits over the threshold.

Finally, in veterinary medicine, as in human medicine, there can be a tremendous burden

3.4
CONVENIENCE EUTHANASIA
RESPONSE
PATRICIA V TURNER

placed on the client caregiver and their family when tending to a chronically ill animal. This can include the need to structure the family's routine and activities around animal treatment or care; additional expenses associated with animal treatment and support; chronic stress and anxiety regarding potential animal pain, suffering or fear of sudden death of a beloved pet; irregular sleep patterns associated with increased bathroom needs of an animal; increased house soiling and associated human hygiene and health concerns; and the physical demands associated with carrying or moving animals that are unable to walk (Christiansen, et al. 2013). What at first glance may appear to be a decision for convenience euthanasia for a treatable condition might actually be the result of the client realistically appraising their personal circumstances. In exclusively considering the patient's medical condition, ignoring the client's needs and circumstances and bullying a reluctant client into a long-term course of treatment, a veterinarian can unwittingly cause both poor animal and human welfare. Some degree of flexibility and compassion should be brought to bear in each case and the clinic must be willing to support the client appropriately to optimise patient care.

||||||||||||||||||||||||||||||||||||||||||||||||||||||||||||||||||

IN the above scenario, it's likely that emotions on both sides will be heightened, even if not always visible. It is recognised that veterinarians and associated professionals that are asked to kill animals are likely to find it stressful – after all, they went into this work to save animals (Rollin 2011b). The exact source of the moral stress may stem from a range of values, for example, the sense of deprivation of the good quality of life that would be available to the animal in the future, or a concern that premature euthanasia goes against the natural order of things. For veterinary surgeons in

the UK who do not wish to euthanase an animal, the professional guidance makes it clear that this is acceptable, stating:

> "where, in all conscience, a veterinary surgeon cannot accede to a client's request for euthanasia, he or she should recognise the extreme sensitivity of the situation and make sympathetic efforts to direct the client to alternative sources of advice."
>
> (RCVS 2015)

Even those who are comfortable with accepting animal death for human benefit, for example to eat, might find it more difficult to justify premature euthanasia of companion animals. It has been suggested that a logical extension of a utilitarian stance such as Singer's could find acceptable a company that provided convenience puppies for people that could be returned for humane killing when they became less fun (lost their utility) – a company that could be called "Disposapup" (Lockwood 1979). However, such a scheme, whilst apparently aiming to increase overall utility in theory, may in practice

△

3.14 A "Disposapup" scheme to replace puppies when owners tire of them would be morally objectionable under most frameworks.
PHOTO ANNE FAWCETT

reduce utility through restrictive systems to breed the puppies, having to enlist people to kill the puppies who do not wish to do so (there are probably not enough psychopaths to operate this scheme) or creating a generally brutalising culture that does not care for these disposable animals or others properly (Sencerz 2011). Others, of course, would object to "Disposapup" on the grounds that it did not respect the intrinsic value of the puppies.

Is the term "convenience euthanasia" itself problematic? The term is often used to refer to (ethically) objectionable euthanasia. In some sense, all euthanasia is convenient and is performed because, ultimately, an animal is no longer wanted alive by an owner, agent for the owner or organisation (Powell 2016). The term itself challenges us to draw out our reasons for distinguishing some euthanasia requests as objectionable and others as unproblematic.

For a discussion of the implications of NOT performing euthanasia, and not refusing it, see chapter 7 on professionalism.

CONVENIENCE EUTHANASIA

What would you do?

You have performed one of two planned operations on Mr R's two-year-old St Bernard, Pierre, to correct his excess facial skin and diamond eye that have caused him continued suffering almost since birth. Pierre has recovered well from his surgery, which was covered by insurance, so you are very surprised when Mr R says he just doesn't think he can put Pierre through it all again. He's emotionally drained and is really angry that the breeder can sell such dogs. He's adamant that he wants euthanasia for Pierre.

How do you respond?

△
3.15 Pierre, a two-year-old St Bernard, has recovered from treatment but his owner is concerned about the prospect of further treatment.
PHOTO ANNE FAWCETT

—
3.4
CONVENIENCE EUTHANASIA
WHAT WOULD YOU DO?

CONVENIENCE EUTHANASIA

What would you do?

Ms A presents a seven-year-old, female, neutered, domestic shorthair cat, Fanfan, that has had a history of good health. Ms A is moving in with a new partner into a rental property that does not allow pets. You suggest rehoming Fanfan, but Ms A shakes her head.

"She's a funny little thing, very shy, and really is a one-cat person. I wouldn't want anyone else to take her on which is why I want to put her to sleep."

How do you respond?

△

3.16 Fanfan is a shy cat whose owner is concerned about rehoming.
PHOTO ANNE FAWCETT

CONVENIENCE EUTHANASIA

What would you do?

Mr C, an executor to the will of a now-deceased client Ms N, presents her two dogs and three cats for euthanasia. While the animals are all healthy, Ms N has specified in her will that any surviving animals are to be euthanased in the event of her death. Mr C presents a copy of the will.

What would you do?

SCENARIO ADAPTED FROM POWELL 2016

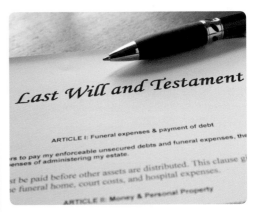

△

3.17 Ms N has specified in her will that her animals are to be killed rather than rehomed after her death.
PHOTO ISTOCK

3.5

Euthanasia or allowed to die?

The following two scenarios contrast different situations where the owners of the animals do not want euthanasia.

SCENARIO 1
EUTHANASIA REFUSAL

▶ You are a veterinarian working in a 24-hour intensive care facility. One of your patients, a 17-year-old Maltese terrier called Teddy, is suffering from multiple diseases including poorly controlled diabetes mellitus, hyperadrenocorticism, benign prostatic hyperplasia, severe (grade 4/4) periodontal disease, severe degenerative joint disease and bilateral cruciate disease. Despite intensive care Teddy remains completely anorexic, and does not move voluntarily. Euthanasia has been recommended on the grounds that he has no quality of life and cannot reasonably be kept alive outside of the hospital environment where analgesia, intravenous fluids and partial parenteral nutrition are administered.

Mr and Mrs K decline euthanasia because they hope that Teddy will die "naturally".

What do you do?

SCENARIO 2
EUTHANASIA REFUSAL

▶ You are a veterinary student on an externship in Thailand with two other students. The veterinarian in charge of the animal welfare organisation you are working for is Buddhist and does not believe in euthanasia. You are asked to draw up a treatment plan for an entire male dog with large, ulcerated, infected tumours on

△

3.18 Teddy has been poorly for a while and is now very sick in hospital. His owners are hoping he will die naturally.

PHOTO ISTOCK

△

3.19 This dog is suffering from a transmissible venereal tumour, a sexually transmitted neoplasm. While it responds well to chemotherapy, this is often not available in areas where the tumour is prevalent.

PHOTO ANNE FAWCETT

the genitals, probably transmissible venereal tumours.

This condition can be treated with chemotherapy but this is not available. The dog is emaciated and in extreme pain. You and your fellow students agree that euthanasia is the most humane option but the veterinarian says that it is not an option.

What do you do?

RESPONSE

STEVEN P MCCULLOCH

▶ It is instructive to analyse these two scenarios side by side, since the contextual circumstances provide insight into ethical analysis of the same overriding problem: the moral issue of a client wishing to prolong the life of a suffering patient. Prior to the analysis, it is worth mentioning that although such scenarios (or "dilemmas") are commonly used in the teaching of veterinary ethics, it is the author's experience that the converse scenario, at least in the context of UK and Hong Kong practice (the two locations where the author has experience), is far more common. The moral problem of presentation of patients that ought *not* to be euthanased, where the client mistakenly believes that they should, is far more common than the moral problem where the client believes that the patient should not be euthanased, when they should (this claim is premised on moral objectivity, or the existence of objectively moral truths, which is also pertinent to the case studies investigated here).

To facilitate analysis, in the following the first case is labelled the "natural death" scenario, and the second case the "Buddhist belief" scenario. The "natural death" scenario is simpler to analyse, and will be considered first here. The analysis of "natural death" is then useful to facilitate examination of "Buddhist belief". The outline of the

scenarios above, how they are interpreted and the ethical analyses highlight the importance of contextualism or situationism in ethics. That is, the *context* or *situation* in which something with potential moral import takes place has a bearing on the moral rightness or wrongness of action. Additionally, the scenarios point to the importance of the *meta-ethical* subject of epistemology in ethics. Are some acts objectively right or wrong, or is morality contingent on the subjective beliefs and attitudes of agents?

The author will assume that the "natural death" case takes place in the UK. Teddy is hospitalised in a 24-hour intensive care facility. Teddy has a number of physical diseases which *contribute* to a negative mental state. Teddy is 17 years old and at least some of his conditions are progressive and incurable. Euthanasia can be described as painless killing (RCVS 2015), or more demandingly as painless killing that is for the benefit of the patient (Broom & Fraser 2007). The author holds that euthanasia is painless killing for the benefit of the patient and this will be assumed in the following analysis. Therefore, two conditions must be satisfied for legitimate euthanasia: firstly, the animal is suffering/has a negative quality of life (QoL)/has a life not worth living; and secondly, the patient cannot be treated such that it will have a positive QoL/a life worth living.

Interestingly, the "natural death" scenario refers to Teddy having *no* QoL. The scenario outline may have been written to mean that Teddy has a negative QoL, but in fact, interpreted literally, means that he has a *neutral* QoL. If Teddy has a neutral QoL, then, based on certain conditions, it is justifiable to accede to the owners' wishes that Teddy die "naturally". The key condition is that Teddy does indeed have a neutral QoL, and not a negative QoL, i.e. that he is not suffering. Thus, the RCVS Code of Professional Conduct for Veterinary Surgeons makes it clear in Section 1.1 that "Veterinary surgeons must make animal health and welfare their first consideration when attending to animals" (RCVS 2015). Therefore, the veterinary surgeon has a primary, overriding duty to the welfare of the patient (McCulloch, et al. 2014). In this context, the fact that Teddy is hospitalised in a 24-hour intensive care facility is important. Arguably, multi-modal analgesia, such as morphine, ketamine and lidocaine infusions, can alleviate Teddy's pain such that he has at least a neutral QoL.

The concept of naturalness is also important. If dying naturally means not being euthanased, at least actively, then modern treatments may mean that Teddy does not suffer during this process. However, if dying naturally also means that Teddy does not receive medication such as analgesics despite being in pain, then he will suffer and have a life not worth living. If the veterinary surgeon reasonably judges that Teddy cannot be treated such that he does not have a neutral or positive QoL, with or without medication, then s/he should recommend euthanasia. If the clients persist with their wishes, since it is the veterinary surgeon's primary responsibility to ensure the welfare of patients under their care, then, after reasonable attempts at persuasion, s/he should insist on euthanasia under the framework of legislation governing the welfare of animals (for instance, the Veterinary Surgeons Act 1966 and the Animal Welfare Act 2006).

The "Buddhist belief" scenario brings a number of further issues to the analysis. A starting point is whether it is possible that any realistic treatment plan can alleviate suffering to provide the dog with a neutral or positive QoL. If there is such a treatment plan available, then pursuance of it should resolve the moral conflict. However, as chemotherapy is not available, the scenario suggests the absence of such a treatment plan. The scenario describes the agent as a veterinary student. This is important for ethical analysis, since veterinary students have a different role and responsibilities

to veterinary surgeons. The veterinary students are working under the direction of the veterinarian in charge of the animal welfare organisation. Qualified veterinary surgeons work with a significant degree of autonomy in clinical decision-making. In contrast, veterinary students, who are presumably on the externship to develop their clinical skills and learn about veterinary practice in a foreign country, do not have such a degree of autonomy.

Consider if the veterinary student were a fully qualified professional – whether a Thai citizen or otherwise – treating the dog in the animal welfare organisation. Consider, additionally, if the person in charge of the animal welfare organisation (assuming this role gives them legal ownership/guardianship) is Buddhist and opposed to euthanasia. In this case, the morally right course of action, for the veterinary surgeon, would to a significant extent depend on the professional and regulatory framework in Thailand. In the "natural death" scenario above, in the UK context the veterinary surgeon has recourse to a legislative framework to euthanase a suffering animal, even against the wishes of the legal owner. In the UK, although sentient animals are regarded as the property of owners, the legislative framework restricts these property rights by permitting, in extremis, euthanasia against the owner's wishes. Therefore, in extremis, the morally right act for a veterinary surgeon is to euthanase an animal against the owner's wishes.

Consider if, in Thailand, there is no such regulatory and legislative framework, and property rights of the owner trump any right of the veterinary surgeon to alleviate suffering. If this is the case, it might be right for the veterinary surgeon to attempt to persuade the client to euthanase the dog. However, it is difficult to defend the claim that the veterinary surgeon should disregard the owner's will and euthanase the dog, thereby breaking either the law or professional regulatory codes. In such cases, the veterinary surgeon should attempt to change the law and regulatory codes to enable such an outcome at a later stage. However, in the meantime, the veterinary surgeon would need to work within the constraints of the legislative and regulatory framework to do their best for patients.

Ultimately, the fundamental moral issues at stake here are (1) quality versus quantity of life, and (2) the perceived rights of humans to actively kill sentient animals. In the "natural death" scenario, the analysis is premised on the idea that non-existence is *better* than a negative QoL. For the animal, a positive QoL constitutes a life worth living, and a negative QoL a life not worth living (FAWC 2009). Religious beliefs add complexity to this concept. The Buddhist believes that euthanasia is morally wrong because s/he believes in the doctrine of transmigration (similarly, Christian doctrine claims euthanasia of humans is morally impermissible due to the sanctity of the immortal soul). Thus, the Buddhist may agree with the veterinary surgeon that a dog is suffering and has a life not worth living. However, according to Buddhist belief, euthanasing the dog destroys the potential for the life (or soul) of the dog to transmigrate into another form.

In moral philosophy, it is broadly accepted that non-religious arguments are more defensible than those based on religious premises. This is because they do not rely on supernatural claims (transmigration in the case of Buddhism, an immortal soul in Christian theology). Hence, one could make the argument that, objectively and in an *ideal* world, dogs suffering with venereal tumours in the absence of accessible treatment ought to be euthanased. Despite this, veterinary practice takes place in the *real* world, and the veterinary practitioner works in the context not only of extant legislative and regulatory frameworks, but of diverse cultural beliefs and attitudes.

The analysis returns to the idea that the moral agent in "Buddhist belief" is a veterinary student. The above analysis suggests it to be problematic, even for a fully fledged veterinarian – perhaps

especially so for a foreigner – to attempt to force his/her moral beliefs on another culture. This does not mean that moral subjectivism (truth cannot hold of moral propositions) and in particular relativism (beliefs are relative to culture) hold. Rather, it points to a distinction between *objective* philosophical argumentation and veterinary practice in the *real world*. The veterinary surgeon must practise with a degree of moral pragmatism. In the case of the veterinary student these points are even more forceful. The veterinary student can – and should – communicate his or her beliefs to the veterinarian in charge. However, it is the role of the veterinarian in charge to be the decision-maker, and with this come the moral responsibilities of decision-making.

Finally, it is instructive to make the point to be wary about stereotyping cultures and religious beliefs when discussing such cases. Just as many Christians do not attend church every Sunday, many Buddhists do not practise scriptural precepts. Furthermore, Buddhists that are opposed to euthanasia are not always opposed to an absolute degree.

During my time in Hong Kong, I was called to treat a recumbent geriatric German Shepherd dog. The patient was suffering from arthritis and peripheral nervous degeneration. Both conditions are progressive and the latter condition meant that he would not walk again. Based on Buddhist beliefs, the client turned down my recommendation of euthanasia and we treated the dog palliatively with analgesics. Two days later I was called to see the same dog. Examination of his perineal area revealed the dog was suffering from fly strike (myiasis). The dog had lived outside all his life and his recumbency had now predisposed him to this disease. After advising that we might be able to remove the maggots, larvae and eggs, but that the condition would recur unless the dog was taken inside, the client requested that I euthanase his dog. This was presumably because of the visual picture of the presence of live maggots eating away at the patient, together with the accompanying putrid stench.

Despite this experience, religious (Buddhist and non-Buddhist) and secular beliefs can sometimes manifest in absolute prohibitions. During 2001, an economic crisis in Hong Kong led to reduced donations to rehoming centres for unwanted pets. Although I cannot verify the claim, I was informed that dogs were starving in a rescue centre due to absolute prohibitions on euthanasia, at least in part due to religious beliefs. Despite this, again, it is important to not stereotype in such cases. In the UK there are large numbers of dogs in rescue centres, some of which have no hope of being rehomed, due to their age, temperament and/or medical conditions. These are not religious but secular organisations, some of which also have absolute prohibitions on euthanasing unwanted animals.

CONVENIENCE EUTHANASIA

What would you do?

A local wealthy widow, Mrs D has always been an animal lover. She wants to set up an animal sanctuary for abandoned animals. She has plenty of resources – space, time and money – and would like you to provide your veterinary services. She makes it clear that she does not believe in euthanasia.

How do you respond?

△

3.20 Mrs D wants to rescue abandoned animals but does not believe in euthanasia.
PHOTO ANNE FAWCETT

3.6

Companion animal hospice

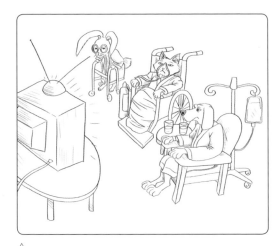

△
3.21

CARTOON RAFAEL GALLARDO ARJONILLA

The rise of companion animal hospices in the USA, UK, Australia and elsewhere is a perhaps inevitable trend given the success of the human hospice movement (Osborne 2009). Hospice care emphasises patient comfort rather than treatment when a cure is no longer expected (Cooney 2015). Hospice care may involve inpatient care in a dedicated environment or treatment of animals in the home environment, and may or may not involve a team of professionals including veterinarians, nurses, social workers and others. The distinction between palliative care and hospice has not been well defined for animal patients (Shanan, et al. 2014, Hewson 2015).

This next scenario considers how to manage the end of life of a cat whose owner wants to hang on to her for as long as possible.

▶   You are working in small animal practice. Miss P presents Sukie, a 13-year-old cat with what has been diagnosed as end-stage chronic renal failure secondary to polycystic kidney disease. Sukie is in poor body condition (body condition score 1/5), is unable to ambulate more than a few steps at a time, has developed urinary and faecal incontinence and cannot eat without assistance.

You believe Sukie has very poor quality of life. Miss P has taken weeks off work to nurse Sukie and provides round-the-clock care, including assisted feeding, cleaning (the cat is wearing disposable nappies) and subcutaneous fluids twice per day.

A number of your colleagues have advised Miss P that the cat should be euthanased on humane grounds, but she refuses to do so, claiming that Sukie has a will to live, still purrs when she pats her and shows an interest in food.

△
3.22 Sukie, a 13-year-old cat with what has been diagnosed as end-stage chronic renal failure secondary to polycystic kidney disease.
PHOTO ANNE FAWCETT

"My mother died of kidney disease," she said. "They certainly didn't put her down."

Miss P is prepared to present Sukie for any treatment you recommend which may extend her life. You have heard that a new pet hospice has just opened up on the other side of your city.

What should you do?

RESPONSE

RICHARD GREEN

▸   The main issues central to this scenario are:

The debate about when the treatment of companion animals can become as much a part of the problem as a part of the solution. As veterinary medicine advances this is clearly an issue that will become more and more relevant with one person's reasonable treatment being another's unnecessary suffering.

The difficulties in assessing quality of life (QoL) in animals and the differences between QoL assessments in animals and people; the difficulties in communicating these differences to pet owners.

The scope for disagreements in the assessment of QoL between concerned individuals, most frequently between owner and vet (and although this scenario concerns a vet's desire to euthanase an animal on humane grounds versus the owner's desire not to, the situation is not infrequently reversed).

The often widely differing views held about the relative merits of a "natural" death versus an assisted, "good" death – euthanasia.

Finally, these issues must be examined within the context of the options available, which in this case are to do nothing – i.e. continue treatment at the practice until eventually a natural death occurs (all the while continuing to discuss Sukie's QoL and using your influence to persuade Miss P as to the benefit of euthanasia);

potentially to refuse further treatment other than analgesia, thus hastening the natural death; to "insist" on euthanasia, either by reporting Miss P to (for example) the RSPCA, or even taking the matter to the police in the same way that one might for a farm animal; or to refer her to the new pet hospice.

A full debate about the methods of QoL assessment in animals is outside the scope of this discussion; however, the salient points are that for animals QoL assessments will always be a best estimate by proxies (usually the vet and the owner), and even more limited than our assessments of non-competent humans where our own experiences are more relevant. Human QoL has been defined differently: one definition is "the difference or gap between the hopes and expectations of the individual and the individual's present experiences" (Calman 1984). In humans there is more emphasis on control and expectation, and more capacity for pleasure when bodily function is compromised, and this may be a useful jumping-off point when discussing why Miss P's mother and cat should be treated differently.

It is important that we do not disregard the ethical as well as clinical merits of treatments we have available. Here prognosis is a key concern. A treatment with the potential to cure or reverse a disease process, or a palliative treatment which can control or manage a disease whilst maintaining QoL, is very different to a treatment which can prolong life without providing the possibility of a cure or the significant control of adverse symptoms – in this case the treatment may actually prolong suffering. The crux of the argument here is of course whether the outcome is an "acceptable" QoL, and the question is "acceptable to whom – the owner, the vet, or ideally, the patient?"

Many owners do feel that the veterinary profession is too quick to offer euthanasia, and the idea that vets can become too comfortable with the notion that euthanasia provides an easy solution has been suggested as one possible factor in the high suicide rate within the profession (for example Bartram and Baldwin 2010). It also raises questions as to whether vets are entitled in any way to (try to) impose their wishes for euthanasia, and most vets would argue that it is their overriding duty towards the animal, above any duty to their client or to themselves, that gives them this right.

In this scenario most (but not all) vets would not elect for a direct confrontation with an owner either by reporting their client, or by trying to enforce euthanasia; however, if there is a strong feeling that the degree of suffering is untenable then this is an option. If so, the vet would be well advised to gain a second (and third) opinion supporting their decision from colleagues, as well as consulting any relevant professional bodies for advice regarding their professional ethical obligations, for example breaching confidentiality when reporting clients. None of these steps would prevent a possible complaint, but they would make a complaint much less likely to be upheld. Although even suggesting to a client that one is considering reporting them is very confrontational, raising this in conversation can be useful in conveying the gravity of the situation.

Continuing to treat Sukie yourself has the advantage of allowing you to be satisfied you are doing the best you can to alleviate her suffering, and to continue to exert your influence to try to change Miss P's mind over euthanasia. However, it may come at a very considerable price, of significant moral stress (another often-cited reason for the high suicide rate in the profession).

If you feel that you are unlikely to be able to influence Miss P to change her mind, and are not going to take a more confrontational approach then the hospice may provide a solution. A pet hospice may offer Miss P a similar, or greater level of care than can be offered by you, since that is their *raison d'être*. Referring Sukie to them would remove the immediate stresses of dealing with this

patient on a day-to-day basis, and for many vets out of sight would be out of mind.

However, for your peace of mind, you would be advised to discuss not only the pertinent details of the case with the hospice, but also their approach to cases.

Do they have a similar outlook to your own, in that they place the pet's QoL at the centre of their management, or do they pursue longevity above all else? Do you feel they are able to offer a similar or better standard of care for a pet in the final stages of life than you can? Hospices are likely to be set up to be better able to provide domiciliary care than most practices, but what about out of hours? Should you (and Miss P) agree an acceptable end point, or at least a framework for deciding one, with the hospice prior to referral?

In the long term, since dilemmas such as this are not infrequent, forming a working relationship with the hospice might be sensible. Clients such as Miss P would be likely to find such an establishment themselves, and having a pre-existing relationship with the hospice would allow you to retain some influence over the case, and might allow you some influence over the working practices of the hospice too.

||||||||||||||||||||||||||||||||||||||||||||||||||||||||||||||||||

OWNERS who care for their animals during life are bound to want to continue with a high standard of care through the terminal period. While hospice care assists both the owner and the veterinary team in the respectful closure of the human–animal bond, not every veterinary team is able to provide this service which requires considerable time and commitment to the medical needs of the patient and the emotional needs of the client or clients (American Veterinary Medical Association 2016).

Much decision-making in this scenario depends on the quality of hospice care available, which may be variable. As alluded to in this case, a disincentive to provide or refer an animal for hospice is the belief that hospice pets are not euthanased. In fact hospice does not exclude nor is it committed to delaying euthanasia where deemed appropriate (American Veterinary Medical Association 2016, Cooney 2015).

Veterinary care has perhaps traditionally been under less pressure to provide palliative care than for humans as euthanasia can be employed to prevent any reduction of quality of life, and treatment costs, during the terminal stages. Traditionally, owners of animals with terminal conditions had three choices: pursuit of aggressive or "heroic" treatment; euthanasia; or limited palliative care followed by euthanasia once quality of life is no longer acceptable (Cooney 2015) (although, as in the scenario above there may be disagreements as to when this point is reached). Animal hospice may provide an appealing alternative to premature euthanasia, or extended suffering which may result from isolating an animal or intensive care, or providing inadequate treatment such as analgesia (Shanan, et al. 2013). Indeed, appropriate hospice seems aligned with utilitarianism (minimising suffering), deontology (respect for the individual patient) and principalism (do no harm – if death is a harm). An owner might also take a contractarian approach, for example "this animal has given me companionship, so I owe it to the animal to maintain his or her quality of life for as long as possible."

As discussed by the author of the scenario, hospice should be provided under stringent conditions, including the ability to provide suitable, round-the-clock analgesia and sanitation (American Veterinary Medical Association 2016).

In a journal issue dedicated to palliative medicine and hospice care, Shearer (2011) notes that both human patients and carers benefit from dedicated end-of-life care (Herbert, et al. 2006, Temel, et al. 2010) and suggests that palliative or hospice care can be sufficient to allow an animal to die naturally without reducing quality of life unacceptably.

COMPANION ANIMAL HOSPICE

What would you do?

You have been asked by your practice principal to set up an ambulatory palliative care service and an in-house hospice for your companion animal clients and others in the area. What safeguards would you put in place to protect the animals from unacceptably poor welfare?

△

3.23 In setting up a hospice service what safeguards would you need to ensure acceptable animal welfare?
PHOTO ISTOCK

Conclusion

Animal death is a critical issue in bioethics and veterinary ethics and consequently our views on it will influence our actions in life and regarding veterinary practices. Animal killing occurs on a large scale and is often systematised and may be invisible.

A number of the scenarios discussed here raise issues about terminology, specifically the varied meanings attached to terms such as euthanasia, slaughter, killing and culling.

Being clear on our views, and being able to defend them, on this of all issues, will help to ensure that we are able to act in accordance with our values and reduce moral stress. Whilst dying may constitute only a small part of an animal's life, almost all societies recognise the importance of preventing poor welfare during this time with strict regulations covering slaughter of animals for food and killing for other reasons. A good death is a desirable outcome for us to provide for all animals whether directly at our hands or not.

References

American Veterinary Medical Association 2016 Guidelines for Veterinary Hospice Care. https://www.avma.org/KB/Policies/Pages/Guidelines-for-Veterinary-Hospice-Care.aspx

Anon 2015 Badger vaccination grant awarded. *The Veterinary Record* 176: 477.

Bartram DJ, and Baldwin DS 2010 Veterinary surgeons and suicide: a structured review of possible influences on increased risk. *Veterinary Record* 166: 388–397.

Bontje DM, Backer JA, Hogerwerf L, Roest HIJ, and van Roermund HJW 2016 Analysis of Q fever in Dutch dairy goat herds and assessment of control measures by means of a transmission model. *Preventive Veterinary Medicine* 123: 71–89.

Broom DM, and Fraser AF 2007 *Domestic Animal Behaviour and Welfare*. CABI: Wallingford.

BVA 2015 Slaughter without stunning and food labelling. British Veterinary Association: UK.

Calman KC 1984 Quality of life in cancer patients – an hypothesis. *Journal of Medical Ethics* 10: 124–127.

Carrington D 2015 Prince Charles letters include strong backing for badger cull. *The Guardian* Wednesday 13 May, 2015.

Cassidy A 2012 Vermin, victims and disease: UK framings of badgers in and beyond the TB controversy. *Sociologia Ruralis* 52: 192–214.

Christiansen SB, Kristensen AT, Sandøe P, and Lassen J 2013 Looking after chronically ill dogs: impacts on the caregiver's life. *Anthrozoos* 25: 519–533.

Cooney K 2015 Offering hospice care for pets. *Veterinary Record* 177: i–ii.

Cousquer G 2013 Everyday ethics. Being a badger's advocate. *In Practice* 35: 350–351.

Dawkins R 2006 *The Selfish Gene: 30th Anniversary Edition*. Oxford University Press: Oxford.

DEFRA 2007 Badgers and Cattle TB. The Final Report of the Independent Scientific Group on Badger TB. House of Commons, HC 130-1: London.

Drew J 2006 *Following Orders*. Bluebell Publishing: Strathaven.

Duerr S, Dohna HZ, Di Labio E, Carpenter TE, and Doherr MG 2013 Evaluation of control and surveillance strategies for classical swine fever using a simulation model. *Preventive Veterinary Medicine* 108: 73–84.

Epicurus 1964 Letter to Menoeceus 124b-127a. In: Greer RM (ed) *Letters, Principal Doctrines, and Vatican Sayings*. Bobbs-Merrill: Indianapolis.

European Council 2009 Council Regulation (EC) No 1099/2009 on the protection of animals at the time of killing. Official Journal of the European Union L 303/1–30.

FAWC 2009 Farm Animal Welfare in Great Britain: Past, Present and Future. Farm Animal Welfare Council: London.

FCEC 2015 Study on Information to Consumers on the Stunning of Animals. Food Chain Evaluation Consortium: Brussels, Belgium.

Food Standards Authority 2015 Animal Welfare Survey of Slaughterhouses. UK.

Fraser D, and Duncan IJH 1998 'Pleasures', 'pains' and animal welfare: toward a natural history of affect. *Animal Welfare* **7:** 383–396.

Gortazar C, Diez-Delgado I, Barasona JA, Vicente J, De la Fuente J, and Boadella M 2015 The wild side of disease control at the wildlife-livestock-human interface: a review. *Frontiers in Veterinary Science* **1:** 1–12.

Gruen L 2014 Death as a social harm. *The Southern Journal of Philosophy* **52:** 53–65.

Herbert R, Prigerson C, Schultz R, and Arnold R 2006 Preparing caregivers for the death of a loved one: a theoretical framework and suggestions for future research. *Journal of Palliative Medicine* **9:** 1164–1171.

Hewson C 2015 End-of-life care: the why and how of animal hospice. *Veterinary Nursing Journal* **30:** 287–289.

Kalof L, and Mason J 2015 *Misothery: Contempt for Animals and Nature, Its Origins, Purposes, and Repercussions.* Oxford University Press: Oxford.

Kasperbauer TJ, and Sandøe P 2015 Killing as a welfare issue. In: Visak T and Garner R (eds) *The Ethics of Killing Animals.* Oxford University Press: Oxford.

Lockwood M 1979 Killing humans and killing animals. *Inquiry* **22:** 157–170.

Lu Z, Mitchell RM, Smith RL, Van Kessel JS, Chapagain PP, Schukken YH, and Grohn YT 2008 The importance of culling in Johne's disease control. *Journal of Theoretical Biology* **254:** 135–146.

McCulloch S, Reiss M, Jinman P, and Wathes C 2014 The RCVS codes of conduct: what's in a word? *Veterinary Record* **174:** 71–72.

McMahan J 1988 Death and the value of life. *Ethics* **99:** 32–61.

Montiel DO, Frankena K, Udo H, Keilbach Baer NM, and van der Zijpp A 2013 Prevalence and risk factors for brucellosis in goats in areas of Mexico with and without brucellosis control campaign. *Tropical Animal Health and Production* **45:** 1383–1389.

National Audit Office 2002. The 2001 Outbreak of Foot and Mouth Disease. National Audit Office: London, UK.

Osborne M 2009 Pet hospice movement gaining momentum. *Journal of the American Veterinary Medical Association* **234:** 998–999.

Palmer C, Corr S, and Sandøe P 2012 Inconvenient desires: should we routinely neuter companion animals? *Anthrozoos* **25:** S153–S172.

Pollard-Williams S, Doyle RE, and Freire R 2016 Influence of workplace learning on attitudes towards animal welfare. *Journal of Veterinary Medical Education* **41:** 253–257.

Powell C 2016 Handling convenience euthanasia. *Veterinary Team Brief.* Educational Concepts, LLC: www.veterinaryteambrief.com

RCVS 2015 Code of Professional Conduct for Veterinary Surgeons. http://www.rcvs.org.uk/advice-and-guidance/code-of-professional-conduct-for-veterinary-surgeons/pdf/

Regan T 2004 *The Case for Animal Rights.* University of California Press: Berkeley.

Rhyan JC, Nol P, Quance C, Gertonson A, Belfrage J, Harris L, Straka K, and Robbe-Austerman S 2013 Transmission of brucellosis from elk to cattle and bison, Greater Yellowstone Area, USA, 2002–2012. *Emerging Infectious Diseases* **19:** 1992–1995.

Rollin BE 2011a *Putting the Horse Before Descartes: My Life's Work on Behalf of Animals.* Temple University Press: Philadelphia.

Rollin BE 2011b Euthanasia, moral stress, and chronic illness in veterinary medicine. *Veterinary Clinics of North America-Small Animal Practice* **41:** 651–659.

Sencerz S 2011 Utilitarianism and replaceability revisited or are animals expendable? *Between the Species* **14:** 81–106.

Serpell JA 2005 Factors influencing veterinary students' career choices and attitudes towards animals. *Journal of Veterinary Medical Education* **32:** 491–496.

Shanan A, August K, Cooney K, Hendrix L, Mader B, and Pierce J 2013 Animal Hospice and Palliative Care Guidelines. https://www.iaahpc.org/images/IAAHPCGUIDELINESMarch14.pdf

Shearer TS 2011 Preface: the role of the veterinarian in hospice and palliative care. *Veterinary Clinics of North America Small Animal Practice* **41:** xi–xiii.

Singer P 2011 *Practical Ethics.* Cambridge University Press: Cambridge.

Temel JS, Greer JA, Muzikansky A, Gallagher ER, Admane S, Jackson VA, Dahlin CM, Blinderman CD, Jacobsen J, Pirl WF, Billings JA, and Lynch TJ 2010 Early palliative care for patients with metastatic non-small-cell lung cancer. *New England Journal of Medicine* **363:** 733–742.

The Animal Studies Group 2006 *Killing Animals.* University of Illinois Press: Illinois.

USDA 2014 USDA Announces Additional Food Safety Requirements, New Inspection System for Poultry Products. United States Department of Agriculture: USA.

Visak T, and Garner R 2015 *The Ethics of Killing Animals.* Oxford University Press: Oxford.

Webster J 1994 *Animal Welfare: A Cool Eye towards Eden.* Blackwell Publishing: Oxford.

Whiting M In press: Could preventing non-stun slaughter in the UK harm animal welfare? *Journal of Animal Welfare Science, Ethics and Law.*

Woods A 2004 *A Manufactured Plague. The History of Foot and Mouth Disease in Britain.* Earthscan: London.

Worboys M 2000 *Spreading Germs. Disease Theories and Medical Practice in Britain 1865–1900.* Cambridge University Press: Cambridge.

Yeates JW 2010 Death is a welfare issue. *Journal of Agricultural and Environmental Ethics* **23:** 229–241.

Yeates JW, and Main DCJ 2011 Veterinary opinions on refusing euthanasia: justifications and philosophical frameworks. *Veterinary Record* **168:** 263.

CHAPTER 4
ANIMAL USE

"…the quality of life for most other sentient animals with whom we share the planet is largely governed by how and where we let them live and what we let them do."
Webster 2005

Introduction

Animal use, in some form or other, is widespread. We use animals as sources of food, clothing, companionship, assistance in life and work, to further scientific knowledge and curiosity, and in cultural and religious contexts. Veterinarians, nurses, technicians and others are charged with the care of animals, within this context of animal use.

But the use of other creatures for our own ends raises some ethical questions, notably:

- How do we justify the use of sentient creatures for our own ends?
- Are these justifications sound?
- What obligations, if any, do we have towards animals?
- What limitations, if any, should be imposed on our use of animals?
- Is our approach to animals consistent?

This chapter incorporates scenarios that explore our use of animals in different contexts. Increasingly, the use of animals by humans is challenged by different groups and individuals. Animal use abolitionists oppose the usage of sentient beings on the grounds that it infringes the right not to be treated simply as the means to another's end. Rather, proponents of this view hold that recognition of the intrinsic value of animals prohibits their use.

First, we begin with a scenario that raises the question of whether we should use animals at all.

SCENARIO

ARE ANIMALS SLAVES?

▶ You attend an animal welfare panel discussion at the national welfare conference. One of the speakers is a self-proclaimed animal activist who argues that animal use is akin to slavery.

Over lunch, one of your colleagues – a companion animal veterinarian – declares that he cannot fault the argument.

"We really do treat production animals as if they are machines, not sentient beings," he says. "It's the reason I became a vegan."

Another colleague – a cattle specialist – pipes up.

"Really? If you followed his line of thinking you'd be out of a job," she says. "Keeping pets is also a form of animal use and, if you like, a type of

◁

4.1

PHOTO ANNE FAWCETT

—
4.0
INTRODUCTION
SCENARIO
ARE ANIMALS SLAVES?

△

4.2

CARTOON MALBON DESIGNS

slavery. Just because you don't eat them doesn't mean you're off the hook."

"I have more influence on animal welfare by buying only welfare certified meat and ensuring that farmers who treat animals well are supported," she continues. "You're out of the market so you have no influence at all."

After some thought, your other colleague responds.

"Well, it's good that someone is prepared to question the status quo."

How might you respond?

—————
RESPONSE
EMMANUEL GIUFFRE

▶ Animal use is justified on economic, cultural and political grounds, and on religious tradition that presumes human dominion over all other living species. (For a brief discussion on religion and animals, see Waldau 2006.)

So entrenched is our use of animals that most humans rarely consider whether the use of animals is morally or ethically justifiable. Over the last three decades, we have seen a burgeoning animal rights movement that has drawn attention to the suffering of animals at the hands of humans. Mainstream discussion on the issue has been dominated, however, by promoting the welfare of animals, as opposed to questioning whether we should be using them altogether.

Our consideration and treatment of animals can be traced back to such influential thinkers as Kant, Descartes and Aquinas, who denied animals' moral worth on the basis of their lack of reason, consciousness or autonomy. In fact, up until the eighteenth century, animals were thought of as nothing more than machines, which avoided any need to consider their welfare, let alone whether or not they should be used (Sankoff 2009).

As the law reflects society, the common law came to mirror this sentiment by relegating animals to the category of "property", a classification which persists today. Distinct from "legal persons", classifying animals as property denies them the ability to bear individual rights and subjects them to ownership by humans.

The capacity to bear rights is considered crucial, as interests protected by a right cannot simply be ignored because it benefits others (Francione 2004, Paper 21, 27). For this reason, some animal rights advocates focus on the property status of animals as the single most influential factor in their continued use and abuse (see e.g. Francione 1995, Wise 2000). In the words of Steven M Wise, "Without legal personhood, one is invisible to civil law. One has no civil rights. One might as well be dead" (Wise 2000, 4).

However, unlike inanimate objects, humans do not have an unimpeded right to do with animals as they please. Anti-cruelty laws first emerged in the United Kingdom in the nineteenth century, prompted by a deepening understanding of the sentience of animals and their ability to suffer (Sankoff 2009, 8). The *Cruel Treatment of Cattle Act 1822* (2 Geo IV c71) deemed it an offence to "wantonly and cruelly beat, abuse, or ill-treat any horse, mare, gelding, mule, ass, ox, cow, heifer, steer, sheep or other cattle". These laws were designed to protect certain animals from gratuitous acts of violence, whilst at the same time permitting humans to continue to benefit from their use.

Nearly 200 years on, this balancing of human interests over those of animals now forms the bedrock of modern animal protection laws, which generally permit animal cruelty if it can be deemed "reasonable", "necessary" or "justifiable" to achieve human ends (Animal Welfare Act 2006). See, for example, the *Prevention of Cruelty to Animals Act 1979* (NSW), section 4 which states:

"(2) For the purposes of this Act, a reference to an act of cruelty committed upon an animal includes a reference to any act or omission as a consequence of which the animal is unreasonably, unnecessarily or unjustifiably: (a) beaten, kicked, killed, wounded, pinioned, mutilated, maimed, abused, tormented, tortured, terrified or infuriated, (b) over-loaded, over-worked, over-driven, over-ridden or over-used, (c) exposed to excessive heat or excessive cold, or (d) inflicted with pain."

Under this balancing act, human needs are given greater weight than those of animals, and it is inevitable given their respective legal statuses: the right of persons to the use of their property will often trump even the most critical interests of the property in question.

Granting nonhumans legal personhood, and accordingly, the capacity to bear rights is not a radical concept. Not all legal persons are human, and throughout history, not all human beings have been considered legal persons. For example, corporations, religious institutions, religious texts and idols, and even a river have been granted legal personality under various common law jurisdictions (Prosin and Wise 2014). The common law once considered married women the property of their husbands; children the property of their fathers; and human slaves the property of their masters. Over time, the law evolved to grant these groups legal personhood, and began the work of eroding the legal, social and political structures that discriminated against them.

Animal advocates often draw comparisons between the exploitation of human slaves and that of animals, and the argument is compelling. As is presently the case with animals, the treatment of slaves was premised on the commodification of another living sentient being – to be worked, traded and disposed of at the whim of "its" human owner. As property, slaves had no intrinsic moral worth; their value was recognised by their owners only in economic terms.

Laws sought to govern the welfare of slaves, but the same concerns arose around balancing the interests of slaves with those of their owners. In certain jurisdictions it was unlawful to kill a slave, but this did not apply to outlawed slaves, slaves killed "in the act of resistance to his lawful owner", or to slaves "dying under moderate correction" (Francione 2004, 25).

> "The justifications put forward by proponents of the animal trade are nearly identical to those put forward by British pro-slave lobbyists in the eighteenth and nineteenth centuries."

Even the justifications for human slavery resemble those put forward by supporters of the use of animals. Take live animal exports, for example, which is one of the more controversial aspects of the animal trade. The justifications put forward by proponents of the animal trade are nearly identical to those put forward by British pro-slave lobbyists in the eighteenth and nineteenth centuries. For the slave trade, justifications included that it:

- supports the local economy;
- supports the health and wellbeing of people in importing countries;
- is necessary, as failure to partake would result in competitors with compromised slave welfare taking over;
- ensures slave welfare, as owners protect, feed and keep traded slaves healthy; and
- is governed by Codes of Practice that result in low and declining rates of slave mortality.

For the common justifications of the animal trade, you need only replace the word "slaves" for "animals" (Cavalieri 2015, Phillips 2015).

The property status of animals is a significant barrier to any higher recognition of their moral worth, in a similar way to how the property status of slaves preserved their oppression. The comparison is not intended to equate the suffering of slaves to that of animals, but to highlight the social injustice of subjugating one class of living sentient beings for the benefit of another.

Questioning the status quo is essential to ending social injustice. It was (and remains) essential in stamping out racism, sexism, heterosexism and all other "isms" which justify the irrational discrimination of one class of people over another. As our knowledge of animal sentience develops, along with our knowledge of their emotional and intellectual complexities, so too speciesism can and should be added to that list.

Purchasing "humane" animal products is put forward as an alternative way to improve the lives of animals. Consumers who purchase higher welfare products must acknowledge, however, that they share in the human-centric philosophy which consents to the exploitation and killing of animals for human purposes, as long as it appears to have been done as humanely as possible. If we were to use the human slavery analogy, it would be comparable to purchasing goods or services derived from slaves, as long as they were treated "nicely" by their masters.

That is certainly not to say that consumers who would otherwise purchase animal products should stop purchasing higher welfare products, nor that we should stop advocating for higher animal welfare standards. Animals benefit from less cruel methods of production, and reducing their suffering is important given society is far from accepting an end to their exploitation or granting them meaningful rights. But we must acknowledge that purchasing animal products reinforces their objectification, and reasserts their status as the property of humans.

"Pet" ownership also reasserts the property status of animals. Purchasing animals for companionship is still a form of exploiting animals for human ends, even if the animal is given an otherwise "good" life. Ethical concerns particularly arise where animals are purchased from pet shops or backyard breeders. While purchased companion animals may be properly cared for by their owners, the exchange too often ignores the deprivation and suffering of breeder animals, the majority of whom are confined in factory farms (see, e.g. Animals Australia n.d.). It also perpetuates unnatural and often debilitating genetic diseases in dogs for aesthetically based "breed standards" (Rollin and Rollin 2015).

Purchasing companion animals from pet shops or backyard breeders ignores the millions of homeless animals kept (and if not rehomed, killed) in shelters. The housing and premature slaughter of millions of otherwise healthy companion animals in shelters are a direct result of commercial over-breeding, and animals being discarded at their owners' convenience. Fostering or adopting animals in shelters is a more ethical approach to acquiring a companion animal.

Animal rights theorists have highlighted the need to reframe the human–companion animal relationship from one of pet "ownership" to one

△ **4.3** Fostering or adopting animals in shelters is a more ethical approach to acquiring a companion animal.
PHOTO ANNE FAWCETT

that recognises the vulnerability of companion animals and the responsibility owed by humans to the animals in their care. A guardianship model – based loosely on the principles that apply to the care of children – would more appropriately redefine the position of "pets", not simply as property, but as individuals owed a duty of care and compassion by their wards (see, e.g. Tischler 1977).

THIS scenario takes an animal rights-based approach, which is consistent with deontology (see chapter 2, section 2.5.2). The animal rights movement is hundreds of years old but became particularly prominent in the late twentieth century. It has been described by some as "the next great social justice movement" (Cao 2015). The contribution of animal rights proponents and animal protectors to animal welfare is acknowledged by animal welfare scientists (for example, see Mellor 1998).

One major concern about a strong rights-based approach is that, if followed to its logical conclusion, it may oppose animal use in any form. John Webster argues that despite these arguments, animal use is acceptable if we attend to animal welfare.

—

RESPONSE

JOHN WEBSTER

▶ "Is animal use akin to slavery?" Emmanuel Giuffre has presented a clear, succinct exposition of the moral arguments that have shaped our attitudes and actions in respect to those animals deemed to be "within our care". In particular, he draws comparisons between the exploitation of human slaves and that of sentient animals. In both cases the individuals have had no moral worth; their value was recognised by their owners only in economic terms. This is a valid moral argument, as far as it goes. Problems arise, however, with its implementation in practice. In brief, he makes a distinction between farm animals and pets. Farm animals will still be regarded as property (aka slaves) but we must treat them more humanely. For pets we could adopt a "guardianship" model with principles similar to those that apply to the care of children; children, I might add, that will not grow up and assume responsibility for themselves.

I wish to take another tack. At the outset, I hope you will agree that, as far as the animals are concerned, it is not what we think but what we do that matters. We may define our role as guardians or slave-masters and their status as persons or merely sentient beings but, to them, these distinctions are meaningless. I acknowledge our many failings in the treatment of the animals in our care. However, in seeking to do things better, I suggest that we may be recruiting the wrong rules of ethics.

What do we actually mean by "ethics"?

In the beginning was the word. Philosophers and others have used the medium of words to create more or less convincing structures upon which to interpret human values, define the rules of morality and set ethical standards. Some non-human animals may have a sense of right and wrong but as a basis for right action, ethics is only applicable to those who have the power of language. Moreover, to simplify Wittgenstein, words cannot be defined by the mental pictures they evoke, only by how they are used. Thus "good" does not exist independently of any good deed. "Slave-master" can only be defined by our actions in respect to the animals in our care.

For those who seek to adhere to the principle of animal rights (I do not include myself in this group) it may be helpful to consider Hobbes' concept of the Leviathan, or Social Contract, which states, in brief:

> "without a common power to keep them in awe, men are in a constant state of war of every man against every man, continual fear, danger of violent death, the life of man solitary, poor, nasty, brutish and short… In pure self-interest and for preservation men entered into a compact whereby they agreed to surrender part of their freedom in order to preserve the rest."

Hobbes' definition of the life of humans who do not embrace the Social Contract reads remarkably like Darwin's definition of survival of the fittest. It can be argued that the animal species that have been most successfully domesticated are those that adapted best to an environment in which the common power is held by the human race. In

4.4 In strictly Darwinian terms, the chicken has proved to be a far more successful species than the tiger.

PHOTO ANNE FAWCETT

strictly Darwinian terms, the chicken has proved to be a far more successful species than the tiger. Personally I would not wish to pursue this argument too far. Let us leave it with the thought that any individual human living in a stable, governed society would be pushed to claim the absolute right to a single freedom, certainly not five!

It should be clear by now that I consider the discussion of animal rights to amount to little more than sophistry, especially when we conclude (without consulting the animals) that, in the matter of rights, some animals are more equal than others. We must recognise that an animal's perception of its own welfare is entirely determined by its own sentience and this is entirely independent of our classification of it as companion animal, farm animal, deserving wildlife (e.g. the tiger) or undeserving wildlife (e.g. the rat). Moreover it is human arrogance to assume that the more "human-like" the species, the greater is

its case to be granted personhood. One of many serious flaws in this argument is the inescapable fact that crows are better problem solvers than chimpanzees.

Albert Schweizer wrote: "The great fault of all ethics hitherto has been that they believed themselves to deal only with the relations of man to man. In reality however, the question is what is his attitude to the world and all life that comes within his reach" (Schweitzer 2010, p. 6). What he recognised is that humans, who have acquired a sense of morality largely (not entirely) through the medium of language, have a responsibility to care for all life that comes within our reach. We are the moral agents. The rest of life, soil, plants and animals, are the moral patients. We have a responsibility to them but they have no responsibility to us. In this sense they are like newborn babies. The responsibility of the moral agent is shared by all humanity because we all, whether omnivore or vegan, depend on the sustainability of the living environment. This obliges us to give special consideration to those to whom we delegate the task of putting our collective responsibility into practice, namely the farmers and stewards of the land. Their prime, practical responsibility is (or should be) to manage the land in ways that are both productive and profitable. While the mass production of meat from cereals and other crops that we could have eaten ourselves is both inefficient and unjust, the fact remains that over 50 per cent of land available for agriculture in the world is made of pastures and range land that can only be harvested by grazing animals. There are also a lot of fish in the sea. In the interests of good planet husbandry, I consider that it is both ecologically unsound and morally self-indulgent to ignore these facts of life.

In regard to our responsibility towards sentient animals in our care, I argue that we should strive to do the best we can in the circumstances that apply to ensure that they are physically fit and feel good during life, and at the end, experience a gentle death. Because I am an advocate for the animals, not a moral philosopher, I make no distinction in this regard between farm animals and pets. As a final thought may I suggest that our love affair with certain animals is not necessarily in their best interests. The two most emotionally disturbed mammalian species are the dog and the horse.

INTRODUCTION

What do you think?

The above scenarios discuss some controversies around animal use:

ONE **What factors would you take into consideration when deciding whether the use of non-human animals for the benefit of humans is acceptable?**

TWO **Which areas of animal use are the most contentious in your opinion?**

THREE **How do your beliefs about the use of non-human animals differ from societal norms?**

△
4.5 Bullriding is one form of animal use.
PHOTO ISTOCK

4.1

Animal use and animal welfare

Animal welfare science, coupled with change in attitudes towards animals, has led to increased consideration of the animal welfare impact of animal care and husbandry. This in turn leads to the question of how we address welfare problems associated with animal husbandry. The following scenario explores two approaches.

△
4.6 Castrating bulls is a controversial husbandry practice.
PHOTO ISTOCK

SCENARIO

CONTRASTING "GOLD STANDARD" AND
"INCREMENTAL IMPROVEMENT" APPROACHES
TO ANIMAL WELFARE CHANGE MANAGEMENT

▶ You are on a committee developing animal husbandry policies. There are frequent, heated debates within the group as to whether some husbandry practices should be banned because of their negative impact on animal welfare, or phased out.

What is your approach?

RESPONSE

DAVID MELLOR

▶ Whenever an animal welfare problem is identified it is natural for those who care to want to correct it. However, the complex interactions between factors such as the particular use to which the animals are put (e.g. farming, competitive sport, companionship, zoo display), current constraints imposed by the circumstances (e.g. free-range or intensive farming, requirement for only elite athletes, long periods alone in the home, restrictive facilities), and limited money to effect remedies, mean that improvements are not always as easily or quickly achieved as would be hoped. On the one hand, this can be a source of great frustration for those who want the solution to be swift and decisive, and who aim for the best outcome for animals immediately. On the other, it may provide spurious relief for those who in fact do not want to change and who therefore cite such complexity as an excuse for taking no remedial actions. Neither position is helpful practically, the first because it usually demands more than can be realistically delivered, and the second because it denies the now well-established, multifaceted drive to improve animal welfare standards locally, nationally and globally (Mellor & Bayvel 2014).

It is helpful to contrast two approaches to animal welfare change management, namely, the gold standard and incremental improvement approaches (Mellor & Stafford 2001).

Gold standard

A gold standard is the best that can be achieved in a particular situation, and it serves as a reference point against which other things of its type may be compared. As a strategy for promoting animal welfare such an approach defines the ideal that is to be required in a particular situation and accepts nothing less than that ideal. Applied in absolute terms, it excludes all of those who cannot meet the requirements. Typically, the demand is for immediate compliance and no concessions are made for practical, financial or other difficulties those affected by the standard might experience in endeavouring to meet it. Common outcomes are resentment, alienation, noncompliance and/or rejection of the imposed standard by animal owners or carers, leading to little or no advance in welfare. Nevertheless, it does have the merit of focusing on what would be the best standards in light of contemporary knowledge.

Typically, many animal rights groups adopt the gold standard approach to serve their objective of changing the way we think about the place of animals in our society. Their major aim is to outlaw any human use that causes animals harm, including, for example, their "cruel" use in laboratories, on farms, for competitive sport, display and even keeping them as pets. They argue that animals do not have the capacity to give or withhold consent to participate in these activities which can, and often do, cause great harm, so that to coerce their participation is unethical. Thus, an extreme position espoused by some members of such groups is that no animal use by humans can be justified.

Although such an absolutist approach does not motivate most animal owners or carers to change

—
4.1
ANIMAL USE AND ANIMAL WELFARE
RESPONSE
DAVID MELLOR

their practices, the activities of such gold standard groups, when lawful, undoubtedly do contribute to positive change in other ways. By strongly challenging the status quo, they question long-standing practices the use of which persists because of tradition, unexamined habits of thought and/or financial or other constraints. Such questioning, supported by drawing attention to the stark contrasts between what "is", illustrated by genuinely serious current problems, and what "ought to be" as exemplified by their gold standard, draws the attention of the public, in particular, to areas of serious welfare concern. Such thinking becomes incorporated as a feature of societal disquiet about our use of animals, and then, at the very least, provides subliminal pressure on groups or regulators committed to incremental improvement to propose and/or implement the largest increments that may be practically achievable at any particular time.

Incremental improvement

This approach is commonly adopted by those who seek to improve the situation we have inherited, the situation as it currently "is". As a strategy it is commonly adopted by animal welfare groups who hold that animal use for human purposes is acceptable provided that such use is humane and, if not completely humane, when the harm can be justified ethically. The justification is usually cast in utilitarian terms such that any harm to animals is traded off against all benefits. However, it is not ethical to engage in *weak utilitarianism*, whereby unsupported assertions are glibly made about the harm being outweighed by the good, or where a genuinely favourable balance is only marginal. What is required is *strong utilitarianism*. To be ethical in these terms, our obligation is to take active steps to ensure that *all* harms are minimised and *all* benefits are maximised, so that the separation between them is the *greatest* that can be practically achieved.

The incremental improvement approach has obvious relevance to minimising the harms in the progression towards a gold standard ideal represented by the welfare of the animals being the best that can be practically achieved under the circumstances of their use. The pragmatic aim is to improve welfare in a stepwise fashion by setting a series of achievable goals, seeing each small advance as worthwhile progress towards the gold standard. This encourages participation, ownership or buy-in by setting reachable targets. These are part of a planned sequence designed to enhance animal welfare progressively. Common outcomes are significant welfare advances, a sense of achievement, willingness to recruit other participants and/or openness to ways of making further improvements. This measured approach, usually adopted by animal welfare scientists, the veterinary profession, animal welfare groups and others, has demonstrably achieved significant progress over at least the last two decades.

Of course, the commitment to incremental improvement must be genuine. It must not consist of mere lip service statements to deflect the pressure to change animal care practices whilst in fact doing nothing. Improvements to animal welfare will always be required for at least two reasons: firstly, because there is a wide spectrum of welfare problems that can arise and it is quite complex to manage all of them in the variable circumstances in which animals are kept; and secondly, because there will always be a need to practically implement the consequences of our continuously advancing knowledge of the ways animals experience themselves and their environments, so that more improvements will be required even when welfare management is already operating at high overall levels.

Some examples

These approaches have often been in conflict (Webster 1994). An excellent example is outlined by Warburton (1998) in a paper entitled, "The humane traps saga: a tale of competing ethical ideologies". This describes the way inflexible gold standard groups, which in this case were not animal rights groups, ultimately stalled attempts to achieve an international agreement designed to progressively improve humane trapping standards worldwide. By insisting that unachievably high standards be applied from the outset, instead of adopting a proposal to consistently work towards achieving far higher standards than then existed, the agreement of the majority of countries to incrementally improve trap standards according to a tight schedule failed to gain full approval and was abandoned.

Yet, gold standard groups have also achieved worthwhile change. Here are two New Zealand examples, both relating to public responses to media broadcasts of cases of farm animals treated in unacceptable ways. First, persistent public pressure by both animal rights and animal welfare groups contributed to the minister of agriculture in 2010 requiring that a planned phasing out of conventional sow stalls must be brought forward by several years to 2015. Second, a public outcry after broadcasts in 2015 of covertly filmed serious mistreatment of some bobby calves in the dairy industry demonstrably led to the rapid establishment of a cross-sectoral group involving dairy companies, meat, pet food and transport companies, the national farmers federation, the veterinary profession and the primary industries ministry. This group then immediately initiated a comprehensive programme to ensure that best practice in bobby calf management would be further developed and credibly implemented countrywide in order to give the public confidence that all of those involved with bobby calves genuinely provide a high level of care.

The incremental improvement approach can be applied to virtually any negative welfare issue that requires attention, apart from wilful ill-treatment or

△
4.7 Sow stereotyping in a stall. Persistent public pressure contributed to bringing forward a planned phasing out of sow stalls.

neglect or the worst consequences of catastrophic weather events. Moreover, its use facilitates discussions between veterinarians and their clients by providing a way for the veterinarian to engender confidence and elicit movement towards solutions when initially there is hesitation or resistance. There are many examples of this working.

Concluding remarks

Incremental improvement provides a way for veterinarians and clients to work cooperatively and constructively *with* each other to achieve agreed goals. This, incidentally, is much more satisfying professionally than repeatedly facing resistance. Although in the wider context the gold standard approach does have some merit, as noted, when dealing one-to-one with clients who require immediate practical solutions, it will usually fail, especially if the tone employed is doctrinaire and moralistic. Incremental improvement is therefore the recommended strategy.

ANIMAL USE AND ANIMAL WELFARE

What would you do?

ONE ⎯⎯⎯ Can you think of an example where the "gold standard" and "incremental improvement" approaches have been used together to improve animal welfare?

TWO ⎯⎯⎯ Can you think of examples where either the "gold standard" or the "incremental improvement" approach has been used to improve animal welfare?

THREE ⎯⎯⎯ Which of the two approaches do you feel has resulted in greater overall animal welfare improvement?

FOUR ⎯⎯⎯ Is there a threshold for animal welfare below which it would be unacceptable to push for incremental improvement? How did you come to this conclusion?

△
4.8 Are there incremental improvements that could be made for these baby rabbits? What would the gold standard for selling rabbits look like?
PHOTO ISTOCK

ANIMAL USE AND ANIMAL WELFARE

What do you think?

You are applying for jobs and have received two offers. One is working in an animal shelter where you deem the standard of animal care to be low, but shelter staff are seeking to improve it. The other is working in an animal shelter which boasts "state-of-the-art" facilities and practices. Which position will you accept? Which ethical framework would you use to justify your choice?

△
4.9 What factors about the level of care offered to animals in a shelter would influence the job you would take?
PHOTO ISTOCK

||||||||||||||||||||||||||||||||||||||||||||||||||||||||||||||||||

THE "gold standard" approach maps best onto a deontological framework, while the "incremental improvement" approach maps best onto a utilitarian framework. As the author suggests, these are not mutually exclusive. The push for the gold standard may create an impetus for welfare improvements. It is with a gold standard in mind that incremental improvements may be made, and new data may alter what is perceived to be the gold standard.

4.2

Food and production animals

The most common form of animal use is in agriculture. Animals are farmed for production of meat, eggs, milk and fibres such as wool. Veterinarians play a critical role in supporting the health and welfare of production animals. One major ethical question raised is whether the veterinarian's primary obligation is to the welfare of the animal, or to the wellbeing of the industry or client they support, and/or consumers of animal products.

Recently, the role of veterinarians in maintaining food security has become a hot topic of discussion. Food security was defined at the World Food Summit as existing "when all people at all times have access to sufficient, nutritious food to maintain a healthy and active life" (World Food Summit 1996).

SCENARIO
LIVESTOCK PRODUCTION INTENSIFICATION

▶ There have been substantial productivity gains in livestock farming during the last 50 years. Pigs and poultry now grow more quickly and efficiently than ever before. Since 1940, the average dairy cow in the USA has tripled its annual milk production to around 15,000 litres a year. There has been a concomitant reduction in the number of cows by two thirds, and the aggregate level of food consumption and manure production per litre of milk has decreased (Roberts 2000). Veterinarians have, on the whole, supported the intensification of livestock farming to increase productivity. Despite this, intensive farming methods have been associated with a range of welfare problems for farm animals. The global human population is projected to increase to 9 billion by 2050, with demand for meat products forecast to double by that time (FAO 2009). In a world of continuing population growth should veterinarians continue to support the intensification of livestock production?

RESPONSE
STEVEN P MCCULLOCH

▶ The veterinary profession has an instrumental role in food security. The profession's position on the future of food and farming will affect the wellbeing of the global human population, the welfare of billions of sentient animals, as well as impacting the living environment for future generations. Food security policy is a complex issue and there are a number of issues the profession should consider prior to forming a position on the debate. Firstly, what impact have intensification and increased productivity had on farm animals to date? Secondly, what is the purpose, both historically and today, of

4.2
FOOD AND PRODUCTION ANIMALS
RESPONSE
STEVEN P MCCULLOCH

intensification? Thirdly, what are the roles and responsibilities of the veterinary profession, particularly in the context of food security and animal welfare?

There has been considerable debate about the impact of intensive farming on the wellbeing of animals (Fraser 2008, Sandøe 2008, Thompson 2008). The intensification of livestock farming has been associated with substantial welfare problems in animals. Increasing productivity has been associated with reduced space contributing to behavioural problems (e.g. fighting in pigs, feather pecking in layer hens) (Bracke & Hopster 2006), rapid growth causing bone abnormalities (e.g. lameness in broiler chickens and turkeys) (Danbury, et al. 2000, Kestin, et al. 2001, Martrenchar 1999), and increased production rates causing metabolic problems (e.g. lameness and mastitis in dairy cattle, osteoporosis in layer hens) (EFSA 2009, FAWC 2009a, 2009b, Webster 2004).

Of course, extensive farming methods can be associated with suffering, particularly by insufficient attention and outright neglect, predation, inclement weather and parasite infestations (Webster 1994). However, since the great majority of animals used by humans are livestock animals, and, worldwide, around 70 per cent of livestock animals are intensively reared, it is probable that the greatest amount of suffering inflicted by humans on animals is caused by intensive farming.

That intensive farming causes widespread and substantial suffering does not necessarily lead to the judgement that veterinarians should oppose further livestock intensification. Can and do animals kept in such systems have, over the course of their lifetime, a life worth living? The Farm Animal Welfare Council (FAWC) recommended that all farm animals should have a life worth living, and that an increasing number should have a good life (FAWC 2009a). The utilisation of FAWC's "Good Life Scale" facilitates ethical analysis by

△

4.10 Injurious feather pecking resulting in marked feather loss is a significant problem on many intensive laying hen farms.
PHOTO SIOBHAN MULLAN

categorising animals as having, overall, a life of net positive value (a life worth living), a life of net negative value (a life not worth living), and a life of significantly higher positive value (a good life) (McCulloch 2015). FAWC's discussion strongly implies that animals reared in intensive systems cannot have a good life:

"It is hard to conceive how certain systems of husbandry could ever satisfy the requirements of a good life because of their inherent limitations. Examples include the barren battery cage for laying hens, and the long-term housing of beef cattle on slats, denied access to pasture."

(FAWC 2009a)

Therefore, two fundamental questions – for the veterinary profession and society more generally – about intensive farming and animal welfare are as follows. Firstly, can animals reared in intensive systems have a life worth living? Secondly, is it morally justifiable for the veterinary profession and society to support husbandry systems that limit quality of life to a mere life worth living, as opposed to a good life?

A further important question is what purpose, historically and today, do intensification and increased productivity serve? In post-WWII Britain, the Agricultural Act 1947 promoted intensification to support self-sufficiency in food production. The impact of modern intensive "factory farming" methods in Britain prompted the publication of Ruth Harrison's *Animal Machines* in 1964 (Harrison 2013). It was public outrage from Harrison's exposé which led the British government to set up an inquiry into livestock farming, led by the eminent zoologist F. W. R. Brambell. The *Brambell Report* acknowledged the link between modern intensive farming and the suffering of sentient animals. The report recommended a research programme that developed into the discipline of animal welfare science, and a government advisory body that developed into the internationally respected FAWC (Brambell Committee 1965). Subsequent to Brambell, influenced by findings in animal welfare science, public pressure and the recommendations of FAWC and other government-advisory bodies, many governments worldwide have banned the most severe forms of intensive farming. These include veal crates (Britain, 1990; EU, 2007); sow stalls (Britain, 1999; EU, 2013);[1] and battery cages (Britain, 2012; barren cages prohibited in EU, 2012).

The 2011 Foresight *Future of Food and Farming* report recommended "sustainable intensification" as a solution to the problem of a growing human population, increased demand for meat and dairy products and climate change (Foresight 2011). Following the Foresight report, the British

government has promoted the sustainable intensification of food and farming (Spelman 2012). However, the concept of and recommendation for sustainable intensification have been contested in the policy arena (FAWC 2012, Garnett & Godfray 2012, Hume, et al. 2011). Elsewhere, I have argued that sustainable intensification is not the right policy for the future of food and farming (McCulloch 2013a, 2013b, 2015). There are serious problems in making an inference from the projected increase in "demand" for meat and dairy and the need to meet this demand within environmental constraints, to the policy of sustainable intensification.

Firstly, the relationship between economic demand and supply is more complex than any simple inference assumes. True demand is affected by supply, via price. Thus, if a policy of sustainable intensification is followed, and greater quantities of cheap food continue to be produced, lower price will indeed promote increased demand. In contrast, a policy which promotes moderate levels of meat and dairy production with a higher price (normal by historical standards) will reduce demand. Thus, true economic demand is not simply a reflection of psychological desire. It may be the case that we would all like to live in great mansions and drive Ferraris. However, since the price of these goods is beyond most of our financial means, there is limited demand for them.

Secondly, the simple inference to sustainable intensification as a solution to food security flies in the face of basic nutrition science. A pig does not convert one unit of plant energy/protein ingested to one unit of animal energy/protein in pork or bacon rashers. The process of converting plant-based energy to meat is inherently inefficient. Energy is "wasted" as heat during the complex metabolic processes of digestion, assimilation and subsequent synthesis of animal protein to consume as meat. For pigs, the conversion ratio

---

1   Gestation crates for sows are permitted throughout the EU for the first four weeks of pregnancy.

—
4.2
FOOD AND PRODUCTION ANIMALS
RESPONSE
STEVEN P MCCULLOCH

is 4:1. For poultry it is around 2:1. Cattle are relatively inefficient converters of vegetable protein to meat, with nine units of plant material ingested for every unit of beef produced (McMichael & Butler 2010). It is worth noting that these inefficiencies are not confined to conversion of energy and protein. It takes an average of 15,500 litres of water to produce 1 kg of beef (Hoekstra 2010). It is forecast that half the global population will be living in areas of high water stress by 2030 (UNDESA 2015). Therefore, policies which promote higher meat and dairy consumption are, morally speaking, problematic.

The dangers of making a simple inference from a growing human population with changing dietary habits to increased "demand" for meat and dairy products to meeting this demand within environmental constraints to "sustainable intensification" should be clear. Let us presume that meat and dairy consumption will indeed more than double by 2050 (Garnett & Godfray 2012). If this is the case, the greenhouse gas produced per unit of meat and dairy must halve, if we are simply to maintain a steady greenhouse gas output from the livestock sector. It seems incredibly optimistic to suppose that science and technology can achieve such a feat within 35 years. Furthermore, the aim is to substantially reduce greenhouse gas emissions, not simply maintain current levels. To illustrate by example, the UK Climate Change Act 2008 mandates a reduction in CO2 emissions of 80 per cent by 2050. Any serious consideration of these figures should provoke genuine scepticism in the veterinary profession that sustainable intensification is an appropriate food policy going forwards.

The third point brings us back to animal welfare. What is the potential for productivity gains in the sustainable intensification of livestock agriculture? Some appreciation of the degree to which livestock intensification has progressed, and the impacts of this on farm animals, helps answer this question. Increasing stocking densities

has resulted in abnormal behaviours and levels of aggression such as tail biting and fighting in pigs and feather pecking and cannibalism in laying hens. Lameness in broiler chickens, which reduces productivity, is primarily due to genetic selection for a rapid growth rate, such that chickens grow so rapidly that their bones cannot support their body weight. Selection for higher milk yields in dairy cows is associated with a range of welfare-related problems, such as lameness. The European Food Safety Authority (EFSA) has recommended selecting for non-yield characteristics to mitigate these problems (EFSA 2009). Modern intensively reared animals are already at their physiological limits of endurance. Webster has compared the dairy cow to a Tour de France cyclist in terms of metabolic demand (Webster 2013). The difference is that the Tour de France lasts only 3 weeks, whilst the cow lactates for 305 days a year, or more in some cases. Additionally,

△

4.11 The only people who work harder than the dairy cow – in a metabolic sense – are polar explorers and cyclists in the Tour de France. But unlike explorers and cyclists, dairy cows lactate year-round.

CARTOON RAFAEL GALLARDO ARJONILLA

the cyclist chooses to engage in such activity, pushing physical limits of endurance. The dairy cow, in contrast, has no choice – a point which is fundamental to the morality of the issue.

Conception rates in dairy cows have reduced by 1 per cent every three years, and it is now only 40 per cent (FAWC 2009b). The declining rate of conception, associated with increased milk yield, is due to exhaustion (Webster 2013). In effect, failing to conceive is a natural response to depleted energy reserves. Of course, the consequences

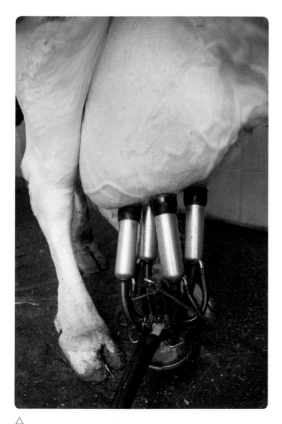

△
4.12 Higher-yielding dairy cows are more likely to be affected by a range of welfare-related problems, such as lameness.
PHOTO ISTOCK

are all too familiar in the dairy industry: the long road to the abattoir. Declining conception rates are associated with increasing cull rates (EFSA 2009, FAWC 2009b).

Thus, livestock intensification is problematic on the following grounds: (1) the policy of sustainable intensification does not logically follow from the food security problem; (2) increased production of low-price meat and dairy food will stimulate true economic demand which will increase production of these products; (3) broadly, the production of meat and dairy is inherently inefficient; and (4) farm animals raised in intensive systems are already beyond physiological limits, manifesting as physical and mental diseases and causing poor welfare. To these problems associated with livestock intensification, a fifth point can be added: (5) society is becoming increasingly opposed to intensive livestock production, demonstrated by consumer behaviour and state legislation prohibiting certain methods of production.

So what are the role and responsibilities of the veterinary profession? Imagine, for the sake of this discussion, that points 1–3 above are, in fact, invalid. That is, for the purpose of food security alone, sustainable intensification is a viable policy for the future of food and farming. At the same time, point 4 is valid: sustainable intensification does cause substantial welfare problems to animals. Based on these conditions, there are three policy positions the veterinary profession might hold: (1) advocate sustainable intensification whether or not it is necessary for the health and wellbeing of a global human population of 9 billion in 2050; (2) oppose sustainable intensification regardless of whether it is necessary for the health and wellbeing of the same human population; or (3) advocate sustainable intensification if and only if it is necessary to sustain the health and wellbeing of the global human population. Now, if one accepts that further intensification and/or measures to increase productivity will, on the whole, cause suffering to animals, should the veterinary profession take position 1, 2 or 3 above?

Position 1, clearly, should be rejected. The veterinary profession has an instrumental role in the provision of safe and nutritious food to the public. Additionally, however, the veterinary profession has a fundamental duty to safeguard animal welfare. If the veterinary profession were to cause animal welfare to decline, when it is not necessary for the health of the human population, it would abdicate its most fundamental responsibility. Position 2 is similarly problematic, although perhaps less so than position 1. To oppose sustainable intensification, even if it were necessary for human health and wellbeing, appears excessively demanding. In effect, the position would prioritise animal welfare over human health and wellbeing. Interestingly, though, this position seems to follow from the UK RCVS Code of Professional Conduct for Veterinary Surgeons which states that the primary responsibility of veterinary surgeons is to animal welfare (RCVS 2015). It follows, therefore, that in the event of genuine conflict between animal and human interests, the veterinary surgeon has a professional obligation to animal welfare, and *not* to promote food production. Despite this observation, society overwhelmingly prioritises human interests over those of animals, and the veterinary profession follows this stance in practice. For this reason, position 2 is both morally problematic and also unrealistic. The rejection of positions 1 and 2 leaves position 3, which has the merit of being more defensible from a moral point of view. In effect, position 3 is based on two premises: firstly, in the final analysis, and in cases of genuine conflict, human wellbeing is more important than animal welfare; and secondly, sentient farm animals have intrinsic moral value, generating direct moral duties to protect their welfare. Indeed, these two premises lead directly to the principle of unnecessary suffering, where it is only morally permissible to cause sentient animals to suffer if it is necessary for human health and wellbeing (Garner 2013, McCulloch 2015). The unnecessary suffering principle is, of course, the cornerstone of animal welfare policy in most of the Western world.

|||||||||||||||||||||||||||||||||||||||||||||||||||||||||||||||||||||||

MUCH of the above discussion hinges on what counts as sustainable. Traditionally, systems were referred to as unsustainable when a resource became depleted so that it was no longer available, or the system itself failed. The term has expanded to encompass impacts on human health, animal welfare and the environment. Broom, et al. argue for a broader definition, in which a system is sustainable "if it is acceptable now and if its effects will be acceptable in future, in particular in relation to resource availability, consequences of functioning and morality of action" (Broom, et al. 2013). A number of scholars argue that animal production is sustainable where human and animal interests (for example in food) are complementary rather than competitive, e.g. where ruminants consume materials that are inedible for humans, as occurs in silvopastoral systems (Broom, et al. 2013, Webster 2013).

Concerns about unnecessary suffering are essentially utilitarian although they are also taken into account in principalist approaches to ethical problems.

FOOD AND PRODUCTION ANIMALS

What do you think?

The term "unnecessary suffering" is commonly used in discussions of animal welfare and anti-cruelty legislation, but a common criticism is that what constitutes unnecessary or indeed necessary suffering is rarely defined.

ONE   How would you define unnecessary suffering?

TWO   How would you define avoidable suffering?

△

4.13 Intensive farming methods have been associated with a range of welfare problems in animals.

CARTOON MALBON DESIGNS

||||||||||||||||||||||||||||||||||||||||||||||||||||||||||

FOOD security is an urgent emerging concern with the human population boom, with some arguing that we will need to broaden the types of animals used as sources. The following scenario explores the ethics of eating insects.

SCENARIO
INSECTS AS FOOD

▶ You are asked to provide an ethical analysis of a large company called "Sect-o-yum". This company is considering diversifying into farming insects and other invertebrates like snails for food. The idea has some perceived benefits, including the fact that the carbon footprint per kilogram of protein could be very low compared to traditional forms of animal protein.

What would you say in your ethical analysis of Sect-o-yum's plans?

△
4.14 Will insects be the protein source of the future?
PHOTO ANNE FAWCETT

RESPONSE
SIMON COGHLAN

▶ The ancient practice of eating insects, known as *entomophagy*, is found around the globe, from Africa to Asia and the Americas. Of the 1 million or so known species of insect, 1500 or so can safely be eaten by humans and livestock. They include various species of grasshoppers, beetles, ants, and wasp, bee and moth larvae. Yet insect eating is largely foreign to Western countries. A 2013 report by the UN's Food and Agriculture Organization (FAO) advances detailed arguments for globally promoting entomophagy (Van Huis, et al. 2013). In this analysis of Sect-o-yum's plans, I shall raise some key ethical questions associated with farming insects (and other small land invertebrates like snails).

Two main ethical reasons for pursuing commercial entomophagy can be found in the FAO report: food security and environmental protection. Malnutrition is a pressing global problem; 9 billion people will need sustenance by 2050. Many edible insects are rich in essential but scarce nutrients like amino acids, iron and zinc. In addition, the feed conversion ratio for insects far surpasses that of beef, pork, farmed fish and even chicken, while the amount of water required is much less. Insect farming promises to generate less damaging greenhouse gas than meat production. Entomophagy may therefore substantially bolster food security around the world and mitigate the disastrous environmental effects of global warming (Van Huis, et al. 2013).

Still, important questions about entomophagy may be raised. Are insect products safe and does insect farming pose a risk of animal-to-human disease transmission (zoonosis)? Will rare species of wild insects become endangered from, say, the escape of mass-produced insects from farms? Unlike traditional wild collection, large-scale insect farming requires contained and intensive systems – a kind of factory farming for invertebrates. But

CHAPTER 4 ANISE

CHAPTER 4 ANIMAL USE

—
4.2
FOOD AND PRODUCTION ANIMALS
RESPONSE
SIMON COGHLAN

will the inevitable application of modern science and technology, combined with global trade pressures, result in the development of practices, such as transgenic insects, with unknown human, insect-welfare and environmental consequences?

The FAO report's authors might reply that such risks should be carefully analysed and, possibly, adequately managed. Industry could devote additional resources toward protecting not only endangered animals like tigers and elephants, but also ecologically vital insect species. In what follows, I put these "extrinsic" moral considerations to one side, and focus on the *intrinsic* ethical value of insects.

At the outset, the objection that entomophagy is disgusting will not do. Intended as a *non*-ethical objection, it invites the rejoinder that, with education and canny marketing, disgust can be overcome. So, what are some ethical objections to entomophagy?

A looming moral question here concerns *sentience* (e.g. Lockwood 1987). In the West, it is now widely acknowledged that the conscious interests of animals carry moral weight. Many of us, however, are likely to feel perplexed about whether insects can feel, have desires or experience pleasure and pain. Insects possess nociceptors (Eisemann, et al. 1984). In mammals, nociceptors convey pain impulses. Insects also respond in some ways to opioids, which in higher animals are associated with pain mitigation. To the objection that such features are compatible with non-sentient detection of noxious stimuli, some scientists reply that insect behaviour reveals a capacity to *learn* to avoid such stimuli (Elwood 2011). Joined with neurophysiological discoveries, such flexible behaviour may suggest that, unlike simpler organisms, insects could be sentient to some degree.

It is often assumed that science is the best way to demonstrate animal sentience. The philosopher Ludwig Wittgenstein (Wittgenstein and Anscombe 1958) disagreed. He advised us to instead carefully examine how language-users *speak* about insects. Wittgenstein said:

"Look at a stone and imagine it having sensations. —One says to oneself: How could one so much as get the idea of ascribing a *sensation* to a *thing*? One might as well ascribe it to a number! —And now look at a wriggling fly and at once these difficulties vanish and pain seems to be able to get a foothold here, where before everything was, so to speak, too smooth for it."
(Wittgenstein and Anscombe 1958, 98; emphasis in original)

We can speak sensibly of a fly being in pain, Wittgenstein says, because the fly can exhibit behaviours that go with our talk of pain. In contrast, there is nothing about a stone that allows the language of pain to find a comparable home. The stone or other object is too "smooth" for pain-language to stick to it. But perhaps Wittgenstein does not completely abolish our perplexity. For while wriggling, struggling flies are very different to rocks (and numbers!), we may nevertheless remain in two minds about applying the language of sentience to insects. Despite our close attention and reflection, our uncertainty about what words and concepts to apply to them may persist. Perhaps insect behaviour just lacks sufficient complexity. How are we to proceed with such uncertainty?

One answer is to adopt a kind of ethical precautionary principle. That is, we could assume that insects *may* be sentient, and so accord their interests *some* weight – perhaps less than other animals – in our moral calculations. Further, we could recognize an obligation to balance the suffering we cause to sentient creatures against the protection of human interests. A utilitarian is likely to favour some sort of precautionary principle. But in any case, since suffering is generally bad, it seems plausible that we ought, as a rule, to avoid causing it. A supporter of rights theory or virtue ethics might well agree. The number of insects poisoned in protecting food crops is vast. Given the huge amount of suffering this potentially entails, it appears that, whatever ethical position we hold, we ought to treat insects more seriously than we do. And if insect farming is likely to cause more suffering than growing crops, there is a strong reason to favour plant-based agriculture.

Under this precautionary principle, we accord some ethical weight to insect interests despite our lack of certainty about their sentience. To illustrate, imagine we are practising archery by firing at a target in front of a thick hedge. If we believed that a cow *might* be behind the hedge, we would have reason not to fire an arrow towards the hedge, even if our belief should turn out to be false. That is because there is some probability that firing the arrow could result in suffering or death. In the same way, we have a reason not to farm insects, because there is some chance – below one but still above zero probability – that suffering will result.

A possible weakness in this argument concerns the words "*in the same way*". For we may well think it impossible to wrong an insect "in the same way" we can wrong a cow, even if it should turn out that insects are sentient after all (cf. MacIver 1948). On this view, the rights of cows, or our duties to them, are in part conditioned by our (virtual?) certainty – *ex hypothesi* missing for insects – that cows are conscious, feeling creatures. In that sense, injuring and killing insects is

less serious than injuring and killing cows, even when, in a peculiar sense, the *probability* of causing a certain amount of harm is judged to be similar in both cases.

Consider now a very different approach to intrinsic value. Environmental ethics speaks of the value of mountains, rivers and special places (e.g. Leopold 1989). It also speaks of the value of living things – an idea famously encapsulated in Albert Schweitzer's (Schweitzer and Lemke 1933) phrase "reverence for life". Philosopher Paul Taylor (2011) later developed a detailed and influential version of *biocentrism*. For Taylor, organisms possessing neither moral agency nor sentience still have an *inherent worth* that derives from their having a good of their own. Organisms have a teleology or goal-directedness that enables them to pursue their own good, even while lacking sentience and intentionality. Unlike stones, living things can be harmed and benefitted.

One objection to ethical biocentrism is that it apparently fails to distinguish between, say, cows, insects and plants. Yet we might feel that insects have greater ethical value than plants. Here is a possible reason. As discussed earlier, insects may not be sentient, but they have behavioural capacities that distinguish them sharply from wheat and soya beans. Small invertebrates can, in a fairly clear sense, *struggle* to escape danger, or *try* to improve their situation. Insects, that is, have a kind of *agency*. Viewers of extraordinary close-up footage in films like *Microcosmos* would be hard pressed to deny it. So, perhaps insects have an ethical value that distinguishes them from plants, even if they are less ethically significant than cows.

Despite all these objections, we may decide that entomophagy is nevertheless justified. We might think, like the FAO, that the arguments concerning food security and environmental protection are so strong that they justify the risks to insect wellbeing, or at least that they make insect farming morally preferable to farming livestock. Of

△

4.15 Are insects sentient?

PHOTO ANNE FAWCETT

course, we might feel the need to be convinced that farming "mini-livestock" causes the same or less harm than plant-based diets. Working that out will require complex cost/benefit investigations.

In closing, I will note an apparent irony in the FAO's largely favourable view of industrialised insect eating. Westerners are typically disgusted by entomophagy because they associate insects with filth, and their consumption with primitive practices and famine conditions. Reality TV contests and some fashionable Western restaurants capitalize on that sort of reaction. The FAO authors suggest, plausibly, that the "disgust factor" is culturally malleable. They claim that educating the public about the marvellous nature of insects and their important ecological roles should blunt this reaction, enabling the social institutionalization of entomophagy.

But perhaps there is a curious ethical tension in this suggestion. For rendering insect eating routine may not, in the long term, boost our *respect*, let alone our *reverence*, for insect life. After all, some may argue, it has proven easier for society to revere free-living or wild animals than to show even minimal respect to livestock like pigs, cattle and sheep. It may, in part, be agriculture itself that blunts our respect for certain animals. That observation might enter into our ethical assessment of Sect-o-yum's plans, even if it is not a decisive objection to them.

||||||||||||||||||||||||||||||||||||||||||||||||||||||||||||||||

MANY debates about animal use invoke the phylogenetic scale, ranking animals according to their overall complexity, with insects much lower on the phylogenetic scale than mammals. As the author points out, a utilitarian analysis may favour use of animals lower on the phylogenetic scale as they may have less capacity to suffer or to be aware of suffering.

Our perception of animals that are sentient has changed over time to encompass "first all humans instead of just a subset of humans, and then also: (a) certain mammals that were kept as companions, (b) animals that seemed most similar to humans e.g. monkeys, (c) the larger mammals, (d) all mammals, (e) all warm-blooded animals, (f) all vertebrates and (g) some invertebrates" (Broom 2016). This corresponds to extending our sphere of moral concern to encompass animals lower on the phylogenetic scale. Thus for example one might argue that despite the fact that insects outnumber the animals higher on the phylogenetic scale, then, their suffering carries overall less moral weight.

ANIMAL USE AND ANIMAL WELFARE

What would you do?

Scientists around the world are working on growing "meat" *in vitro*. You are working for a food company seeking to develop a new source of protein that is environmentally sound, safe for consumption and a replacement for traditional meat. How would you decide whether to invest in insects or *in vitro* "beef"? How would you promote this to your customers?

△
4.8 This beef is cultured in a laboratory from muscle cells isolated from a tissue biopsy.
PHOTO DAVID PARRY, PA WIRE

4.3

Idiosyncracies in animal usage

One of the issues around animal use is our application of inconsistent treatment of animals. Some we treat as cherished companions, others as a source of food and still others as vermin – yet we may struggle to justify this varied treatment.

According to John Webster, "Our actions towards other animals – whether we care for them, simply manage them or seek to destroy them – is defined not by their own sentience but by how we categorise them in terms of their *extrinsic value* (i.e. their value, or otherwise, to us)" (Webster 2005).

The inconsistency in our approach to different species has been raised previously, notably by authors such as James Serpell (1996) and Hal Herzog (2010).

One question often raised is why is it that we cherish some animals as pets, cultivating intimate and reciprocal relationships with these animals, yet we feed other – equally sentient – animals to these pets? A number of people attempt to avoid this inconsistency by feeding their pets a vegan diet, a topic explored in this next scenario.

_____

SCENARIO
VEGAN PET FOOD

▶ Mr N presents you with Simba, a two-year-old male neutered cat, and Nisse, a five-year-old female German Shepherd dog, for routine health checks and vaccination. When you ask about dietary history, Mr N explains that both pets are on a vegan diet, as he feels it is unethical to feed his pets any form of animal protein. On physical examination you determine both animals to be underweight (body condition score 2/5), and both have skin conditions. You believe this may be due to dietary deficiency.

What do you do?

_____

RESPONSE
RICHARD GREEN

▶ If you genuinely believe (either with or without evidence) that the animals' diet is causing their current health problems, then as a veterinary surgeon with a primary obligation to safeguard your patients' welfare, you have a duty to try to influence this situation. However, it would be worth – possibly before exploring with your client the ethical aspects of feeding pets vegan diets – trying to determine to the best of your ability how likely it is that the diets are genuinely responsible for the health issues. Is there any particular element of the diet whose deficiency is potentially

△

4.17 Although dogs may enjoy eating vegetables there are ethical concerns around both feeding dogs and cats meat and giving them a vegan diet.
PHOTO ISTOCK

responsible, and are there any tests to confirm this? A failure to demonstrate a direct connection may not diminish your suspicions, nor avoid the conversation, but showing a cause and effect could significantly strengthen any argument you may have, and may also suggest a treatment if the owner is not willing to consider a radical change in diet.

The questions underlying this dilemma are the ethical issues surrounding feeding pets, on the one hand with foods derived from (primarily) intensively farmed animals, and on the other hand feeding a diet which is "unnatural", but which does not result in harm to (food) animals (and arguably results in less harm to the environment). It is also worth bearing in mind that there is a significant body of opinion (including many vets) that holds commercial meat-based pet food diets responsible for a number of health problems, including dental disease and obesity-related illness, as well as more esoteric disorders.

The dilemma may best be examined as a series of questions – questions for yourself about the case, and questions to put to the owner:

Why is the owner feeding a vegan diet? This is perhaps most likely to be because of ethical and welfare concerns about the animals raised to make conventional diets, but is worth checking. If your client is vegan for its supposed health benefits, and is extending that rationale to his pet, the discussion would be more likely to focus on the provision of appropriate species-specific diets, and one would not expect the client to be as resistant to change if this were best for their animal. If there is resistance, perhaps the owner would consider a dietary trial of a more conventional food for a period of time to see if the signs are alleviated?

If the owner is feeding his pets their vegan diet out of concern for the welfare of the contents of conventional pet food, and there is a significant chance the diet is causing the clinical signs, then the dilemma becomes a question of comparative harms: the welfare loss caused by the poor nutrition of the pets versus the (perceived) negative welfare balance of animals ending up in pet food. This would be a utilitarian argument; however, one might argue using a relational theory of ethics that owners should have a greater duty of care towards animals with whom they have a relationship, than ones with whom they don't. One might also make an argument that if one could source meat-based pet food from high welfare farms, that might provide a compromise.

If one's client is morally opposed to eating meat at all, from a more deontological or rights-based view, then (other than heroically refraining from pointing out the inherent irony in keeping a carnivore for a pet – highlighting the faults in one's clients' logic may be satisfying but is rarely conducive to negotiating solutions) one's best course of action might be to go back to basics and explore with them the ethical reasons why one would or wouldn't want to feed one's pet a vegan diet. An ethical matrix would be an ideal tool for this purpose, as it would allow both vet and owner to examine their reasoning in an objective and critical fashion.

Major stakeholders would be the pet(s), the owner, the vet and the potential contents of the pet food. Depending on the reasons for the owner's veganism, environment might also be added to this list. Ethical arguments for feeding a vegan diet might be that the welfare cost to those animals reared to be eaten by pets is greater than the welfare cost incurred by a pet on an inadequate diet (this would depend on the degree of health issues directly attributable to that diet) – this is a utilitarian argument primarily. If the diet is not responsible for any health problems, then that argument is even stronger. Autonomy of owner and (arguably) food animals is better respected by allowing the vegan diet, as is the wellbeing of the owner. Whether the wellbeing of the food animals is better or worse as a result would depend on one's opinion of their welfare while being farmed, and whether it is better for them to have lived a life, or not to have existed.

Ethical arguments against a vegan diet might arise because of a clear link between a pet's ill-health and its vegan diet, or simply because of a different weighting in the welfare benefits/costs to the pets and the food animals from the vet and owner. This is why an ethical matrix can be very useful to inform discussion, but cannot provide a definitive answer. It would be fairly safe to assume a loss of the pets' autonomy in feeding a vegan diet – it is unlikely cats or dogs would choose a non-meat-based diet over a meat-based one, and one could argue that their "telos" as a carnivore is compromised by feeding a non-meat-based diet.

Professional ethics will also have an impact on the view of the vet at least. The vet in this scenario has a greatly increased duty towards the animals in his or her care over the animals going into pet food. As a slight counter to this, professional ethics do advise that vets should attend to their clients' needs, but still never at a cost to animal welfare.

As with so many areas in veterinary medicine, the outcome is likely to be a compromise with the eventual outcome depending as much upon the vet's skill as a negotiator as a clinician. However, if as clinician you are sure that the primary cause of your patients' ill-health is their diet, then your primary obligation is to work towards improving that situation.

|||||||||||||||||||||||||||||||||||||||||||||||||||||||||||||||||||

THE author in the previous scenario might be seen as employing a principalist approach: seeking to do no or minimal harm (non-maleficence), doing good by promoting the health of the animal (beneficence), respecting autonomy (in this case

of both the client's decisions as well as the animal's preferences and dietary needs) and being just to all parties.

Recent studies of commercial vegan diets available at the time of publication found that the majority did not meet the nutritional requirements of dogs or cats (Gray, et al. 2004, Kanakubo, et al. 2015). As obligate carnivores, cats in particular may suffer from dietary deficiencies (for example deficiencies of vitamin A, cobalamin – vitamin B12, niacin – vitamin B6, phosphorus, calcium, vitamin D and/or taurine) and associated clinical signs.

Different arguments may be advanced based on different starting points. For example, one may argue that the feeding of a commercial diet of any kind can only be ethically justified when dietary analysis and feeding trials show that these diets provide for the animal's nutritional requirements and do not lead to deficiencies, or at the very least are better than the alternatives. Such studies require significant funding and given that vegan pet food is such a small proportion of the market there may not be funding to carry out such trials. An alternative argument may be made that such foods must at least be better than available alternatives, such as scavenging or feeding human leftovers.

The dilemma of the vegan companion animal owner represents what has been described as a "tragic trade-off" between two sacred values: protecting the wellbeing of the companion animal in question, and protecting the wellbeing of other animals and the environment. If the client has adopted a vegan diet for ethical reasons, they may rightly be concerned that their companion animal may undo or cancel out the benefits or at least reduced harm of their own dietary choices (Rothgerber 2013).

A tragic trade-off represents a true or irresolvable/insoluble ethical dilemma.

"If feeding one's pet meat/fish is perceived as necessary for their good health and animal-based diets are perceived as unethical because they contribute to cruelty to animals and to environmental decay, the vegetarian's dilemma pits harming one's beloved pet with harming other animals and the environment, both of which may constitute absolute prohibitions offering no simple solution."

(Rothgerber 2013)

The dilemma for the client could be solved for some with the availability of a reliably complete, palatable vegan pet food.

IDIOSYNCRACIES IN ANIMAL USAGE

What do you think?

In the wild, dogs and cats are predators and will hunt and kill other animals for food. How might the feeding of meat to pets be morally different?

△
4.18 Does the natural predatory behaviour of wild dogs justify our feeding of farmed meat to pet dogs?
PHOTO ISTOCK

IDIOSYNCRACIES IN ANIMAL USAGE

What do you think?

It has been argued that predation results in significant suffering (McMahan 2010). If predatory animal species could gradually be replaced by herbivorous species, without ecological disruption resulting in more harm than would be prevented by ending predation, would you support this? Why or why not?

△
4.19a–b Would eliminating predators reduce suffering?
PHOTOS GRAEME ALLAN

4.3

IDIOSYNCRACIES IN ANIMAL USAGE
SCENARIO
PEST OR PATIENT?

|||||||||||||||||||||||||||||||||||||||||||||||||||||||||||||

MANY species are treated differently and have variable legal protection depending on the context in which they find themselves being used by humans. In this next scenario we explore one such case.

SCENARIO

PEST OR PATIENT?

▶   You are working in a rural mixed practice during a rodent plague. On your way into work you bump into Ms S, one of your clients, who tells you that she is concerned her pet rats, Scabbers and Harry, are unwell and would like to bring them in to see you.

Your colleague overhears the conversation.

"Surely you aren't going to treat those things?" she says. "We're overrun with rats and mice, their food is probably attracting more, and they've probably got diseases. I would refuse to see them."

What should you do?

△
4.20 Can we justify treating pet rodents while trying to eradicate them elsewhere?
PHOTO ANNE FAWCETT

RESPONSE

RICHARD GREEN

▶   From a common-sense point of view, it does not appear to make sense to treat two pet rats whilst you or others make very significant efforts to eradicate many more elsewhere. However, by the same logic, does it make sense to treat sick dogs and cats, when healthy ones are destroyed in pounds or shelters, or indeed, to use valuable resources treating pet animals at all, while countless humans suffer worldwide?

Indeed, there will be very many people, vets included, who think that is where the answer to this question should end, although they will probably not be reading this book.

However, there are a number of other ethical and legal considerations which could also bear upon this dilemma, which essentially reduces to the question: can it be ethically acceptable to treat two very similar populations in very different ways?

Looked at from a welfare perspective, the suffering of the "pest" rats is no different than that of the pet rats, so if one took a purely utilitarian or deontological stance, then it would seem the two different populations deserve equal treatment, whether that be their nurture or destruction. In such a case there would therefore be no ethical grounds for a veterinarian to treat one rat in one way and another rat in another way. Your colleague may be right.

If one were to produce an ethical matrix for each population of rats, and to rate the relative impacts on the different stakeholders – rat, pet owner, the wider human population suffering from the rodent plague – then one would be able to score the two populations differently by allowing a greater value for the impacts on the non-rat stakeholders than for the rats, thus giving greater weight to the extrinsic rather than the intrinsic value of the rats.

But are there other ethical frameworks that would allow for the different treatment of these rat populations?

Hobbes proposed a contractual theory of ethics (Sorell 1996) whereby obligations to those with whom we have "contracts" are greater than to those whom we don't (Hobbes was considering humans when he proposed this, but one can argue that the theory could be extended more widely), and a number of ethicists have proposed a "relational" theory, to account for our intuitive views that our obligations to those (humans and animals) with whom we have a close relationship are greater than to those with whom our relationship is more distant, or non-existent.

The special nature of the relationship between certain animals and humans may be viewed in microcosm (the relationship between individual pets and their owners) or in macrocosm (the "contract" domesticated animals have entered into with the humans who domesticated them). In either situation, those subscribing to a contractarian or relational view will hold that our responsibilities to these groups of animals are greater than to their non-domesticated, un-owned cousins.

Veterinary professional ethics can be read as supporting a relational or contractual theory of ethics in this case. The RCVS code of conduct holds that a vet has a greater duty of care to an animal presented to him or her for treatment, than to a wild or un-owned animal (RCVS 2015), and there would be significant grounds for a charge of unprofessional conduct for the veterinarian choosing to refuse treatment for Scabbers and Harry, especially if they are already registered as patients with the practice, regardless of the external context.

Companion animal vets, and to a lesser degree, large-animal vets by the very nature of their business model exist to service (or at least subsist by servicing) these "special relationships", and for them to choose not to do so in this case would run counter to their core business. One might go so far as to argue that a refusal to treat the rats in this case would, especially if the client was registered

with the practice, be in breach of its contract with respect to consumer laws, regardless of any considerations of professional conduct.

Legally, animals falling under the direct control of people are due a duty of care under the Animal Welfare Act 2006 in the UK (HMSO 2006), and also by legislation in many other countries. As such they are afforded rights not due to wild animals, which are protected, mostly to a lesser degree, by other legislation.

In this case, you would be well advised to ignore your colleague, even if you can see their point of view.

||||||||||||||||||||||||||||||||||||||||||||||||||||||||||||||||||||||

THE term "pest" denotes a particular relationship to humans, typically a negative relationship in which the pest is perceived to pose a threat to human health, safety or economic wellbeing (Littin, et al. 2004). Control of pests is generally justified on utilitarian grounds. Strictly speaking, the benefits of any control regime (reduced economic losses, reduced rate of disease transmission, or protection of endemic flora and fauna) should outweigh the costs (Beausoleil & Mellor 2015).

Use of the term has been criticised, particularly by proponents of compassionate conservation, because it is an emotive, value-laden term that scapegoats one species and fails to address the complex role of this and other species in the ecosystem.

Once an animal is identified as a member of a so-called "pest" species they may be viewed with a negative bias, as described by Serpell (1996):

"...negative personification is particularly prevalent in human attitudes to predators, pests and scavengers – species that either compete with us directly, or which survive off the surpluses of human culture. Highly emotive, anthropomorphic language is often used to describe such species: they are spoken of as filthy, disreputable, gluttonous, sly, ruthless, evil, cowardly, blood-thirsty and savage. They are portrayed as dangerous and despicable enemies of society that invite nothing but hatred and loathing. Needless to say, these grossly inaccurate epithets serve nicely to justify their callous extermination."

(Serpell 1996)

A relational or contractarian approach as discussed in this case can help overcome this issue around inconsistency in our approach to rats, but does it go far enough? One question raised is whether we have a duty to review our approach to pest species in the light of our relationships to pets from the same species.

IDIOSYNCRACIES IN ANIMAL USAGE

What do you think?

ONE_____Which animals in your area have differing levels of legal protection depending on their relationship to humans?

TWO_____What is your view on the welfare experienced by the animals in these different settings?

THREE_____How might the different levels of protection be ethically justified?

FOUR_____Which forms of protection are inadequate in your view?

△

4.21 Foxes are often portrayed as wily, sly, bloodthirsty and ruthless – characteristics which may be used to defend their inhumane treatment by some.
PHOTO ANNE FAWCETT

△

4.22 Rabbits appearing in magic shows have higher levels of welfare protection than wild rabbits in areas where they are considered a pest.
PHOTO ISTOCK

4.4

Animal use across cultures

Animal use varies between contexts and cultures. There may be a conflict between exercising tolerance of religious and cultural differences and maintaining standards of animal welfare or principles of treatment, as discussed in the following scenario.

△

4.24 The traditional bits used are particularly harsh.
PHOTO GLEN COUSQUER

△

4.23a−b You are called to assess the welfare of mules used for tourist treks. They are thin, dehydrated and overloaded with luggage.
PHOTO GLEN COUSQUER

SCENARIO

MULE TREKKING

▸   You are asked by a well-known and reputable trekking company to undertake an animal welfare audit on their pack animal supported expeditions in the Moroccan High Atlas. In return they are offering you a 10-day trek.

They have recently received a number of complaints from clients concerned that mules are being exploited in order to provide them with a holiday. The company is anxious to have an independent evaluation of the situation from an equine specialist with experience of Morocco. The complaints suggest that the mules used were thin, dehydrated and overloaded with luggage that they had to carry up steep trails. The mules were also worked in traditional bits and subject to rough handling by their young − mostly adolescent − handlers. The accompanying photograph shows the bits to be particularly harsh.

In preparing to undertake this audit you are concerned about how you are going to assess and evaluate the welfare of these animals. What norms should you refer to? According to what standards should you measure your findings?

What should you do?

RESPONSE
GLEN COUSQUER

▶   As an experienced veterinary professional you are confident that you can assess the extent to which a mule is dehydrated. Body condition scoring and the weighing of loads carried would also appear to be objective assessments. Advising on the acceptability of any such measurements is a value judgement, however. What constitutes an acceptable means of communicating with and controlling a mule can also be characterised as a value judgement.

In pondering how you can arrive at a value judgement that recognises and respects the different cultures and value systems of those who come to know the mule, you recognise that the mule is a member of a host community, whereas you, together with the tourist and trekking company, are "visitors". You will therefore be judging the mule's welfare from outside of the community.

Wrestling with this problem, you are forced to consider whether you lean towards moral relativism or whether you believe that there are universal values.

According to Steven Lukes (2008) there are two ways of thinking about morality and moral norms. The first is the descriptive view, taken by the external observer who, observing that these norms vary from society to society, culture to culture or even group to group, concludes that they can only be judged from within the said society, culture or group. According to moral relativists, there is "no unique viewpoint from which moral norms are rationally compelling and universally

binding" (p. 21), for the "authority of moral norms comes not from reason but from religious authority say, or tradition or custom or convention, and its scope of application is local and time bound" (p. 23). The second way of thinking about this issue is the practical or normative view. This maintains that "morality is single not plural" (p. 18) and can be justified rationally in such a way that it would be compelling for all relevantly similar moral agents.

The appeal of moral relativism has seduced and confounded many nation states during their struggle to apply universal democratic principles within increasingly multicultural societies. Benhabib (2002, 87–89) describes how the US legal system has found itself accepting cultural justifications as a valid defence for crimes such as rape and homicide because in the defendant's own culture these would have been more acceptable. Benhabib deplores this rigid view of culture and the law's failure to extend equal protection to all citizens:

> "The cultural defence strategy imprisons the individual in a cage of univocal cultural interpretations and psychological motivations; individuals' intentions are reduced to cultural stereotypes; moral agency is reduced to cultural puppetry."
>
> (Benhabib 2002, 89)

A similar example, this time pertaining to animal welfare, is provided by Thiriet (2004), who describes how the Australian legal system has chosen to allow Aboriginals an exemption from animal cruelty legislation where their hunting practices are deemed to be traditional.

In both these examples, the authors refer to the unwillingness of the ruling classes to take a strong line and perhaps be seen to oppress a minority group. Benhabib (2002, p. 89) decries a situation in which "white liberal guilt is pitted against the 'crimes of passion' of Third World individuals", whilst Thiriet (2004, p. 176) observes that the "imposition of Western values of conservation is regarded as a new type of dispossession, and the same would inevitably apply if Western animal welfare values were similarly imposed".

The examples cited illustrate the tendency in some quarters to believe in moral relativism. It would appear, however, that what we are actually witnessing is a reluctance to judge others and criticise behaviour and practices that we believe to be wrong. Indeed, as Martha Nussbaum (2000) points out, "much of the appeal of relativism rests on confusing it with tolerance of diversity, respecting the ways of others" (p. 49). Lukes (2008) remarks on the "inconsistency of asserting the relativity of all moral principles and then proclaiming the moral principle of tolerance as a universal principle and moreover one backed by reasoning" (p. 37), for as he sees it "tolerance … presupposes such negative judgement or condemnation. To tolerate something or someone is to abstain from acting against what one finds unacceptable" (p. 37).

The difficulties lie in (1) making an unbiased, objective, neutral and therefore universal judgement on what is good or bad, right or wrong and (2) finding a way to study, understand and critique social practices in a non-confrontational way.

In seeking a way forward, you draw inspiration from Jacques Derrida's deconstructive approach. This allows you to recognise that the distinction made between visitor and the host communities is empirically inaccurate. It is a false dualism, a binary opposition that overlooks the fact that, both within and across these communities, there are similarities and differences. It also fails to recognise that these communities are very much part of a network and that tourism is constantly blending and challenging these differences.

You therefore set out to deliver a balanced and insightful contribution to discussions about how these pack animals should be treated. By fostering new insights and by facilitating dialogue and discussion over these animals' needs, you are able to progress and develop practice. This is significantly aided by you collecting extensive video footage to document and support your findings. These videos are reviewed on trek with the mule owners, allowing further understanding to be acquired. This spirit of engagement then allows alternatives to be identified and explored in partnership with the communities themselves.

The muleteers tell of how their ancestors endured great hardships when travelling across these mountains. Treks involving whole days without water and remote camps that offer no grazing were not uncommon. Where these might have been acceptable for raiding parties, these communities are now at peace and treks are undertaken, not as part of a military campaign but in the name of tourism. It also emerges that tourism has greatly increased the number of mules living and working in the mountains. The steep valleys, however, are unable to provide adequate food for the mules and this has to be bought in. Consequently, many mules do not receive adequate roughage; this in turn can lead to dental problems in older mules. Young muleteers receive no training in how to handle their mules and are reliant on cruel traditional bits. In both cases, it is the lack of alternatives that accounts for the problem. Once this has been recognised, the trekking company is in a position to help organise and subsidise the bulk purchase of fodder for the mules and arrange for training in natural horsemanship that reduces the communities' reliance on traditional bits.

In summary, your contribution shifts from making an objective assessment of the welfare of the mules working in tourism to recognising that welfare is not simply something which is measured but should be discussed. This involves it being deconstructed and reconstructed in partnership with the local communities. A renegotiated understanding of welfare can then be cultivated.

4.4

ANIMAL USE ACROSS CULTURES
RESPONSE
GLEN COUSQUER

||||||||||||||||||||||||||||||||||||||||||||||||||||||||||||||||||||||

THIS scenario demonstrates that it is possible to demonstrate humility and respect for another culture whilst providing education and guidance. Such an approach is consistent with virtue ethics. In this case we would expect the invited expert to demonstrate compassion for humans and animals, including the muleteers, truthfulness in providing accurate information about the welfare compromise involved and alternatives available, and integrity and conscientiousness in ensuring that a proper welfare assessment is carried out and that the result is not influenced by the offer of a free trek.

To this end the virtuous professional would identify the potential conflict of interest in accepting a trek. They would further recognise that the company, and wider industry, do not have access to animal welfare experts and cannot afford access to expertise. The holiday here is therefore not a bribe or inducement, but entirely necessary for evaluation of animal welfare in situ. Detailed reporting on animal welfare concerns would be impossible otherwise.

For more discussion about conflict of interest, see the scenario in chapter 7 by Andrew Knight regarding "Samples and freebies". For further discussion about animal welfare versus cultural and religious interests, see the scenario "Cultural rights vs animal welfare" by James Yeates in chapter 14.

ANIMAL USE ACROSS CULTURES

What would you do?

You are on holiday visiting a veterinary friend abroad. Your friend lives in a small village just along from the local slaughterhouse. Your friend offers to take you round there, as they know the owners well. Whilst there you find out that most cattle arrive in the afternoon and are then starved of food and water and tethered on bare concrete overnight prior to slaughter. You witness the casting of cattle onto the ground, and then killing being carried out with multiple stabs and cuts to the throat. Death comes slowly. What do you do?

△

4.25 How would you approach an "off duty" visit to a slaughterhouse abroad where conditions for protecting public health and animal welfare are far from ideal?
PHOTO ISTOCK

4.5

Experimentation on laboratory animals

Experimentation on animals in laboratories is typically overseen by ethics committees. While policies may vary, many committees assess study proposals in the light of the "3Rs": replacement, reduction and refinement. The following scenario explores the application of the 3Rs.

SCENARIO
CAUSING DELIBERATE HARM

▶ As the veterinary surgeon responsible for the welfare of the animals in a research establishment, part of your job is to advise on, and monitor, the use of animals in experiments. In general, you are aiming to lower the severity in any experimental work while maintaining the scientific objectives.

As a result of legislation and institutional policy (e.g. as a member of the Ethics Committee) you are asked to assess a research project that involves mapping pain pathways in rats. Pain is a common and serious clinical problem in humans and animals. Deliberately causing pain in healthy animals to study pain is against the veterinary ethic of protecting the welfare of animals, even if it may lead to better pain relief for both.

How should you advise the scientists and reconcile your own internal conflicts?

RESPONSE
DAVID MORTON

▶ Pain research is always a difficult ethical area and veterinarians can make an important difference partly because they, together with the animal care staff/technicians, can recognise when an animal is not normal, as well as the degree of that deviation from normality. This could be used as an indicator for the degree of pain and distress an animal is suffering (NB animals in pain are nearly always also in distress of some sort). Despite the deliberate intention to cause pain and suffering, which is in itself a serious ethical concern, from a utilitarian viewpoint it is only justifiable if one can

△
4.26 There can be a conflict between the veterinary ethic of protecting the welfare of animals and infliction of pain on animals in experiments.
PHOTO ANNE FAWCETT

be reasonably certain that there will be important gains, for example in knowledge about pain relief treatment.

The key here is to minimise the animals' suffering whilst maintaining the scientific objective. Occasionally, this may not be possible, in which case the vet has to argue for the animal as that is their first priority, as if they do not do it, who else will do so? It is up to the scientists to argue their case and then for the Ethics Committee members to decide on what to advise or permit as part of the continual harm:benefit analysis.

The potential benefit of such experiments to our understanding of pain mechanisms and pain therapy needs to be shown, and that the work is likely to provide useful data. In terms of harm there should be an upper limit so that the level of pain must never be debilitating, for example as shown by inappetence and drastic weight loss, profuse vocalisation, neglect of pups, muricidal behaviour and abnormal social interactions – such as extreme aggression to other animals and humans.

Should researchers take into account the phylogenetic scale? If so they must seriously question the use of dogs, primates and dolphins as subjects in neuropathic pain models, especially when the evidence in pilot experiments provides no significant differences between higher or lower animals in the phylogenetic scale. Furthermore, painful experiments can be justified only if a significant number of animals or humans are likely to benefit.

The standard ethical framework to evaluate animal experiments, introduced by Russell and Burch in 1959, is known as the Three Rs: replacement, reduction and refinement, and this can be applied in pain research.

Replacement: Are there any *in vitro/in silico/ ex vivo* alternatives for the work that do not use conscious animals? Has the work been done before? Have the scientists performed a systematic review?

Reduction: The number of animals involved must be kept to a minimum, using just that needed to establish statistical significance (power analysis) and the lowest number of experimental groups. Pilot experiments may sometimes yield minor and statistically non-significant differences but may still be added to a main study and provide a basis for abnormality scoring. From an ethical point of view, it is questionable whether small intergroup differences justify a larger-scale study, particularly for this type of research.

Refinement: Can the work be performed in such a way as to reduce suffering? For example, can the most painful part of experiments be performed under a general anaesthetic? Measures should be taken to ensure that the animal is exposed to the minimal pain necessary for the purposes of the experiment. If possible an animal presumably experiencing chronic pain should be treated for relief of pain, or should be allowed to self-administer analgesic agents or procedures, as long as this will not interfere with the aim of the investigation or as long as such treatment can be taken into account in some way. Humane end points should be established to minimise suffering of animals.

The duration of the experiment must be as short as possible. It is often possible to test the degree of pain more accurately by providing an environment in which abnormal behaviours are more pronounced in some way. For example, how the animal interacts with cage "furniture and toys" in an enriched environment may give more clues on the impact of the experiments on the animals than if they were in a barren environment.

Other approaches

In order for other groups to avoid repeating a similar experiment the work should be published and the methods used to evaluate the levels of pain and distress should be included in any publication.

In this example the research group are human clinicians or have close relationships with pain clinics. It would seem logical that there should be some cross-fertilisation of ideas, experience and techniques between the human and animal study groups.

||||||||||||||||||||||||||||||||||||||||||||||||||||||||||||||||||||||||||||

A humane end point is defined as "the earliest indicator in an animal experiment of (potential) pain and/or distress that, within the context of moral justification and scientific endpoints to be met, can be used to avoid or limit pain and/or distress by taking actions such as humane killing or terminating or alleviating the pain and distress" (Morton 1999).

Humane end points are established prior to the study taking place. For example, in a study involving pancreatic tumour induction in mice aiming to test an anti-neoplastic agent, a humane end point may be when the tumour reaches a certain size or stage. The mouse would be euthanased when it reached this pre-determined end point, regardless of whether the research protocol has been completed.

Another example of a humane end point would be an agreed percentage of weight loss – for example 15 per cent loss of bodyweight as measured at the beginning of the study. As animals in these studies become closer to these end points, measurements may occur more frequently according to the study protocol.

In studying pain pathways in rats, a certain degree of pain and distress may be intrinsic to the experiment. However, study design can minimise pain and distress by measuring potential pain and establishing non-clinical parameters such as hormone level changes, biochemical parameters, physiological parameters or genetic up or down regulation as indicators of pain or distress.

In such a study, a humane end point may prompt removal of a painful stimulus or source of pain, and/or administration of analgesia.

Such an intervention may disadvantage scientists, particularly where end points occur before the study protocol is complete. This is why it is important that animal care and ethics committees are independent and thereby not influenced by the aims of the study.

Experimental use of animals is largely justified on utilitarian grounds: the costs to the animal (in terms of behavioural restriction, discomfort and pain) are weighed against the benefits to humans. This is reflected in research applications where costs to an animal are weighed against expected benefits to humans. However, in many countries including the UK the government has also decreed that no licences will be granted for procedures on great apes, regardless of any potential benefits. The latter maps onto deontology, in that such apes cannot be used as a means to an end in any circumstances (Home Office 2014).

Those opposed to animal experimentation typically argue that it infringes on the rights of animals to bodily integrity and liberty, and/or that it is wrong to use an animal as a means to an end.

CHAPTER 4 ANIMAL USE

—
4.5
EXPERIMENTATION ON LABORATORY ANIMALS
WHAT DO YOU THINK?

EXPERIMENTATION ON LABORATORY ANIMALS

What do you think?

You are asked to review a licence application for an experiment involving animals investigating the regenerative potential of heart muscle following damage such as might occur during a "heart attack". In total, 2000 zebra fish, 3500 day-old mice and 3000 adult mice are proposed to be used over five years. Each animal will have a small area of heart muscle damaged under general anaesthesia, be assigned to one of four treatment or control groups and be monitored. The researchers state in their application that "these models of heart injury carry a *substantial* severity rating and there is a risk of an animal showing signs of premature heart failure after surgery".

ONE _____ What additional information would you need to properly assess this application?

TWO _____ What refinements might be possible?

THREE _____ Regardless of the legal position, what elements of this application make you more or less sympathetic to it?

△
4.27 Zebra fish are commonly used in experimental scientific procedures.
PHOTO ISTOCK

4.6

Personal values vs professional responsibilities

Veterinarians may need to consider whether they work within a context of a form of animal use they find objectionable, as discussed in the following scenarios.

_____
SCENARIO
LIVE EXPORT. RESPONSES:
CLIVE PHILLIPS AND JOY VERRINDER

△
4.28 Millions of cattle raised in Australia are shipped to the Middle East each year.
PHOTO ISTOCK

_____
SCENARIO 1

Background information
Millions of cattle and sheep are sent from Australia to the Middle East each year, a journey of some 10–14 days. En route, the animals will face challenges of heat stress, ammonia accumulation, lack of feed and overcrowding. Exporter companies are required to employ a veterinarian

and/or stockperson to accompany long haul shipments of cattle, who meets with the captain of the ship regularly to discuss the animals' welfare, and they should also complete a voyage report, which identifies the mortality and documents any problems during the voyage. If cattle mortality exceeds a threshold of 0.2 per cent, there is a government investigation. Government theoretically has the power to stop shipments of animals by any exporter.

Scenario

You are employed by a major live export company, Tradco, as a veterinarian to accompany cattle on long haul shipments to Egypt. A ship is loaded in Fremantle in August with 500 Angus steers, which have been trucked from Victoria and held overnight before the morning. Up until the equator the weather is acceptable, but as you approach the Gulf of Suez the wind drops and the temperatures climb to over 40°C. No advance weather report was obtained by the company's captain that could have predicted such temperatures. With limited ventilation capacity on the vessel, the cattle close to the engine room are obviously stressed by the heat, with open-mouthed panting and copious salivation. You arrange for some hosing of the cattle with seawater but, given the risk of extra humidity further increasing heat stress, when their condition does not improve after a short period you desist from this activity. You advise the captain not to enter the Gulf until the temperatures have dropped. He answers that he has a schedule to keep to and must continue.

The next day, as you approach the Suez Canal at the end of the Gulf, you find three cattle dead and you notice that the internal temperature on the ship is 46°C. You advise the captain not to enter the Canal, but to wait out in the Gulf where there is at least some breeze. He insists on continuing and does not accept your report, which includes the mortalities. You remonstrate with him

but to no avail. By the end of the voyage, 10 cattle have died and relations between yourself and the captain are even more strained. Shortly after you enter port, temperatures in the region decline and you realise that a delay in entering the Gulf would probably have saved the lives of several cattle. You are aware that any complaints to the company may jeopardise your position. You wonder whether to make a complaint to the company.

RESPONSE

▶ You first advise the captain that your duty to the animals, to your profession and to the company as ship's veterinarian requires you to report the high mortality. You then contact the company Chief Executive and advise that there was high cattle mortality on the voyage and that you believe that some deaths would have been avoidable if the ship had waited at the entrance to the Gulf and even more if the ship had not entered the Suez Canal. You advise that the continuation of the voyage was contrary to your advice and that the Australian Standards for the Export of Livestock require an advance weather report to be obtained that, had it been obtained, might have averted the disaster. You indicate that there will now be an Australian Government investigation of the causes of the mortality and ways of preventing it in future, which will consider your report. You end by stating that you are concerned about the many impacts of these journeys on the animals and that you would like to meet to explain these and to help develop alternatives as soon as possible.

SCENARIO 2

△
4.29 The increasing demand in Asia for dairy products has created a growing demand for high-quality dairy heifers.
PHOTO MICHAEL QUAIN

Background information

The increasing demand in Asia for dairy products has created a growing demand for high-quality dairy heifers. As there is a shortage of these in most Asian countries and milk prices in Australia and New Zealand are low, dairy farmers in Australia and New Zealand have developed a lucrative trade in heifers sent by ship to ports in China, Pakistan and other countries with potentially high-output dairy farming industries. As there are insufficient numbers of cattle available in these two countries, heifers are also being sourced from South America, where there is less control of their welfare before and during the voyage to China.

Scenario

You are a veterinarian with a client who is a dairy farmer in Victoria. The client has increasing debt problems and is receiving increasingly low milk prices relative to the cost of production. You hear that there is a market for some of his dairy heifers in Pakistan. However, there have also been reports of cows sent there suffering with little food, inadequate care and high mortality rates.

You wonder whether you should inform him of this opportunity to sell heifers to an exporter who is shipping high-quality heifers to Pakistan.

RESPONSE

▸ You consider that it is important to maintain your integrity by respecting the well-being of animals and working toward the fairest outcome for all. By not informing the farmer, he may still hear of the opportunity, or sell his cows to someone else who is exporting cows to Pakistan, and remain unaware of, or ignore, the welfare issues. You advise him of the opportunity to market his cattle in Pakistan and strongly recommend he does not send or sell his cattle for this purpose as it would damage their welfare in the short and long term and may also impact on his and dairy farmers' well-being due to growing public awareness and concern for dairy cow welfare You indicate that you would like to work with him to find alternatives to his continual unprofitable situation, rather than move towards less ethical solutions.

SCENARIO 3

Background information

The live export trade in sheep to the Middle East provides a ready source of meat, particularly for religious festivals in which the meat is shared with the purchaser's neighbours and poor people in the region. The journey is long, about 12 days, and there are often welfare problems en route, such as the build-up of ammonia, high temperatures, inability of some sheep to eat and contagious diseases such as salmonellosis.

Scenario

Sheep are in surplus in Australia as a drought has been decimating farming land for nearly five

△
4.30 Pustular dermatitis can spread quickly in the closely confined conditions on a ship.
PHOTO ANDREA TURNER

years. A shipment of sheep has been sent to Saudi Arabia but has been rejected at the port because the Saudi veterinary inspector says that he has detected a high prevalence (6 per cent) of scabby mouth (pustular dermatitis). Although another country, Bangladesh, has already said that it will take any sheep rejected, the shipping and export companies would lose millions of dollars in lost revenue and the sheep would have further to travel. You are accompanying the shipment on behalf of the shipping company in the role of veterinarian and are asked to determine the extent of scabby mouth in the sheep. You assess a sample of each deck and come to the conclusion that approximately 7 per cent of the sheep are infected. The captain of the ship meets with you and asks you to consider that some were not very serious cases and it should be reported as 5 per cent, this being the legal limit for sheep to enter Saudi Arabia in live shipments. You wonder whether to agree to the captain's request.

RESPONSE

▶ You explain to the captain that your professional standards require you to report an accurate prevalence of scabby mouth on this voyage, regardless of how mild or serious, and to falsify the records would damage your integrity. You indicate that you will request a report on the incidence of scabby mouth at the pre-export inspection and compare it with your detected levels. You contact the Saudi veterinary inspector to inform him of your concern about the impact of a further voyage to Bangladesh on the welfare of the sheep, and ask if he would join you in recommending that the importing company accept the sheep anyway, as you both have detected only a small percentage with scabby mouth over the current rather arbitrary limit. You request that, if the sheep are transported further to Bangladesh, you want to accompany them, rather than flying home from Saudi Arabia as planned.

POSTSCRIPT

▶ Many believe the live export of animals is unethical, as it imposes prolonged suffering on sentient beings during transport and eventual slaughter (Phillips & Santurtun 2013).

Veterinarians therefore have a professional responsibility to consider whether they should work to minimise the harm while the live export trade exists or refuse to participate in it. Working within the trade to help animals currently in need, becoming fully informed of the issues, and using this knowledge as evidence to stakeholders and governments to argue the case for moving toward alternatives is one way. As a professional body, veterinarians could alternatively oppose the trade by refusing to participate in it, in which case stock persons may replace veterinarians on board the ships. Either way veterinarians are in a strong position to work with industry and governments to replace industries and practices that cause animal suffering with more ethical alternatives.

All the major ethical frameworks can be used in a complementary way to identify whether the trade is ethical: the deontological perspective based on duty to universal principles such as respect for the lives and well-being of all sentient beings, the utilitarian viewpoint to achieve the greatest happiness or well-being for all concerned, Rawls' justice as fairness perspective with the greatest benefit to the least advantaged, care ethics focusing on compassionate relationships, as well as virtue ethics to apply universal principles consistently to become the best we can be.

Conclusion

Animals are impacted by our uses of them, with some forms of animal use more controversial than others. As society changes, the conditions under which and the ways in which we use animals are consistently reviewed. What once appeared to be respectable practices – such as veal crates, sow stalls, ear cropping and tail docking – may later come to strike us as deeply objectionable. Veterinarians and allied health professionals need to be aware of the implications of the use of animals by ourselves, our clients and wider society, and understand that different forms of animal use may be more or less acceptable to society. We must also consider what duties and obligations flow from the way we keep and use animals, and how best to bring about improvements.

ANIMAL USE ACROSS CULTURES

What would you do?

You are approached by Mr F, the owner of a "puppy farm", to be their vet. He is keen to move away from some of the bad practices that have blighted what he sees as a legitimate activity. Nevertheless, he keeps his 67 bitches in individual kennels and lets them out in groups to exercise in a fenced-in area each day, with limited human contact. When puppies are born he encourages kennel staff to spend extra time with them but admits this is in addition to doing other jobs. Will you accept Mr F as a client?

△
4.31 In some countries many puppies are bred in "puppy farms".
PHOTO ISTOCK

References

Animal Welfare Act 2006 Office for Public Sector Information. UK.

Animals Australia N.d. Where do puppies come from? http://www.animalsaustralia.org/puppies

Beausoleil NJ, and Mellor DJ 2015 Advantages and limitations of the Five Domains model for assessing welfare impacts associated with vertebrate pest control. *New Zealand Veterinary Journal* **63**: 37–43.

Benhabib S 2002 *The Claims of Culture: Equality and Diversity in the Global Era*. Princeton University Press: Princeton.

Bracke MBM, and Hopster H 2006 Assessing the importance of natural behavior for animal welfare. *Journal of Agricultural and Environmental Ethics* **19**: 77–89.

Brambell Committee 1965 Report of the Technical Committee to Enquire Into the Welfare of Animals Kept Under Intensive Livestock Husbandry Systems. HM Stationery Office: London.

Broom DM 2016 *International Animal Welfare Perspectives, Including Whaling and Inhumane Seal Killing as a W.T.O. Public Morality Issue*. Springer International: Cham, Switzerland.

Broom DM, Galindo FA, and Murgueitio E 2013 Sustainable, efficient livestock production with high biodiversity and good welfare for animals. *Proceedings of the Royal Society of London B: Biological Sciences* **280**.

Cao D 2015 *Animal Law In Australia*. Lawbook Co. Thomson Reuters: Sydney.

Cavalieri P 2015 For an expanded theory of human rights. In: Armstrong S and Botzler R (eds) *The Animal Ethics Reader*. Routledge: London.

Danbury TC, Weeks CA, Chambers JP, Waterman-Pearson AE, and Kestin SC 2000 Self-selection of the analgesic drug carprofen by lame broiler chickens. *The Veterinary Record* **146**: 307–311.

EFSA 2009 Scientific report on the effects of farming systems on dairy cow welfare and disease. *The EFSA Journal* **1143**.

Eisemann CH, Jorgensen WK, Merritt DJ, Rice MJ, Cribb BW, Webb PD, and Zalucki MP 1984 Do insects feel pain?–A biological view. *Cellular and Molecular Life Sciences* **40**: 164–167.

Elwood RW 2011 Pain and suffering in invertebrates? *Ilar Journal* **52**: 175–184.

FAO 2009 How to Feed the World in 2050. Food and Agriculture Office of the United Nations: Geneva, Switzerland.

FAWC 2009a Farm Animal Welfare in Great Britain: Past, Present and Future. Farm Animal Welfare Council: London.

FAWC 2009b Opinion on the Welfare of the Dairy Cow. Farm Animal Welfare Council: London.

FAWC 2012 View on Sustainable Intensification. Farm Animal Welfare Council: London.

Foresight 2011 The Future of Food and Farming: Challenges and Choices for Global Sustainability. The Government Office for Science: London.

Francione G 1995 *Animals, Property and the Law*. Temple University Press: Philadelphia.

Francione G 2004 Animals – property or person? *Rutgers Law School (Newark) Faculty Papers*, Paper 6.

Fraser D 2008 Animal welfare and the intensification of animal production. In: Thompson PB (ed) *Ethics of Intensification: Agricultural Development and Cultural Change*, 167–189. Springer: Breinigsville.

Garner R 2013 *A Theory of Justice for Animals: Animal Rights in a Nonideal World*. Oxford University Press: Oxford.

Garnett T, and Godfray C 2012 *Sustainable Intensification in Agriculture. Navigating a Course through Competing Food System Priorities*. Food Climate Research Network and the Oxford Martin Programme on the Future of Food: London.

Gray CM, Sellon RK, and Freeman LM 2004 Nutritional adequacy of two vegan diets for cats. *JAVMA-Journal of the American Veterinary Medical Association* **225**: 1670–1675.

Harrison R 2013 *Animal Machines*. CABI: Wallingford.

Herzog H 2010 *Some We Love, Some We Hate, Some We Eat: Why It's So Hard to Think Straight about Animals*. Harper.

HMSO 2006 Animal Welfare Act (c.45).

Hoekstra AY 2010 The water footprint of animal products. In: D'Silva J and Webster J (eds) *The Meat Crisis*. Earthscan: London.

Home Office 2014 Guidance on the Operation of the Animals (Scientific Procedures) Act 1986.

Hume DA, Whitelaw CBA, and Archibald AL 2011 The future of animal production: improving productivity and sustainability. *The Journal of Agricultural Science* **149**: 9–16.

Kanakubo K, Fascetti AJ, and Larsen JA 2015 Assessment of protein and amino acid concentrations and labeling adequacy of commercial vegetarian diets formulated for dogs and cats. *JAVMA-Journal of the American Veterinary Medical Association* **247**: 385–392.

Kestin S, Gordon S, Su G, and Sørensen P 2001 Relationships in broiler chickens between lameness, liveweight, growth rate and age. *The Veterinary Record* **148**: 195–197.

Leopold A 1989 *A Sand County Almanac*. Oxford University Press: Oxford.

Littin KE, Mellor DJ, Warburton B, and Eason CT 2004 Animal welfare and ethical issues relevant to the humane control of vertebrate pests. *New Zealand Veterinary Journal* **52**: 1–10.

Lockwood JA 1987 The moral standing of insects and the ethics of extinction. *Florida Entomologist* **70**: 70–89.

Lukes S 2008 *Moral Relativism*. Profile Books: London.

MacIver AM 1948 Ethics and the beetle. *Analysis* **8**: 65–70.

Martrenchar A 1999 Animal welfare and intensive production of turkey broilers. *World's Poultry Science Journal* **55**: 143–152.

McCulloch SP 2013a Agriculture, animal welfare and climate change. In: Wathes CM, Corr SA, May SA, McCulloch SP and Whiting MC (eds) *Veterinary & Animal Ethics: Proceedings of the First International Conference on Veterinary and Animal Ethics*, 84–99. Wiley-Blackwell: Oxford.

McCulloch SP 2013b Sustainability, animal welfare and ethical food policy: a comparative analysis of sustainable intensification and holistic integrative naturalism. In: Potthast T and Meisch S (eds) *Climate Change and Sustainable Development: Ethical Perspectives on Land Use and Food Production*, 175–180. Wageningen Academic: Wageningen, The Netherlands.

McCulloch SP 2015 The British Animal Health and Welfare Policy Process: Accounting for the Interests of Sentient Species. University of London.

McMahan J 2010 The meat eaters. *New York Times*: New York.

McMichael AJ, and Butler AJ 2010 Environmentally sustainable and equitable meat consumption in a climate change world. In: D'Silva J and Webster J (eds) *The Meat Crisis,* 173–189. Earthscan: London.

Mellor DJ 1998 How can animal-based scientists demonstrate ethical integrity? In: Mellor DJ, Fisher M and Sutherland G (eds) *Ethical Approaches to Animal-Based Science*, 19–31. ANZCCART: Wellington, New Zealand.

Mellor DJ, and Bayvel ACD 2014 Animal Welfare: Focusing on the Future. Scientific and Technical Review, 1–358. Office International des Epizooties 33.

Mellor DJ, and Stafford KJ 2001 Integrating practical, regulatory and ethical strategies for enhancing farm animal welfare. *Australian Veterinary Journal* **79**: 762–768.

Morton DB 1999 Humane endpoints in animal experimentation for biomedical research: ethical, legal and practical aspects. In: Henriksen CFM and Morton DB (eds) *Humane Endpoints in Animal Experimentation for Biomedical Research*, 5–12. Royal Society of Medicine Press: London.

Nussbaum MC 2000 *Women and Human Development. The Capabilities Approach*. Cambridge University Press: Cambridge.

Phillips CJC 2015 *The Animal Trade*. CAB International: Wallingford.

Phillips CJC, and Santurtun E 2013 The welfare of livestock transported by ship. *The Veterinary Journal* **196**: 309–314.

Prosin N, and Wise SM 2014 The Nonhuman Rights Project: coming to a country near you. *Global Journal of Animal Law* **2**: 1.

RCVS 2015 Code of Professional Conduct for Veterinary Surgeons. http://www.rcvs.org.uk/advice-and-guidance/code-of-professional-conduct-for-veterinary-surgeons/pdf/

Roberts M 2000 US agriculture: making the case for productivity. *AgBioForum* **3**: 120–126.

Rollin BE, and Rollin MD 2015 Dogmatisms and catechisms: ethics and companion animals. In: Armstrong S and Botzler R (eds) *The Animal Ethics Reader*. Routledge: London.

Rothgerber H 2013 A meaty matter. Pet diet and the vegetarian's dilemma. *Appetite* **68**: 76–82.

Sandøe P 2008 Re-thinking the ethics of intensification for animal agriculture: comments on David Fraser, animal welfare and the intensification of animal production. In: Thompson PB (ed) *Ethics of Intensification: Agricultural Development and Cultural Change,* 191–198. Springer: Breinigsville.

Sankoff P 2009 The welfare paradigm: making the world a better place for animals? In: Sankoff P and White S (eds) *Animal Law in Australasia*. The Federation Press: Sydney.

Schweitzer A 2010 *The Wisdom of Albert Schweitzer: A Selection*. Philosophical Library, Open Road Integrated Media: New York.

Schweitzer A, and Lemke AB 1933 *Out of My Life and Thought: An Autobiography*. JHU Press: Baltimore.

Serpell J 1996 *In the Company of Animals: A Study of Human–Animal Relationships*. Cambridge University Press: Cambridge.

Sorell T 1996 *The Cambridge Companion to Hobbes. Cambridge Companions to Philosophy*. Cambridge University Press: Cambridge.

Spelman C 2012 Speech to Food and Drink Association on "Secure and Sustainable Food – The Rio+20 Challenge".

Taylor PW 2011 *Respect for Nature: A Theory of Environmental Ethics*. Princeton University Press: Princeton.

Thiriet D 2004 Tradition and change – avenues for improving animal welfare in indigenous hunting. *James Cook University Law Review* **11**: 160–178.

Thompson PB 2008 *The Ethics of Intensification: Agricultural Development and Cultural Change*. Springer: Breinigsville.

Tischler J 1977 Rights for nonhuman animals: a guardianship model for dogs and cats. *San Diego Law Review* **14**: 484–506.

UNDESA 2015 The United Nations World Water Development Report 2015: Water for a Sustainable World. https://www.unesco-ihe.org/sites/default/files/wwdr_2015.pdf

Van Huis A, Van Itterbeeck J, Klunder H, Mertens E, Halloran A, Muir G, and Vantomme P 2013 Edible Insects: Future Prospects for Food and Feed Security (No. 171, p. 187). Food and Agriculture Organization of the United Nations (FAO).

Waldau P 2006 Animals and religion. In: Singer P (ed) *In Defense of Animals – the Second Wave*, 69–83. Blackwell Publishing: Oxford.

Warburton B 1998 The humane traps saga: a tale of competing ethical ideologies. In: Mellor DJ, Fisher M and Sutherland G (eds) *Ethical Approaches to Animal-based Science*, 131–137. Australian and New Zealand Council for the Care of Animals in Research and Teaching: Wellington.

Webster AB 2004 Welfare implications of avian osteoporosis. *Poultry Science* **83**: 184–192.

Webster AJF 2013 *Animal Husbandry Regained: The Place of Farm Animals in Sustainable Agriculture*. Earthscan, Routledge: London.

Webster J 1994 *Animal Welfare: A Cool Eye towards Eden*. Blackwell Publishing: Oxford.

Webster J 2005 *Animal Welfare: Limping towards Eden*. Blackwell Publishing: Oxford.

Wise SM 2000 *Rattling the Cage: Towards Legal Rights for Animals*. Perseus Publishing: Cambridge.

Wittgenstein L, and Anscombe GEM 1958 *Philosophical Investigations*. Blackwell: Oxford.

World Food Summit 1996 Rome Declaration on Food Security.

CHAPTER 5

VETERINARY TREATMENT

Introduction

Veterinary treatment involves interventions to prevent disease and suffering, diagnose and manage or cure disease in animals. Over the previous decades there have been significant advances in veterinary medicine and surgery. New pharmaceuticals, treatment protocols and diagnostic modalities (such as digital radiology, high-quality sonography, computed tomography and magnetic resonance imaging) are increasingly available. Veterinary patients can benefit from cancer chemotherapy, stem cell therapy and minimally invasive surgical procedures such as laparoscopy.

In addition, in many countries there has been an increase in the availability of specialist treatment. In the USA, for example, the number of active board-registered diplomats grew by 15.46 per cent between 2006 and 2009. Between 2008 and 2009 alone the number of specialists in internal medicine increased by 11 per cent (American Veterinary Medical Association 2010).

While the aim of improved veterinary treatment may be laudable, with the intention to benefit both animals and owners, with all advances come increased costs. Training and continuing education, pharmaceuticals, disposables, equipment

and maintenance all cost money, and in the private practice setting, the client typically pays. In charity and non-government organisation (NGO) settings, limits may be placed on the nature and extent of treatment that can be provided to individual animals.

There are also potentially welfare costs to animals. Veterinary treatment may involve isolation of animals or separation from conspecifics or owners, physical restraint, unpleasant auditory and olfactory stimulation, and potentially painful procedures, all of which may lead to fearful behaviour and even a conditioned avoidance response in some animals (Dawson, et al. 2016).

With improved technology comes the dilemma – just because we *can* treat an animal with a certain condition, does it mean we *should*? In relation to the use of 3D-printing of prostheses for animals, surgical specialist Noel Fitzpatrick urges consideration of ethical issues:

"The bottom line now is that anything is possible, if you have a nerve and blood supply. That means that we now have a line in the sand: not what is 'possible' but what is 'right'. In the past it was just the case of if it wasn't possible, you'd move to euthanasia."
[Fitzpatrick, interviewed in Finan (2016)]

As in human medicine, variation in the practice of veterinary medicine is a major issue: similar patients with similar conditions may be treated

◁

5.1a–b
PHOTOS ANNE FAWCETT

differently (Djulbegovic, et al. 2015) not just because of differences between veterinary teams but also potentially because of the status of the animal.

Veterinary treatment raises a number of ethical issues, some of which we will discuss in depth in this chapter:

- Should we embrace all treatment advances? How do we assess these from an ethical standpoint?
- Is it in the best interests of the animal to pursue treatment and if so, should we seek curative or palliative treatment?
- How can we assess quality of life and determine treatment end points in animal patients?
- How do we navigate barriers to treatment?
- What are the costs and benefits of different interventions to different stakeholders? Is there potential for conflict and if so, how do we resolve this?
- To what extent do we treat a problem or symptom that is inherited, and to what extent do we attempt to address the underlying cause?
- How can members of the profession ensure that veterinary treatment is as accessible as possible, while running a sustainable business?

5.1

Barriers to treatment

There are numerous barriers to treatment: lack of resources, costs, the knowledge base of the veterinarian and client, current evidence or lack thereof. Another is patient considerations including the temperament of the patient and impact of examination, diagnosis and treatment on the individual animal, as explored in the following scenario.

SCENARIO
TREATING AN AGGRESSIVE CAT

▶ Kaiser is a five-year-old, male, domestic medium hair cat with a history of severe aggression. Mr H, the owner, has been bitten several times transporting Kaiser to your clinic, staff have been bitten and scratched trying to handle Kaiser, and even when placed in a carrier he launches. At home he behaves well – unless the veterinarian makes a house call, in which case he reverts to his unhandleable status.

So aggressive is Kaiser that staff are very reluctant to book appointments for him. Mr H does not appear to recognise the risk the cat presents to others.

At a previous visit, during which you had to gas Kaiser in an induction box, you diagnosed neoplasia. Mr H indicates he is keen for you to undertake chemotherapy which, for this condition, will involve regular visits for administration of cytotoxic drugs as well as blood tests.

What should you do?

△

5.2 Treatment of aggressive cats requires additional ethical considerations.
PHOTO ISTOCK

RESPONSE
ANDREW GARDINER

▸ Kaiser behaves well at home but unfortunately seems locked into a fear-aggressive behaviour pattern triggered by veterinary visits. As his vet, I am placed in a dilemma with regard to my conflicting obligations to Kaiser, my practice colleagues and Kaiser's owner. I need to try to adopt a common-sense approach which is fair to all. In one sense, it seems wrong that Kaiser should be treated differently (or not treated at all) just because he is terrified, particularly as he seems to be a "normal" cat at other times and can be assumed to enjoy the pleasures of his feline life with his caring owner. However, this could be a somewhat utopian view, given the difficulties encountered with Kaiser and the prospect of treating this particular disease.

I decide to try to work through the four guiding concepts of principalism which might help me produce a potential road map for dealing with Kaiser's disease.

(1) Non-maleficence
Whatever action I take with Kaiser, I do not want to make his situation worse; this is the principle of doing no harm. Given Kaiser's history, there is a possibility I could do him harm. He is a sick cat, so subjecting him to the great stress of veterinary visits, which would need to be frequent, may be inadvisable. If Kaiser were to struggle during the placement or removal of an IV line, cytotoxic fluids could leak and cause a tissue slough which would be painful and very difficult to treat and would seriously affect his welfare. The specifics of treating his disease need to be seen alongside his entrenched fearfulness at the vet.

Doing no harm also applies to myself, my veterinary colleagues and Kaiser's owner. Someone could easily get injured while trying to treat Kaiser. The option of doing nothing, however, goes against his owner's wishes, who is aware that the underlying disease can be treated. It is clear that the principle of non-maleficence alone does not provide me with an answer, but cautions me that whatever I do, I must try to avoid making the situation worse.

(2) Beneficence
I should try to promote good for Kaiser, his owner and my colleagues. Either treatment or some form of palliative care could achieve this, but only if it satisfies the criterion of non-maleficence above. That could be challenging. Is it possible to argue that *not* treating Kaiser for neoplasia is being beneficent? I am not sure. I feel that, to satisfy this principle in the spirit in which it is intended, I actually need to *do* something, otherwise it is really no different from non-maleficence. I need to do something which has the potential to add to Kaiser's welfare in the short and long term, and also helps my client, who wants his cat treated. The prognosis for the disease can be good.

(3) Autonomy
If Kaiser had the choice, I infer from his extreme behaviour that he would prefer not to attend the veterinary clinic or have treatment. So not treating Kaiser would respect his autonomy and relieve him of stressful visits and handling. This is an interesting proposition that could be worth discussing with his owner, given the extreme nature of Kaiser's behaviour. However, this interpretation of autonomy is very frequently disregarded in veterinary medicine (and areas of human medicine such as paediatrics) – because most companion animals would probably prefer not to attend the veterinary clinic. Kaiser's autonomy after successful treatment could be excellent, so he could benefit, even if he did not understand or appreciate this. This is the basis on which most veterinary treatment takes place.

The autonomy of Kaiser's owner is respected by attempting treatment. Autonomy of clinic staff could be respected, by allowing staff to decline to be involved in Kaiser's treatment if they so wished.

(4) Justice

One could argue that it is just that Kaiser receives treatment, so long as the cost of treatment is outweighed by the benefits, and that it would be unjust to withhold it just because of his fearful behaviour. This assumes treatment is in Kaiser's best interest and that we treat him as we would any other cat. It is also just that Kaiser's owner is treated the same way as other owners of cats with neoplastic disease, and given the same choices, and that he is not prejudiced because he happens to own a very fearful cat. Following the principle of justice, we should be fair and treat similar cases in a similar manner (Anzuino 2007).

Rather like the ethical matrix, I find that working through the guiding principles does not provide me with a ready answer but it has opened the dilemma up slightly and added some context to an ethical decision which seems to hover between utilitarianism and deontology (including a sense of Kaiser's "right to have treatment").

My aim was to provide a "road map" for Kaiser's treatment, whatever that may be, bearing in mind that, at home, Kaiser appears to be a happy and relaxed cat. I therefore suggest to the owner that we form an informal "ethical committee" to oversee Kaiser's treatment. The committee would comprise myself, two members of practice staff who are willing to be involved in Kaiser's care, and two members of Kaiser's human family (or friends). As such, Kaiser's response to treatment and its stresses could be monitored. Actions to reduce stress would be taken, including:

- Acclimatisation to a squeeze cage at home, in which he would always be brought to the practice. Use of cat-appeasing pheromones could

be tried. Alternatively, a top-loading carrier with soft bedding that Kaiser can hide under may make it easier to inject him with sedation in a safe manner.
- Direct handling of Kaiser without sedation would be banned.
- The use of reversible sedative drugs, if these can be used safely, to facilitate handling and treatment, delivered through the squeeze cage.
- Arranging treatment sessions for quiet periods, always using the same staff and routines, and with no waiting in the waiting room.
- Agreeing criteria for when treatment should be discontinued and switched to palliative care or watchful observation of his untreated condition.

The situation would need to be kept under review but I would be happier to keep Kaiser under my care, rather than have him go to another practice.

△

5.3 In this scenario the veterinary team must find a way to deliver treatment while being aware of everyone's safety.
PHOTOS ANNE FAWCETT

MANY patients are fearful in a veterinary context, but a very limited number may exhibit extreme behaviour. It is assumed, in the above scenario, that Kaiser's behaviour is extreme. It is also assumed that a single intravenous therapy is the only potential definitive treatment. Increasingly there are a range of alternatives available including total oral chemotherapy.

One thing that must be assessed on a case-by-case basis is to what extent fear can be mitigated, for example by low-stress handling and judicious use of sedation. There may be an option, for example, for the owner to administer some agents at home.

The precautions taken in scenarios such as these may vary with work health and safety legislation applicable in different jurisdictions. Veterinary patients cannot consent to veterinary intervention, although some may tolerate it better than others. Unlike many (though not all) human patients, they cannot assess a positive trade-off between short-term discomfort, stress or pain associated with treatment (such as that incurred by restraint) and long-term benefits (remission or cure of disease, relief of pain and discomfort in the longer term, improved wellbeing and so forth).

Such considerations are not limited to treatment of severe diseases. The same considerations apply when assessing common interventions, for example castration of male cats. Often our decision to intervene is made on a cost:benefit or utilitarian analysis, but for animals who are excessively fearful or aggressive the costs could easily exceed the benefits.

For some veterinarians, this is more a practical than an ethical issue. Fearfulness in veterinary settings is common. However, it is important to reflect on how we manage fear in patients. The principalist approach stresses the need to minimise costs or harms, which may require a flexible approach and – as suggested above – a review of standard operating procedures.

BARRIERS TO TREATMENT

What would you do?

Mr F rings the surgery to ask you to visit Hercules, his aggressive Arabian stallion, who has a nasty gash to his leg. The last time you visited Hercules you took the head nurse with you and it was a very stressful experience for all concerned. Now your head nurse states categorically that she doesn't want to have anything to do with Hercules and he doesn't think it's fair for you to ask other members of the team to be involved. What options are open to you? What would you do?

△
5.4 Hercules is difficult to get close to even when he is not in pain.
PHOTO ISTOCK

5.2

Individual versus group health

Depending on the context in which the veterinary team is working, the patient may be an individual animal (for example, a companion animal), a herd or a population of animals. There is potential for conflict between the interests of an individual animal and the interests of a wider population or herd. For some farmed animals such as fish and chickens the groups are so large, and the value of the individual is so low, that treatment is almost exclusively undertaken at the group level, with sick individuals being culled. When pigs, cattle and sheep become sick they may still be individually treated, despite much of the overall management involving the whole herd.

Companion animals are almost always treated based on their individual needs but even though

they may be the only animal in a household they can still be considered part of a wider community, particularly for disease transmission purposes. As shown in the below scenario, treatments that impact on the wider population can be further complicated by the interests of the client and may result in rejection of, or resistance to, treatment recommendations.

SCENARIO
VACCINATION REFUSAL

▶ Ms G brings her new puppy, Bramble, to you for a check over. She's very clear that she doesn't want Bramble to be vaccinated as she's heard (now discredited) scare stories about the human Measles, Mumps and Rubella (MMR) vaccination. You presume she could easily afford the vaccinations. In truth, you know the incidences of the diseases you vaccinate against are very low in your area, thanks partly to decades of vaccination of the pet population, although parvovirus is common in small pockets around the country where vaccination rates are lower.

What should you do?

RESPONSE
JAMES YEATES

▶ Some elements of this case focus on the balance between individuals and the "greater good". The ethical basis for respecting consent is partly human "rights" to property but also partly because of the potential societal risk in owners not trusting vets, and therefore not presenting their animals for treatment. Similarly, the benefit of vaccination is partly to the individual but also partly to the animal (and owner) population from the herd immunity provided (especially as

△

5.5 Sheep are usually managed in large groups and may be treated on an individual or a flock level.
PHOTO ISTOCK

∧

5.6 Deciding whether to vaccinate or not must take into account the fact that dogs are social animals and usually part of a local community, with implications for disease transmission.
PHOTO ANNE FAWCETT

no vaccination is perfect, although some provide much better protection than others).

Consent

In general, owners have some legal rights to determine what happens to their animal (within certain bounds). In particular, there is an assumption that (non-emergency) treatments should not usually be given to an owned animal without the owner's consent. This is partly because of the animal's legal status as property. For most of us, this would rule out performing vaccination without consent.

This does not mean we cannot try to persuade the owner by providing information and clinical advice. While valid consent should be free from influence, it should also be informed.

The owner has brought in her new puppy for a health check, suggesting she has some confidence that a veterinarian is able to carry out a comprehensive clinical examination. Building on

this level of trust, you may be able to identify areas of agreement (perhaps on other topics) to establish rapport.

The veterinarian may wish to talk about the different types of evidence available. For example, the now discredited Measles, Mumps and Rubella (MMR) vaccine scare stories in popular media have been addressed by careful consideration of information from much higher quality peer-reviewed papers. The veterinarian may wish to talk about the quality of the evidence that they have gathered over their time as a practitioner. They will have relevant information about local disease patterns; they can give an indication as to the frequency of incidents under the reporting scheme for adverse reactions; they can discuss the likelihood of serious disease in an individual pet from an outbreak of parvovirus. It may be relevant to discuss how a vaccine works and how unlikely (and clinically trivial) an adverse event is likely to be. Perhaps the owner can appreciate this by reflecting on their own immunisations.

As an alternative to regular vaccination, some veterinarians perform antibody titres. However, there remains uncertainty about correlation between antibody titres and the degree of protection. What happens to the measured level of protective antibodies with or without vaccination? How do we measure protection conferred by vaccines that rely more on cell-mediated immunity? In any case, this is not relevant in an unvaccinated puppy like Bramble.

The veterinarian might refer to evidence-based vaccination guidelines such as the World Small Animal Veterinary Association's Guidelines for the Vaccinations of Dogs and Cats (Day, et al. 2016). Among other things, the guidelines state that veterinarians should "aim to vaccinate every animal with core vaccines", of which the parvovirus vaccine is one.

An alternative would be to rely on simply treating the disease if acquired. Veterinarians could highlight that "it's easier (and cheaper) to prevent than to treat parvo". The veterinarian may honestly stress the negative welfare impact, plus severe morbidity and high mortality, associated with parvovirus infection.

In this scenario, strong beliefs are unlikely to be altered in a single conversation and subsequent presentations to the practice serve as an opportunity to refresh the debate. To be open and non-judgemental in a clinical discussion often requires more than a 10-minute consultation. Perhaps an invitation to an evening practice event would be a useful forum for debate. If it is clinically appropriate and relevant to the area, a presentation on the sudden emergence of a cluster of clinical cases of haemorrhagic gastroenteritis may serve as a useful catalyst. Of course the beneficial impact of the presentation relies on its influence taking effect before Bramble contracts parvovirus.

Free-riding

Nobody sane would question the value of vaccination at the population level (for diseases that have a reasonable prevalence otherwise). The issue here is that in some areas there may be sufficient population-level "herd" immunity to protect Bramble. However, parvovirus outbreaks still occur (for example, see www.diseasewatchdog. org), and diseases such as distemper are emerging in unvaccinated populations (Norris, et al. 2006). Even so, it is possible that Ms G can get away with not vaccinating.

But perhaps there is a more radical position. If herd immunity is high, then perhaps it is actually in Bramble's interests not to be vaccinated: the small risks of vaccine reaction may outweigh the also small risks of contracting the disease. In fact, if we are concerned about the overall welfare in an animal when assessing the risk of a low-prevalence disease, a cost:benefit analysis could even suggest that we should *advise* against vaccination. (We should be mindful that Bramble is an unvaccinated puppy, and reducing vaccination frequency carries different risks to withholding the first vaccination in a puppy) (Day, et al. 2016).

The paradox here is that if we did that for every animal, then there would not be herd immunity – in which case, the cost:benefit ratio for each animal would change and we should then advise vaccination. This risks population disease levels yo-yoing up and down, causing significant suffering.

Such paradoxes are familiar from the literature on Game Theory, such as the Prisoner's Dilemma. These create major headaches for proponents of direct act utilitarianism (although they generally refer to contrived cases). Fortunately, many can be resolved by coordination: if we can legitimately coordinate all decisions then we can come up with an optimal rule for general application – an example of rule utilitarianism that would advocate vaccination by all.

This still leaves us with a dilemma about Ms G. We may say that there is a general rule that everyone should follow, but Ms G can shirk this rule without any significant risk (and so can her friends as long as there aren't many of them). In fact, on the above logic, one could even argue that she *should* break the rule. Doing so means Bramble avoids the risk of vaccine reaction while still being well protected from parvovirus due to herd immunity. Effectively, act utilitarianism and rule utilitarianism lead to different conclusions. But if we defend the rule in terms of overall utility, it seems inconsistent to object to breaches that increase utility even more.

This suggests that we need some other approach to defending the rule than rule utilitarianism. Perhaps we can frame it as a concept of fairness: it seems unfair that Ms G gets away without paying (and Bramble gets away without vaccination risks), when other owners (and animals) carry the risk. However, it seems wrong to counteract that fairness by making Bramble and Ms G worse off. Many concepts of fairness aim at benefitting the worst off, not harming the best off for no benefit to others.

Instead, perhaps we can frame this in some concept that does not focus on the consequences (but which otherwise looks very similar). This perhaps needs to be framed as a personal responsibility of each and every owner to undertake particular actions (that include reasonable vaccination). For example, the Golden Rule or Kantian categorical imperative would both suggest that Ms G should act as she would want all other owners to do. On this basis she should not shirk her responsibility even if fulfilling it has no effect or, for her dog, a negative effect.

Conclusions

So you can legitimately advise Ms G to vaccinate. You can even perhaps inform her that it is part of her responsibility (although without exercising undue influence).

△
5.7 Canine parvovirus causes severe, often fatal disease in puppies and dogs, but can be prevented with vaccination.
PHOTO ANNE FAWCETT

Non-complicity

As a final thought, if Ms G does refuse vaccination, do you have a moral responsibility for any subsequent disease? The answer is yes if you gave wrong advice about the risks of the disease. But the answer is no if you were constrained by the owner's legal right to refuse treatment.

Nevertheless, you could ensure you are not complicit, while perhaps legitimately influencing the owner's decision. For example, having a practice policy that requires all adult animals to be vaccinated before admission for elective surgeries might mean you are not later responsible for unnecessary disease transfer. As a patient of your practice, therefore, Bramble will need to be vaccinated to receive any inpatient care. This may influence Ms G's decision to vaccinate.

||||||||||||||||||||||||||||||||||||||||||||||||||||||||||||||||||||||||||||||||||

IN this scenario the author refers to types of utilitarianism, notably act utilitarianism and rule utilitarianism. These are two of numerous subtypes of utilitarianism that we did not cover in the introductory chapters simply as they aren't frequently referred to in the veterinary literature, but some explanation will assist here.

Act utilitarianism is what is often most commonly thought of when we talk about utilitarianism as espoused by Jeremy Bentham and others including John Stuart Mill. In short, an act utilitarian holds that an act is morally right if – and only if – it maximises happiness or wellbeing. An act is judged by the consequences of that act alone.

In contrast, rule utilitarianism holds that an action is right if it conforms to a moral norm or rule which leads to the greatest good. Like act utilitarians, rule utilitarians are interested in consequences – just not the consequences of a single act. Rather, they hold that if a rule or moral norm generally maximises happiness or minimises suffering, it is a good rule. A rule utilitarian believes that following rules that generally lead to outcomes where the good is maximised has better consequences *overall* than making exceptions for individual cases.

In this case, act utilitarianism may justify not vaccinating Bramble, but this is only good if everyone else in the community vaccinates their dogs. Rule utilitarianism supports vaccination of Bramble if it increases group immunity.

It is critical in such cases to review current evidence. Canine parvovirus vaccines carry very little risk to dogs, and thus it may be argued by some veterinarians that the decision to vaccinate is non-problematic. But, as discussed, not all vaccines are equally effective and some carry higher risks of adverse reaction. Decision-making around such cases can be more challenging. For example, some feline leukaemia and rabies vaccines have been associated with feline injection-site sarcoma, an aggressive neoplasm. However, a recent review concluded that "vaccination of cats provides essential protection and should not be stopped because of the risk of feline injection-site sarcoma" (Hartmann, et al. 2015).

Instead, steps can be taken to mitigate harm, for example vaccinating animals as often as necessary (based on available evidence) but as infrequently as possible. Explaining this rationale to Ms G may satisfy her that the veterinarian is not simply recommending vaccination as a means of making money. Other harm-minimising steps include use of non-adjuvanted, modified-live or recombinant vaccines where possible, avoidance of the interscapular region as an injection site, and post-vaccination monitoring (Day, et al. 2016, Hartmann, et al. 2015).

INDIVIDUAL VERSUS GROUP HEALTH

What do you think?

ONE How much responsibility do we have for the general health of our local population, versus that of our individual patients?

TWO If we consider an owner's right to refuse vaccination, what other rights should we consider in such cases?

△
5.8 Free-roaming cats may spread infectious diseases such as feline immunodeficiency virus (FIV) through fighting and other forms of close contact.
PHOTO ANNE FAWCETT

5.3

Costs and benefits of treatment

Veterinary treatment incurs welfare costs to patients, however trivial this may be. The decision to treat is often made on a cost:benefit analysis, although in some cases the benefits to the patient may be minor. Furthermore, there may be a gap between the proposed and actual costs and benefits of treatment, as in the following scenario.

SCENARIO

PRE-ANAESTHETIC SCREENING

▶ Your practice has a policy of recommending pre-anaesthetic screening for every procedure requiring a general anaesthetic on the grounds that this may reveal underlying disease which would preclude anaesthesia or necessitate a different anaesthetic protocol.

You admit a dog for a routine desexing procedure, but there has been a miscommunication. One team member does not alert the other members of the team that the owner has elected for a pre-anaesthetic screening blood test until the dog is anaesthetised. The dog was quite fearful and difficult to anaesthetise in the first place, and had to be restrained by several team members for induction.

Another team member says, "Let's go ahead. It's peri-anaesthetic screening anyway."

What should you do?

RESPONSE

CHRIS DEGELING

▶ Pre-anaesthetic screening (PAS) remains controversial in both human and veterinary medicine. Using diagnostic tests to screen asymptomatic individuals for the presence of underlying or "subclinical" disease is fundamentally different from testing those who have clinical signs. Intuitively, the possibility of identifying and treating biological dysfunction before it causes disease appears to offer great benefits, but screening has the potential to do more harm than good. This is because the use of diagnostic technologies is rarely completely benign – tests are interventions, and in the case of screening they are interventions performed on individuals in which a relevant clinical abnormality has not been detected. Like paediatricians, veterinarians often practise in conditions of an information deficit. Young children and non-human animals cannot provide an account of their symptoms, and parents and animal owners can sometimes fail to notice a significant clinical sign. Yet perceptions about a lack of information are not

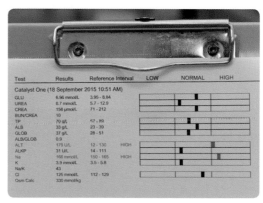

△

5.9 What are the reasons for pre-anaesthetic screening and how do you respond when results are abnormal?

PHOTO ANNE FAWCETT

5.3
COSTS AND BENEFITS OF TREATMENT
RESPONSE
CHRIS DEGELING

a priori sufficient grounds to justify screening – to hold this position would invite a regress where we constantly test every healthy individual for every conceivable condition. For screening practices to be considered ethical, clinicians who propose interventions that do not address a clear clinical need for patients have an added responsibility to ensure the likely benefit of testing outweighs any possible harms.

Judging the utility and ethical acceptability of screening by weighing up the potential harms and benefits might seem to be relatively simple, yet it is actually quite difficult. In the first instance experience with "over-diagnosis" in human medicine indicates that the harms and benefits caused by screening are often counterintuitive and difficult to recognise (Moynihan, et al. 2012). Reflecting on the veterinary scenario presented, aside from the injury caused by venepuncture, taking a blood sample from an animal almost always requires some form of physical or chemical restraint. However minor these interventions may seem to a busy veterinary clinician, they do harm to the animal. The number of attempts at venepuncture needed to obtain a useable sample and the amount of pain and stress each of these interventions causes the animal can compound these harms. Further harms are possible if the test result initiates further fruitless investigations of what turns out to be a "pseudo-disease". In contrast to the harms, the benefits of testing seem obvious. But claims that screening tests provide benefits for the animal and their owners often rest on the value of reassurance (for owners and clinicians) and sets of assumptions about what would have happened to the health and welfare of the animal if the test had never been performed.

Screening: understanding the potential for benefits and harms

To help understand how harms and benefits typically cash out from screening interventions let us consider the potential impacts of the different test outcomes. If the PAS tests are normal then they have not provided a benefit to the animal, only the harms associated with venepuncture and restraint. The only conceivable benefits gained through normal PAS test results are accrued by the owner and clinician if they provide a sense of reassurance. However, any reassurance gained must be tempered by the knowledge that screening blood tests do not rule out all of the subclinical conditions that can affect any anaesthetic. From this it is possible to argue that any benefit we might assign to a negative screening result actually only addresses a sct of conditions created by the decision to perform the test. The test might provide a sense of reassurance, but it has not informed a change in anaesthetic practices. The course of events surrounding the animal's care has not been altered by the test, except that the animal has suffered additional harms.

In contrast, if the test results point to an abnormality then it is possible the test has provided benefit to the animal, but only if subsequent diagnostic investigations "uncover" a treatable disease that was highly likely to have a detrimental effect on the health and welfare of the animal, such as to shorten or diminish its quality of life. If the test abnormality initiates further investigations of what turns out to be pseudo-disease, then every subsequent diagnostic intervention could cause further harm to the animal. Discovering pseudo-disease, discovering a disease that cannot be treated or discovering a disease that is unlikely to affect an animal's health or welfare does not confer benefit on the animal or its owner – unless an otherwise futile treatment plan is changed or the owner gains some sense of control or solace from knowing what is likely to happen in the future.

If we discount the value of some limited reassurance for owners and clinicians, from the range of possibilities described above it becomes clear that the ethical permissibility of pre-anaesthetic screening in companion animal practice depends on a value judgement about how much the harms caused can be justified by obtaining potential benefits for a few individuals. Evidence as to the substantive benefits of pre-anaesthetic screening is mixed, with some studies suggesting that it has little influence on veterinary anaesthetic practices (Alef, et al. 2008), while others note that the test results cause a modification in protocol in upwards of 4 per cent of canine and 9 per cent of feline cases (Davies & Kawaguchi 2014). Notably, in this more recent study, the final outcome measure was a change in protocol in response to an abnormal test result; the external validity of taking such measures was not necessarily established by subsequent investigation for the presence of clinically significant disease. Given that clear evidence of substantive benefit for a significant number of screened animals is lacking, we should be wary of assumptions that seeking some reassurance about the biological function of otherwise clinically healthy animals can justify the individual and cumulative harms instantiated by establishing and encouraging pre-anaesthetic screening programmes.

So how should we handle this dog booked in for desexing?

Based on the analyses above it would seem that the appropriate response to the clinical scenario presented is straightforward. The stress caused to the dog by physical restraint, and the potential for injury to the animal, veterinary staff, or both, preclude any attempt to obtain a blood sample without first applying some form of chemical restraint.

If possible the owner should be contacted so a new course of action can be agreed to.

Hopefully the idea of anaesthetising a dog to obtain a blood sample to test whether he or she is a suitable candidate for a general anaesthetic would seem problematic, to even the most guideline-driven practitioner. While it is possible to argue that the dog could be allowed to recover consciousness while the test was run, and then anaesthetised again for surgery, it is difficult to see how such a course of action aligns with the animal's interests. Utilising two induction processes magnifies the risk of complications. Alternatively a blood sample for testing could easily be obtained while the surgical procedure is being undertaken. However, given the history, if the clinician were so concerned about their patient it is arguable that providing benefit through instituting fluid support should be a higher priority.

But given this dog may be unlikely to tolerate a fluid line and could be expected to become quite agitated during its removal, taking any further interventionist steps is also highly questionable. Screening in the circumstances described is both clinically questionable and morally untenable if the test result is highly unlikely to change a clinical decision, let alone provide any benefit for the animal. Assuming that the owner still wishes to have their dog desexed, performing a screening test should not take primacy over nor in any way inhibit or delay the careful and considered care of this difficult patient through surgery and anaesthetic recovery. Especially given the limited scope for benefit from further intervention, screening this dog seems to be a self-serving and futile measure.

The social and economic context in which screening decisions are made also matters

Although we typically consider any effort towards disease prevention to be both clinically and ethically laudatory, it is important that the interests of the patient (and potentially their

5.3
COSTS AND BENEFITS OF TREATMENT
RESPONSE
CHRIS DEGELING

"As much as we might seek to always offer the best standard of care in veterinary practice, what veterinary patients receive is often determined by the value the owner ascribes to the animal, and the level of resources owners are willing and have available to allocate to their animal's care."

owners) are always given primacy in making screening recommendations and undertaking screening procedures. As much as we might seek to always offer the best standard of care in veterinary practice, what veterinary patients receive is often determined by the value the owner ascribes to the animal, and the level of resources owners are willing and have available to allocate to their animal's care. Even as screening can mitigate some risks for a small number of individuals, it potentially exposes many other animals and their owners to significant physical, psychological and financial harms, and potentially wastes resources that could be more usefully allocated elsewhere. More testing will pick up more abnormalities and lead to more interventions (and more veterinary income), but this does not easily or even necessarily translate into better health and welfare for veterinary patients.

While arguments are promulgated by industry stakeholders that veterinary professionals typically "under-service" their clientele leading to under-diagnosis, it behoves practitioners not to test indiscriminately but to choose to test wisely. If experiences in human medicine in general, and paediatrics in particular, are any guide, increasing the scale of screening asymptomatic and otherwise healthy animals is likely to promote circumstances where even the best-intentioned practitioner might unwittingly end up providing too much medicine for the good of their patients (Coon, et al. 2014). That said, the risks of under-diagnosis and over-diagnosis are both real, so screening should not be a default position, but the result of a decision actively made by the owner. At a minimum, veterinarians who offer screening must ensure that their clients are able to make decisions that are informed by knowledge of the possible benefits and also the costs and risks for them and their pets.

AS the subheading suggests, the author performs a cost:benefit or utilitarian analysis of this scenario. If indeed the purpose of the test is simply pre-anaesthetic screening, to determine if the patient is a suitable anaesthetic candidate or determine whether the procedure should be modified, the author has convincingly argued that the test cannot be justified after commencement of anaesthesia.

There is scope for debate here, as some would argue that the detection of severe illness in 1 in 100 animals is enough of a "good" to outweigh the costs to the other 99 animals.

Much of this analysis hinges on the reasons for pre-anaesthetic testing. If such tests are performed routinely, without much consideration, it is hard to see their value.

However, there may be multiple reasons for performing such tests and these should be explored. Cats and dogs are adept at masking disease and may have severe subclinical disease. On this basis screening for subclinical disease may be easier to justify than it is in human patients.

If it is impossible to draw blood from a patient until they are anaesthetised, it may still be useful to have the results early in the procedure. For example, if the packed cell volume is low a transfusion may be given and the plan to perform surgery aborted, but anaesthesia may facilitate other diagnostic modalities. If the patient is azotaemic, fluid therapy may be suitably adjusted.

Pre-anaesthetic blood tests may also be used to determine if any post-operative analgesics such as non-steroidal anti-inflammatory drugs are contraindicated, and thus modify an analgesic protocol. Pre-anaesthetic tests may also be used in animals with known pre-existing conditions to determine the patient's current status and potentially aid in prognostication.

The harms of restraint, fear and discomfort associated with a blood test may be mitigated by pre-medicating or sedating animals. In addition, the results of a test should be interpreted in the light of a thorough clinical examination and history. The interpretation of test results without this information, or based on a hastily conducted physical examination, may be compromised.

In such cases, based on the author's response, it may be justifiable to proceed with the test. One would expect the veterinary team to be clear with the owner about the purpose or purposes of the test.

COSTS AND BENEFITS OF TREATMENT

What do you think?

There has been a recent media storm over the use of hormonal treatments for infertility in dairy cattle. Infertility is a common problem in dairy cows but most causes do not appear to result in negative experiences, such as pain, for the affected cows. Detection of the problem usually involves per-rectal palpation of the reproductive tract and may also require a blood sample or other forms of imaging.

ONE — Who are the stakeholders in this issue?

TWO — What are the costs and benefits of hormonal treatments to improve fertility in dairy cows?

THREE — How can some of the costs be mitigated?

FOUR — How important would regulating hormonal use be to developing a solution acceptable to all stakeholders?

△
5.10 Fertility testing in a dairy cow.
PHOTO SIOBHAN MULLAN

5.4

Transplants and transfusions

In human medicine, suitable candidates may volunteer to donate blood, tissue or whole organs to aid family members, friends and even strangers. In such cases procedures are established to ensure that donors are not coerced and freely consent to the procedures.

Animals cannot consent to such procedures, which are generally invasive and may involve sedation, anaesthesia and major surgery with attendant risks of complications. Particularly in relation to whole-organ donation, the term "source" is used more frequently in the literature than "donor". This is because, unlike the donor in human medicine, the source animal cannot consent to nor refuse the procedure and therefore does not have autonomy which is valued in deontological and principalist approaches.

While organ transplantation is reasonably rare in veterinary practice, it is common in experimental settings. As transplantation techniques are refined, and anti-rejection drugs developed, transplantation may become a more accessible and viable treatment option. The following scenario explores the subject of renal transplantation in cats.

SCENARIO
ORGAN TRANSPLANT
– RENAL TRANSPLANT IN CATS

▶   You work in a referral practice. A new surgeon who joins the team has experience in performing renal transplants in cats, and is keen to offer this service. Under this proposal, source animals would be sourced from the pound and screened for infectious disease. The owner of the recipient cat must adopt the source cat.

The team is divided about this. Those in favour of offering the service argue that it will extend the life of cats that would otherwise die, and allow source animals who may not otherwise be adopted to be placed in homes. Those against claim that the welfare of source animals is impacted by an invasive procedure, to which they cannot consent, and which therefore is unethical.

How should you respond?

◁

5.11 Renal transplantation in cats currently involves harvesting a kidney from a source cat.
CARTOON
RAFAEL GALLARDO ARJONILLA

RESPONSE
DAVID MORTON

▶ First, a rather broad issue that applies across animal ethics. Why is the value of a cat's life so important that we try to go to such great lengths to save the life of the sick recipient cat as well as the life of the source animal? While it is ethically acceptable to many that animals die prematurely, e.g. for humans to eat, it is not true for all animals in all contexts. Is that simply a matter of taste, or is it that cats are intrinsically different in some material way, e.g. they have claws and fur, unlike most of the animals we eat? Or are cats put into another category, being morally different in some relevant way? Or is it all a matter of human culture and preference and does it have some ethical underpinning? Putting these issues aside, let's deal with the current scenario and the recipient cat first.

The cat requires some form of treatment for its condition if it is to regain a reasonable quality of life. While there are treatments available other than transplantation, they do not seem to offer the same long-term outcome. All treatments will require some follow-up medication to either control the failing kidney, or to prevent rejection after transplantation. Let us assume that they have an equal impact on the welfare of the animal in terms of pain, discomfort and distress. The transplant has the advantage that it may "cure" the cat, providing that there is no underlying condition that will persist and cause the transplanted organ to fail also. The transplanted cat seems therefore to have a distinct advantage, as complications notwithstanding, its quality of life may even return it to a relatively healthy "normal" state, as often happens with successful transplants in humans. We also have to assume that the surgeon is competent at their job and that the rejection control measures work long term, or at least for a reasonable length of time. But what would that time be? Would that

depend on the age of the cat at the time of surgery and would age and other co-morbidities be an exclusion criterion for the selection of a recipient?

Other factors that might play into consideration are the ability of the owners to continue treatment with lifelong immunosuppressive therapy (a common reason for failure of treatments is a failure to take or give the prescribed medicines), financial limitations, the personality of the cat to cope with treatment, the effectiveness of treatment (no treatment is 100 per cent guaranteed) and the degree of mismatch between the organ and recipient. Currently there is unlikely to be sufficient data to answer all these questions, so providing complete information in a comprehensible way when obtaining informed consent from the owners becomes a crucial issue. Moreover, training owners to look out for signs of early rejection, and signs of other disease due to long-term immunosuppression, becomes important so that early rescue therapy can be started. Potential complications of renal transplantation include post-transplant malignant neoplasia, with lymphoma reported in 22.5 per cent of renal transplant recipients (Durham, et al. 2014), retroperitoneal fibrosis (Wormser, et al. 2013), hyperacute, acute and chronic rejection, hypertension and neurologic signs including seizures (Pressler 2010). In addition, immunosuppression required to prevent organ rejection may be associated with recrudescence of latent infections such as toxoplasmosis.

Now let's consider the source animal which is healthy and does not require any treatment. It will have to undergo screening for health and tissue typing before extensive surgery to remove one of its kidneys. One would assume that this would happen at roughly the same time as the transplantation so that organ preservation was not an additional complicating factor. Thereafter the source animal will have only one kidney and if all goes well this does not seem to be a serious physiological problem, other than for any other aging animal in that environment. Other than immediate treatment for pain and infection, it will not require any ongoing treatment. As the source cat is not from the same household as the recipient it will be expected to settle down with new owners in a new environment etc., in addition to being with a "strange" cat. This could be evaluated before using that particular animal. The scenario assumes that the owners will take good care of the source animal, that is ongoing, and will continue to treat it as well as they do the recipient animal, or at least as their own.

An alternative approach would be to euthanase the source animal, thereby releasing two kidneys for transplantation, thus saving the lives of two recipient animals. That might involve staggering the two removals to await a second recipient. As the source animal comes from a pound, it is likely to be more of a financial transaction for the pound owners to help fund care of the other animals there. This might maximise the outcome and save many more lives than a single recipient cat.

Another approach could be to mimic some human situations and wait for a cat to be brought into the surgery that had, or was going, to be euthanased anyway, for example because of a road traffic accident, or behavioural problems such as aggressiveness.

A third option would be to breed cats specifically for this type of therapy with a suitable universal tissue type and in good health, in the same way as breeding cats for research.

Finally, what about the veterinary surgeons involved? Have they received adequate training, are they competent in the surgery and aftercare for what will be an uncommon operation? As it is likely to be very expensive, is it being seen as a way of earning "easy" money or as an intellectual challenge?

|||||||||||||||||||||||||||||||||||||||||||||||||||||||||||||||||||||

IN the majority of cases where renal transplantation is performed, it is defended on utilitarian grounds: the recipient (if treated successfully) enjoys greater quality of life, the source cat is adopted and removed from a potentially harmful shelter environment and the owner benefits from the improved health of their animal. One might add that the surgical and medical teams derive satisfaction from performing a challenging procedure and the practice benefits from income. One could argue that a successful renal transplant results in the greatest good for the greatest number of stakeholders. This is a typical utilitarian cost:benefit analysis.

This of course relies on an accurate prediction of outcome, which is something of a gamble between unpredictable deterioration due to renal disease versus short-term effects on wellbeing followed by the possibility of a longer life in better health, of unknown duration – subject to the absence of serious complications (Yeates 2014). A consequential justification for renal transplantation relies heavily on analysis of risks, which requires reviewing current evidence about potential complications. Some cats with chronic kidney disease can be managed medically for an extended period of time, a period which could exceed the lifespan of a cat receiving a renal transplant (Sparkes, et al. 2016).

An assessment of risks to the recipient alone may lead to the assessment that the procedure is ethically unsound on utilitarian grounds.

The use of a source cat introduces a second gamble, in that the selection of this cat is based on a gamble that it will have a better life than it might otherwise (for example, if it remains in the shelter) (Yeates 2014). This assumes that the adopter is scrupulous and ethical and will not neglect the source cat or have it euthanased.

The risk of complications may be reduced by placing strict selection criteria (for example an upper age limit, the absence of co-morbidities and so on) on potential recipients and behavioural screening of the source cat to ensure that adoption of this animal does not cause stress to the recipient or the source cat. In addition, detailed screening of owners to ensure compliance may be helpful in eliminating owners who are less dedicated, but cannot guarantee good treatment of the animals by the owner post-operatively. The harvesting of a kidney from a cat during non-recovery is an alternative which a utilitarian might justify on the grounds that the source cat then does not experience post-operative pain and this eliminates the risk of any stress caused by adopting an additional cat into the household.

However, this does not eliminate risk of complications. Currently it is impossible to ascertain that a recipient or donor is 100 per cent free of infectious disease, which may only become apparent when the recipient is immunosuppressed and can lead to life-threatening complications.

The use of purpose-bred animals as source cats may be justified on the grounds that this reduces the risk of introducing unknown infectious diseases.

The key ethical objections to transplantation are based on deontological or principalist approaches. The deontologist would object to renal transplantation on the grounds that it involves treating one stakeholder (the source cat) as a means to an end (treatment of the recipient). Similarly, breeding cats for the purposes of using them to source organs involves treating them as a means to an end and is not acceptable from a deontological viewpoint. It does not respect the integrity of these animals.

A principalist objection to renal transplantation in cats would be based on an argument that it violates the principle of non-maleficence, in that harm is inflicted on a healthy patient (the source animal) and potentially the recipient (due to the risk of life-threatening complications, being in some cases more immediately life-threatening than the disease being treated); also it does not respect the autonomy of the source animal and it sacrifices the health of the source animal for the benefit of the recipient, and is therefore unjust. According to James Yeates, the source animal is doubly disadvantaged by society because it has been first relinquished and second harvested to benefit a well-kept recipient (Yeates 2014).

In order to try to address these concerns, guidelines for feline renal transplant usually specify that equal consideration should be given to the interests of the source and recipient animals. This is one reason why the purpose-breeding of source animals is frowned upon: there is an assumption that the harm of invasive surgery and organ harvesting will be offset by adoption where the cat will be otherwise "better off", or in some cases by euthanasia if the animal cannot be rehomed.

In addition, existing policies such as that of the New Zealand Veterinary Association (New Zealand Veterinary Association 2010) and that of the Royal College of Veterinary Surgeons (currently suspended) stress the importance of informed consent of the owner, with particular reference to potential complications of both the transplant itself as well as adoption of the source cat into the household.

According to utilitarian, deontological and principalist analyses, it is hard to justify use of a source cat for renal transplantation in cats. In fact, a utilitarian argument may be made for euthanasing the cat with chronic kidney disease, and adopting the potential source cat.

Alternatively, if the use of a source cat could be eliminated, for example by cultivating renal tissue *in vitro* without requiring tissue from other animals, many ethical concerns are eliminated although the risks to the wellbeing of the recipient must still be considered and may be grounds alone for avoiding this procedure.

TRANSPLANTS AND TRANSFUSIONS

What do you think?

ONE——— **Does whole-organ transplantation differ from blood or tissue sourcing?**

TWO——— **What information would you seek prior to deciding whether or not to offer whole-organ transplantation?**

THREE——— **If you agree that transplantation is acceptable, would you place any conditions on this? Why?**

FOUR——— **If you do not agree that transplantation is acceptable, how will you address the welfare of cats with otherwise terminal renal disease and the concerns of their owners?**

Organ & Tissue Donation
Share your life...
1998
USA
32

△

5.12 Human organ donation, even when it occurs after death, is only permissible for volunteer donors in many countries, whereas live source cats cannot consent to give up their kidney.
PHOTO ISTOCK

The following scenario explores the ethical aspects of blood transfusions in animals.

SCENARIO
BLOOD TRANSFUSION

▶ You are working in a small general practice. A six-month-old German shepherd cross, Buddy, is presented following a motor vehicle accident. The dog, weighing around 20 kg, is in lateral recumbency with a temperature of 35°C and a respiratory rate of 80 breaths/minute. There is marked abdominal pain. You perform abdomino-centesis and confirm haemoabdomen. The dog's PCV is 18. An ultrasound examination shows extensive free fluid in the retroperitoneal space. You suspect an avulsed kidney leading to haemorrhage from the renal vessel.

Neurological examination reveals an absence of deep pain sensation in the hind limbs. In order to stabilise the dog you need to perform a blood transfusion. This may alleviate the acute anaemia, but the animal may have a poor prognosis.

What should you do?

RESPONSE
VANESSA ASHALL

▶ In this case a young dog is in a critical condition due to acute internal blood loss. Whilst a blood transfusion might benefit the animal in the short term and allow stabilisation for surgery (if required), the neurological exam indicates that there may also be a serious spinal injury with the potential for permanent paralysis. It is important to quickly decide whether the transfusion is to be given, but the severity of the spinal injury may not be known for some time. Therefore, the immediate question here is whether it is justifiable to

△

5.13 Following a serious road traffic accident blood transfusion may be considered.
PHOTO ISTOCK

administer a blood transfusion to an animal who you believe may not ultimately recover.

Identifying ethical concerns

The use of blood and blood products for animal transfusions creates unique ethical concerns. If a drug could be used whose cost and effectiveness were the same as blood itself then this scenario would be no different to many others concerning the short-term treatment of an animal with a poor prognosis. However, animal blood is a biological product which has been produced at some cost to another animal. A central ethical dilemma which is particular to transfusion (and transplant) decisions is the justification for causing harm to one animal in order to help another. I will focus on this aspect of the given scenario although other more commonly encountered issues such as cost will also be relevant to the final decision.

Identifying stakeholders

When we consider the stakeholders involved here the existence of a donor animal is an unusual

addition to the more commonly encountered animal patient, veterinary surgeon and owner. If banked blood products are used the blood bank will also be a stakeholder. I include the donor animal because I consider the removal of blood from a dog to be morally important; I am working on the assumption that dogs are capable of being harmed and that the removal of a quantity of blood from a dog could be considered potentially harmful. These assumptions are important because when making decisions about using animals for the benefit of others, such as in animal research, ethical arguments are sometimes made that certain species are not capable of being morally harmed or that the procedures proposed are not harmful.

Information and evidence required

The information which we need to make our decision is therefore how much harm may be caused in producing the blood or blood products and what benefits might result from the transfusion. The processes involved in collecting blood for transfusion to animals vary according to the species (Davidow 2013) and to the techniques and skill of the collection personnel. Whilst it is sometimes possible to collect blood from a relaxed and conscious donor, blood is also collected from donors who are sedated or anaesthetised with attendant risks; some protocols require the administration of intravenous fluid therapy. Bruising, bleeding or faintness could result from blood withdrawal and non-medical harms such as fear and distress are also relevant (Ashall 2009).

At present UK vets can choose to purchase banked canine blood or to find a donor for blood collection at their practice. Vets should be inquisitive about the welfare of animals who donate to blood banks; however, the potential distance which blood banking creates between animals who donate blood and those who receive it means that vets using banked blood products are reliant upon information provided by the blood banks about the welfare of their donors. Equally, vets who choose to collect blood in practice may sometimes find it difficult to argue against transfusion on donor welfare grounds such as when an owner is insistent that their other dog should be used as a donor. It should also be remembered that what is best for donor welfare could be at odds with other concerns such as the cost of obtaining blood or blood products and this problem would need to be honestly explored.

In other countries vets might need to consider the welfare of colonies of blood donors or the biological manipulation of animals to produce specialist blood products. If information about the harm which may be caused to a blood donor is not available or is not properly considered this could result in a factually incorrect and therefore invalid ethical justification.

We should also consider the likely benefits of transfusion. Evidence must be sought as to the likely outcome of the proposed treatment, be it through personal experience, reliable advice or literature. A recent retrospective study suggests that "surgical intervention for treatment of hemoperitoneum, regardless of etiology, resulted in discharge from the hospital for 70 of the 83 (84%) dogs" (Lux, et al. 2013). However, we also know that the animal may not ultimately survive due to its spinal injury and if this should be the case, are there any benefits to giving the transfusion? Without the transfusion the animal is unlikely to survive at all so that the extent of its spinal injury may never be known. The transfusion might be viewed as "giving the animal a chance" by extending its life for further testing which may be assumed to be a benefit provided the dog will not be suffering. There may also be non-medical benefits to this transfusion; the owner will get to spend a little more time with the dog and will be satisfied that they had done everything they could if euthanasia is ultimately suggested.

These benefits to the owner are also relevant to the ethical justification.

Making a considered judgement

As we have identified the importance of considering harm and benefit in this scenario, we might think of appealing to utilitarian theory to help make our decision; however, this approach has potential limitations. For example, it could lead to the justification of sacrificing or "farming" a few individuals to provide a supply of blood for many others of the same species (Harris 1975).

Deontological or rights theory does not support this argument for humans whom we recognise as having moral and legal rights to make decisions concerning their own bodies. The principle of autonomy is therefore vital in justifying human blood donations since it is the donor's informed consent which justifies the infliction of harm by a medical professional in order to collect blood for the benefit of another.

Animals are not legally recognised as having equivalent rights and furthermore dogs cannot consent to the donation procedure. The principle of autonomy in our scenario is therefore difficult to apply to animal donors; however, even in humans arguments are sometimes made for non-autonomous donations.

I find the four principles are a useful framework for considering the ethical justification of animal transfusions. My consideration of harms and benefits above is broadly equivalent to the two principles of non-maleficence and beneficence. Autonomy as it relates to the human stakeholders is reliant upon a proper understanding of the relevant information. This translates to the need for each stakeholder to be careful in communicating all pertinent information, such as the source of the blood and implications for the donor, alongside a realistic assessment of the likely outcome of transfusion to all decision-making parties.

And finally this framework highlights the principle of "justice" or fairness which can help to moderate some of the unappealing outcomes of a pure utilitarian approach. Ultimately, calling for donor and recipient to be treated as moral equals seems fair when balancing benefits and harms in non-autonomous individuals of the same species. Practically we should consider how fairly the collection and distribution of blood is organised; how are blood donors sourced, are there any benefits to them and might they be in a position to benefit from a blood transfusion one day? Considering the principle of justice allows us to consider the fair distribution of benefits and raise concerns over the use of animals as blood donors who do not have an equivalent quality of life to the blood recipients, or who will never genuinely benefit directly or indirectly from their involvement.

Alongside the four principles of biomedical ethics Beauchamp and Childress raise the importance of certain moral virtues (Beauchamp & Childress 2001). When justifying animal blood transfusions using this framework it is important to recognise the value of two key virtues in particular: trust and conscience. The contemporary process of producing and using animal blood products can involve many stakeholders who each need to take responsibility for considering the ethical implications of how they manage their own role. They must also be able to trust that other stakeholders are doing the same.

In conclusion I believe that an argument could be made for transfusing the dog in this case using a principled approach. Given that the benefits to the dog are likely to be short lived, the fully informed owner should know that there are considerable benefits to them in the transfusion being given. The source of the blood for the transfusion will also be a deciding factor here because only blood produced with very little harm to a donor dog who is equally respected could be justifiably used.

UNLIKE whole organs, blood can be regenerated in a donor and thus long-term harm from such a procedure is less likely; however, there are still associated risks. Increasingly, evidence-based guidelines are available for clinicians considering transfusion which may be helpful in decision-making. For example, some guidelines state that blood transfusion should be avoided in terminal animals (Pennisi, et al. 2015). The difficulty may lie in determining whether a patient, for example a cat with anaemia due to feline leukaemia virus, is terminal or not.

TRANSPLANTS AND TRANSFUSIONS

What do you think?

Develop a policy for the administration of transfusions and consider the following:

ONE — **From where/whom would you source the blood? Do you have an order of preference?**

TWO — **What would be your criteria for eligibility to receive a transfusion?**

THREE — **What would be your criteria for excluding some animals from receiving a transfusion?**

△

5.14 When is it appropriate to perform a blood transfusion on veterinary patients?
PHOTO ANNE FAWCETT

5.5

Actual and potential patients

In the previous scenarios we considered forms of treatment such as transplant and transfusion which impact multiple animals – both the source and recipient. But other treatments in veterinary practice impact multiple animals, including reproductive technologies such as cloning (see chapter 15, on changing and cloning animals). In the following scenario we explore issues around addressing the interests of actual and potential animals.

SCENARIO
SPAYING A PREGNANT CAT

▷   Mrs B, a farmer, brings in a heavily pregnant feral cat to be spayed. She says they are overrun with cats and are lucky to have caught this one. Tamara, the vet nurse on surgical duties that day, has a responsibility to admit the cat for surgery, and assisting with the operation. She is not happy about spaying such a heavily pregnant animal, and the consequent death of the kittens. The other nursing colleagues feel the same way and offer to hand-rear the kittens if they are delivered alive.

What should you do?

RESPONSE
PETER SANDØE AND SANDRA CORR

▷   Whatever choice is made, the veterinarian will be under a legal obligation (in most Western countries) to get consent from Mrs B, since legally the cat belongs to her.

Since the cat is heavily pregnant, the surgery could be considered more risky, and therefore the vet may recommend waiting to spay the cat until she has delivered the kittens. However, this could prompt Mrs B to decide to have the pregnant cat put down – a solution which may be equally distressing for all concerned.

It is not clear what the vet nurses object to most: that the cat is not being allowed to have the kittens (and be spayed later), or that the spay will go ahead without an attempt to save the kittens.

One option is for Mrs B to sign over the cat to the practice, and for a vet nurse to adopt the cat, allowing the cat to give birth and rear her kittens. The cat can be spayed after delivering the kittens, and either returned to Mrs B or rehomed after the kittens have been weaned.

These options raise difficult questions, such as the following:

(1)   The cat is feral, and may not take well to being confined – what if she disappears to give birth?

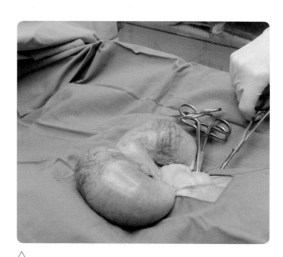

△
5.15 Spaying pregnant animals terminates the pregnancy. This cat is being spayed late-term.
PHOTO ANNE FAWCETT

(2)   The cat may have a difficult birth, causing her to suffer or even die (during, or later, as a result of complications). The cat may require an expensive caesarean – who pays for that?

(3)   If the kittens are hand-reared by the nurses, they will be deprived in terms of learning and normal social functioning from not having had a normal cat mother during the first weeks of their lives – who takes responsibility for that?

(4)   Volunteering to hand-rear kittens may seem like a good idea at the time, but this is very hard work, and subsequently the involved people may not bother, may find it too hard, or may not do it properly with bad consequences for the kittens – who will ultimately have to bear the responsibility?

(5)   The vet nurses may not be able to find good homes for the cat or kittens – and it may be necessary, ultimately, to have them put down – is that acceptable, and who takes responsibility?

Alternatively, the veterinarian may decide to do exactly what she or he has been asked to do by Mrs B. This decision by the veterinarian may be based either on convenience, or on a concern about poor outcomes for the mother and/or kittens – why risk problems and trouble if there is an easier solution? This approach may be unwise, however, as it may alienate the vet nurses, with detrimental effects on staff morale and working relations at the veterinary clinic. If the decision is made to spay the cat, then the veterinarian should try to change the nurses' minds about letting the kittens survive based on an ethical argument.

This ethical argument may take as its starting point the premise that there are already too many surplus kittens for which it is difficult to find homes. On this basis, it can be argued that instead of bringing more kittens into the world, the vet nurses could direct their good efforts towards helping find good homes for some of the existing surplus cats. Good homes are difficult to find, particularly for and kittens adult cats, and the argument could be made that rehoming some of the many abandoned cats that already exist may ultimately increase cat welfare more, and save more cat lives, than bringing more kittens into the world to compete for the good homes. (And of course there is no guarantee that the kittens not yet born will end up in good homes, or have good lives).

The suggested argument is very much in line with utilitarianism or other forms of consequentialism, which consider the consequences of an action, rather than the action itself. This is a line of ethical thinking that may be held by many, but certainly not all, of those who care about animal welfare. Some will adhere to a kind of animal rights or virtue based view that disagrees with the "emotionally detached" approach of consequentialism. They may argue that the unborn kittens have a moral right not to be killed that cannot be overruled. Of course veterinarians and vet nurses cannot save all kittens, but they may – according to a rights view – still have an obligation to save those for which they have an immediate concern, such as the unborn kittens of the cat brought in by Mrs B.

So our suggested approach is based on encouraging ethical discussion between the veterinarian and the vet nurses. While this may not lead to full agreement, it enables each stakeholder's view to be expressed and acknowledged, facilitating mutual respect, even if the outcome is an agreement to disagree.

Overall, if one accepts the premise that there are already more abandoned cats and kittens in existence than can be found good homes for, and that a complication-free birth for the mother and the subsequent thriving of healthy kittens who find

CHAPTER 5 VETERINARY TREATMENT

5.5
ACTUAL AND POTENTIAL PATIENTS
RESPONSE
PETER SANDØE AND SANDRA CORR

good homes cannot be guaranteed, then the logical course of action is for the vet to spay the cat (simultaneously terminating the pregnancy), in line with the request of Mrs B.

We are aware that this recommendation relies on controversial consequentialist premises – and therefore also recommend that the opposing view be recognised as part of a respectful ethical discussion.

||||||||||||||||||||||||||||||||||||||||||||||||||||||||||||||||||||||||||

THIS scenario requires weighting the interests of an actual animal (the mother), as well as surplus cats already in existence, as well as potential animals (foetuses). We are taking the position that a foetus is a potential animal, although this is contentious. The authors perform a cost:benefit analysis clearly weighted in favour of the interests of actual animals.

There is some overlap here with our discussion on conscientious objection in section 10.1 in the chapter on education and training.

ACTUAL AND POTENTIAL PATIENTS

What would you do?

Twin pregnancies in horses commonly lead to early gestational death, late-term abortion or delivery of small foals with retarded development and high susceptibility to infection. Late-term abortion and birth of twins are associated with dystocia, trauma to the reproductive tract and subsequent poor fertility. Of twins that do survive to term, many die in the neonatal period. As a result, twin pregnancies are often reduced by reducing the pregnancy to a single foetus (Tan & Krekeler 2014). You diagnose a twin pregnancy in a mare owned by a hobby farmer. You believe it is approximately 15 days post-ovulation, with most twin pregnancies manually reduced between days 16 and 17. When you offer to reduce the pregnancy to save one foal, the client is horrified. "Absolutely not," she says. What do you do?

△
5.16 Most twin pregnancies of horses result in problems but occasionally twins can be successfully born and raised.
PHOTO ISTOCK

(180)

5.6

Inherited disease

The risk of inherited disease in some animals may be known or unknown by breeders, with some breeding practices more likely to increase the risk of inherited disease than others.

Management of inherited diseases involves a number of ethical issues, including genetic counselling, selection of breeding animals, removal of animals from the breeding programme and whether or not the veterinarian should actively influence breed standards.

Veterinarians are increasingly called upon to prevent rather than simply treat inherited diseases, as explored in the following scenario.

△

5.17 Irish setters can be affected by Haemophilia A, a genetic disorder.
PHOTO ISTOCK

SCENARIO

MANAGEMENT OF INHERITED CONDITIONS

▷   You have known Mr and Mrs D for years. They've been great clients and really cared for their dogs in a way that you've admired – you're not sure you would have been able to make the sacrifices they did to ensure Tilly, their black Labrador, had her diabetes well controlled.

Following Mr D's redundancy they have decided to turn their passion into a small income through showing, breeding and selling Irish setter puppies. Having chosen a male and two females their first litter arrives uneventfully. However, after a while some of the puppies don't seem quite right with small swellings and mild lameness. You recommend testing for Haemophilia A, an X-linked genetic disorder known in Irish setters. To your disappointment the sire of the litter is diagnosed in the test as affected and the dam is a carrier for a mutation known to cause the disease.

How should you counsel Mr and Mrs D?

RESPONSE

IMKE TAMMEN

▷   The breeding of animals requires breeders to have a sound understanding of animal genetics and to apply this knowledge ethically to have "good" outcomes for individual animals and the population or breed as a whole. Increasingly, the general public is concerned about the impact of animal breeding decisions on animal welfare and health. Veterinarians are often asked to provide advice on breeding matters relating to animal health and welfare, especially in the context of treatment and management of inherited diseases.

Haemophilia A is an inherited coagulation disorder caused by deficiency in Factor VIII, which has been identified in many dog breeds (OMIA: 000437-9615 (n.d.)), as well as in many other species (OMIA: 000437 (n.d.), OMIM: 306700 (n.d.)). The disease was first described in Irish setters in 1946 (Graham, et al. 1949) and at least one mutation has been identified to cause the disease in this breed (Lozier, et al. 2002).

5.6
INHERITED DISEASE
RESPONSE
IMKE TAMMEN

In the following, specific issues relating to this scenario are discussed and some general principles that relate to ethical management of inherited diseases in animals will be highlighted. In relation to inherited diseases in animals, affected animals are often used as models for the corresponding human disease and dogs with Haemophilia A have been used in this context (Lozier & Nichols 2013). However, this issue is not investigated in this scenario.

Key issues

Regulations and recommendations relating to the management of inherited diseases exist, but vary between different countries/states, are not consistent between breeds or species and are undergoing constant change. Using the framework of an ethical matrix/principalist approach (Beauchamp & Childress 2013, Mepham 1996), issues impacting on wellbeing (beneficence and non-maleficence), choice (autonomy) and fairness (justice) in relation to key stakeholders (existing animals, Mr and Mrs D, Irish Setter Breed society, the Irish setter population and the general public) should be explored jointly by the veterinarian and the owners of the animals. In these discussions, the veterinarian should provide information about clinical signs, preventive measures, treatment options (and their costs) and long-term prognosis as well as providing detailed information about the genetics of the disease. This should include information on the predicted outcomes of different matings, especially as X-linked diseases are not very common. The veterinarian should strongly recommend that the Ds contact their breed organisation to report the occurrence of this inherited disease. Considerations relating to the management of inherited diseases should be made within the context that all animals (and humans) are expected to be carriers of multiple deleterious recessive mutations (Nicholas 2010).

Immediate treatment and diagnostic decisions

As the male dog is affected and several of the puppies appear to be affected, the disease in these dogs needs to be treated, to minimise harm. Recent studies have shown that the clinical course of the disease can vary from life threatening to very mild (Aslanian, et al. 2014, Barr & McMichael 2012). Progress in treatment of inherited coagulation disorders together with preventive measures have resulted in good long-term prognosis for some affected animals (Aslanian, et al. 2014, Barr & McMichael 2012). Considering the Ds' previous commitment to controlling a chronic disease in their pet dog and the fact that their male Irish setter has reached breeding age with no apparent clinical signs of Haemophilia A, treatment of the affected puppies has a reasonable prognosis. However, depending on various issues – including severity of clinical signs and response to treatment – euthanasia of affected animals might be considered.

The second female and all puppies should be tested to have accurate information on the disease status (disease genotype), preferably not only with a coagulation assay but with a direct DNA test that accurately identifies carrier animals (Barr & McMichael 2012). As with all direct DNA tests, which test only for known mutations, the possibility that more than one mutation can cause the same clinical signs (i.e. genetic heterogeneity) needs to be considered in the interpretation of the test results.

Future breeding and sale of animals

Based on the results of the diagnostic tests, Mr and Mrs D will be able to make more informed choices about daily management, selling and breeding of their animals. According to the guiding principles it is important to not create more affected dogs (non-maleficence) and to fully inform buyers of the genetic status of the animals (autonomy and justice).

In the following plan, X_h and X_H correspond to the disease and normal allele on the X chromosome, respectively, and Y corresponds to the Y chromosome, which does not contain the gene.

Genotypically normal animals (i.e. they are not affected by the disease and do not carry a disease-causing mutation: $X_H X_H$ or $X_H Y$) can be sold or bred without restrictions.

Heterozygous female animals ($X_H X_h$) are clinically normal but breeding with these animals should be avoided as there is a high risk that affected puppies are born in every mating. Nevertheless, these dogs would be great pet animals.

The use of a Punnett square (Diagram 1) assists in predicting in more detail the outcomes of such matings.

(A)	X_H	Y
X_h	$X_H X_h$	$X_h Y$
X_H	$X_H X_H$	$X_H Y$

(B)	X_h	Y
X_h	$X_h X_h$	$X_h Y$
X_H	$X_H X_h$	$X_H Y$

△

DIAGRAM 1 Punnett squares for the mating of a carrier female ($X_H X_h$) with a normal male ($X_H Y$) (A) and a carrier female ($X_h X_H$) with an affected male ($X_h Y$) (B) for recessive X-linked Haemophilia A. In a mating between a female carrier and a normal male, on average, a quarter of the puppies will be male and affected, a quarter will be male and normal, a quarter will be female carriers and the final quarter will be normal and female. In a mating of a female carrier and an affected male, on average a quarter of the puppies will be male and affected, a quarter will be male and normal, a quarter will be female carriers and the final quarter will be affected and female.

Buyers should be advised of the carrier status and it might be advisable to spay the dogs to ensure that they are not accidentally bred in the future.

Homozygous affected females ($X_h X_h$) or affected males ($X_h Y$) are very likely to develop clinical signs and are at risk of excessive haemorrhage especially in the case of trauma or surgery. These animals would require special management and ongoing treatment. Any sale of such animals, even after disclosure of the disease status, appears problematic and is very likely to damage the seller's reputation. A simple Punnett square analysis can again identify predicted outcomes of various matings: interestingly, the mating of an affected male dog with a normal bitch results in 100 per cent clinical normal offspring. All male puppies will be normal ($X_H Y$). However, all of the females will be carriers ($X_H X_h$). The results from a mating between a carrier female ($X_H X_h$) and an affected male ($X_h Y$) were already discussed above and are very likely to result in the birth of affected puppies. In a mating of an affected female ($X_h X_h$) with a normal male ($X_H Y$), all female puppies will be carriers ($X_H X_h$) and all male puppies will be affected ($X_h Y$), and in a mating between an affected male ($X_h Y$) and an affected female ($X_h X_h$) all offspring will be affected.

Often cross-breeding is recommended as a safe measure to breed from carrier animals, and Irish setters are crossed to create "designer dogs" such as Golden Irish and Irish Doodles. However, as this is an X-linked recessive disease, breeding of a carrier female with a dog from a different breed is still expected to result in affected male puppies being born (see Diagram 1a).

This would mean for the Ds that they should not breed again from the bitch that has already given birth to affected puppies and that they should consider breeding their second bitch only if she tests homozygous normal for the disease. The use of the affected male for breeding with a homozygous normal female (of any breed) would result in puppies that are not affected by the disease. If measures are taken to ensure that all female offspring of such a mating are not used for future breeding and their genetic status is disclosed in any sale, than this could be ethically permissible. The Ds would need to check with their breed society if showing of dogs that tested positive for Haemophilia A is possible. Unfortunately, their plans to turn their passion into a small income through showing, breeding and selling of Irish setter puppies have got off to a bad start. However, if their second bitch is tested homozygous normal, informed breeding decisions can allow them to still achieve their goals.

Reporting to the breed association

Inherited diseases are familial, i.e. they not only affect individual animals but also to various degrees related animals. By reporting to the breed society (and to the breeder that the foundation dogs have been purchased from) that Haemophilia A has been diagnosed in these dogs, management of the disease in related dogs and the wider Irish setter population becomes possible, and the birth of further affected puppies can be prevented if appropriate measures are implemented. The veterinarian should advise the Ds to contact the breed society, or if they decline to do so (e.g. for fear of discrimination, loss of reputation), respect the autonomy of the Ds but in view of the wellbeing of the Irish setter population, ask for permission to contact the association and report the occurrence of Haemophilia A in Irish setter dogs without providing client details.

The breed society has various options on how to manage the disease on a population level. Further research might be needed to identify the frequency of the disease allele in the population and to assess the effective population size as this would impact on a management plan. As DNA testing is a possibility, a recommendation of DNA testing of related animals, the inclusion of DNA testing results in the studbook, and an education programme about this inherited disease among their breeders should be immediate measures.

Conclusion

Historically, inherited diseases in livestock and companion animals have been under-reported. We have today effective tools to investigate emerging inherited diseases, to develop DNA tests and to implement effective strategies to manage these diseases on a population level. Therefore, proactive and transparent approaches that minimise negative effects for individual animals and the population are required, to reassure animal owners as well as the general public that animal breeding has a strong focus on animal welfare.

Effective and ethical management of inherited diseases often requires case-specific solutions. Severity of the disease, mode of inheritance, frequency of the disease allele in the population, accuracy of available tests, population size and structure, as well as the underlying views on the moral status of animals and/or their position in the *sociozoological* scale (Arluke & Sanders 1996) are some of the factors that need to be considered. Inherited diseases with known modes of inheritance and/or a known disease-causing mutation (and therefore the possibility for DNA testing), as described in this scenario, are relatively easy to manage, and the genetic testing should be implemented to enhance, not limit breeding choices (Bell 2011). However, to implement effective solutions, cooperation and change are required, not only by individual breeders but by whole breed societies.

POOR breeding choices are associated with welfare problems not only in companion animals but production animals also. For example, an overemphasis on selection for milk yield in the first lactation has been associated with negative impacts on fertility and traits linked to sustained fitness (Webster 2013). It will take at least five years before the effects of modern selection indices with emphasis on robustness and improved lifetime performance become apparent.

In this and the above scenario, it is evident that excess emphasis on traits that we consider desirable has a welfare cost, the burden of which is borne by the animal and potential owners.

INHERITED DISEASE

What do you think?

"One easy way to prevent a heritable condition in a given breed is to ban any more breeding from these animals."

ONE   **How would you describe the ethical value of having "breeds"?**

TWO   **What arguments are there for and against this statement?**

THREE   **What is your view on this proposal?**

△
5.18 What is the impact of having distinct breeds on people and animals?
PHOTO ANNE FAWCETT

5.7

Alternative treatment modalities

The term "conventional treatment" refers to treatment that is generally accepted and used widely by healthcare professionals. Alternative and complementary treatment modalities are not as widely accepted or used. Nonetheless, clients may demand such treatments, which can raise ethical problems as discussed in the following scenario.

SCENARIO
HOMEOPATHY AND PROFESSIONAL CONDUCT

▸ Your clients, Mr and Mrs M, arrive in great distress with Pompidou, their 10-year-old border collie, who is walking with difficulty around in circles and has had his head on one side for the past 4 days. He has been reluctant to go for walks and has vomited four times. On examination, the ataxia, strabismus and nystagmus lead you to diagnose vestibular syndrome, with no reason to suspect this is anything other than the idiopathic peripheral form with a good prognosis.

You reassure the Ms and advise them on supportive care for Pompidou; the syndrome will likely resolve itself without treatment. You offer to prescribe some anti-sickness medication, maropitant, for Pompidou to help him feel less nauseous. The Ms instantly get very angry at you for "trying to pollute Pompidou with pharmaceutical chemicals". They insist upon a homeopathic remedy for the vomiting as they have already been treating him with homeopathy at home. Concerned about your suggestion of pharmaceutical treatment, they immediately leave the clinic.

A few days later you call the Ms to see how Pompidou is doing. They had taken Pompidou straight from your surgery to a homeopathic vet, who had initially hospitalised Pompidou and given him homeopathic treatments and acupuncture. Now Pompidou has made a full recovery, "no thanks to you", the Ms allege that "you did nothing to help Pompidou", that you should have offered homeopathy and they are considering reporting you to your professional regulator.

How should you respond?

RESPONSE
MARTIN WHITING

▶ Professional communication is paramount in cases like this. It is very easy to immediately resort to discrediting the homeopathic veterinarian to the Ms in order to regain your standing. The Ms have challenged your competency and decision-making process and it is common to take that personally; the threat of being reported to a

△

5.19 Older dogs may suffer from vestibular syndrome, causing their owners concern, and some to seek "alternative" treatments.

PHOTO ISTOCK

professional regulator is also incredibly stressful. Such threats should always be taken seriously, even if they have no merit, and so ensuring your records and notes are meticulous is important. Of course, any veterinarian should keep contemporaneous and complete clinical and client records at all times.

First, consider if there are grounds for a professional complaint. Such threats should give us pause to reflect upon our actions to find where there could be room for improvement. The veterinarian approached the case appropriately and after clinical examination determined that Pompidou did not require further treatment or diagnostic investigation other than supportive care. The treatment for nausea is an appropriate course of action for the wellbeing of Pompidou, where the feeling of nausea in vestibular syndrome can be substantial. Unfortunately the veterinarian was not given the opportunity to discuss further treatment options before the Ms left the practice. However, by following up the case a few days later the veterinarian has demonstrated due diligence in the care of Pompidou. It is highly improbable that a professional complaint would be upheld.

From the perspective of the client, the second veterinarian took a much more aggressive approach to Pompidou's care. Immediate hospitalisation and instigation of therapy conflict with the assessment of the primary veterinarian. It is perfectly possible that both courses of action were appropriate, if Pompidou was on the boundary of clinical severity and the second veterinarian thought it more appropriate to admit for treatment. The client may not fully understand this difference of professional judgement and to them it could appear that the primary veterinarian was negligent in their treatment while the second veterinarian did "all they could". It is important not to over-interpret the anger of the client and assume that this has come from the second veterinarian. There is not enough information to comment on the professional conduct of the second veterinarian in terms of their communication. The main potential issue relating to the professional conduct of the second veterinarian is that they took over care of Pompidou without making contact with the primary veterinarian. This could be hugely detrimental to Pompidou as the second veterinarian will not know, for certain, if Pompidou has already received medication or had diagnostic tests performed, and will not have discussed their concerns with the primary veterinarian. Although, again it is important not to jump to conclusions; the Ms may not have informed them about the initial visit.

In the case of the Ms, it is very unlikely that a conversation about your decision to use evidence-based medicine over homeopathy will be fruitful, the clients are too angry. However, there are points that arise from this case which are worthy of thought to prepare us for future cases.

First, the Ms stated they instigated homeopathic treatment on their own before they brought Pompidou in. It is worth checking in your local jurisdiction the rules that relate to owners "self-medicating" their pets. In countries like the UK, it is not lawful for non-veterinarians to diagnose or treat animals; this is covered by the Veterinary Surgeons Act 1966. Some types of homeopathy may be exempt from this but it is worth familiarising yourself with your local laws regarding non-veterinarians treating animals.

The key topic that needs to be considered, however, is the use of homeopathy in an era of evidence-based veterinary medicine and if such practice impacts on the professional standing of a scientific discipline. The Australian government recently conducted a large-scale call for evidence on the use of homeopathy in humans and concluded that there is insufficient evidence to demonstrate homeopathy as an effective treatment for the reported clinical conditions (Optum 2013). One veterinary systematic review was unable to conclude the efficacy of homeopathy in veterinary

use (Mathie & Clousen 2014). MEPs in Europe have assigned 2 million Euros to investigate homeopathy in farm animals (Anon 2011). While in the UK, the Veterinary Medicines Directorate issued a warning that they were clamping down on veterinary homeopathic remedies to ensure that they only claim to be medicinal if they can produce the safety and efficacy data for their use (VMD 2010). The BVA added to the concern by highlighting the dangers of the public self-treating animals with homeopathy: "unauthorised products may at the least be ineffective and at worst could cause harm because serious life-threatening diseases may go undiagnosed" (VMD 2010). This statement gets to the core of the problem with homeopathy; there are too many unknowns and no reliable evidence of their benefit. Animal welfare is the veterinarian's primary concern and when an animal's welfare is compromised, to select a remedy that has no proven efficacy (and no data on its contraindications) and no scientific proof of its mode of action seems to run contrary to the professional duty to do no harm.

Informed consent is another key element with homeopathy. To gain informed consent for a course of treatment, the risks, benefits and potential harms must be explained to the client, the client must understand them and then agree on the course of action that is most agreeable to them and their animal. This information usually comes from drug labels and large-scale studies of impacts of interventions. As stated above, these do not exist, or are inconclusive, for homeopathy. Informed consent also requires veterinarians to explain all appropriate treatments that are available, and their associated risks and benefits. If homeopathy is included in this, then it will require the veterinarian and the client to opt for a treatment that has no demonstrable effect and unknown risks and harms, over a medication that has known values. This becomes hard to justify both professionally and logically.

—
5.7

ALTERNATIVE TREATMENT MODALITIES
RESPONSE
MARTIN WHITING

Homeopathy could represent a conflict between clients' wishes and one's professional obligation to animal welfare. While clients may demand certain types of treatment for their animals, the professional veterinarian is not under any obligation to provide that treatment if they deem it unsuitable for that patient. Professional autonomy is not compromised by client demand. The laws regarding the treatment of people and the treatment of animals differ greatly, and this may be unknown to the general public. While the public may be able to treat themselves and their children with homeopathy, they may, in the same jurisdictions, be limited in what they may administer to an animal without a veterinary prescription.

The UK established the RCVS by Royal Charter so as "to improve the veterinary art which had been theretofore practised generally by ignorant and incompetent persons, which had been long and universally complained of" (RCVS 1844). The objective was to provide academic rigour and robust professional principles to what became the art and science of veterinary medicine. The veterinary profession is mixed in its views on homeopathy, but governing bodies and universities adhere to, and teach, the principles of evidence-based veterinary medicine so as to be sure that the animals under the veterinarian's care receive the efficacious treatments they need. It is hard to defend the choice of a treatment that has no evidence base or scientific validity to its alleged mode of action. While it is possible to proceed with medications where the mode of action is unknown, this should only be the case when the benefits, harms and contraindications are well documented.

Almost all veterinary professional organisations have issued statements of a disapproving nature with regards to homeopathy's use in an era of evidence-based veterinary medicine. There appears to be no evidence to support its use, other than clients' demand. Yet, national and international veterinary regulatory bodies do not have a unified approach to the control of products that may have a substantial impact on the welfare of animals. In Sweden, veterinarians are prohibited from prescribing homeopathy, but the public are not. In the UK, the public are prohibited from prescribing homeopathy but the veterinary profession are not. Horses racing in Finland must not have homeopathy up to 96 hours before a race due to fears of unfair influence. It seems that the jury is undecided on how the complex issue of homeopathy regulation needs to be tackled. Either it has the potential to cause an effect (positively or negatively) and therefore it needs to be pulled into the domain of veterinary therapy and be controlled and regulated (or prohibited). Or it offers no effect at all on animals, and thus needs to be prohibited on the basis of a fraudulent trade that wastes clients' money that could otherwise be used to help animal welfare.

To continue to practise non-evidence-based veterinary medicine under the guise of a profession undermines the professional and scientific standing of veterinarians across the world.

||||||||||||||||||||||||||||||||||||||||||||||||||||||||||||||||||||||||||||||

EDUCATIONAL institutions such as universities and training colleges tend to equip future members of the veterinary team with a sound knowledge base in conventional treatment (even if the evidence base in some cases may be scant) but do not generally provide training in alternative and complementary modalities.

As the author points out, it isn't the fact that the particular treatment modality is alternative that is the issue – rather, it's the lack of evidence and scientific justification in a science-based profession that is problematic. Whilst such a treatment may seem to "do no harm", an ineffectual treatment may allow the underlying disease to progress, ultimately harming the patient.

ALTERNATIVE TREATMENT MODALITIES

What would you do?

Mr H, an organic farmer, has a flock of 70 sheep. The lambs are now four to eight weeks old and many do not seem to be thriving and some have poor coats and diarrhoea. You confirm a diagnosis of coccidiosis and consider treatment options. Mr H is very committed to organic principles and wants to try oregano oil treatment as he has heard it might help. Organic regulations require you to make a case to get a derogation to treat with conventional coccidiostats. How would you respond?

△

5.20 Treatment of organic sheep may require additional considerations of owner views and compliance with regulations.
PHOTO ISTOCK

For another perspective on homeopathy, readers are directed to the scenario in section 10.5 in the education and training chapter.

5.8

Quality of life

The goal of much veterinary treatment is to maintain or improve quality of life. "Quality of life" (QoL) encompasses all experiences that animals have and can inform a value judgement such as whether an animal's life is worth living, or avoiding. The assessment of quality of life can be challenging. There are tools such as the five freedoms, the five domains and systems for estimating quality of life over time (Wolfensohn & Honess 2007), but none allows us direct insight into the experience of the animal whose quality of life we are assessing. Furthermore, practical application is challenging as currently there is no consensus or agreed minimum QoL below which the decision should be made to euthanase an animal.

Thus there is scope for disagreement about assessment of QoL, in terms of both overall assessment as well as methodologies used to assess QoL. The following scenario explores the use of QoL assessment tools in companion animal practice.

SCENARIO
QUALITY OF LIFE ASSESSMENT IN PRACTICE

▶   Your busy mixed practice has got so big that, to your great relief, you are now able to employ a Clinical Director who is responsible for maintaining and improving clinical care of animals and clients. They suggest that a formal QoL assessment would be useful to use within the practice, both to highlight animals with poor welfare, or at risk of poor welfare, and to monitor progress of clinical patients and assess patient outcomes. There is some resistance amongst your staff who say they do this every day as part of their job already, without a formal framework or recording system.

How should you proceed?

RESPONSE
PATRICIA V TURNER

▶   This is an interesting scenario that deals with a tool (a formal QoL assessment) that may be used potentially to enhance the welfare of clients' pets. The scenario also touches upon a common issue in veterinary practice management – that of dealing with a change. The latter issue will be dealt with in the context of establishing a welfare-friendly practice.

Making the changes necessary to become a more welfare-focused practice goes far beyond

△

5.21 What factors should we take into account when assessing the quality of life of companion animals?

PHOTO ANNE FAWCETT

stocking packages of hay in the front office for rabbits or having separate entrances for canine and feline patients. It is a way of practising veterinary medicine that must be embraced by all hospital personnel – from partners to veterinary associates to veterinary nurses to managers to front office personnel. It will be a significant change for the practice to incorporate routine QoL assessments into client visits; however, the fundamental goal of every veterinary practitioner should be to enhance the QoL of their patients, so the concept should not be difficult to "sell" to others in the practice. This represents an opportunity for clinic owners and personnel to discuss their current methods of conducting client visits and communications to ensure consistency as well as evaluating potential methods, such as the use of standardized forms, to improve their approach to enhancing patient wellbeing. Some training may be necessary to ensure that clinic personnel as well as clients understand the purpose of adopting this new assessment tool. Ultimately, it is important to ensure buy-in of all practice personnel if the tool is to be effective in enhancing animal wellbeing.

QoL assessments are used to evaluate animal wellbeing, which can be explained to the client as enjoyment of life, and they can be applied to any veterinary species. The purpose of these tools is to assist with making more objective end-of-life decisions as well as decisions about patient care and treatment, and to make a conscious effort to serially evaluate overall animal welfare as part of a regular wellness examination. It is important to note that QoL assessments may not be the sole determinant for making animal treatment or end-of-life decisions, as other factors, such as human safety, financial situation of the client, and client social support network, may also be important in decision-making. Bearing this in mind may help to alleviate fears or concerns that practice personnel or clients may have about a new process driving decision-making.

△

5.22 "What sort of physical activity does he or she enjoy?" Many practitioners routinely incorporate aspects of an informal QoL assessment into their overall examination.

PHOTO ANNE FAWCETT

△

5.23 QoL assessment provides a summary of perceived changes in animal wellbeing over time, something that may otherwise be difficult for the veterinary nurse or practitioner to appreciate, especially if changes are subtle.

PHOTO ANNE FAWCETT

Many practitioners routinely incorporate aspects of an informal QoL assessment into their overall examination, which may be used to inform the recommendations that they make to the client. However, a formal assessment tool reminds clinic personnel to keep animal wellbeing at the forefront of their care (part of ethical decision-making), it may help to better integrate care of a patient in a multi-person practice in which the same people may not examine the patient at each visit, and it may help the client to better understand the medical needs and progression of any condition of their pet throughout its life.

Clinics can always create a new QoL assessment tool; however, there are several QoL assessment tools available online that can be used as a template or modified to suit a practice or specific situation (for example, in the UK practices can undertake training in the use of the PDSA's

PetWise MOTs). Each has its own merits and should be discussed by hospital personnel to ensure that they are completed and used consistently. More than one assessment tool may be necessary to cover all contingencies. For example, a short generic questionnaire that focuses on daily pleasures and enjoyment of a companion animal may be given to the client to complete in the waiting room prior to each annual wellness examination. This requires the client to really focus on the needs and wants of their pet and to think about whether there have been any physical or behavioural changes in the animal since the last visit. When incorporated into the medical record and reviewed annually, this brief QoL assessment also provides a succinct chronological summary of perceived changes in animal wellbeing over time, something that may otherwise be difficult for the veterinary nurse or practitioner to appreciate, especially if changes are subtle. A more detailed and specific questionnaire may be developed and used to monitor animals that are undergoing treatment to objectively assess response to care or

5.8
QUALITY OF LIFE
RESPONSE
PATRICIA V TURNER

progression of disease, such as diabetes mellitus, osteoarthritis or neoplasia.

As with any change in practice, it is critical to assess the use of a formal QoL assessment tool(s) over time through both client and personnel feedback and to make any needed adjustments. A successfully integrated tool will enhance practice quality and consistency, assist with ethical decision-making for patient care and add value for clients.

|||||||||||||||||||||||||||||||||||||||||||||||||||||||||||||||||||||||||||||

QoL assessment is often informal, and may rely on intuition. There may be disadvantages to this approach, including poor ability to predict outcomes from a consequentialist or utilitarian perspective, and failure to appreciate the patient's perspective (which maps onto autonomy in the principalist approach).

Yet this is not the only situation where the veterinary team may be called upon to assess the subjective state of an animal. Increasingly, validated pain scoring systems are utilised to assess pain in veterinary patients and facilitate administration of analgesia (Epstein, et al. 2015).

QUALITY OF LIFE

What do you think?

ONE _____ What parameters do you consider when assessing QoL in animals?

TWO _____ How useful do you think a formal QoL assessment checklist would be?

THREE _____ How would you approach a situation where a colleague or client disagrees with your QoL assessment?

The Gift

△

5.24 People who care for animals consider their animal's QoL, but does formal assessment aid decision-making?

CARTOON FRANKO, GRRINNINBEAR.COM.AU

Conclusion

Veterinary treatment is fundamental to veterinary practice and as such requires particular ethical consideration. Veterinarians are not in practice to harm animals, so deciding how, when and importantly why to treat animals, when very often these treatments inflict at least short term-harms, is essential. Communicating with owners will usually be required to enable a shared decision to emerge that is acceptable to both parties, who each have, amongst other things, the animal's interest at heart.

References

Alef M, Von Praun F, and Oechtering G 2008 Is routine pre-anaesthetic haematological and biochemical screening justified in dogs? *Veterinary Anaesthesia and Analgesia* **35**: 132–140.

American Veterinary Medical Association 2010 New AVMA Market Data Reveals Trends in Veterinary Medicine, Pet Care and Women's Career Choices. PR Newswire, 9 March.

Anon 2011 MEPs approve funding for farm animal homeopathy pilot project. *Veterinary Record* **169**: 274.

Anzuino K 2007 Everyday ethics. *In Practice* **29**: 234.

Arluke A, and Sanders C 1996 *Regarding Animals*. Temple University Press: Philadelphia.

Ashall V 2009 Everyday ethics. *In Practice* **31**: 527.

Aslanian ME, Sharp CR, Rozanski EA, De Laforcade AM, Rishniw M, and Brooks MB 2014 Clinical outcome after diagnosis of hemophilia A in dogs. *JAVMA-Journal of the American Veterinary Medical Association* **245**: 677–683.

Barr JW, and McMichael M 2012 Inherited disorders of hemostasis in dogs and cats. *Topics in Companion Animal Medicine*. **27**: 53–58.

Beauchamp TL, and Childress JF 2001 *The Principles of Biomedical Ethics*. Oxford University Press: Oxford.

Beauchamp TL, and Childress JF 2013 *Principles of Biomedical Ethics, Seventh Edition*. Oxford University Press: New York, Oxford.

Bell JS 2011 Researcher responsibilities and genetic counseling for pure-bred dog populations. *The Veterinary Journal* **189**: 234–235.

Coon ER, Quinonez RA, Moyer VA, and Schroeder AR 2014 Overdiagnosis: how our compulsion for diagnosis may be harming children. *Pediatrics* **134**: 1013–1023.

Davidow B 2013 Transfusion medicine in small animals. *Veterinary Clinics of North America-Small Animal Practice* **43**: 735–756.

Davies M, and Kawaguchi S 2014 Pregeneral anaesthetic blood screening of dogs and cats attending a UK practice. *Veterinary Record* **174**: 506.

Dawson LC, Dewey CE, Stone EA, Guerin MT, and Niel L 2016 A survey of animal welfare experts and practicing veterinarians to identify and explore key factors thought to influence canine and feline welfare in relation to veterinary care. *Animal Welfare* **25**: 125–134.

Day MJ, Horzinek MC, Schultz RD, and Squires RA 2016 WSAVA Guidelines for the vaccination of dogs and cats. *Journal of Small Animal Practice* **57**: 4–8.

Djulbegovic B, Hamm RM, Mayrhofer T, Hozo I, and Van den Ende J 2015 Rationality, practice variation and person-centred health policy: a threshold hypothesis. *J Eval Clin Pract* **21**: 1121–1124.

Durham AC, Mariano AD, Holmes ES, and Aronson L 2014 Characterization of post transplantation lymphoma in feline renal transplant recipients. *Journal of Comparative Pathology* **150**: 162–168.

Epstein M, Rodan I, Griffenhagen G, Kadrlik J, Petty M, Robertson S, and Simpson W 2015 2015 AAHA/AAFP pain management guidelines for dogs and cats. *Journal of the American Animal Hospital Association* **51**: 67–84.

Finan V 2016 TV supervet says scientific advances including 3D printing of body parts mean he can save ANY animal but it's not always the right thing to do. *Daily Mail Australia*: MailOnline.

Graham JB, Buckwalter JA, Hartley LJ, and Brinkhous KM 1949 Canine hemophilia; observations on the course, the clotting anomaly, and the effect of blood transfusions. *Journal of Experimental Medicine* **90**: 97–111.

Harris J 1975 The survival lottery. *Philosophy* **50**: 81–87.

Hartmann K, Day MJ, Thiry E, Lloret A, Frymus T, Addie D, Boucraut-Baralon C, Egberink H, Gruffydd-Jones T, Horzinek MC, Hosie MJ, Lutz H, Marsilio F, Pennisi MG, Radford AD, Truyen U, Möstl K, and Diseases EABoC 2015 Feline injection-site sarcoma: ABCD guidelines on prevention and management. *Journal of Feline Medicine and Surgery* **17**: 606–613.

Lozier JN, Dutra A, Pak E, Zhou N, Zheng ZL, Nichols TC, Bellinger DA, Read M, and Morgan RA 2002 The Chapel Hill hemophilia A dog colony exhibits a factor VIII gene inversion. *Proceedings of the National Academy of Sciences of the United States of America* **99**: 12991–12996.

Lozier JN, and Nichols TC 2013 Animal models of hemophilia and related bleeding disorders. *Seminars in Hematology* **50**: 175–184.

Lux CN, Culp WT, Mayhew PD, Tong K, Rebhun RB, and Kass PH 2013 Perioperative outcome in dogs with hemoperitoneum: 83 cases (2005–2010). *JAVMA-Journal of the American Veterinary Medical Association* **242**: 1385–1391.

Mathie RT, and Clousen J 2014 Veterinary homeopathy: systematic review of medical conditions studied by randomised placebo-controlled trials. *Veterinary Record* **175**: 373–381.

Mepham TB 1996 Ethical analysis of food biotechnologies: an evaluative framework. In: Mepham TB (ed) *Food Ethics*, 101–119. Routledge: London.

Moynihan R, Doust J, and Henry D 2012 Preventing overdiagnosis: how to stop harming the healthy. *BMJ* **344**: e3502.

New Zealand Veterinary Association 2010 9i. Renal transplantation in cats. http://www.nzva.org.nz/policies/9i-renal-transplantation-cats

Nicholas FW 2010 *Introduction to Veterinary Genetics*. Wiley-Blackwell: Chichester.

Norris JM, Krockenberger MB, Baird AA, and Knudsen G 2006 Canine distemper: re-emergence of an old enemy. *Australian Veterinary Journal* **84**: 362–363.

Online Mendelian Inheritance in Animals (OMIA) N.d. Faculty of Veterinary Science, University of Sydney. MIA Number: OMIA 000437. http://omia.angis.org.au/

Online Mendelian Inheritance in Animals (OMIA) N.d. Faculty of Veterinary Science, University of Sydney. MIA Number: OMIA 000437-9615. http://omia.angis.org.au/

Online Mendelian Inheritance in Man (OMIM) N.d. Johns Hopkins University, Baltimore, MD. MIM Number: 306700. http://omim.org/

Optum 2013 Effectiveness of Homeopathy for Clinical Conditions: Evaluation of the Evidence. Overview Report. National Health and Medical Research Council, Australian Government.

Pennisi MG, Hartmann K, Addie DD, Lutz H, Gruffydd-Jones T, Boucraut-Baralon C, Egberink H, Frymus T, Horzinek MC, Hosie MJ, Lloret A, Marsilio F, Radford AD, Thiry E, Truyen U, Möstl K, and Diseases EABoC 2015 Blood transfusion in cats: ABCD guidelines for minimising risks of infectious iatrogenic complications. *Journal of Feline Medicine and Surgery* **17**: 588–593.

Pressler BM 2010 Transplantation in small animals. *Veterinary Clinics of North America-Small Animal Practice* **40**: 495–505.

RCVS 1844 The Royal Charter of the Royal College of Veterinary Surgeons. 8 March.

Sparkes AH, Caney S, Chalhoub S, Elliott J, Finch N, Gajanayake I, Langston C, Lefebvre HP, White J, and Quimby J 2016 ISFM Consensus Guidelines on the diagnosis and management of feline chronic kidney disease. *Journal of Feline Medicine and Surgery* **18**: 219–239.

Tan DKS, and Krekeler N 2014 Success rates of various techniques for reduction of twin pregnancy in mares. *JAVMA-Journal of the American Veterinary Medical Association* **245**: 70–78.

VMD 2010 Press Release: Alternative pet remedies: government clampdown. https://www.gov.uk/government/news/alternative-pet-remedies-government-clampdown

Webster AJF 2013 *Animal Husbandry Regained: The Place of Farm Animals in Sustainable Agriculture*. Earthscan, Routledge: London.

Wolfensohn S, and Honess P 2007 Laboratory animal, pet animal, farm animal, wild animal: which gets the best deal? *Animal Welfare* **16**: 117–123.

Wormser C, Phillips H, and Aronson LR 2013 Retroperitoneal fibrosis in feline renal transplant recipients: 29 cases (1998–2011). *JAVMA-Journal of the American Veterinary Medical Association* **243**: 1580–1585.

Yeates JW 2014 Ethical considerations in feline renal transplantation. *Veterinary Journal* **202**: 405–407.

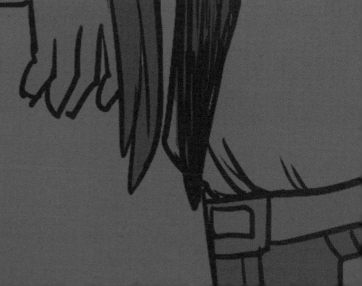

CHAPTER 6
MONEY

Introduction

Much of veterinary practice all around the world operates within the private sector. The practices are businesses, the owners (often vets) are business people and the clients have to pay for veterinary products and services.

The affordability of veterinary treatment can have a direct impact on the welfare of animals and the wellbeing of their owners. Even when veterinary practice is undertaken by governmental or charitable organisations it is not free from monetary constraints, requiring ethical considerations to decide on the appropriate distribution of available resources.

Unlike in human healthcare, euthanasia is an option if treatment is not affordable. While this may be humane, euthanasia of an animal with an ultimately treatable or manageable condition can be a source of moral stress in veterinarians (Rollin 2011). At the same time, it is not sustainable for veterinarians to personally fund treatment for all animals whose owners cannot afford to pay the bill.

In this chapter we will consider scenarios where owners can't afford treatment, how motivations to make profits and treat animals fairly can be reconciled and where a charity must justify its policy on the treatments it will provide.

6.1
Owners unable to afford veterinary care

Trying to find an appropriately ethical treatment path for sick animals whose owners are financially constrained will be familiar to many veterinarians. Yet it seems fair, reasonable and in keeping with our professional obligations that such animals are treated. In 1935 Principal and Dean of the Royal Veterinary College, Sir Frederick Hobday, made this point at the reopening of the RSPCA's Liverpool Animal Clinic:

> "We assert emphatically that the animal of a poor man when ill has just as much moral right to proper diagnosis and treatment of its ailment as the animal of the rich..."
>
> (Anon 1935)

At the time, most companion animals – particularly those belonging to the poor – were treated by laypersons in facilities such as the People's Dispensary for Sick Animals of the Poor (Gardiner 2014).

"Middle-class approval for the treatment of working-class pet animals opened up a field of practice that would later grow exponentially as a highly successful branch of private medicine – the birth of the small animal clinic" (Gardiner 2014).

But the problem of poor clients persists. In a more recent survey in the UK a majority of the 58 practising vets reported that the most common ethical dilemma they encountered concerned clients with limited finances (Batchelor & McKeegan 2012). The vets in this survey produced a median stress rating of 7 out of 10 for cases where financial limitations impact on treatment – a high score, but lower than their stress ratings for euthanasia of healthy animals and for situations when owners want to continue treatment despite poor welfare. The authors speculate that this relatively lower stress score:

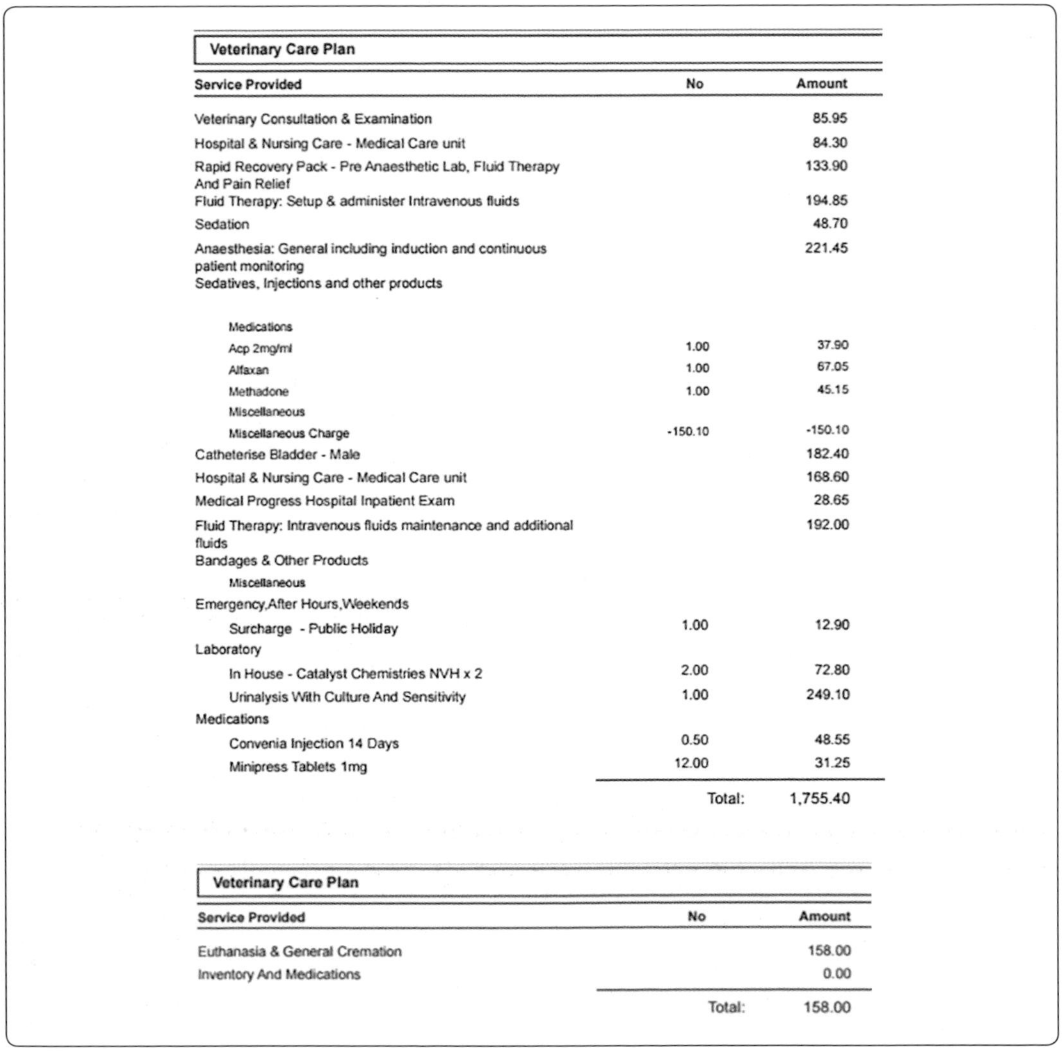

Veterinary Care Plan

Service Provided	No	Amount
Veterinary Consultation & Examination		85.95
Hospital & Nursing Care - Medical Care unit		84.30
Rapid Recovery Pack - Pre Anaesthetic Lab, Fluid Therapy And Pain Relief		133.90
Fluid Therapy: Setup & administer Intravenous fluids		194.85
Sedation		48.70
Anaesthesia: General including induction and continuous patient monitoring Sedatives, Injections and other products		221.45
Medications		
Acp 2mg/ml	1.00	37.90
Alfaxan	1.00	67.05
Methadone	1.00	45.15
Miscellaneous		
Miscellaneous Charge	-150.10	-150.10
Catheterise Bladder - Male		182.40
Hospital & Nursing Care - Medical Care unit		168.60
Medical Progress Hospital Inpatient Exam		28.65
Fluid Therapy: Intravenous fluids maintenance and additional fluids		192.00
Bandages & Other Products		
Miscellaneous		
Emergency,After Hours,Weekends		
Surcharge - Public Holiday	1.00	12.90
Laboratory		
In House - Catalyst Chemistries NVH x 2	2.00	72.80
Urinalysis With Culture And Sensitivity	1.00	249.10
Medications		
Convenia Injection 14 Days	0.50	48.55
Minipress Tablets 1mg	12.00	31.25
	Total:	1,755.40

Veterinary Care Plan

Service Provided	No	Amount
Euthanasia & General Cremation		158.00
Inventory And Medications		0.00
	Total:	158.00

△

6.2 Veterinary treatment may be unaffordable to some.
PHOTOS ANNE FAWCETT

"might be related to the acceptance of an unpleasant but common reality, availability of methods to work around this (for example, doing an operation at a reduced fee) or possibly that financial constraints are seen as the client's responsibility and are as a result of their actions (as opposed to the other scenarios that were more directly related to the veterinary surgeons' actions)."

(Batchelor & McKeegan 2012)

The next two scenarios consider different cases where owners have limited finances available for veterinary treatment.

SCENARIO
OWNERS CANNOT AFFORD TREATMENT

▶ You have been presented with Sasha, a three-year-old intact female Labrador. She has weight loss and lethargy, but appears alert and happy. The clients, Mr and Mrs B, are a couple in their late 50s who are generally dedicated to their animals, but of limited financial means. Nevertheless, they agree to screening blood and urine tests, thoracic and abdominal radiographs and a biopsy. The results confirm early-stage cancer. The main treatment options include: immediate euthanasia; doing nothing until Sasha's quality of life deteriorates, probably in several months, and then euthanasia; or chemotherapy. The latter would have to be performed by a specialist colleague at a referral practice. Her chemotherapeutic regime for this condition has a fairly high probability of remission, which normally lasts about 6–18 months, for most of which Sasha could be expected to have a good quality of life. Just over 20 per cent of dogs on this protocol have survived 2–2.5 years. However, the chemotherapy is associated with a range of unpleasant side effects, which can be managed to varying

degrees using medications. The clients would need to be able to pay £1,500–2,000 in fees, and would need to regularly transport Sasha to the hospital for intravenous injections and blood monitoring, as well as medicate Sasha at home and nurse her through any possible side effects. The clients appear frightened and somewhat confused as you explain all of this. They care a lot about Sasha and seem keen to proceed, but don't think they can afford it.

What should you do?

RESPONSE
ANDREW KNIGHT

▶ Once appropriate empathy has been expressed and deficiencies of understanding have been corrected, you can progress to addressing the ethical dilemma this case presents. On the one hand Sasha would clearly benefit from treatment, and may suffer poor welfare

△
6.3 Mr and Mrs B care for Sasha but may not be able to afford expensive veterinary treatment.
PHOTO ISTOCK

—
6.1
OWNERS UNABLE TO AFFORD VETERINARY CARE
RESPONSE
ANDREW KNIGHT

and eventually die without it. At three years of age she is only young, and would normally enjoy many more years with her caring human family. Even with her neoplastic diagnosis, chemotherapy has a high probability of giving her 6–18 months of remission, with a generally good quality of life. Hopefully any side effects should be largely manageable with medication.

On the other hand, this situation is placing the clients under considerable financial pressure. Much as they care about Sasha, and would doubtless enjoy spending future months to years with her, extra nursing duties notwithstanding, they may not actually be able to afford her treatment. Regrettably, it appears they have not taken out pet insurance, which would probably have covered treatment costs. They would also face additional duties and perhaps some level of distress, when nursing Sasha. Nevertheless they seem keen to proceed if they can.

Sasha and her owners are the main interested parties, although as a caring veterinarian you probably also have a strong personal and professional interest in ensuring your patient's welfare is appropriately safeguarded: that her life is preserved without suffering for as long as possible, and that when this is no longer possible, she is humanely euthanased. Other parties, notably the veterinary practice, also have interests with varying degrees of legitimacy, such as in maximising revenue generated through treatment. However, the essence of this case is the dilemma posed between the need for treatment, and the inability (despite their keenness) of the clients to pay for it.

The next advisable step would be to see if sources of financial assistance are available that might assist the clients, and your patient, and hence reduce or eliminate this dilemma. As decided by the practice owner or manager, your practice might have a policy of allowing payment via instalments for long-standing clients, or in certain types of cases. You might wish to check whether the referral veterinarian could offer such a plan. In some locales charities can assist animals in general, or specific breeds. Clients may or may not be financially tested to determine eligibility. CareCredit is a US lending organisation that provides credit to eligible clients to pay for medical, and veterinary, treatment. It would be worth seeing whether any such organisation exists in your region. With sensitivity, you might ask whether friends or family could be called on to assist, and might raise the potential of fundraising efforts such as online "crowdsourcing", and in particular, "crowdfunding", which have, on occasion, assisted others. However, if none of these options are available, or within the ability of the clients to achieve, then the fundamental dilemma remains.

Cases such as these often tempt veterinarians to reduce or eliminate their charges. However, while clearly benefiting the patient and client immediately present, such choices may adversely impact future patients and clients. As McCulloch (2011) put it, "Supererogatory acts such as this

△
6.4 Veterinarians may be in the position of euthanasing animals with treatable conditions.
CARTOON RAFAEL GALLARDO ARJONILLA

can be criticised for encouraging irresponsible behaviour (e.g. unable to afford veterinary fees, no insurance) and might also be construed as unfair to those who pay the normal fee." In the worst-case scenario, overly sympathetic actions or policies could result in the practice becoming financially unviable, ultimately leaving the patients and clients with no veterinarian to serve them. Hence, an appropriate balance should be struck. In some practices this is achieved by having a limited charitable fund, funded partly from practice profits and also client donations, with clear eligibility policies to prevent abuse, and a monthly limit to safeguard the practice finances and other resources.

Ultimately, if sufficient funds could be found to treat Sasha, the choice that would best protect her interests and those of the clients, veterinarian and practice, would be to treat her as described. If at some point she became refractory to treatment, and the neoplasia recurred and progressed to the point where she was undergoing significant suffering with a poor prognosis for recovery, then the most ethical choice would seem clear: Sasha should then be euthanased to prevent further suffering (and also distress for the clients, and the veterinary staff treating her). The same choice would also apply if the funds could not be found to treat Sasha, after exhausting all the options described above. Sasha is currently alert and happy, and should be left to enjoy her life, and the time with her human family, and vice versa, for as long as she has a good quality of life. When her quality of life declined to the point where she was undergoing significant suffering with a poor prognosis for recovery, then she should be euthanased.

Euthanasing Sasha when these criteria are met would effectively provide the "greatest good for the greatest number" of stakeholders, which represents the most common form of utilitarian ethical decision-making. However, if using a rights-based ethical framework, some might argue that Sasha has a "right" to life. Conversely however, it could be argued that Sasha has a "right" to be spared serious ongoing suffering. A key question is what Sasha would choose for herself, assuming she was sound of mind and competent to do so. Of course the degree of detail to which Sasha could actually contemplate and choose between alternate futures is unknown, and in any case unable to be communicated. Veterinarians often have to make decisions or recommendations in the best interests of the patient, and, to the greatest extent possible, other stakeholders with legitimate interests such as clients, despite having to deal in probabilities, rather than certainties. Indeed, this burden is one of the key responsibilities of the veterinarian.

||||||||||||||||||||||||||||||||||||||||||||||||||||||||||||||||||||||||||||

SOME of the ethical concerns relating to treating cancer in the veterinary setting are discussed by Moore (2011), including gaining informed consent, use of resources that could be used by humans, euthanasia, unproven therapies, client communication and addressing concerns of the veterinary team. Although these concerns also apply to other treatments, some forms of cancer medicines are perhaps unique in their ability to produce serious short-term side effects. Moore stresses the importance of proper staging and evaluation of the cancer before any decisions are made about treatment, except when the owner cannot afford to, or does not want to, treat their animal under any circumstances.

In the above case there may be a conflict between the interests of the client and the interests of the animal. Managing such conflicts is difficult if the veterinarian sees their role as advocating in the best interests of the animal.

One way of eliminating or reducing conflict is discounting services, explored in the following scenario.

SCENARIO

DISCOUNTING SERVICES

▶ You are a veterinarian in a small animal practice. On the same day you are presented – by coincidence – with two dogs with broken femurs, following trauma, owned by two different owners. One belongs to an older widow who lives alone on a pension, the other belongs to a young couple who have no children. Neither client has an insurance policy that will cover the costs of the treatment. You estimate the fee for treatment to be $3000 for each dog, but suspect that the older woman cannot afford to pay this treatment.

What should you do?

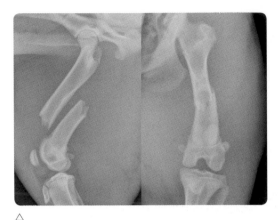

△

6.5a–b Do we ever treat patients with the same injury differently due to their circumstances?
PHOTO DR JOHN CULVENOR,
NORTH SHORE VETERINARY SPECIALIST CENTRE

RESPONSE

PETER SANDØE AND SANDRA CORR

▶ In this case the veterinarian has to balance concerns and interests relating to at least five different parties: (1) the first dog, (2) its owner, the older widow, (3) the second dog, (4) its owners, the young couple (which we for the sake of simplicity will here just consider to be one party), and (5) the veterinarian her- or himself.

Both dogs are urgently in need of treatment. Assuming that they are not very old dogs, it must be in their interest to receive the best available treatment (the "gold standard"), which would be internal fixation, usually with a bone plate. Although there would be some initial discomfort, most animals will use the leg well within a few days of appropriate fracture repair, and are expected to ultimately make a complete recovery back to normal function.

In most Western countries it will be a legal offence for an owner not to seek treatment for a dog with a fracture; and it will be a legal offence

for a veterinarian not to administer first aid or pain relief – even if the owner is not able to pay. However, even in affluent Western countries, a dog does not have a right to the best veterinary treatment. The owner may decide to have the dog put down, or the veterinarian may decide only to offer euthanasia or a secondary standard of treatment if the owner cannot pay for the best treatment.

This situation, where dogs are not entitled to treatment other than first aid or euthanasia, unless someone is willing to pay for it, is of course something that can be discussed from an ethical point of view. We will, however, not pursue this discussion here but can refer interested readers to the other cases where this issue is in focus. Instead, our focus in the remainder of the response will be on the veterinarian's responsibilities to the two clients.

For a start it is important to recognise that although the veterinarian may make assumptions about the financial situation of the clients, she or he cannot actually know about the means and preferences of the two clients unless the clients tell them. Therefore the veterinarian should offer the older lady *all* the treatment options for the dog, including the best treatment, and explain the costs of each, without prejudging the situation. However, the discussion can be phrased very tactfully, for example, "The ideal way to fix this is x, which costs y, but there are a few other options too…" Similarly, with the younger couple, the veterinarian cannot assume that they are able and willing to pay for the best treatment; and the veterinarian should, unless the couple from the start indicates a willingness to pay "whatever it costs" for the best care, mention other and less costly options, including euthanasia.

One outcome may indeed be that the young couple requests the best treatment, but the elderly lady says that she cannot afford to pay $3000, although she very much wants to keep her dog, who is her dear friend and companion, and without whom she would feel very lonely indeed.

A possible solution may be for the elderly lady to seek treatment for the dog from a charity, such as the Blue Cross or PDSA. However, such organisations do not operate everywhere, and even if they do, there are strict rules as to who is eligible based on income or benefits and even postcode.

If there is no charity available or if the elderly woman, despite not being able to pay what it costs to have her dog treated, is not eligible, then ultimately the problem is the old lady's and not the veterinarian's. And it is important to recognise that the old lady could have prevented the problem by taking out pet insurance. So in a way she has not been acting responsibly in relation to her dog.

However, most veterinarians will want to try to help and may even feel an obligation to do so, especially if the older woman is an old and loyal client of the practice. Situations like this one can contribute significantly to increasing stress levels in the working life of many veterinarians, and it is therefore advisable to try to develop a policy for dealing with such situations.

If the veterinarian decides to try to help the old lady beyond what is legally required (offering first

aid or euthanasia), there are a number of options available – apart from suggesting to the old lady to sign the dog over to someone who will pay for the treatment:

(1) To fix the fracture in a cheaper way that "should" work, but may not have as predictable an outcome, e.g. using an external skeletal fixator instead of a bone plate, which has an increased risk of complications with a femoral fracture.

(2) To have a less experienced surgeon operate (so they can gain experience) with the informed consent of the owner (e.g. a first opinion vet instead of a specialist).

(3) To reduce costs, for example, by taking fewer x-rays post-operatively, which could compromise the outcome if a technical mistake is not then spotted.

(4) To offer limb amputation, which is often cheaper than repairing a fracture.

(5) To offer a payment plan, e.g. have the owner pay half the bill at the time, and then pay the rest in instalments over an agreed time.

(6) To reduce the cost overall, without compromising care.

The first four suggestions will all mean that the veterinarian will have to compromise on the desire to give animals in her or his care the best possible treatment. This is another huge cause of stress to veterinary professionals, who often see it as failing the animal, and it can undermine the person's pride in her- or himself as a professional. Furthermore, with the first three options, there is a possibility that things may go wrong; and then the veterinarian may be held responsible for not providing better treatment. Therefore many veterinarians would be unhappy with these options.

This then leaves the two "financial" solutions – to offer a payment plan and/or reduce the costs. These solutions may incur both economic risks and losses in terms of getting paid less than the normal rate. If the veterinarian does not own the clinic, but only works there, it will of course be necessary to involve the practice owner. In that case an agreement may be made that the veterinarian does the work on unpaid overtime so that she or he covers some of the extra costs on her or his own.

Whether or not it is a good idea for the vet/practice to reduce costs and thereby lose money for the benefit of clients who are unable to pay will depend on a lot of things – not least how often this occurs. Most practices undertake a certain amount of pro bono work, as vets are generally highly motivated to treat ill and injured animals, and this contributes to a sense of wellbeing and satisfaction, where the economic bottom line is not the sole consideration. However, if helping poor clients causes the veterinarian economic difficulties, then it will not be sustainable and will then, conversely, contribute significantly to increased stress and unhappiness for the veterinarian.

If the veterinarian decides to reduce the cost without compromising care and thus offer the treatment to the old lady at a substantially reduced price there is also the danger that the young couple, if they find out, will think that they have been treated unfairly. They may use this as a basis for complaint. Already many clients think that veterinary bills are too high and complaints to this effect is another factor giving rise to stress and unease among veterinarians.

So if we should offer advice on the case it is probably to go for option 5 – to offer a payment plan. This gives the old lady a chance to give her dog the best care. It imposes an economic risk on the veterinarian, but she or he avoids several other potential stressors, i.e. compromising on professional standards; feeling bad about not helping the old lady and her dog; or having to face other clients who think that they have been treated unfairly.

||||||||||||||||||||||||||||||||||||||||||||||||||||||||||||||||||||||||||||||

BOTH of the scenarios just discussed support the notion that the ultimate obligation for animal welfare, over and above any legal requirements to euthanase or provide first aid treatment, lies with the owner. However, as the authors note, veterinarians commonly engage in activities to help people who cannot afford treatment, either to identify sources of financial support, or by taking a financial hit themselves.

In a US survey, veterinarians admitted to discounting veterinary products and services for certain groups. For example, 87.8 per cent of respondents discounted services and products for staff members, a practice noted to be common in other industries such as retail (Kogan, et al. 2015). Other groups receiving discounts included friends and family members of staff, professional colleagues, Good Samaritan or stray cases, good clients who have fallen on hard times, senior citizens, military personnel or good neighbours of the practice. In addition, services were frequently discounted for rescue, shelter and working animals including service animals. Other reasons for discounting included working within an owner's limited budget to treat a sick animal, long-term medications and long-term hospitalisation.

The key reasons given for discounting were to provide the best possible care for the animal, and to do everything possible for an animal. However, 43.9 per cent felt that clients expected discounted services and 34.4 per cent felt clients expected discounted products, while 32.1 per cent felt that clients expected free services and 16.6 per cent felt that clients expected free products.

Managing public expectations around charging can be difficult. As one practitioner commented: "It sometimes feels like we can't win. If we put profits into charitable causes it's because we overcharge; if we aren't there 24/7 for free, it's because we don't care" (Davidson 2015).

Ad hoc discounting (combined with missed charges) can be a significant drain on practice revenue, not uncommonly accounting for a 5–10 per cent loss of total turnover (Felsted 2014). This is particularly problematic if the discounter is doing so without authorisation from the practice owner. There may be a perception by the discounter that the fee schedule is unfair. Thus "Non-owners are… more accepting of the level of fees charged in the practice if the overall practice is run fairly; for example, employees are paid well and treated well; clients are treated well; the value and quality of services are apparent; and employee pay is tied to performance" (Felsted 2014).

The opportunity to provide pro bono services is increased in successful practices, and the opportunity to be involved in pro bono work may be appealing to potential employees and a useful means of attracting and maintaining skilled veterinarians on staff (Kogan, et al. 2015).

As we've seen, there is the possibility for veterinary practices to act as a redistributor of the resources available for veterinary treatment through a number of means. For example, through making slightly more profit on common, relatively inexpensive items, such as vaccinations and worming, vets may be able to subsidise treatments for sick animals, particularly of owners with limited financial means. But, is this fair? And who should decide when or by how much to increase or reduce charges?

The issue of fairness, or distributive justice, in the veterinary setting has received much less attention than for human healthcare, perhaps because treatment for companion animals is seen as a luxury commodity for owners and for farm animals as a business transaction. One of the main models of distributional health justice, where the overall resource is shared out based on need, might be appropriate for state or charitable veterinary practices but charging a flat rate fee, regardless of the animal's problem, to fund such a scheme in private practice would seem unfair to many clients.

However, some clients might accept a small inequity in charging structure to benefit the neediest. Rarely is this made explicit though, and such a "scheme" is usually administered on an ad hoc basis with little oversight. Greater transparency could serve to promote a public sense of both practitioner honesty and societal responsibility, and ultimately it could help fuel a debate to determine exactly what form of distribution was seen to be fair.

△
6.3 Many veterinarians seek out opportunities to help others by performing pro bono and volunteer work for organisations such as Pets in the Park.
PHOTO LINDA WORLAND @CLIQUE PHOTOGRAPHY

—
6.1
OWNERS UNABLE TO AFFORD VETERINARY CARE
SCENARIO
DROUGHT-STRICKEN FARM

OWNERS UNABLE TO AFFORD VETERINARY CARE

What would you do?

Consider how you would approach the situation when Ms T brings her three-year-old Great Dane Princess to you at 10 pm. Both are in distress and you suspect Princess has gastric dilatation, although at this stage she is still standing. When you quote her a cost of $500 for initial stabilisation and a possible $1000 for surgery Ms T completely breaks down crying and says, "There's no way I can get that sort of money. What on earth should I do?"

ONE　　　What are the particular features of this case that make it difficult to deal with?

TWO　　　Consider what information you need to decide the best course of action.

THREE　　What is your ethical justification for this decision?

△

6.6 Large breeds are predisposed to gastric dilatation.

PHOTO ANNE FAWCETT

Veterinarians, nurses and technicians may be confronted by situations in which they have concerns about how clients have allocated resources in a way that negatively impacts on animal welfare, such as the following scenario.

———
SCENARIO
DROUGHT-STRICKEN FARM

▶ Your clients, Mr and Mrs Q, have been farming sheep for 30 years and have never seen a drought like it. It's not just the effect of this year, but comes on top of the last couple of dry years too. They estimate that 20 per cent of their ewes have died already and the remaining sheep are now very thin. The Qs have had a bad time of it and are now seriously out of cash to truck feed and water in. Mrs Q sobs that she just doesn't know what to do for the best, but thank goodness they've already paid for their children to go through private school and university. They only need to consider themselves now.

What should you do in this situation?

———
RESPONSE
JOHN BAGULEY

▶ Tannenbaum (1995) has suggested four fundamental approaches to your role as a veterinarian: a healer of animals, a friend and counsellor to clients, an economic manager and herd health consultant, or a business manager. There is inherent conflict in these approaches as you consider what is best for the animal, what is best for the client and what is best for your veterinary business.

Whatever approach or approaches you take to your career or individual scenarios and ethical dilemmas, it is vital to remember that your first

△
6.8 Drought has both welfare and economic impacts.
PHOTO ANNE FAWCETT

priority is animal welfare. You must always be the advocate for the animal. In most jurisdictions the priority of animal welfare is an administrative or legal obligation and beyond that it is your ethical obligation. The public has put their trust in you, has accepted your scientific knowledge of animals and expects you to promote animal welfare. Anything less is a betrayal of that trust and jeopardises your standing as a professional.

In this scenario your scientific knowledge of animals and farming systems provides you with an opportunity to ensure that the welfare of the remaining sheep on this property becomes the priority and that decisions are made which are in the best interests of these animals within the constraints of their circumstances. You cannot change what has already happened but you can try to minimise poor animal welfare outcomes in the future.

Again, your first priority is for the remaining sheep; be the advocate for these animals and encourage your clients to put the welfare of these

sheep first. Talk them through the options as a friend and counsellor, provide them with advice about each option and about possible financial support so that they make the best economic decision for their farm. Most importantly, take some time to ensure that, if they continue farming or return to farming, you assist them to develop a drought plan that triggers appropriate action well before this level of animal suffering.

In addition to seeing this scenario through Tannenbaum's considerations of your role as a veterinarian, this scenario also highlights how a deontological approach and a consequentialist approach to ethical decision-making can sometimes work together.

Firstly, it emphasises the priority of animal welfare for veterinarians and the five freedoms model in assessing the welfare of farm animals. One of the main criticisms of a deontological-type approach to ethical decision-making is the problem of competing rights or rules and whether some rights or rules are more important than others. "Thou shall not kill" – but what happens when we are forced to kill in self-defence or in the defence of others? Animal welfare is your priority and animal welfare expects freedom from thirst, hunger, discomfort and disease but what if the clients cannot afford feed or treatment? The public and the legislators expect you to be the advocate for the animal.

Arguably, in this scenario you can use your priority of animal welfare and your scientific (and business) knowledge and skills to understand all stakeholders and create a solution that ensures the best possible outcome for the animals concerned given the circumstances.

Finally, this scenario most likely represents an ethical dilemma where the solution is fairly simple but the implementation may be very difficult depending upon the attitude of your clients. Scientific knowledge and practical competences provide the foundation for your veterinary work,

6.1

OWNERS UNABLE TO AFFORD VETERINARY CARE
RESPONSE
JOHN BAGULEY

your personal values and understanding of ethics will guide how you approach this work and, in many cases, your ability to empathise, communicate and negotiate will make your veterinary work worthwhile from not only a personal, but also an animal welfare and professional perspective.

|||||||||||||||||||||||||||||||||||||||||||||||||||||||||||||||||||||||||

HERE, the sheep are suffering, but so are the farmers. They can't currently pay for the necessary care, but some may argue that they could have done more in the "boom" years when they spent money away from the farm and perhaps did not properly plan for such difficult times. In one farming sector in the UK there is a long-standing joke that the farm buildings are falling down around the newly built swimming pool. Managing risk and uncertainty is a key part of farming and although droughts are often seen as a "natural" disaster it has been argued that in drought-prone areas significant improvements to welfare can be made through the use of "straightforward management changes, such as improved planning for extended dry periods and drought" (Petherick 2005). Regional legislation may also support this principle, where allowing animals to die from hunger or thirst is not permissible.

OWNERS UNABLE TO AFFORD VETERINARY CARE

What do you think?

ONE _____ How does a failure to plan affect owners' culpability for welfare problems experienced by their animals in the future?

TWO _____ How should any additional culpability of owners affect your own actions in such cases?

△

6.9 Is failure to plan properly a relevant ethical consideration?
PHOTO ISTOCK

6.2
Unowned animals

The scenarios so far have considered the relative responsibilities of the owner and veterinary practice for treatments. Unowned animals, such as stray animals and wildlife, present a unique challenge as there may be no party responsible for the treatment costs. Yet there may be a legal obligation to provide first aid, treatment or euthanasia. Such interventions all incur expenses, usually borne by veterinarians or veterinary practices.

In our next scenario we will consider the ethical implications when there is no owner to be responsible for an animal's treatment costs.

SCENARIO
STRAY DOG HIT BY CAR

▶   After the clinic is closed a member of the public phones to ask to bring in a small dog that they saw being hit by a car. They presume the dog is a stray as they've seen it hanging around their street for a while. On examination the dog is pale, tachycardic and at the very least has a nasty de-gloving injury of the left forelimb, which could involve a fracture. The dog's identity cannot be determined; it is not microchipped, tattooed or wearing a collar. If the dog were owned, and money were no object, you would expect that following several general anaesthetics and protracted care the dog would make a good recovery.

What should you do?

RESPONSE
PETER SANDØE AND SANDRA CORR

▶   The veterinarian may avoid the call if the call is diverted to an emergency service or other out-of-hours provider that covers that clinic's night duty. If the veterinarian takes the call she or he would, in many Western countries, be obliged to see the dog – even if it is not clear whether the person calling is willing to pay for the treatment. The veterinarian may tell the person calling about charities, such as the RSPCA in the UK, who may pay for the treatment of the dog – however in some cases, for example with the RSPCA, the charity has to give permission *before* the animal is seen by a vet.

We will here assume that the veterinarian sees the dog, and that there is no reason to think that a charity will pay for the treatment. Thus the veterinarian has a case where she or he has to balance concerns and interests relating to at least

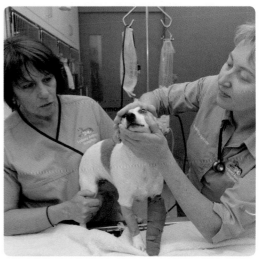

△
6.10 What limits do you place on treatment of an unowned animal?
PHOTO ANNE FAWCETT

four different parties: (1) the dog, (2) the person who brought in the dog, (3) a potential owner of the dog and (4) the veterinarian her- or himself and possibly also the practice owner, if the veterinarian is not self-employed.

Firstly, let us consider a potential owner of the dog. It is not known whether there is someone out there who misses the dog, or whether the dog was abandoned. However, even if there is an owner who wants the dog back, this person seemingly has not been a responsible dog owner. Firstly, the person has not provided the dog with identification allowing it to be reunited with them, should it run away (although the dog may have had a collar with a name tag which has subsequently been lost). Secondly, it seems from the information provided by the person calling in that the dog has been hanging around for a while, although the veterinarian does not know whether this information is to be trusted.

So the veterinarian will usually make an effort to find out whether the dog has a legally responsible owner, who would be willing and able to pay for the dog to be treated.

Regarding the person who brings in the dog, the veterinarian could with some justification expect the person to share some of the burden for taking care of the dog – for example by covering the treatment costs, or at least some of them. However, in practice, this will rarely happen. Most ordinary people seem to take for granted that veterinarians have a special duty to take care of sick animals, which goes far beyond what can be required of other people. In a way this is positive in that it shows that people hold veterinarians in high regard and view them as moral authorities. However, it is also potentially very stressful and burdensome for the veterinarian to be made responsible for the plight of animals that others have abandoned, or are not willing to take responsibility for. Also, since the person bringing in the dog claims to have seen the dog hanging around

in the street for a while, she or he could have taken action earlier to take care of the dog, for example by notifying the RSPCA or dog warden, and thereby have prevented an accident from happening. Furthermore there is a real danger that people may misuse veterinarians to take care of their own sick or injured animals under the pretence that the animal does not belong to them.

From a legal point of view the obligations the veterinarian has towards the animal may vary, even among rich Western countries. In some countries the veterinarian may be allowed to put the dog down if, given the circumstances, it seems likely that the dog in question is a stray dog. In other countries the veterinarian may be required to administer first aid, so that the dog would be assessed and stabilised. The dog would likely be given IV fluids, analgesia (painkillers) and would have the leg injury flushed and bandaged for support (but not x-rayed). A decision on how to proceed is usually made within the following 24 hours.

We will here, for the sake of argument, assume that no owner shows up, and that the person who brought in the dog has withdrawn. This leaves the veterinarian with the decisions about what to do – to treat the dog in anticipation of finding an owner; to treat the dog and subsequently adopt or rehome it; or to euthanase the dog. If the veterinarian considers rehoming, the end result may be that the dog recovers, finds a good home and that the new owners pay the costs of treatment – or at least part of them. However, there are a lot of uncertainties here and the case may give rise to both high costs and stress to the veterinarian. On the other hand, if the veterinarian decides to avoid the hassle and the risk of further costs and just euthanases the dog, this may be quite stressful too. For many veterinarians, the killing of an animal with a potential for a good life may appear utterly wrong. And of course an owner may subsequently appear looking for the dog and may make a complaint against the veterinarian for killing the dog.

Here there seems to be a real conflict between two philosophies concerning the value of animal life. For some veterinarians, the killing of an animal is not a moral issue. What matters is whether the animal lived a good life and was killed humanely. This view may be termed "animal welfarism". It contrasts with a view according to which saving an animal's life matters in its own right – a view that may be termed "value of animal life". This view could take the shape of an animal rights view, according to which, other things being equal, it is morally wrong to kill an animal that is able to go on living a good life. It could also take the shape of a form of consequentialism, which ascribes value to the avoidance of killing of animals, but allows a trade-off with other values.

So the veterinarian must make up her or his mind about the moral bottom line. If the veterinarian endorses animal welfarism, euthanasia will in most cases be a relatively straightforward solution in relation to the animal. However, if the value of animal life is endorsed, other options will have to be considered. One important consideration here will be whether there is any real prospect of rehoming the dog. This will depend on a number of factors, including how friendly, well-behaved and young the dog is, what sort of breed it is etc. For example, if it looks well looked after, and is very friendly, it seems more likely that there will be an owner looking for it, so it may be kept for several days. If it is thin, mangy and trying to bite, the dog is more likely to be euthanased, and that decision made sooner.

Cases like the one described here can be costly to veterinarians, and indirectly part of the costs may be borne by normal clients who through their fees enable the veterinarians to undertake a certain amount of pro bono work for the benefit of unowned or stray animals. This may seem unfair, and will be a particular problem where charities do not exist to share some of the responsibility of care with veterinarians.

It would not be appropriate for us to give advice on the main issue here, which is whether the veterinarian should engage in efforts to treat and subsequently rehome abandoned or stray animals that are brought into the clinic, or whether the animals should be euthanased sooner rather than later. Rather our advice is for the veterinarian to have a clear policy that is socially and economically sustainable, and that is in line with the considered moral view of the individual veterinarian regarding the value of animal life.

||||||||||||||||||||||||||||||||||||||||||||||||||||||||||||||||||||||||||

THIS scenario raises many potential costs of treating stray animals to veterinarians and veterinary practices. But are there other potential benefits? For example, one might consider that treating stray animals contributes to one's caseload and knowledge bank. When confronted with a similar ethical dilemma, the philosopher Bernard Rollin observed that, "Some veterinarians view unowned animals as 'continuing education from God', and treat the animals as they would a client's animal" (Rollin 2006). He goes on to suggest that one strategy in ensuring all parties benefit from such cases is that local papers may cover "before" and "after" stories about treatment of these animals, resulting in "extraordinary publicity that could not be bought" (Rollin 2006).

Of course such an approach assumes that the stray animal caseload is not overwhelming, and publicity does not generate enquiries from pet owners seeking discounts.

A recurring theme in some of these scenarios has been the need for a practice policy to determine, in the cold light of day rather than the heat of the moment, exactly how the practice deals with implicit or explicit requests for free or discounted services.

What do you think?

ONE_____What features should be included in a practice policy to decide the principles of any pro bono work undertaken by the practice?

TWO_____Which ethical principles would you want to underlie your practice policy?

THREE_____What would be the key points of your policy?

△

6.11 Some practices have policies relating to pro bono work.
PHOTO ISTOCK

6.3

Pet insurance

In many countries it's possible for pet owners to insure their pet for veterinary costs. In the UK 15 per cent of pet owners have insurance, some 2.5 million people, who made 800,000 claims in one year (ABI 2014). The average cost of each claim rose by 52 per cent from 2007 to 2012 (ABI 2013). In contrast it is estimated that only around 1 per cent of pets are insured in the USA (Embrace 2013).

Usually the insurance companies are profit-making enterprises operating in a competitive market themselves. Their business centres on estimating risk and already there are usually larger premiums for certain species, for example horses compared to cats, or certain types of animal, e.g. large compared to small dogs, based on the greater risk of incurring more expensive veterinary treatment. For many owners insurance provides peace of mind and adequate cover if the need arises. However, a range of possible ethical issues surrounding veterinary insurance cover have been identified.

Potential ethical issues with insurance companies:

- Unfair treatment of clients through over-zealous exclusions both prior to, and after clinical conditions arise, particularly long-term problems.
- Unfair use of "excess" payments.
- Confusing terms and conditions.
- Poor customer service, such as quibbling over forms, late payments and unavailability for client contact (Taylor 2012).

Potential ethical issues with veterinary surgeons:

- Conflicts of interest when veterinary surgeons promote one insurance company.

- Over-inflation of treatment prices by veterinary surgeons for insured animals.
- Over-treatment of insured animals by veterinary surgeons (Watkinson 2009).

Unlike for other insurance areas the risk models for veterinary insurance are relatively crude and the companies are seeking better ways to estimate risk. The ethical issues of one way that companies may be able to do this are discussed in the next scenario.

SCENARIO
ACTIVITY TRACKER FOR DOGS

▶   Mr V is one of your most technophilic clients and he's often telling you about the latest gadget he's got his eye on. He is considering an app for monitoring aspects of his own health and asks your advice about an advert for a similar system for his dog, Sasha. He would be able to monitor Sasha's exercise, including when the dog walker is taking her out. The company, PetStepz, has even teamed up with an insurance company so that if Sasha does enough exercise his insurance premium is reduced.

How should you respond?

RESPONSE
JAMES YEATES

▶   On the face of it, the immediate consequences would be expected to make us in favour of this device. Sasha may become healthier, the owner saves money – win-win.

Privacy
One set of difficulties come from the use of information.

We tend not to think of animals as having rights of confidentiality or privacy. Animals may need places where they feel safe, secure and hidden – but our rationale for such provisions is usually based on the utility of such facilities, for example to avoid suffering, rather than a value of privacy per se when the animal is otherwise unaffected (or else birdwatching would be an inexcusable voyeurism). It is with the information about people that we might be concerned.

Mr N's data may be kept, exploited and/or shared by PetStepz. How much exercise your dog gets is probably not sensitive personal information, but nevertheless it is information about the owner that others could use to advertise certain products to Mr N (or, perhaps unethically, to change his personal health premiums). But Mr N, if fully informed by the company, has implicitly consented to them having his data and to however they plan to use them. Unless one is being very paternalistic about Mr N in a way that goes beyond our veterinary remit, one would say that he is best placed to make that decision in his own

△
6.12 Would owners of less active animals be penalised under a new pet insurance scheme?
PHOTO ANNE FAWCETT

interests and/or has the power to waive the right to privacy.

His dog walker's privacy may be another matter, especially if anyone is in a position to identify the data as relating to him/her. Mr N could use it to check up on him/her. Again, if they are informed then perhaps this is acceptable. He/she may sooner not be checked up on, but it seems a fair part of the "contract" between them that Mr N can be sure he is getting what he pays for.

Fairness

The use of the data by an insurance company raises an issue of fairness if it means Mr N will be paying less than other people (or those people then choose not to have pet health insurance) or if the insurance company declines to fund certain treatments for other animals (e.g. obesity-related conditions).

Fairness can be considered as based on what people deserve, and one element thereof is what effort they have "put in" – just as we might feel a worker should get more money for grape-picking for a whole day than for a half day. On this basis it could be argued that Mr N paying cheaper premiums is fair.

Other owners who do not have the app may end up paying more. That may seem unfair, insofar as they are being disadvantaged by Mr N's new gadget. However, arguably, such actuarial corrections simply correct a previous unfairness in which Mr N's premiums for his healthy well-walked pet were subsidising other owners who did not take such care.

Fairness can also be considered in terms of the "pattern" of outcomes, such as the distribution of costs and benefits. It seems unfair to those dogs who are not exercised enough to also not be insured because the premiums become too high, or if those animals do not receive adequate treatment because the insurance company uses the data not to set premiums but to limit what they

pay out for. Effectively, these animals are subject to a double jeopardy of getting neither the preventive nor the remedial care they need. One concept of fairness (proposed by Rawls) suggests that one should avoid anything that makes the worst off even worse off, even if it helps the better off. So, the improvements for Sasha are perhaps not enough to justify the disbenefits to these worse-off animals.

Unfortunately, one of the problems of fairness is that there are lots of different concepts (as there are for morality; indeed many map onto one another) and they are going to conflict. In particular, concepts of fairness that are based on the process (e.g. fair exchanges or getting-what-you-deserve) are likely to conflict with concepts that are based on the outcome (e.g. equal distributions or getting-what-you-need).

This is perhaps especially clear if one considers that the concept of animal ownership, while not something that I would argue against, is inherently unfair: the idea that one individual can "own" another seems likely to lead to problems in trying to be fair to both.

Overall utility

However, we could add an additional consequentialist argument based on the reasonable expectation that this "progress" may be expected to, in the long run, incentivise those other owners to take better care of their animal. Based on a probabilistic concept of benefit we might think that the risks of being harmed from altered insurance are outweighed by the chances of having better care. This suggests that for those animals, the introduction of this app is a net good thing on a utilitarian approach.

Arguably, however, this does not help them from a Rawlsian concept of fairness insofar as the probabilities do not matter – if some are made worse off and some benefitted, then what matters is those worst off. But this assumes a

"God's-eye" view of our decision-making. We are not able to know which animals are benefitted and which are harmed. As such, we are benefitting all animals. Indeed, if we could not do anything that risks making life worse for some individuals, then we would be unable ethically to give many – or perhaps any – veterinary treatments that are beneficial on a probabilistic basis, including vaccination, any medicine, surgery or maybe anything at all. All veterinary medicine involves risk.

Conclusions

So perhaps it is not quite win-win. It is perhaps another "win" for the other animals who have a chance of being better cared for (albeit also with a chance of not getting such good veterinary treatment). The "lose" affects the other owners, but arguably that is their choice and better than Mr N paying for their laziness. As such, you can be pleased with Mr N, and maybe even get a monitor for yourself.

PET INSURANCE

What would you do?

One of your clients, Ms D, is considering insurance for her small cross-breed dog, Tippi. She has heard of the high costs of some life-saving treatments but is also aware that her dog is probably a low risk for veterinary treatments based on the breed and lifestyle. Ms D is wondering whether she would be better off putting the same amount of money away in a bank account for veterinary emergencies. She also wonders whether if something happened to Tippi before she'd built up a large deposit you would be able to accept monthly payments too? How do you respond?

△

6.13 Would you advise an owner of a low-risk dog to take out pet insurance?
PHOTO ANNE FAWCETT

6.4
FAIR DISTRIBUTION OF HEALTHCARE RESOURCES
SCENARIO
SCOPE OF SERVICE OF A CHARITY

6.4
Fair distribution of healthcare resources

Large healthcare organisations have difficult decisions to make about how to spend their resources. This next scenario considers the implications of an ethical policy proposed by a large veterinary charity.

SCENARIO
SCOPE OF SERVICE OF A CHARITY

▶ You are on the ethics committee of a pet charity. The organisation provides subsidised veterinary services for the financially disadvantaged and acts as a shelter. One of the goals of the organisation is to work against pet overpopulation by promoting neutering of pets. To this end, all animals rehomed by the organisation are neutered.

Recently there have been several cases where owners claiming to have no money demanded a caesarean section, and offered to pay off the account once the puppies were sold. A number of staff feel this goes against the philosophy of the organisation. They have proposed a policy whereby caesareans will only be performed if the owner consents to having the animal neutered at the same time.

What should you do?

RESPONSE
RICHARD GREEN

▶ In this scenario, the purpose of the ethics committee is to decide if this policy is ethically acceptable for adoption by the charity.

△

6.14 How should an animal welfare charity manage clients who rely on their discounted services to breed animals?
PHOTO ANNE FAWCETT

This will depend on a number of factors which may not all immediately be apparent, including: the charity's own vision, mission and values; the beliefs and values of the charity's trustees, staff and volunteers (which one hopes would be in alignment, but this may not always be the case); any veterinary policy such as this must be compliant with the national professional code of conduct for veterinary surgeons (and the charity would be well advised to run their policy past the conduct department for opinion prior to its adoption); and policies must also be acceptable to the majority of the public and media. Charities are totally dependent on the public for financial support, and a policy which is ethically sound but which flies in the face of public opinion or media approval will be potentially catastrophic.

The ethics committee would thus be advised to evaluate the proposed policy against all of these criteria, and to weigh up any potential deficiencies or problems against the potential benefits. It is of huge benefit to involve not only managers and key decision makers from the charity in evaluating a policy such as this, but also some of the staff likely to deal with the consequences of such a policy on the ground. It is all too easy to make what seems like a perfectly reasonable decision to adopt a policy such as this on a broadly utilitarian platform, but it is vital to consider what staff should and would do in practice when faced with a bitch urgently requiring caesarean section at 2 am presented by an owner who will not consent to a spay, and with no resources to go elsewhere. To such end it is useful when writing such policies to include some real case-based scenarios to facilitate their implementation.

Perhaps the primary concern for a welfare charity with a remit to provide subsidised veterinary services and to reduce pet overpopulation, will be to assess the impact on animal welfare of such a policy.

A simple ethical matrix would serve very well, and can be extended to look at the wellbeing, autonomy and justice of: the charity, its supporters, its staff, the individual animals affected (bitches requiring caesarean and their pups) and the pet population as a whole, with specific regard to those pets which end up in shelters through oversupply.

Clearly a key part of the evaluation will be to solicit veterinary opinion as to whether it is acceptable to spay bitches at the time of caesarean, and whether this has a significant (detrimental) welfare impact on either the bitch or the puppies when compared to a caesarean without spaying. If there is any significant impact, then the policy becomes very much harder to justify and implement for a number of reasons, although there are also arguments to be made that a bitch requiring one caesarean would be more likely to require subsequent ones, so that spaying her at

the time of the first would benefit her own welfare too in the long term.

Some people will also argue that spaying bitches is a mutilation and that preventing them from breeding compromises their autonomy, but this argument is far more wide-reaching than just this one scenario, and one that this animal welfare charity has already considered and discounted in its general neutering policy.

Problems arising if caesarean-spay does compromise individual welfare would be:

- Veterinary professional bodies tend to place far greater emphasis on vets' responsibilities to the welfare of the animal under their care than to others outside their sphere of responsibility.
- It would be very hard for a welfare charity to justify any deliberate and significant compromise of individual animal welfare, even for the "greater good".
- Both public opinion and the media are far more likely to focus on the individual than the bigger picture, so there is considerable risk to reputation.

If we assume (for the sake of argument) that the effect on both bitch and puppies of caesarean-spay is minimal, then the dilemma shifts largely to a conflict between the charity's mission to reduce pet overpopulation by preventing bitches from breeding again (and by the educational message that having such a policy sends), and owners unwilling to consent to caesarean-spay, but unable to afford private veterinary care.

Arguments for the charity adopting a hard line on such policies would be that, as long as the conditions are set out clearly and applied equally, then there is no significant moral difference between this and, for example, a private vet requiring payment for the operation. The veterinary professional body should be content as long as first aid and analgesia are always unconditionally offered. The

—
6.4

FAIR DISTRIBUTION OF HEALTHCARE RESOURCES
RESPONSE
RICHARD GREEN

policy also fits with the mission to reduce overpopulation, and is agreeable to staff, and one hopes, supporters. If there is no significant detrimental effect on the bitch or puppies of caesarean-spay, then there is unlikely to be any adverse publicity directly associated with the policy – although see below.

The areas where the policy may struggle ethically are: in obtaining informed consent from owners – most definitions of informed consent specifically preclude any form of coercion, which this might appear to be; and the dilemma which will face staff at 2 am – the intractable owner with the bitch requiring urgent caesarean – the unstoppable force and the immovable object. In this case, the main losers could be the bitch and the puppies, who are the pawns between the two, and the staff on duty may find themselves risking significant compromise to the bitch's (and puppies') welfare in order to enforce the policy – a high-stakes game of poker.

Even staff who happily support the policy in daylight hours may find it hard to offer a choice of euthanasia or "analgesia and find somewhere else" (and this clearly might not play well on social or other media) and for this reason, it is often prudent to rehearse such scenarios before they arise.

In this instance, one solution might be to agree to perform the caesarean without spay to alleviate the immediate welfare issue, but then to "ban" the client from further access to the charity's veterinary services.

And that is another, separate ethical dilemma.

ONE of the most interesting ethical issues of this case is the question of whether the charity should use its resources to influence people's actions in this way. Perhaps the charity should just be there to provide care as required, and deemed fit, by the people it serves. In that case the onus is on the clients, and the expectation perhaps is that they would only ask for services deemed acceptable by their wider society. Or conversely, it could seem unethical for a charity *not* to try to aim for long-term improvements through ensuring that its policy objectives are met. The level of restriction imposed by charities on their services is very variable and depends not only on their respect for personal autonomy but also pragmatically on the level of resource they have compared to demand.

Often such policies are developed on the basis of utilitarian or cost:benefit reasoning, but there is an expectation – particularly of a charity – that such policies are fair and just to stakeholders. Quite often when such policies are criticised, a deontological or rights framework is invoked – for example, refusal to perform a service may be seen as infringing the rights or autonomy of the client and staff.

FAIR DISTRIBUTION OF HEALTHCARE RESOURCES

What would you do?

You work for a veterinary treatment charity that provides services to clients on low incomes. Currently the charity will fund medical and surgical treatments that are costed at less than $1000 when there is a "reasonable" chance of the animal having a good quality of life for at least 2 years. The charity is always under financial pressure but last year it significantly overspent as income was down and more people used the service. Each veterinarian has been asked to contribute three types of treatment that they think could be cut.

ONE _____ Which types of treatment do you think might be candidates for the charity to cut?

TWO _____ Which elements of veterinary care would you not like to see sacrificed?

△
6.15 Children from a South African township have brought their puppies in for free veterinary care including vaccination.
PHOTO ISTOCK

6.5

Financial conflicts of interest

Financial conflicts of interest can occur when our secondary or private interests (for example, to make a profit) take primacy over our professional and public duties (for example, to work in the best interests of our patients and clients). Conflicts of interest in general are discussed in section 7.2. The need to remain profitable in order to be able to help other animals is important. But just

△
6.16
CARTOON RAFAEL GALLARDO ARJONILLA

6.5

FINANCIAL CONFLICTS OF INTEREST
SCENARIO
PROFIT SHARING

how profitable is it acceptable to be? And when could our motivation to make money become, or be seen to be, a conflict of interest with our primary role of helping our patients? The following scenario involves a financial conflict of interest.

SCENARIO

PROFIT SHARING

▶ You own a small animal practice employing three assistant vets. You have always employed people on a straight salary and have been making reasonable profits. A business advisor suggests to you that you could make more money if you employed your vets on a profit-sharing basis, whereby they would receive a small fixed salary but then an additional percentage of their takings. Your advisor says it could be a win-win situation as the best vets will earn more than they currently do, and you will also increase your profits.

What should you do?

△

6.17 Some practices run a profit-sharing scheme for vets and other employees.

PHOTO ANNE FAWCETT

RESPONSE

JAMES YEATES

▶ The ethics of this may depend on what it achieves. Let us differentiate three things:

(a) It may increase the amount of patients that each vet treats.
(b) It may increase the amount of treatment that each vet gives per patient.
(c) It may increase the amount each vet charges per treatment.

(a) seems a legitimate thing to reward – to a point. If one vet simply stays late while another is work-shy, then it seems fair to reward the former. This can go too far, for example if it leads to vets working when they are tired or stealing cases (or changing the name on the practice management system), but these eventualities can be avoided both in practical terms and ethically insofar as these represent constraining principles (just as laws on inheritance can be abused by people killing their aunts but are constrained by laws on murder). Plus, perhaps there are reasons why it would be unfair to pay differential amounts (e.g. if one vet was coping with a disability or did more work which is invoiced for a lesser amount, such as charity or shelter work which is discounted by the clinic).

(b) seems riskier. It is hoped that vets are making clinical decisions based on what animals need (and what owners can afford). As such, anything that biases that decision-making could be considered unethical. In particular, this seems unethical if it would encourage vets to "overtreat" animals, providing them with treatment that is not in their interests. Again, this could be argued to be an ethical constraint: vets have a responsibility not to let it influence their decision-making (but this seems somewhat naive). The danger of overtreatment suggests one case where a profit-sharing payment scheme to increase (b) might be legitimised: where vets are systematically undertreating animals when owners could afford to pay for treatments that would be in the animals' interests.

(c) seems even riskier. It is hoped that vets would charge a "reasonable" amount. One might argue that the responsibility to ensure they receive value for money is the buyer's. But this seems hard to justify in veterinary care, when owners lack the knowledge to distinguish service quality or even to meaningfully compare prices (they can assume oranges are similar from all vendors, or at least know what type they like, but be unable to meaningfully compare treatments for cruciate rupture). So vets have a responsibility to charge "fairly". Again, perhaps using profit-sharing as a means to offset undercharging could be legitimate (especially if the increased turnover went into charity work rather than partners' profits) but "overcharging" would not be.

Before concluding, it is worth considering a couple of practical matters.

Firstly, some ethical vets may actually end up working less hard, providing less treatment or charging less because they are worried about their own biases and therefore (over)compensate.

Secondly, returning to the question of fairness, there is a danger that differentiating pay on a profit-sharing basis would actually mean giving more money to the unethical employees – those who overtreat or overcharge, and perhaps – to a degree – those who overwork (depending on whether the motivation is to help more animals and clients, or to make more money). This not only seems unfair but may lead to those more ethical vets leaving or losing morale (from a profit-based perspective, good riddance, but nevertheless from an ethical point of view that is unfair).

Thirdly, and conversely, it could be the case that if a profit-sharing scheme is not introduced then otherwise good vets may leave the practice

and go to other, less scrupulous competitors. If so, perhaps there is a pragmatic reason for having a profit-sharing scheme to ensure that your more ethical practice is not outcompeted – but this argument is at risk of dangerous misuse, for as we have already seen such schemes may favour unethical veterinary activity.

Finally, of course, clients are unlikely to be pleased to hear that their vet is incentivised to provide as much treatment and charge as highly as possible and you may find that overall it has a negative effect on your profit margins, especially in comparison to marketing yourself as not having a profit-sharing scheme – unlike your unscrupulous competitor.

Conclusion
The risks of the scheme seem sufficiently likely to outweigh the benefits (apart from your own profit). Unless in the specific case you can genuinely be sure (and monitor) that this is not the case, then it seems safer ethically to avoid the scheme.

||||||||||||||||||||||||||||||||||||||||||||||||||||||||||||||||||||||||||||||

THESE profit-sharing schemes are becoming increasingly common in some areas of veterinary practice. In line with James Yeates' views expressed above Glen Cousquer also felt such schemes to be ethically dubious, suggesting that vets' "remuneration should be fair, but should not tip the balance in favour of exploitation of vulnerable clients and insurance companies" and that "it may be appropriate to consider other reward schemes that uncouple pay from work undertaken, with rewards being recognised through reviews of an individual's CPD budget, holiday entitlement or the staff social fund" (Cousquer 2011).

There has been a surge in interest in what motivates people in different types of jobs. Time and

"Our clients absolutely need to trust that we have their animals' best interests at heart."

again, particularly in professional roles, money is not found to be a strong motivator, and in these cases providing monetary incentives can actually have negative consequences in the long term by reducing intrinsic motivation, enjoyment of the job and productivity (Pink 2009).

As we are operating in a sub-free market, due to the inequality of knowledge of the transacting parties we need to be particularly mindful of our need both to act with integrity, and to be seen to be doing so. Our clients absolutely need to trust that we have their animals' best interests at heart. In the UK at least, veterinarians are one of the most trusted professions, coming third behind pharmacists and opticians with 34 per cent of people completely trusting vets and another 61 per cent generally trusting them (BVA 2015). To lose that position would be devastating, making communicating with clients difficult for individual vets and with potential knock-on consequences for animal welfare.

FINANCIAL CONFLICTS OF INTEREST

What do you think?

ONE _____ What aspects of veterinary practice would it be ethical for practice owners to reward staff for?

TWO _____ How should you decide how much profit is reasonable for a practice to make?

△
6.18 Veterinary practice is about more than selling.
PHOTO ISTOCK

Conclusion

In private as well as public practice, financial considerations will affect the care animals receive. There is often pressure to discount services for clients or provide pro bono care for some animals. These can be stressful experiences for the veterinary team and in some cases a practice policy governing such requests may help. In private practice there is a requirement to remain profitable in order to be able to help other animals. However, care is needed to ensure that a conflict of interest derived through financial motivation does not negatively affect animal welfare.

References

ABI 2013 Insurers pay out over £1.2 million every day to treat sick cats and dogs. Association of British Insurers.

ABI 2014 We are claiming cats and dogs – pet insurance claims on the rise. Association of British Insurers.

Anon 1935 An RSPCA Liverpool animals clinic. *Veterinary Journal* **91**: 187.

Batchelor CEM, and McKeegan DEF 2012 Survey of the frequency and perceived stressfulness of ethical dilemmas encountered in UK veterinary practice. *Veterinary Record* **170**: 19.

BVA 2015 Public Trust in the Veterinary Profession. British Veterinary Association: London, UK.

Cousquer G 2011 Principled profit-sharing? *In Practice* **33**: 142–143.

Davidson J 2015 Forever defending vet practice as a business. Vet Times. https://www.vettimes.co.uk/forever-defending-vet-practice-as-a-business/

Embrace 2013 Facts & Statistics on the US Pet Insurance Industry. Embrace Pet Insurance.

Felsted K 2014 Controlling discounts and missed charges. *In Practice* **36**: 371–373.

Gardiner A 2014 The 'dangerous' women of animal welfare: how British veterinary medicine went to the dogs. *Social History of Medicine* **27**: 466 487.

Kogan LR, Stewart SM, Dowers KL, Schoenfeld-Tacher R, and Hellyer PW 2015 Practices and beliefs of private practitioners surrounding discounted veterinary services and products. *The Open Veterinary Science Journal* **9**: 1–9.

McCulloch S 2011 Everyday ethics. *In Practice* **33**: 297–298.

Moore AS 2011 Managing cats with cancer: an examination of ethical perspectives. *Journal of Feline Medicine and Surgery* **13**: 661–671.

Petherick JC 2005 Animal welfare issues associated with extensive livestock production: the northern Australian beef cattle industry. *Applied Animal Behaviour Science* **92**: 211–234.

Pink D 2009 *Drive: The Surprising Truth about What Motivates Us*. Riverhead Books: New York.

Rollin BE 2006 An ethicist's commentary on soliciting client contributions to a fund raiser. *Canadian Veterinary Journal-Revue Veterinaire Canadienne* **47**: 1158.

Rollin BE 2011 Euthanasia, moral stress, and chronic illness in veterinary medicine. *Veterinary Clinics of North America-Small Animal Practice* **41**: 651–659.

Tannenbaum J 1995 *Veterinary Ethics: Animal Welfare, Client Relations, Collegiality*. Mosby-Year Book: St Louis.

Taylor N 2012 The Potential Pitfalls of Pet Insurance. https://orchardvetgroup.com/2012/02/22/the-potential-pitfalls-of-pet-insurance/

Watkinson M 2009 Why I'm ashamed to be a vet: a shocking exposé of the profession that puts pets through 'painful and unnecessary treatments to fleece their trusting owners'. *Daily Mail*.

CHAPTER 7
PROFESSIONALISM

Introduction

Being a professional comes with some special expectations in society. Besides competency in the technical aspects of the job, trustworthiness and other virtues are usually seen to be part and parcel of being a professional.

According to Eliot Freidson, author of *Professionalism, The Third Logic*:

> "Professionalism may be said to exist when an organised occupation gains power to determine who is qualified to perform a defined set of tasks, to prevent all others from performing that work, and to control the criteria by which to evaluate performance."
>
> (Freidson 2001)

Many talk of a sort of professional–social contract: society allows professions to self-govern and determine criteria for entering the profession, as long as the profession conducts itself ethically. Hence veterinary practice around the world is generally a regulated and respected profession, with only those veterinarians who attain and maintain specified standards of care being allowed to practice. This is a unique privilege which, if abused, could conceivably be withdrawn by society.

> "...professionalism is a set of institutions which permit the members of an occupation to make a living while controlling their own work. That is a position of considerable privilege. It cannot exist unless it is believed that the particular tasks they perform are so different from those of most workers that self-control is essential."
>
> (Freidson 2001)

According to the Australian Council of Professions (1997), "A profession is a disciplined group of individuals who adhere to *high ethical standards* and uphold themselves to, and are accepted by the public as possessing special knowledge and skills in a widely recognised body of learning derived from research, education and training at a high level, and who are *prepared to exercise this knowledge and these skills in the interest of others*" (emphasis added).

Problems with professions

The concept of a profession is open to criticism. Common criticisms include:

- The fact that a profession has a monopoly on providing a service or services makes this open to abuse, for example, members of a profession might seek to maximise their profits while failing to provide appropriate benefit to consumers.
- Because there is a restriction on competition, since only members of the profession can provide the same service, this lack of competition may not be sufficiently motivating and professionals may be inefficient, unreliable and costly.

7.1
CARTOON AILEEN DEVINE

(225)

7.0
INTRODUCTION

- Because professionals are free to make professional judgements, they may be prone to error.

These arguments have been used, for example, in recent debate about "anaesthesia-free dentistry" offered by non-veterinarians. Despite numerous risks, anaesthesia-free dentistry is sold as less risky for pets because anaesthesia is avoided. Providers have argued that vets abuse their monopoly by providing costly treatment. It's an argument that is believed by some members of the public, despite sound reasons for providing anaesthesia (American Veterinary Medical Association 2016).

Codes of conduct and professional codes of ethics are formulated, in part, to justify trust in professions.

They provide guidelines which, if followed, prevent professionals from taking "selfish advantage" of their monopoly.

They provide guidelines for appropriate behaviour and also form a basis for disciplinary action.

Professional ethics are a set of rules or principles created by a profession to ensure that professionals do the right thing, i.e. act ethically. How is that judged? Professional ethics need to be aligned with the predominant social ethic. If it is generally held in society that it is right to recognise and respect private property, the professional code of ethics should be consistent with that. Similarly if the predominant social ethic holds that individuals have a right to privacy, the professional code should be consistent with that.

What would happen if professions could create their own code of ethics without acknowledging the predominant social ethics?

- It could mean that the professional code of ethics purely served the interests of the profession.

△

7.2 Maintaining professionalism is challenging in the face of workplace stress.
CARTOON MALBON DESIGNS

- Professionals might be able to justify price-fixing or protection of members by other members, thereby serving the needs of the profession ahead of those of the clients or the public.
- This would erode public trust and the ultimate outcome may be society withdrawing the profession's ability to self-regulate.

One increasingly common feature of being allowed to conduct veterinary practice around the world, and similar to medical practice, is the requirement to swear an oath, agreeing to abide by certain overarching guiding principles. The World Small Animal Veterinary Association (WSAVA) collated information about the oaths sworn in 15 countries. Of these, all included reference to promoting animal welfare, very often first in the oath and cited as the primary responsibility of veterinarians in two oaths (UK and Iran); all included requirements to act ethically; and 10 oaths included a promise by veterinary surgeons to continue professional learning throughout their careers (WSAVA 2013). Subsequently the WSAVA developed a "Global Veterinary Oath" they deemed suitable for all veterinarians:

THE WSAVA GLOBAL VETERINARY OATH
"As a global veterinarian, I will use my knowledge and skills for the benefit of our society through the protection of animal welfare and health, the prevention and relief of animal suffering and the promotion of 'One Health'. I will practice my profession with dignity in a correct and ethical manner, which includes lifelong learning to improve my professional competence."

(WSAVA 2014)

In this chapter we will consider scenarios relating to professional conduct, including conflicts of interest, indiscretions and professional obligations.

7.1

Confidentiality

Client confidentiality or professional secrecy is one of the cornerstones of veterinary practice. Like other professionals, veterinarians are required to ensure the confidentiality shared by clients (Lachance 2016). In society individuals may reasonably expect to have a right to privacy, including privacy of information shared with professionals. It is important because it signifies the trust between the veterinary team and client which is necessary for both parties to be able to make the right decisions about an animal's welfare. The requirements for client confidentiality are usually spelled out in professional guidance, for example, in the UK in the Veterinary Nurse (RCVS 2015a) and Veterinary Surgeon (RCVS 2015b) Codes of Professional Conduct. In the first scenario we consider a possible breach of client confidentiality through an online social media post.

△
7.3
CARTOON AILEEN DEVINE

SCENARIO

SOCIAL MEDIA CONFIDENTIALITY

▶ You are a veterinary student on clinical placement at a local veterinary practice. You regularly keep in touch with friends and colleagues via Facebook. Another veterinary student and close friend, on placement at the same veterinary practice, has publicly posted a comment to Facebook about their recent experience assisting with surgery. The post reads:

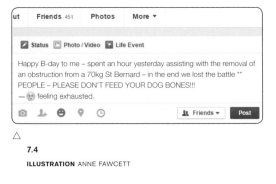

△
7.4
ILLUSTRATION ANNE FAWCETT

What should you do?

RESPONSE

JASON B COE

▶ Social media offers a relatively new form of communication within society, where it is argued that boundaries can become blurred between one's personal and work life, posing risks for veterinary staff, veterinary practices and the veterinary profession (Coe, et al. 2011, Weijs, et al. 2013).

Understanding and managing the inherent risks of using social media that exist for all members of the veterinary profession are important in preserving the profession's reputation and, in turn, clients' trust.

In this scenario a veterinary student comes across a post to Facebook by a classmate. In the post their classmate expressed publicly a recent experience of losing a surgical patient due to a gastrointestinal obstruction that resulted from a client feeding their dog bones. Nowhere in the post is there an indication that the client provided consent to the student to disclose publicly information about the surgical case. As such, the scenario poses an ethical dilemma for the student reading the post in how they should respond. Some of the options that exist for the student who has come across the post include ignoring the post, speaking to their classmate about removing or modifying the post, or alerting management at the clinic to the post.

In choosing a path forward it is important first to identify the stakeholders affected by the situation. In this scenario, consideration must be given to the effect the post has on the veterinary student who made the post, the client involved, companion animals in general, the veterinary practice and its members, the veterinary profession and society as a whole. Next, one needs to consider the ethical principles of non-maleficence, beneficence and autonomy in relation to the effects the post has on each of the stakeholders.

In posting the comment, it is possible that the student veterinarian intended to promote good (i.e. beneficence) for companion animals. More and more, Facebook and other social media are used by individuals and organisations, including veterinary practices, to provide public messaging and education. It is very likely the student veterinarian intended, at least in part, for their post to act as a public message advising people not to feed their dog bones, with the intention of benefiting other animals. In addition, Facebook has been found to offer a form of social support for its users (Hampton, et al. 2011); therefore, it is possible the student veterinarian gained some individual benefit from the post through the opportunity to vent their frustration to friends at having lost a patient in their care. Although the intention and outcome of the post carry the potential for good, the broader effects of the post, including possible harm to other stakeholders, need to be weighed.

At first glance this post may appear to pose little risk to the client because it does not specifically identify the client or patient by name; however, it is possible that information available on the veterinary student's Facebook profile can be pieced together along with details of the case in such a way that the client involved, and likely others, are able to identify the client or patient, resulting in a breach of client confidentiality (Australian Medical Association 2010).

Although people, including veterinary students, are entitled to choose the content they post to their own Facebook profile, this autonomy is limited insofar as societal laws and professional standards permit (Coe, et al. 2011). In many jurisdictions, breaching veterinary–client confidentiality is considered professional misconduct (Ontario Government 2011, RCVS 2015b). Veterinary professionals and staff have an obligation not to reveal information about a patient or client except in various prescribed situations, such as information that is disclosed with the consent of the client, information arising from reportable diseases, when acting upon a court order (such as a subpoena or summons), information that is for the benefit of human or animal health and, in some jurisdictions, information relating to suspected animal abuse (Douglas C. Jack, personal communication, 18 June 2015).

Unless one of the exemptions is present, information cannot be shared beyond members of the veterinary profession. Further, clients reasonably trust that the details of their interaction with members of the veterinary profession are confidential (Weijs, et al. 2013), unless consent has been obtained from the client. Therefore, by posting information that breaches client confidentiality,

this student exceeds the boundaries of autonomy and their entitlement to post what they choose. A breach of client confidentiality brings potential harm to the client (e.g. loss of trust, feeling publicly admonished), the veterinary student (e.g. reprimand, disciplinary action) and the individual(s) ultimately responsible for ensuring that client confidentiality is maintained within the veterinary practice (e.g. reprimand, loss of trust). Although in this scenario some good may come from cautioning the public against feeding their dogs bones, alternative ways to educate the public exist. Therefore, any benefit that may exist from this post for companion animals does not outweigh the harm that the breach to client confidentiality would bring.

A breach of client confidentiality also poses potential harm to the student's peers including the veterinary practice and veterinary profession they represent, as well as society as a whole. Trust forms the foundation of effective healthcare relationships (Mechanic 1998). Veterinarians and veterinary personnel including students have important roles in managing the veterinary profession's reputation in a manner that preserves the public's trust. The veterinary student in this scenario is a representative of the veterinary practice where they are on placement and, in turn, the veterinary profession; therefore, the student's breach of the client's confidentiality is a violation of the client's trust in the practice and its members, and more broadly, undermines society's trust in the veterinary profession overall. As a result, the post poses harm to the veterinary practice (i.e. loss of clients' trust) and the veterinary profession (i.e. loss of society's trust).

In this scenario, the harms to the client, the veterinary practice, the veterinary profession and to society as a whole outweigh any benefit to companion animals or individual benefit gained by the student making the post. In addition, the student has exceeded his or her right to post what he or she chooses by breaching societal and professional expectations for client confidentiality. Based on this assessment, the student observing the post needs to take the actions necessary (e.g. discuss with their classmate, notify management) to have the post removed or modified to eliminate the breach of client confidentiality. The student reader may also take the opportunity to suggest to their classmate alternative approaches for educating the public about the dangers of feeding their dogs bones. In this case, it would have been relatively simple for the veterinary student to seek out the consent of the dog's owner. In addition, the student veterinarian in this scenario should be cautioned that using privacy settings to protect who views the post would still not be an acceptable option for sharing details about the clinical case without the client's consent (i.e. breach of client confidentiality). Precedence suggests online disclosures, including those protected by privacy settings, are accessible to evidentiary discovery in courts of law (Dhawan 2009).

To prevent future harm, this scenario offers the veterinary practice the opportunity to pursue a discussion with staff about client confidentiality and the boundaries that exist in posting details about clients and patients online without client consent. The conversation should include a discussion with staff about privacy settings and how they are insufficient to legally protect information posted online from evidentiary discovery. Furthermore, information protected with privacy settings is often still available to a large audience of individuals. For example, a recent study involving veterinarians identified that participants on average had 225 Facebook friends (Weijs, et al. 2014). When privacy settings are managed to allow friends of friends to see a post, the scope of the audience for that post becomes even greater (e.g. 225 Facebook friends each with a mean of 225 Facebook friends could result in a

post reaching 50,625 Facebook users). Discussing the discrepancy that can exist between the intended audience for a post and the reality of how far information posted to social media can travel is an important consideration in managing the potential risks of social media. In a recent study exploring veterinarians' use of and attitudes toward Facebook (Coe 2014), a number of participants were able to provide real-life examples of the negative consequences that can arise from the public posting to social media of certain content by veterinary staff members. As shared by one veterinarian participating in the study:

"A technician had various photos posted relating to patients. The photo album privacy setting had not been changed… A client who was a FB friend of another person in the practice found these photos and felt they were highly inappropriate… The client brought a case against the technician with the licensing board."

Being aware of the risks associated with social media and the ways that breaching client confidentiality can happen on social media will provide staff members who represent a veterinary practice and ultimately the veterinary profession with the information needed to manage the potential risks and repercussions of posting client- or patient-related information online. In addition, proactively promoting a practice culture where staff informally regulate one another's use of social media offers a process from within the veterinary team to immediately identify and rectify potential breaches of client confidentiality in order to safeguard the public's trust in the veterinary profession.

||||||||||||||||||||||||||||||||||||||||||||||||||||||||||||||||||||

IN this scenario, Coe has taken a utilitarian approach and highlighted a number of issues, including the negative consequences of a breach of client confidentiality. However, preserving confidentiality does not mean work-related matters must stay at work. It can be acceptable to talk about clients and patients when they cannot be identified in any way. Using a pseudonym for a patient may not be enough to prevent recognition and any given time frame could also act as a clue and make recognition more likely. It may also be that the information you have posted elsewhere, such as the place of work, will add up to possible identification. Guidance for doctors reminds them that in order to maintain confidentiality "you must ensure that any patient or situation cannot be identified by the **sum** of information available online" (Anon 2010; emphasis in original).

Two elements about the motivation of the poster in this case were clear. First, the poster was compelled to try to educate the public about the risk of exposure to bones, which has led in this case to a fatal manifestation of an ultimately preventable condition. The second, perhaps less obvious, is an apparent need for the poster to share what has been a traumatic experience.

CONFIDENTIALITY

What do you think?

ONE   Assuming the intention of the poster in the above scenario is to prevent intestinal obstruction in other animals, what alternative approaches might you suggest?

TWO   Veterinarians, nurses, technicians and animal carers are often exposed to trauma and many cases have poor outcomes. What are more appropriate ways for the poster to express their feelings about the case?

THREE   How can health professionals post online about cases without breaching client confidentiality?

△
7.5 The temptation to post one's current status on social media can be problematic for professionals.
CARTOON RAFAEL GALLARDO ARJONILLA

|||||||||||||||||||||||||||||||||||||||||||||||||||||||||||||

OTHER problematic elements of professional conduct are raised by social media and online posting. Those related to relationships between veterinary team members are discussed in chapter 11 and here are three further examples:

"Will you be my friend?"

△
7.6 How would you respond if a client sends a friend request on social media?
ILLUSTRATION ANNE FAWCETT

Maintaining good relationships with true friends, whilst protecting oneself from calamitous exposure to the wider online audience, is the art of successful management of social media. Accepting clients, or even former clients, as friends blurs the professional/personal boundary. It's very important that a veterinary team member's relationship with clients is professional. Whilst it may not always seem like it when dealing with difficult clients, there is a power imbalance in this relationship – team members have knowledge about, and influence over, the care of the client's animals. Keeping a professional distance from clients protects them (and team members) from exploitation. For some time, doctors and teachers in the UK have both been advised to keep

their private and personal lives separate by politely declining requests from patients and pupils, both current and former, to become a "friend" on social media (BMA 2011, NUT 2010). However, this is much less clear-cut when an existing "friend" on social media becomes a client.

"I'm a bit hungover today"

△
7.7 The person posting this may have been joking, but not everyone will interpret this generously.
ILLUSTRATION ANNE FAWCETT

How would you feel about a client knowing you were out late the night before their pet comes in to see you in the surgery? We've all felt a bit off-colour at work (whether self-inflicted or not) and usually make a decision about whether we can still perform our duties safely before coming in to do our job. In one case a government worker in the UK tweeted that they were hungover at work and was subsequently vilified in the press who printed her photo for millions to see. She argued that the tweets were private to her Twitter followers but in a landmark case the Press Complaints Commission found the tweets were in the public domain and that this information was of wider interest as it affected her ability to do her job (Press Complaints Commission 2011).

Whilst Twitter is perceived as a more public forum than Facebook, there was a case of a vet student jokily posting on Facebook that they were struggling to perform their anaesthesia checks due to being hungover. Unfortunately, a (Facebook) friend of a (Facebook) friend's pet was

> "Keeping a professional distance from clients protects them (and team members) from exploitation."

undergoing surgery at the practice that day. Whilst in this case no harm came to the pet, and it is not clear whether the student's condition affected their ability to act competently, broadcasting this doubt to a large number of people resulted in a protracted period of explanation by both the student and the practice. It was considerably more harmful than making a light-hearted comment to a few friends after work.

"Let's post some funny pictures"

Everyone loves a funny picture and, more than most workplaces, veterinary team members have access to all sorts of unusual sights. Posting funny pictures of people or animals may bring the poster, the practice and the profession into disrepute. A group of seven doctors and nurses

in the UK were suspended after they posted on Facebook pictures of themselves playing "The Lying Down Game" in a range of places around their hospital (Fleming 2009). A student nurse in the USA was expelled from her course, along with three other students, for posting a picture of herself posing with a placenta. In itself the picture was pretty inoffensive, along the lines of "this is what I do in my practicals" rather than "look how I can fool around with body parts". However, the nursing tutor at the university stated that her "demeanor and lack of professional behavior surrounding this event was considered a disruption to the learning environment and did not exemplify the professional behavior that we expect in the nursing program" (Anon 2011). Similarly, posting images of patients recovering from anaesthesia, for the purpose of humour, may be seen as exploiting their vulnerability.

Posting photos can also be problematic if the animal can be identified and client confidentiality is breached. Even if there is a good reason to post a picture, such as requesting help from other professionals about the care of the animal, it is still important to maintain client confidentiality. In such a case it would be more appropriate to send an email, or communicate via a password-protected forum.

△
7.8 Posting "funny" pictures can backfire.
PHOTO ANNE FAWCETT

CONFIDENTIALITY

What would you do?

You are working at an animal shelter. An eight-year-old Golden retriever is surrendered by a family whose circumstances have changed, such that they can no longer care for the dog. They express their wish that the dog be rehomed. On admission to the shelter the veterinary team find that the dog is suffering from severe arthritis and make the decision to euthanase the animal. Not all team members agreed with the assessment. You notice later that one staff member who had been in charge of looking after the dog all weekend has posted the dog's image on social media. What do you do?

△

7.9 It can be challenging to resist the urge to vent on social media, particularly in relation to topics one is passionate about.
PHOTO ANNE FAWCETT.

CONFIDENTIALITY

What would you do?

You attend a continuing education workshop and take photos at a luncheon which includes wine-tasting. You post a group photo (which includes yourself) of veterinarians toasting the tutor with champagne on your Facebook page. A week later, a loyal client makes a complaint about the appearance of a surgical wound in their pet. Your employee performed the surgery. The complaint includes a photo of their pet's wound and a link to the group photo, and states that your employee "must have been drunk" at the time of surgery to produce the result. How should you respond?

△

7.10 The availability of wine at a continuing education lunch causes unforeseen problems.
PHOTO ISTOCK

CONFIDENTIALITY

What would you do?

You are a veterinary student on clinical place-ments. You keep in touch with colleagues via Facebook. One of your fellow students posts about the experience of collecting semen from a dog at a practice that undertakes a significant amount of the region's canine artificial insemina-tion. The post shows a photo of the student grin-ning, with a small amount of blood on her gloved hand. "OMG FAIL...I made his penis bleed and my supervisor had to take over. LOL". How should you respond?

△
7.11 There are different ways you can respond to a social media post in person and online.
PHOTO ISTOCK

‖‖‖

IN the following scenario we will consider a case where clients are in conflict with each other. Before thinking about conflict resolution the veterinary team must decide whether client confidentiality will be breached by bringing the parties together. To pass on confidential informa-tion, such as details about a patient's treatment or client personal or financial details, to a third party would be considered unethical. However, just letting it be known that you might know of the opposing party, and importantly, asking per-mission from each party to bring them together, will not necessarily breach client confidentiality.

SCENARIO
PATIENT KILLS ANOTHER PATIENT

▶ You are working in a companion animal prac-tice on a busy Saturday morning. One of your clients, Mrs L, rushes into the waiting room, distraught, with her cat Chloe who is bleeding profusely. As you perform your examination and administer oxygen Mrs L explains that a man was walking his dog past her house, on a lead. The cat hissed and the dog lunged, grabbed the cat and shook it. Both owners attempted to pull the animals apart and both were bitten and scratched in the process.

Chloe has a number of injuries including a broken neck. You euthanase her on humane grounds.

You ask if Mrs L knew the man with the dog.

"I didn't get his details," she sobs. "I just raced down here as soon as I could get Chloe out of that dog's mouth. I wish I knew. He'll pay for this."

As you step out of the room to bill up the con-sultation, Mr R, another client, rushes into the waiting room. He has scratches all over his fore-arms and his dog, Oscar, has multiple lacerations around the muzzle and face.

△

7.12 When one client's animal injures another there are many ramifications.
PHOTO ISTOCK

"My dog was attacked by a cat," he says. "Can you see Oscar now?"

You strongly suspect that this is the dog that killed the cat.

What should you do?

———

RESPONSE

GLEN COUSQUER

▶   Who is the protagonist in this situation?

This question is a good starting point for reflection for it reveals much about how people tend to respond when faced with such a conflict. Did you feel inclined to identify one of those involved as the protagonist, the other as the antagonist? Would you approach such situations with the assumption that someone is to blame and that there is a right and a wrong? How does this then apply to the

animals involved – that is to say – how does it apply in the field of non-human relations?

Conflict is a fundamental part of everyday life. As a veterinary professional and as a member of a community, one has to consider how one responds to, and manages, conflict. Dealing with conflict is, however, challenging at the best of times, particularly when powerful emotions hold sway.

So how does one approach such matters?

According to Kraus (1993), "... the psychological egoism and shortsightedness of individuals, which work in tandem to cause conflict and noncooperation in the state of nature, must be attributable to intrinsic human nature. Without some coercive apparatus to introduce external incentives for cooperation, intrinsic human nature will continue to produce conflict and noncooperation." In recognising this, it becomes clear that your role cannot be limited to dealing with the medical needs of those involved. It must take into account the need to address the conflict and will, ideally, seek to establish a cooperative situation.

Conflict resolution

Since the end of the Second World War, conflict resolution has established itself as an academic discipline (Ramsbotham, et al. 2011); it is to this body of work that we turn for insights into what the clinician should do in this situation.

David Hume (2003 [orig. 1739]) saw morality as a product of the emotions, placing empathy (which he called sympathy) at the top of the list. Hinckfuss (1982) argues that, for this reason, "morality always inhibits the rational resolution of conflict whenever it is used within a dispute because decisions are thus made on the basis of false beliefs." He further argues that: "Using morality as a device within the resolution of conflicts is like using a brick as a toothpick. If you want to be rid of the fibre between your teeth and you do not want broken teeth, then throw the brick away, and think of how best you can rid yourself of the fibre without it." This allows him to argue for the rational resolution of conflicts. Hinckfuss (1982) proposes that:

"the rational resolution of conflict involves the following

(a) sorting out any conceptual confusions between them relevant to the conflict;
(b) finding out the facts of the case relevant to the conflict and,
(c) if it is still necessary, devising ways of solving their mutual problem.

The object of requirements (a) and (b) is to eliminate the possibility of a dispute continuing when there is no conflict of interests, but merely a belief on the part of one or more of the disputants that there is such a conflict."

The words of Donald Nightingale (1976) sum up the matter:

"When a conflict situation is defined in terms of absolutistic values or in terms of ideological principles, parties have little room to manoeuvre. Beliefs about human rights, moral precepts and ideology cannot be sacrificed piecemeal to an opponent. There is an all-or-nothing quality to such conflict situations, which makes resolution difficult."

So how can these insights into rational conflict resolution help here?

A way forward
In the first instance, further conflict should be avoided by ensuring that those involved receive the treatment they need and do not meet in the waiting room. Oscar and his owner Mr R should be ushered straight into a free consulting room. Mr R can then be provided with the materials

required to clean his own injuries whilst waiting for you to attend to Oscar.

Given that it is a busy Saturday morning, that Chloe has been euthanased and that Mrs L probably needs to seek medical attention for her own injuries, it would not be inappropriate for Mrs L to be walked straight out to the car park. Arrangements could be made for her to return to pay the bill at a later date, booked to coincide with a follow-up appointment during which bereavement support could be provided and the facts of the case further clarified.

Mr R could similarly be encouraged to return for a follow-up appointment at which support could be offered and the facts of the case revisited.

On this particular Saturday morning, further conflict has been averted. There is, however, clearly a need to cooperate to address the wider situation and work for the greater good.

With this in mind and, once the facts of the case have been explored with each owner, it could be suggested that they help in the production of a practice newsletter or even an article for the local paper. This should be proposed as a joint project with clear expectations set that the two owners work together in an attempt to help others avoid such traumas in the future.

With the consent of the owners, each owner could be encouraged to produce an account of the event, detailing their own experiences as well as what they think may be the others' views on what happened. These accounts can then be exchanged and you can arrange to discuss these differing perspectives with each owner. Recognition that all four individuals have been traumatised and that both owners have regrets and are sorry can help.

Providing any conceptual confusion is addressed and, providing the facts of the case are clarified, this gives an opportunity for ways to be found to solve the problem to the satisfaction of both parties. In order to realise this – Hinckfuss's third stage of conflict resolution – it is suggested that the two owners be invited to meet and further discuss what happened and what lessons can be drawn.

At this meeting, the task confronting the disputants is to move beyond conflict and to work together to reinvent the situation. They must in the words of Hinckfuss (1982) "devise different means of satisfying their mutually consistent fundamental desires, other than via the secondary desires which brought them into conflict in the first place". This involves the recognition that "the fact that secondary desires are in conflict does not entail that there is inconsistency between desires at a deeper level".

Thus Mrs L and Mr R can come to recognise that, fundamentally, they both want their pets to live free from fear and conflict. They also realise that this ideal is only possible if owners understand each other and work together to do what they can to avoid conflict. They find themselves willing to work with the practice team to produce material for the practice newsletter and website. This then leads to them helping the practice redesign their own waiting room and producing signs that help owners develop a better understanding of their responsibilities. Further down the line, this experience leads the practice to fund training in bereavement counselling and conflict resolution for members of their nursing team.

CONFIDENTIALITY

What do you think?

Client confidentiality may be breached by receptionists on the phone in the waiting room, leaving notes visible to other clients and by simply chatting to others.

ONE **Can you think of examples where client confidentiality has been breached?**

TWO **How serious would you consider these examples?**

THREE **What could have been done to avoid breaches of confidentiality?**

FOUR **What ethical justification could you give for taking or not taking these actions?**

△
7.13 Even talking to clients in the reception area can lead to breaches of confidentiality.
PHOTO ISTOCK

7.2

Conflicts of interest

A conflict of interest is defined as:

(1) A situation in which the concerns or aims of two different parties are incompatible;
(2) A situation in which a person is in a position to derive a personal benefit from actions or decisions made in their official capacity. (Oxford Dictionaries 2016)

We usually talk about primary interests or public duty, for example the duty of a veterinarian as a professional with a primary endeavour to benefit animals and society, and secondary or private interests, for example personal benefit such as making a profit. Conflicts become particularly toxic when subsidiary or secondary interests take primacy.

An example of (1):

- A conflict between the interests of an animal (for example, to live a good-quality life and die a humane death) and the interests of the owner (for example, to prolong suffering to keep an animal with them; to maintain an animal in a situation that is not good for the animal's welfare – e.g. animal hoarding).

Examples of (2):

- A veterinarian breeding and selling brachycephalic dogs, and performing surgery to correct brachycephalic airway syndrome;
- A specialist veterinarian presenting a talk on the treatment of a medical condition by a pharmaceutical or pet food company that sells a particular treatment of a medical condition;
- A health professional drawing a commission based on number of sales, regardless of whether these are in the interests of the patient;

△
7.14 Treating a sporting animal that one has a financial interest in is a conflict of interest.
CARTOON MALBON DESIGNS

• A veterinarian treating sporting animals that he or she has a financial interest in, or bets on.

There are alternative definitions of conflict of interest, for example:

"A conflict of interest is a conflict between public duty and private interests which could influence the performance of official duties and responsibilities. A reasonable perception of a conflict of interest is where a fair-minded person, properly informed as to the nature of the interests held by the decision maker, might reasonably perceive that the decision maker might be influenced in the performance of his or her official duties and responsibilities."

(Australian Health Practitioner Regulation Agency 2013)

Maximising personal gain is not consistent with professional ethics, according to Freidson:

"There can be no ethical justification for professionals who place personal gain above the obligation to do good work for all who need it, even at the expense of some potential income… it is not profit itself which is unethical, for all workers must gain a living: it is the *maximisation* of profit that is antithetical to the institutional ethics of professionalism."

(Freidson 2001; emphasis in original)

Conflicts of interest may be real (where there is an actual conflict of interest between a person's private interests and public duties); potential (where a person has private interests that could potentially conflict with their public duties); and/or perceived (where someone might perceive that a person's private interests could improperly influence the performance of their public duties). Conflicts of interest, unmanaged, undermine public trust in the profession, which is the foundation of professionalism.

A conflict of interest itself is not inherently negative. However, the term itself is problematic because it is perceived as a pejorative term: there is a presumption of inappropriate behaviour. It also suggests that primary and secondary interests are always in conflict, which may not be the case. The literature on conflicts of interest tends to focus on financial conflicts and ignores non-financial interests such as the prospect of fame, e.g. publication, invitations to speak at conferences based on a non-funded study outcome and so on.

In response to these criticisms, Cappola and Fitzgerald suggest scope for use of the term confluence of interest when primary and secondary interests align (Cappola & Fitzgerald 2015).

In the example of the veterinarian breeding dogs and performing surgical procedures, the interests of those animals are compromised (the vet knows they can perform the surgery and charge for this so their secondary interest in profiting may take precedence over their duty to the welfare of the animal). This raises the question: are veterinarians part of a profession or an industry?

In this case the interests of the client (in having a "cute" animal) and the interests of the vet (making money) are sustaining a population of sick animals – the costs of which are being borne by the animals and the owners.

A professional uses their craft and skill to benefit society. A professional acting out of duty does so because they have a duty to the animal. So they will still use their skill to treat that animal; however, they would take steps to avoid suffering being passed onto future generations (e.g. by desexing that animal, and in their control by not breeding from those parents). In such a case, drawing money from performing surgery is a confluence of interests, because it is not being done primarily for commercial gain.

A similar argument might apply around pre-scription. It is argued that veterinarians have a conflict in both prescribing and dispensing drugs, because we profit directly from our recommenda-tions. If we are dispensing drugs as a service to the client, for example, convenience, and they are aware there are alternatives (e.g. taking a script to the pharmacy), there is no conflict. If we dispense drugs to profit from these, and do not offer alterna-tives, there is a potential conflict of interest.

Williams (2002) listed the interests of veteri-narians as:

(1)   Acceptance of the use of animals (with a spectrum of views on the moral standing of animals);
(2)   An interest in preventing and relieving animal suffering;
(3)   Maintaining health, wellbeing and productiv-ity of animals;
(4)   Earning a living and maintaining positive relationships with clients (adapted from Wil-liams 2002).

There is scope for conflict here, for example between 1 and 2 (where animal use causes suf-fering); and 3 and 4 (where the interests of the animal conflict with the interests of the client).

Thus for example, in the case of animal hoard-ing, a veterinarian may have a very good client who is an animal hoarder. In reporting that client to the relevant authorities, the veterinarian risks losing that client and associated income.

To what extent is it acceptable to work with the client (to try to change their treatment of animals, through education and so on) and where should a vet draw the line, for example when does one report a client?

In the next scenario we explore a perceived conflict of interest.

SCENARIO
TREATMENT OF OWN ANIMALS

▶   You are a veterinarian working in a mixed animal practice. You are catching up socially with some friends. One friend, a fellow veterinarian, says that she recently performed two cystotomies on her cat to remove bladder stones. Five stones were retrieved in the first procedure. Post-opera-tive radiographs did not reveal any other stones; however, the cat re-obstructed and a stone was visualised in the distal urethra. This was removed successfully. Another friend, who is a medical doctor, appears shocked.

"I can't believe you operated on your own pet," he says. "I would never operate on a family member – it's against our ethics."

"I would have got a colleague at least to do the second cystotomy, as I would have blamed myself for the complication," said another friend.

How should you respond?

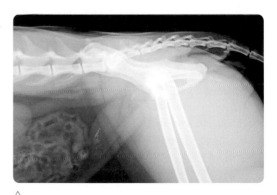

△
7.15 A fellow veterinarian performed two cystotomies on their own cat to remove uroliths. It is considered unethical for doctors to treat members of their own family – should the same apply to vets?
PHOTO ANNE FAWCETT

RESPONSE

JANE JOHNSON

▶   The principal ethical issue raised by this case has to do with a veterinarian treating their own animal. And the response of the medical doctor to this is understandable as, except in special circumstances, it is widely regarded as unethical for doctors to treat members of their own family. However, on closer examination there may be important differences between the situations in veterinary and in human medicine that mean drawing parallels here may not be helpful.

The main worry in human medicine is that a patient's care might be compromised if they are treated by a medical professional to whom they are related. The reasons why it might be compromised have to do with epistemological challenges, patient autonomy and emotional/psychological attachments. Although discussion often focuses on the latter, it is worth taking some time to unpack these other two areas, as they will also be important to deciding how to respond to the situation.

Part of the worry in treating family members is that it may be difficult to get the kind of patient knowledge which is essential to inform appropriate treatment. For instance, a doctor is unlikely to have the relevant patient history to hand when a family member asks for advice, and the patient might not be forthcoming about important elements of their condition, pre-existing treatments and so on, especially if these are sensitive. For example, they might not want the doctor to know of a termination, an STD etc. A doctor is also less likely to undertake a proper examination of the patient as it might cause embarrassment for both parties or be difficult in the context where advice is sought (e.g. over the phone, at a barbecue etc.). Once advice is given and treatment dispensed, neither the physician nor the patient is likely to notify the patient's regular doctor, compromising the patient's medical record as an accurate reflection of their medical history.

Concerns regarding autonomy and informed consent also arise when doctors treat family members. Depending on the nature of the relationship, patients may not feel empowered to fully discuss treatment options with the doctor, to express the values which inform their medical decision-making or to decline a recommendation. As noted earlier the context in which advice is sought can impact treatment. In this case the process of informed consent would likely be compromised if discussions occur at a family event or over the phone. Again depending on the nature of the relationship, doctors may simply comply with a patient's wishes (e.g. for an antibiotic) rather than interrogate and challenge these wishes.

A frequently cited reason for doctors not to treat family members involves the worry that physician objectivity might be clouded by the emotional and psychological dimensions of their relationship to that person; that clinical decisions might be influenced, either consciously or unconsciously, in a way that compromises patient wellbeing. For instance, perhaps serious symptoms might be ignored or simply missed because of the gravity of what those symptoms might mean for an individual to whom the physician is related. To further their own needs and interests, a physician might also selectively present treatment options to the patient. On the flip side, the personal nature of the relationship might actually enhance the level of care accorded to a patient. Some people argue that this is precisely what is missing from modern medicine – caring concern, thoughtfulness and genuine attentiveness to the patient and their needs and values.

Both the epistemological worries and issues over autonomy articulated in the previous paragraphs seem less troubling when it comes to a veterinarian treating their own animal than when a physician treats family members. Even if a vet does not routinely treat their animal, they are still likely to know the relevant patient history, and the animal patient is not in a position to intentionally conceal their previous treatments, other conditions and so on. Providing the vet has access to the required equipment, doing an examination of the animal should not be a source of embarrassment or difficulty. Since the veterinarian is also the owner it seems more likely that any treatment they undertake will be reported back to the animal's regular veterinarian, if the vet does not always treat their own animal, so the integrity of the patient record will be maintained. Similarly, patient autonomy and informed consent are not compromised in the case of a vet treating their own animal because it is owners rather than animals that negotiate and make decisions over treatment. However, what a vet might lack in this circumstance is the kind of sounding board furnished by a dialogue between a veterinarian and an owner.

The most relevant point of comparison between the medical and veterinary situations involves the emotional relationship with one's own creature which might mean one is too close to make appropriate and objective clinical judgements. A veterinarian might read too much into symptoms or alternatively gloss over them by virtue of the stake they have in their implications. At the same time, an emotional investment in the animal one is treating might ensure an attentiveness and level of care that would be otherwise absent. Many animals also have negative experiences and fear going to the vet; presumably if the vet were also your owner this would not be an issue.

Therefore, in reply to the physician, the rightful and well-grounded ethical prohibition over treating members of one's own family which exists in medicine should not necessarily apply in the practice of veterinary medicine. It might, however, be prudent for a vet treating their own animal to undertake this in consultation with a colleague who could act as a sounding board for diagnoses and proposed treatments.

CONFLICTS OF INTEREST

What would you do?

You realise that the next client waiting to see a vet is Mrs G, the hospital doctor who's been treating your wife for her breast cancer. The reception notes just say the dog is "unwell". There are other vets in your practice but no-one is expected to become free for about 20 minutes. What would you take into consideration when deciding whether to see Mrs G? What would you do?

△

7.16 Should you see the next client who is your wife's doctor?
PHOTO ISTOCK

||||||||||||||||||||||||||||||||||||||||||||||||||||||||||||||

OUR next scenario discusses conflicts of interest that may arise through receipt of gifts, samples or "freebies".

SCENARIO
SAMPLES AND FREEBIES

▶ You are invited to attend a continuing education evening about obesity by a sales representative for a pet food company that is launching a new brand of diet food.

What should you do?

RESPONSE
ANDREW KNIGHT

▶ In an ideal world, all clinical decisions, including decisions about which drugs, diets and products to prescribe, would be purely rational and evidence-based, to maximise optimal patient outcomes. The reality, of course, is that this ideal is subverted to varying degrees by factors ranging from relatively benign to considerably less so. More benign examples include lack of certainty about a diagnosis, or of certain evidence about a proposed treatment's efficacy, in which case decisions are made on the basis of probabilities rather than certainties; and genuine financial limitations of owners, which result in the selection of a suboptimal, but cheaper, treatment option. Less benign influences on the prescribing process include overemphasis on the interests of the treating veterinarian or practice in maximising profit, or on ease of treatment (e.g. minimising hospitalised caseloads over the weekend), or on trialling a therapy with less evidence of efficacy, in which a clinician has a particular interest. And a particularly prominent

area of concern – which is related to this case – is the attempted subversion of treatment decisions by the commercial interests of companies supplying pharmaceuticals, diets or other veterinary products.

Of course such companies exist partly to create and distribute products that make a major contribution to animal health and welfare, and in doing so make important, positive contributions to society. However, they also have a duty to their shareholders to maximise profits, which occurs when their products are used as widely as possible – regardless of the degree to which their products are actually clinically superior to alternative treatment options. This can create a strong interest within such companies in influencing prescribing decisions. After all, these companies do operate within a competitive environment. They succeed or go bankrupt in a corporate version

△

7.17 Continuing education events are often sponsored by pharmaceutical or food companies, but should you avoid these potential conflicts of interest?
PHOTO ISTOCK

of the "survival of the fittest", based on a combination of the effectiveness of their products, and perhaps even more importantly, the extent to which they can successfully influence clinicians to use them, and clients to request them.

Companies supplying pharmaceuticals, prescription diets or other healthcare products are known to seek to influence prescribing decisions in a variety of ways. Through sponsorship or gifts they seek to build relationships and influence with clinicians. They may sponsor continuing education events, and even the travel and accommodation costs of clinicians. They may also arrange such events themselves, such as in this example relating to obesity. Such sponsorship is very significant, particularly within the world of human healthcare, where the financial stakes are even higher. As D'Arcy and Moyniham (2009) reported, "The pharmaceutical industry is an extremely important source of funding for continuing medical education – 35% of the estimated US$9–14 billion that industry spends each year on pharmaceutical marketing goes towards educational support." Companies may also target students, as occurs when suppliers of prescription veterinary diets offer discounted pet food to students in veterinary schools.

Even more disturbing is the subversion of scientific evidence by such companies. It is well understood within the scientific world that studies of the effectiveness of a new treatment are more likely to show a positive result when funded by companies with a commercial interest, than when funded by independent sources such as government agencies, charities or universities. Such studies more often have favourable efficacy results and overall conclusions, and are less likely to show evidence of harm, than non-industry-sponsored studies (Lundh, et al. 2012).

Goldacre (2009, 2012) has described at length the various methodological manipulations that occur within such studies that predispose them to outcomes more likely to be favourable to the industry funder. This problem is pervasive within science. In a survey of scientists randomly sourced from databases maintained by the National Institutes of Health's Office of Extramural Research, 15.5% of all 3247 respondents reported changing the design, methodology or results of a study in response to pressure from a funding source (Martinson, et al. 2005).

Tempting though it might appear at first glance, the solution to problems such as these is not to ban all industry involvement in scientific studies or educational events. As D'Arcy and Moyniham (2009) stated, because sponsorship of continuing educational events is so substantial, "If pharma-sponsored education is no longer allowed, we may witness tomorrow's doctors practicing yesterday's medicine." Similarly, a great deal of scientific work relating to the development of new therapeutics would not occur, without industry sponsorship.

However, we must recognise that the primary interests of industry are not in advancing science or patient welfare, but in advancing their commercial competitiveness. Accordingly, we must demand absolute transparency with respect to their generation of scientific results, and we must subject claims about the safety and efficacy of their products in scientific studies or educational events to very rigorous critical scrutiny. In this particular case relating to the obesity presentation, you should attend if possible, but you should closely scrutinise the claims made about the efficacy of the new company diet, and you should examine the evidence in support of those claims, paying particular attention to the methodological design of any supporting studies.

In order for such increased scrutiny and critical review to become more firmly embedded within the culture of medicine – both human and veterinary – scientists and policymakers must be further educated about the nature and extent of

7.2
CONFLICTS OF INTEREST
RESPONSE
ANDREW KNIGHT

this pervasive problem, and about how to critically assess evidence, and particularly study methodologies, for sources of bias, and about how to minimise these through good experimental design. Such training should be included within the curricula of veterinary schools, and should be made available through continuing education to veterinarians.

||||||||||||||||||||||||||||||||||||||||||||||||||||||||||||||||||

WHILST being aware of, and concerned to reduce, any possible conflicts of interest may help to mitigate negative outcomes there may be some benefits to associations with pharmaceutical or drug companies. If veterinary practices are not aware of new treatments they will not be used, even if they are the best available. Companies, partly driven by self-interest, aim to increase awareness of their products through advertising and other promotional efforts. Free samples may enable discounted treatments for financially compromised clients, or allow a greater understanding of a product, for example food or routine treatments, if they are used on the veterinarian's own animals.

CONFLICTS OF INTEREST

What do you think?

ONE What do you think are the most important ethical considerations for whether there should be restrictions on sponsorship of student, veterinary and nursing training events by pharmaceutical or food manufacturers or other veterinary companies?

TWO Which, if any, restrictions would you support?

△

7.18 Would you consider restrictions on sponsorship of veterinary training events? **PHOTO** ISTOCK

Virtue ethics is a useful framework through which to examine conflicts of interest. The virtuous veterinarian is trustworthy, and therefore we would presume would be open and transparent about potential conflicts of interest. In addition the virtuous veterinarian has integrity and we would assume would resist the temptation of freebies or gifts for personal benefit if they are aware of the potential influence or perceived influence these may have.

Many organisations have responded to concerns about conflicts of interests by adopting policies. For example, the Association of American Veterinary Medical Colleges has developed "Guiding Principles and Considerations: Ethical Interaction between Schools/Colleges of Veterinary Medicine and External Entities". The policy, under review at the time of going to press, acknowledges that:

> "Gifts and services to faculty, staff, students and student organisations may take many forms, such as pens, food and beverages, backpacks, items of clothing, books, and sporting/concert tickets. Such gifts have been shown to significantly influence the attitudes and preferences of the receiver, regardless of the value or nature of the gift."
>
> (Association of American Veterinary Medical Colleges 2011)

The document goes on to suggest strategies to mitigate or eliminate potential and real conflicts of interest, including development of a funding consortium with no single company acting as a dominant contributor, clear and open disclosure of funding arrangements to students by faculty, use of generic rather than brand names of drugs in teaching and so forth.

One of the practical challenges is that the practice of giving "freebies" is widespread and insidious, and many veterinarians, nurses and technicians aren't aware of the potential conflict of interest these represent.

CONFLICTS OF INTEREST

What would you do?

Your practice has a sales deal with a major drug wholesaler. In exchange for using the wholesaler's products exclusively, you are offered free bonus stock that you can resell for full price. What should you do?

△

7.19 The offer of bonus stock may be tempting.
PHOTO ISTOCK

7.3

Personal and professional values

We've already seen that there are particular expectations of professionals, but veterinarians work as part of a team. One of the challenges lies in maintaining professionalism when there are differences of opinion as to the appropriate course of action. Some differences of opinion are based on differences in personal values and convictions about what a professional should do in a given situation.

△

7.20 How can you align your conscience with your professional duties when asked to euthanase a healthy goat?
PHOTO ISTOCK

SCENARIO
CONSCIENTIOUS OBJECTION

▶ You are a senior veterinary nurse in a mixed practice working with an experienced veterinarian. You have built up a rapport with your colleague over many years and he has stated that you are his most trusted employee.

You have noticed over the past six months that every time the veterinarian performs euthanasia, he requests that you assist, often commenting that he wants to make sure that the procedure goes smoothly.

On one particular occasion, he is required to euthanase Isolde, a three-year-old pet Pygmy goat, because the owners cannot find her a home. Clearly conflicted about whether this is the best course of action, he has been talking with the client for some time before he calls you out to the vehicle to assist.

You do not wish to do so.

What should you do?

RESPONSE
PATRICIA V TURNER

▶ There are several considerations for this scenario, which will be considered separately. These include: euthanasia of a healthy animal, conscientious objection to euthanasia by a staff member, the double-edged sword of being highly technically skilled (and thus overused for certain procedures), offloading of unpleasant tasks to other staff, and loyalty to an employer or practice.

Pygmy goats serve few agricultural purposes, and while some may be kept for milking, as a companion to other livestock or as a guard animal, hobbyists largely keep these inquisitive and highly intelligent animals as companions. Thus, the issue at stake is not slaughter of an animal for food consumption or because it is diseased or dying, but rather convenience euthanasia of a healthy pet. Many individuals, including veterinarians, staff and members of the public, find the concept of killing healthy companion animals to be morally objectionable.

The fact that the veterinarian in this scenario has been talking to the client for "some time" before calling for assistance suggests that the veterinarian may also be experiencing conflict about the task at hand. It is never inappropriate to make one's ethical views known to one's workmates or employer; however, these conversations tend to come up over the course of time, as one works with and gets to know others in a practice environment.

Suddenly announcing a conflict just before a critical procedure is to occur is awkward for the employer and client, potentially jeopardising the veterinary–client relationship, and it may also create a welfare risk for the animal in question, especially if the veterinarian decides to proceed and is unable to carry out the procedure on his own without assistance.

Quietly offering up suggestions for managing the animal's care or future to the veterinary colleague may be more appropriate, depending upon the relationship that the nurse has with their employer, especially if there are new opportunities that may not have been considered during the previous veterinary–client discussion. On the other hand, outwardly voicing moral objection to the procedure, while creating immediate social tension, may be what is needed to force the client (and the employer) to think through her request more closely. It is highly likely that if the client is unable to place this one animal in a home, there may be other animals in a similar situation in the future, unless further measures are taken to better control the profligacy of her goats.

Immediately following this appointment, the nurse must have a candid discussion with this veterinarian to thoroughly explain their actions and viewpoint. Convenience euthanasia is likely to

have come up as an issue many times over the course of the staff member's time with this clinic and there was likely ample opportunity to express their moral stance or at least to engage in a discussion prior to this event, thereby circumventing a stand-off in the client's presence. However, it may be that the staff member was not confident in bringing this issue forward previously or wished to avoid offending their employer by disagreeing with him publicly. It may be that the subject was broached previously but this was not listened to.

To sidestep the appearance of insubordination, it is critical that the issue of moral objection to convenience euthanasia be discussed openly. Depending on the outcome of the conversation and the openness of the employer to further discussion, this may also be an item of broader interest for discussion at the next hospital staff meeting. A more forward-thinking practice will recognise that clients who request convenience euthanasia are generally not ones that contribute long term to the financial soundness of a practice and that convenience euthanasia conducted on a "one-by-one" basis does little to address the root causes of animal overpopulation. Whether the nurse's employment is in jeopardy after this scenario will in part be influenced by how tactfully the situation was handled with the client, how well the discussion is handled with the employer, as well as local labour laws governing small business management. In many jurisdictions, the employer would not be considered to have just cause for firing an employee over this issue.

The long-term outcome of this situation will depend on the openness of the veterinary owner(s) to discussions by valued and trusted staff members and the availability and willingness of other personnel to assist with technical aspects of euthanasia. Because euthanasia can create an intense emotional burden for veterinarians and staff, it is important to share the workload over multiple individuals. This also ensures that competent technical back-up support is available in the event that personnel who routinely assist are away for planned or unplanned absences. It may be that other staff members are not morally conflicted by convenience euthanasia, are quite willing to assist and were only waiting to be asked.

Further, in many practices, veterinarians conduct euthanasia procedures without technical support and it may be that none is needed for the rare times that this situation arises in the future. While the veterinary nurse may lose some standing in the eyes of their employer, and thus "status" in the eyes of other clinic personnel, equally, they may be respected more than ever for declaring the conflict and remaining true to their ethical view. In the long run, they will be happier and more satisfied in their work if not placed in a position of moral conflict.

The final issue to consider for this scenario surrounds loyalty to a long-term employer and whether this should trump loyalty to one's ethical viewpoint. It is important to emphasise that euthanasia is not just an unpleasant task, as there are many unpleasant tasks in a veterinary hospital that must be done on a daily basis, such as cleaning kennels etc. Euthanasia is a special responsibility of veterinary personnel and because of the irreversibility of the consequences and the public trust shown by tasking this to veterinarians, it must be considered seriously. With the responsibility of killing humanely comes the burden of knowing that it was conducted for appropriate reasons.

Forcing oneself or an employee to carry out tasks that are morally objectionable is unethical and can result in long-term physical and mental health issues. Thus, it cannot be conceived of as immoral to resolve personal ethical conflict at the cost of employer loyalty. If the employer in this circumstance is unwilling to consider any alternatives for this nurse, the nurse should carefully reflect on whether this practice is one in which they can continue to work.

||||||||||||||||||||||||||||||||||||||||||||||||||||||||||||||||||||||||||

IN her response Pat Turner is clear that being true to one's ethical beliefs is important, and as far as possible should be respected by one's employer. Human nurses are sometimes asked to participate in treatments they find objectionable, such as those relating to abortions, procedures against a patient's wishes or caring for technologically supported brain-dead persons. In some instances guidance has been provided as to when and how nurses may conscientiously object (Trossman 2014), but it may still remain unclear (McHale 2009) and has been the subject of litigation against nurse employers (Stein 2011).

This approach may be seen as a rights-based view, where an employee is seen to have a right to personal values and a right to conscientious objection. As such an employee should never be forced to perform a task that violates such rights.

PERSONAL AND PROFESSIONAL VALUES

What would you do?

Your colleague has been treating Ms V's horse Magic for grass sickness, a disease of unknown aetiology that causes wasting, difficulty swallowing and colic. Magic has been intensively nursed intermittently and has certainly suffered greatly at times. The prognosis for acute grass sickness is poor but for more chronic cases has improved over recent years with some horses returning to full work. However, you are concerned that the treatment of Magic may not be in his best interests. Your colleague has gone on holiday for two weeks and left you in charge of the case. What should you do?

△

7.21 You are not sure whether continuing to treat Magic for grass sickness is in his best interests.

PHOTO ISTOCK

7.4

"What would *you* do if it was your animal?"

In this next scenario we consider how a veterinary professional should respond to a question commonly posed to all professionals, whatever their line of work: what would you do?

___

SCENARIO
WHAT WOULD *YOU* DO IF IT WAS YOUR ANIMAL?

▶ Mrs F brings Tokyo, an elderly cat, in to see you again. Tokyo has been under your care for the last few weeks since she became ill with signs of chronic renal failure. She initially improved with treatment but is now inappetent and has barely moved in the last 24 hours. You discuss taking Tokyo in for fluid therapy and other supportive treatments again, but also broach the subject of euthanasia with Mrs F. She knows it's been coming but really isn't sure what to do. She asks what you would do if it was your cat.

How should you respond?

___

RESPONSE
MARTIN WHITING AND
ELIZABETH ARMITAGE-CHAN

▶ This very common scenario seems to divide the profession in their approach to client and patient management. The problem revolves around the dilemma of whether the veterinarian should answer the client honestly or does this create a fear within the professional of biasing the client's freedom of choice? Sadly, such a scenario tends to occur, similarly to Mrs F, at a time

△

7.22 Tokyo is an elderly cat with chronic renal failure, but her owner wants to know what you would do if your cat were in the same situation.
PHOTO ANNE FAWCETT

of deciding between exquisitely painful options, regarding either welfare or financial considerations, so preparatory forethought of a resolution is advantageous to effective case and client management.

One critique of such a scenario seeks to divide the question into its many facets (McCulloch 2012). Is the client asking a purely clinical question, namely what is the best course of action for the best clinical outcome? This relies solely on the veterinarian's knowledge of the disease and treatment options and appeals strongly to the gold-standard treatment within evidence-based veterinary medicine. The second facet may be a welfare question; the client is asking for the veterinarian's expert knowledge of welfare to provide guidance as to what to do for the cat that is going to be best for the cat's welfare. The third element of the question could be a purely practical one relating to husbandry, ongoing care and finances. It is worth bearing these three elements in mind if

you decide to answer Mrs F's question in a way that would be constructive for her informed decision-making. But first we need to decide if even answering this question is the right thing to do.

Some veterinarians believe it is morally wrong to answer such a question for the client. To them, a response to this question can represent an overly paternalistic approach to case management where the client dissociates from the decision-making process and it can present a strong feeling of bias. To give the veterinarian's opinion on such a case can then make it very hard for a client to decide anything differently; to do so requires the client to say that the opinion of their veterinarian is wrong. Anecdotally, some veterinarians are concerned that to provide their own personal feelings on a case can lead to them becoming too emotionally invested and, in terms of their self-protection, they feel it is better not to provide an answer. Consequently, there has been a gradual reluctance to provide an answer; it is not uncommon to hear clinicians say "our professional ethics guidance prohibits us from answering that question". Such a response has caused upset amongst clients and is perceived to be unhelpful, to the point of leading to complaints (Mullan 2012).

When considering the influence that a veterinarian has over their client's decision-making it has been proposed that a principle of reasonableness is applied (Yeates & Main 2010). In this scenario, it is possible that the client is specifically asking to be influenced in their decision-making. In this interpretation it would be appropriate to answer their question in a way that helps the client realise what they, themselves, wish to do. Yeates suggests it would be illegitimate to use this moment to influence the client into a course of action that is not in their, or the animal's, interest. There is merit to this consideration. The client has asked a genuine question and is seeking an answer to their question, irrespective of what their motive may be. To deny them an informative response would be to deprive them of an element of the wealth of knowledge, experience and expertise of the professional before them. This is a key point. The veterinarian is a professional person; this entails not only providing the client with the factual nature of the veterinary world, but also providing them with the professional interpretation of these data. The professional identity of a veterinarian is not reduced to a "walking Wikipedia" of veterinary information, but rather as an expert able to provide interpretations of the vast wealth of data in a format suitable for the client to understand and decide what they wish to happen to their companion. To refuse to answer such a question, for fear of undue influence or coercion, is to sell the profession short of its true caring expertise. Yet *how* to deliver this information to the client must give us pause.

The pre-eminent veterinary ethicist Bernard Rollin explains his view on the use and abuse of Aesculapian authority in an exceptional paper (Rollin 2002). Rollin notes that society holds those with the power to heal in special regard; they do so with such reverence that one physician commented, "I can get almost anyone to do whatever I tell him or her." The power of the words of a healer is immense upon those who are in need, and such power is also heralded by the veterinarian. Thus, the ability to influence clients who are in such need as Mrs F asking for advice on the fate of Tokyo should not be underestimated.

So how should such a power be used? All authors cited agree that client decision-making and animal welfare are to be upheld; the veterinarian's power to influence should be used sparingly, with caution and to promote welfare and client decision-making. In human medicine, authors have argued that neither a direct response to the question nor a neutral disclosure of clinical options constitutes ethically responsible care by the professional (Minkoff & Lyerly 2009). The emotional power of Mrs F asking such a personal question implicitly reflects her trust in the veterinarian's

judgement, but without giving a direct response or remaining neutral, what is left but a lie?

To resolve this scenario effectively an appeal to the narrative theory of ethics is most enchanting. Chapter 2 describes the narrative theory as one in which moral values are imparted through the narrative of a story. The factual disclosure to the client of what a veterinarian would do with their animal in any given situation, or if the veterinarian were deciding on behalf of another client's animal, is of little value to the client who may then become dissociated from the problem and defer responsibility and decision-making away from themselves. A narrative approach to this scenario can help convey the message of animal welfare, and potentially the veterinarian's preference, in a form to which the client can relate.

The client can then transpose their understanding of their animal into the story to glean what they wish to. Such a process is one of client empowerment; it enables them to understand an analogous scenario, draw out the underlying key principles of the decision-making process and then use those skills and apply them to their own scenario. The Aesculapian authority, or Yeates' influence, may be borne out in the details of the story that can resonate with the client's understanding of their animal to assist them in making the decision they want to.

In this scenario, the veterinarian may reply with an example of their own animal, for example of Bailey, a 13-year-old Labrador. "Bailey was a highly active Labrador having been a working dog for the early part of his life, he was never happier than when he was out on walks or running through the woodland retrieving anything he thought I might be interested in receiving. But when Bailey in old age developed very severe arthritis in his back following an accident he had when he was younger, he began to lose his ability to do the things that he enjoyed. He could not go for the walks he loved, he struggled with stairs or when jumping over trees, he even began to struggle getting out of his bed. Bailey's condition could have

7.4
"WHAT WOULD YOU DO IF IT WAS YOUR ANIMAL?"
RESPONSE
MARTIN WHITING AND ELIZABETH ARMITAGE-CHAN

△
7.22 The veterinarian shares the story of his own dog, Bailey, a 13-year-old Labrador.
PHOTO MARTIN WHITING

been partially managed; he could have had strong analgesia to alleviate some of his pain but even with this, he still was not able to do the things he had previously enjoyed. He could not do the things that made him flourish. Had Bailey been a lap dog, happy to sit in his bed all day long or be lifted and carried around, then perhaps the analgesia would have been a perfect treatment option for him. But Bailey was too independent for that. We made the decision that although he was still a wonderful dog for us, he was himself only a shadow of what he wanted, every day was a struggle for him and he had lost his *raison d'être* with no hope of return."

This story conveys to the client the process by which the terminal fate of Bailey was decided. The focus was on Bailey's welfare and it was not about what the owner wanted. The decision was made purely based on the nature of Bailey and what he "enjoyed" in life and how he flourished. His illness had taken that from him. A different dog, with the same problem, may only suffer a minimal welfare

compromise. But when we considered Bailey for who he really was, then the difficult decision of euthanasia became the only option for his best interests.

The idea of the narrative ethic is that the client can then take the story of Bailey and see how Tokyo fits with that. Would repeated admission to a veterinary hospital, away from her familiar home and being handled by unfamiliar people with the gravest feeling of malaise associated with renal failure fit with her lifestyle? Perhaps Tokyo is still able to enjoy the things she "loves" after the fluid therapy – maybe this is limited to sitting in front of an open fire, or lying in the sun. Perhaps Tokyo is a real hunter, who hates being indoors, does not like to use a litter tray and struggles to eat the specialised renal diet.

Assisting the client to transpose the narrative story of the veterinarian's familiar pet to the scenario they face themselves can do little more than contextualise and empower them into making the decision that is most appropriate for them, in their life, with their pet that they know intimately. Mrs F has, at this point in time, invited you into her "inner circle" of support and guidance (Meyer, et al. 2012). Through encouraging her to identify and rely on her intrinsic values, and promoting shared decision-making, rather than refusing to answer her question or providing a substituted judgement, the veterinarian has created an outcome for all that is a nurturing process of shared decision-making.

||||||||||||||||||||||||||||||||||||||||||||||||||||||||||||||||||||||||||||||||

AS the authors point out, there are many potential reasons that clients would ask "what would you do?" – and understanding those reasons is important in the shared decision-making process. The approach outlined here maps well onto the virtue ethics approach, where one appeals to the professional's honesty, wisdom and integrity to discover the best way forward.

Consider the following scenarios.

"WHAT WOULD *YOU* DO IF IT WAS YOUR ANIMAL?"

What would you do?

Mr G calls you out to see his 12-year-old daughter's pony Herbie who has been stiff and off his food. You diagnose laminitis and suspect Cushing's disease. You know that Herbie means the world to Mr G's daughter and you presumed that they would want to treat the laminitis and explore the possibility of Cushing's. Mr G takes you aside and explains that his business has been losing money and he's in a lot of debt. He asks you what you would do. How do you respond?

△

7.24 Herbie's owner is financially stretched, and asks you what you would do in his situation.

PHOTO ANNE FAWCETT

7.5

Beyond the call of duty?

Veterinary practice is changing. It was once the case that veterinarians were on call 24 hours per day, 7 days per week. This has changed considerably – many factors including dedicated emergency centres, corporatisation and an increased awareness of the need for so-called work–life balance mean that veterinarians are not always expected to be "on duty".

△

7.25 "Veterinary medicine is a way of life – a ministry in the true sense of the word – so I'm a veterinary surgeon 24 hours a day, 7 days a week, 365 days a year..." according to Dr Jamee Harris, DVM.

CARTOON SUHADIYONO94

Eliot Freidson believes that being a professional brings an onerous but necessary burden of expectation:

"Professional ethics must claim an independence from patron, state, and public that is analogous to what is claimed by a religious congregation. This conclusion may seem extravagant when attached to the average weary professional who puts in many more hours a week than most other workers, who has become accustomed to a standard of living and a level of respect that are becoming increasingly difficult to sustain, and whose workload constantly presses toward routinisation, the cutting of corners, and the loss of any pleasure from work. Give or take here or there, that is the condition of all of us who work. But those with the status of professions in advanced industrial nations are still in a position of symbolic as well as economic privilege."

(Freidson 2001)

However, expectations around what it means to be a veterinary professional vary. Veterinarian James Harris, for example, believes that "veterinary medicine is a way of life – a ministry in the true sense of the word – so I'm a veterinary surgeon 24 hours a day, 7 days a week, 365 days a year, and I'm available to do what I think is appropriate, and to help people and my colleagues at any time because I think service is man's most noble activity" (Llayton-Bennett 2013).

Others believe there can and needs to be protected time off-duty to recharge. In some studies, for example, long work hours and on-call time were contributing factors in psychological morbidity in veterinarians (Fritschi, et al. 2009, Shirangi, et al. 2013).

To what extent should professionals work "beyond the call of duty"?

SCENARIO
LAME SHEEP

▶ You are driving past a field of sheep when you notice that several are so lame as to be grazing resting on their front knees. Otherwise, the sheep look in reasonable condition and the field appears suitable for them. You now work in a practice solely treating companion animals and do not know whom the sheep belong to or who their vet is.

What should you do?

△
7.26 What obligations do you have driving past a field where many sheep are so lame they are grazing on their knees?
PHOTO SIOBHAN MULLAN

RESPONSE
PETE GODDARD

▶ My response to this scenario will pose a number of questions rather than identify an obvious solution.

The actions that could be taken range from doing nothing at all to stopping the car, jumping over the fence and treating the sheep then and there – or actions equivalent to that (extreme) effect! I believe that most people would not go to those lengths but what action might be contemplated?

Given that lameness is a major welfare issue for sheep and that certain flocks have particular problems with lameness (some struggle with an incidence of around 10 per cent, possibly more), it is unlikely that you would take action at every affected flock you drove past. Since up to 3 million sheep in the UK could be lame at any one time you may not get very far on your journey in some areas! So in this case, where you have observed that other features of the flock seem reasonable, at least superficially, it is likely that you would do less, rather than more, according to your conscience.

In a parallel situation where a doctor was witness to a non-emergency situation, they would have no obligation to act in every case; the severity and the immediateness with which it was felt action should be taken are important considerations. If someone collapsed in the street, a passing doctor would help, but they probably would not stop to aid someone with rheumatoid arthritis or a limp. In an emergency, a passing doctor would help with basic life support but for specialist input, care would be discharged to a more appropriate colleague with the itinerant doctor passing on relevant information to them. There may also be professional indemnity issues for "good Samaritan" acts, especially if intervening without invitation could cause litigation problems. So here we come full circle to the scenario in question.

Thus the question remains, as a vet, regardless of the type of work you are engaged in, do you have a special moral responsibility through your training and expertise, to take action when you suspect an animal welfare issue exists? As noted earlier in this book, veterinarians swear an oath on joining the profession, which states in the UK that animal welfare is the primary consideration – but it does not mention the reality of duties towards owners, financial and professional. It was pointed out that animals are also property in law and, although vets may have a degree of moral authority, they have no power to force owners to treat animals in a certain way. In this case, as a vet treating companion animals, would you consider that you are closer in position to a concerned member of the public than a sheep veterinarian in this case? After all, these sheep are not under your care. The public, if so minded, may inform the police or an animal welfare protection charity about the sheep and then likely feel responsibility has passed from their hands.

Assuming that you do feel a moral duty to do something, action is needed but what should be done and what could be achieved? You could call in at the farm and confront the owner with your concerns. The reaction this direct approach evokes may be challenging and the likelihood of this may deter you from pursuing this course of action. If you did so, however, would the fact that you were a vet enhance your argument? The fact that you treated small animals could actually weaken your position if a more detailed discussion ensued; you may certainly not be up-to-date with the latest advice regarding foot trimming for example and would be in a poor position to make any comment about exactly what to do. Importantly, what you would not know prior to following this line is the situation of the farm. Could it be that the farmer was fully aware of the issues and struggling to get the situation under control? There may even be a comprehensive lameness control plan with veterinary input in place but perhaps it is yet to have a marked effect, recognising the difficulty of controlling lameness in some situations even with the best advice.

Perhaps the farmer has serious personal or financial problems. These would not absolve him or her from proper animal care but you are not to know that steps are not in place to help them and their sheep. Your arrival on the doorstep, unprepared for what might follow, would be particularly unwelcome in these circumstances and may even precipitate some catastrophic response. In essence, how is one to balance the moral responsibility towards both a farmer in distress and the welfare of the sheep? A multi-agency approach would usually be needed to resolve this and your intervention may derail an existing initiative.

An alternative option could be to write to the farmer expressing concerns; you would need to consider what you would do if you did not receive a response or on passing the field again if there were no improvement. Also some of the aforementioned responsibilities towards the farmer would equally apply.

An option many might consider would be to contact a veterinary practice serving the local farm animal community and identify the vet responsible for the flock. Even if it was not a client of that particular practice, they would probably know who it was. Having said that, the extent of veterinary involvement with sheep flocks varies widely and may be as little as being associated on an annual basis with the flock's required health plan. Even so, there has likely been at least minimal contact between the farmer and the nominated practice and this could be exploited potentially to raise the concern. However, you could end up in a more convoluted position if the practice "looking after" the farm in question said that they were aware of the problems but the farmer refused to entertain their proposed solutions, despite much effort. You would then have a new task in order to try to persuade the practice (with or without your continued involvement) to take things further. They might say that it was felt that the lameness on the farm was pretty much typical for the area. What would you do then?

A final active approach, as noted above in relation to a member of the public, would be to contact an animal protection charity, the police or the local government veterinary office. As a vet, regardless of your specialty, you may find that the organisations are able to discuss your concerns on a professional basis and may even be aware

of the farm in question. Having taken this last approach, you may consider that your moral duty has been discharged – but how would you feel on driving past the field a week or two later if nothing seemed to have changed?

Or you could drive on past.

||||||||||||||||||||||||||||||||||||||||||||||||||||||||||||||||||||||||||||||||

THIS scenario raises the question of what is considered a "supererogatory act". Supererogatory actions are those that are considered "beyond the call of duty". There is debate in the literature about what constitutes a supererogatory act, or indeed whether they can be said to exist at all. For example, some would argue that a bystander who drags a child out of a burning building, saves an old man from drowning or intervenes when they see a thug kicking a dog is in each case going beyond what we might expect. They are doing something morally good, but not something they are obligated to do. We are all familiar with news stories in which ordinary people are called "heroes" because they have performed such acts.

But when those heroes are interviewed, they will often respond that they were doing "what anyone would have done". Similarly, anti-supererogationists argue that it is absurd to divide "good" actions into those that are "required" and those that are "good" (Heyd 2015).

The author takes a consequentialist approach, evaluating the possible consequences of different courses of action – some with the potential to cause harm to stakeholders such as the farmer, the veterinary profession or themselves.

One thing to consider is that professionals are held to a higher standard of conduct than members of the public. One argument is that the knowledge that we have (for example about animal welfare) obliges us to act when we observe that something is not right. Again, this maps best onto virtue ethics or deontology.

BEYOND THE CALL OF DUTY?

What would you do?

You are on holidays with your family after a long break when the traffic comes to an unexpected stop. You hear that in front of you a horse float has overturned and one horse has bolted. A crowd has gathered and a police vehicle arrives. You have a long day of driving ahead. What would you do?

△
7.27 Up ahead a horse lorry has overturned. What would you do?
CARTOON RAFAEL GALLARDO ARJONILLA

BEYOND THE CALL OF DUTY?

What would you do?

It's your first night off-call in a week. You have just sat down to dinner when a friend contacts you in distress: his dog is having difficulty breathing. You know from conversations with your friend that the dog has had long-term mitral valve degeneration and you believe that the dog is likely in terminal congestive heart failure. Your friend says, "I am not sure if this is the end, and I feel more comfortable with you than my regular vet."

What would you do?

△
7.28 You sit down to dinner on your first night off-call when a distressed friend calls.
PHOTO ANNE FAWCETT

7.6

Challenges to professionalism

Veterinarians and allied health professionals may experience occasional requests from clients or colleagues that challenge our professionalism. Determining how to respond is not always easy, particularly when social as well as professional values may be breached, as in this next scenario.

SCENARIO

IS IT OKAY FOR A PROFESSIONAL TO TELL A "WHITE" LIE?

▶ You are working in companion animal practice. Mr and Mrs G present you with their 10-year-old terrier that has been hit by a car and sustained a femoral fracture. You explain all the treatment options, but the Gs decide to have the dog put down on financial grounds. They request that the whole family be present, including their teenage children.

Before the children arrive, Mrs G asks for a quiet word with you.

"The kids will be really upset," she says. "Could you please explain to them that this is the only option for Mack, as I don't want to traumatise them even more."

You know that one of the kids is thinking of applying to veterinary school.

What should you do?

RESPONSE

PETER SANDØE AND SANDRA CORR

▶ The veterinarian is asked by one of the two owners of the dog to tell an outright lie to the owners' children. They want the veterinarian to tell the children that the only option is to put the dog down, while in reality it would be easy to return the dog to good health by performing a simple (but costly) orthopaedic procedure. Allegedly the purpose is to spare the children from being upset (or angry) that the dog will be put down because the parents are not able or willing to pay for it to be treated. At the same time the lie will likely make the life of the parents less difficult, enabling them to avoid potential conflict with their children about the decision.

Whether in principle it can be morally acceptable to tell a lie is a divisive issue in moral theory. Famously, the German philosopher Immanuel Kant argued that telling a lie is never justified, even when telling the truth may have disastrous consequences for an innocent person, as in Kant's own example of a potential murderer inquiring about the whereabouts of his potential victim. Few – if any – contemporary ethicists would follow Kant that far, but some, subscribing to a rights or a virtue-based theory, would claim that, at least prima facie, it is always wrong to tell a lie to another person. So on this view telling lies should be avoided unless there are strong counterbalancing arguments, such as protection of life.

Adherents of utilitarianism or other forms of consequentialism, on the other hand, would argue that whether or not telling a lie is wrong will depend entirely on the consequences, direct or indirect, of deviating from truthfulness. For utilitarians and other consequentialists there are therefore no direct moral arguments against lying. However, there are some good pragmatic arguments as to why it is generally not prudent to lie. Firstly, it is so difficult and demanding to lie without being found

△
7.29 Is it ever acceptable for a professional to tell a "white lie"?
CARTOON MALBON DESIGNS

out, because a lie has endless logical ramifications that are difficult to predict and control, whereas telling the truth requires much less thinking. Secondly, and related to the first point, there is a great risk of being exposed as a liar, and this may have severe social and personal consequences.

We will argue that there are strong consequentialist reasons why the veterinarian in this and other professional contexts should not engage in deliberate lying. If our arguments are accepted, they will also be of relevance to most adherents of the view that it prima facie is always wrong to tell a lie. The latter may accept consequentialist arguments, but still require stronger reasons for accepting lying as being morally legitimate. They would therefore agree with our conclusion that even based on valid consequentialist arguments it is wrong for the veterinarian to tell a lie.

The starting point for our argument is that it is an important asset for veterinarians to be seen as trustworthy. If clients don't trust their veterinarian

life will be difficult for both parties, both psychologically and economically: psychologically because a lack of trust will undermine the confidence of both parties, and economically because as a result, costly regulation may be required and/or litigation may become widespread.

The main premise of our argument is that if the veterinarian deliberately lies, it is bound to undermine trust in the veterinarian, both directly and indirectly. Firstly, one or more of the children may find out that they have been lied to and this may undermine their trust in veterinarians and other professionals such as medical doctors. Even the client who asked the veterinarian to tell a lie may lose trust: if it is possible to make the veterinarian tell a lie that easily, why then believe the veterinarian in other contexts? And the veterinarian doesn't actually know whether the *other* parent condones the idea of telling a lie to the children.

Indirectly, telling a lie in this case may undermine the moral stance of the veterinarian, and even if it doesn't, it may have psychological costs in terms of a guilty conscience and potential demotivation. Furthermore the veterinarian has an obligation to her or his colleagues; and if one veterinarian is caught in telling a deliberate lie and the story spreads, this may undermine confidence in the profession as a whole. This is part of the reason why the veterinary profession in the UK and other countries aims to uphold a code of veterinary practice where truthfulness is an important element. Among the five "principles of practice" listed by the British Royal College of Veterinary Surgeons Code of Professional Conduct for Veterinary Surgeons are that veterinarians should maintain "honesty and integrity", as well as "independence and impartiality" (RCVS 2015b).

We therefore recommend that the veterinarian tells the parents that she or he can't say that this is the only option for Mack because it's simply not true. The veterinarian can suggest that the parents have that conversation with the children in their absence, and can agree that they will not, as long as they are not directly asked, say anything to contradict the parents.

If the children stay with the animal when it's put down, and ask questions, it's easy for the veterinarian to direct the attention back to the animal, or make an ambiguous comment such as "I'm sorry, there's nothing I can do for Mack" – which is true, as the owner has made the decision not to pay to treat the dog, and is different from "there's nothing that can be done".

In the previous paragraph we have, based on consequentialist reasons, argued that the veterinarian can be economical with the truth as long as it prevents the children from unnecessary upset, and does not contribute to undermining the trust in the veterinarian or the veterinary profession. Adherents of other moral positions may take a more restrictive view on this.

CHALLENGES TO PROFESSIONALISM

What would you do?

Mrs E, an elderly client of yours with a retriever and two cats, is now very poorly. The senior partner in the practice usually makes the house visits to treat her animals but this time you need to go. Your boss tells you, "Every time I go to Mrs E's house she makes me promise I will euthanase her three animals so they can all 'travel to heaven' together when she dies. I'm not going to do it, I'll find new homes for them, but she doesn't need to know that. If she asks I want you to promise too, otherwise it'll only upset her." What will you do when Mrs E asks you to make the same promise? What will you tell your boss?

△
7.30 What would you do if your boss asked you to lie to Mrs E?
PHOTO ISTOCK

|||||||||||||||||||||||||||||||||||||||||||||||||||||||||||

VETERINARIANS are authorised to sign certain documents, and sometimes receive requests to be less than transparent in the declarations, as in the following scenario.

SCENARIO
COMPROMISING REQUEST

▶ Mr and Mrs T come to visit you with their 12-week-old Beagle puppy Max for vaccination. They mention that he has been very quiet, off his food and vomiting for the last 24 hours. During the consult, he has bloody diarrhoea. Instead of vaccination, you recommend a parvovirus test, a full blood count and electrolytes, and intravenous fluids.

You estimate the costs at around $500 initially with an additional $150 for each additional day of hospitalisation, and more if imaging or other treatment is required.

The Ts are very concerned about costs. Mr T confers with his adult son on the phone, then authorises treatment – on the condition that the consultation is forward-dated until tomorrow. His son has applied online for pet insurance for Max.

What should you do?

RESPONSE
CHRIS DEGELING

▶ Mr and Mrs T are asking you to commit a deception in order to facilitate the provision of urgent medical care. From the description provided it would seem that the Ts were unaware that their puppy Max was so unwell when they entered the consultation, and they were completely unprepared for this turn of events. Suddenly finding themselves to be both morally and

△
7.31 The owners of this beagle puppy ask you to alter his medical record.
PHOTO ANNE FAWCETT

financially responsible for the health and welfare of a sick animal, the Ts have resorted to giving you an implicit ultimatum: commit fraud so they can shift the costs for Max's potentially lifesaving treatment to an insurance company, or funding for the benefits of this treatment will be withheld.

You are being placed in a situation where a moral imperative to attend to the urgent medical needs of a patient is in conflict with your obligations to the legal demands of the pet health insurance system and maintaining the good standing of your profession. The prima facie moral argument against perpetrating a deception is that lying is wrong and unethical, even when the intention is to assist another in desperate need. Yet it can be also be argued that bending the rules to ensure that Max receives the best care possible is a valid form of patient advocacy, an act of mercy and perhaps even a veterinary professional obligation. While veterinary practitioners who are committed to putting their patients first may find ethical arguments supporting deception compelling, it is also important to note that there is a clear financial incentive to fudge the patient's record because the veterinary practice will directly profit from providing the required services.

At first glance the Ts' behaviour appears to be manipulative, but this does not provide sufficient grounds to refuse their request. In the privately funded US human healthcare system the sort of deception the Ts are asking you to commit is commonly known as "gaming the system". In a published commentary on a similar hypothetical veterinary case, Glen Cousquer (2011) argues that a standard utilitarian approach which weighs the benefits for Max, his owners and the veterinary practice of funding services through deception against the harms caused to the insurance system is too simplistic. I agree. However, while Cousquer's argument rests on the assumption that honesty and veracity are essential to the maintenance of trust and a larger public good, for me this rationale is not completely satisfying because it does not clearly explain if or why a veterinarian still owes something to Max, an animal in need of veterinary care.

Alternative approaches

The central ethical problem posed by gaming health insurance systems has been usefully explicated by Matthew Wynia and colleagues (2000) as a conflict between two competing perspectives:

- That the provision of healthcare should be viewed as a profession-based fiduciary model.
- That the provision of healthcare should be viewed as a market-based contractual model.

So should a professional healthcare provider such as a veterinarian obey what Tavaglione and Hurst (2012) have described as *the internal morality of medicine* and above all, aim to provide the best care possible to their patients, which, in some circumstances, might require them to bend or break rules? Or are healthcare suppliers only required to provide good information and a list of services, and patients (veterinary or otherwise) simply get what they (or their owners) are willing to contract and pay for?

Contractual models

Contractual approaches to healthcare preclude the ethical permissibility of gaming the system, and can be supported through both consequentialist and duty-based arguments. These arguments end up sounding quite similar but there are subtle distinctions in what they take to be the morally relevant features. If we limit our focus to veterinary cases, from a consequentialist viewpoint it is possible to argue that if gaming of the insurance system becomes a common practice it may undermine trust in professional standards, professional conduct and contractual systems, which will eventually cause harm to insurers, veterinary healthcare providers, animal owners and broader society. Untoward consequences such as upward pressure on insurance premiums and stricter limitations on coverage may make contracting veterinary care less reliable and more unaffordable, and may be ultimately self-defeating. Notably, contractual arguments are often embedded within veterinary professional norms. Veterinary ethicists such as David Main (2006) have argued that as long as veterinarians offer reasonable recommendations for the animal's future care (which may include immediate euthanasia), it is the owner who has the moral responsibility to ensure their pet receives appropriate care, and the legal responsibility to prevent unnecessary suffering. Unless an offer is accepted and a contract is entered, then the veterinarian is only compelled to act if they have overriding animal welfare concerns.

In contrast, duty-based objections to gaming health insurance systems either rest on claims that: (1) deceptive practices rely on a form of lying that contravenes common morality; or (2) gaming is an egregious breach of the healthcare provider's duty towards contractual justice (Tavaglione & Hurst 2012). In the case of lying, as Cousquer (2011) points out, honesty is also morally important because veracity or being truthful is considered to be essential to a health professional's integrity, and the effective conduct of healthcare provider–patient/client relationships. Contractual justice demands that each party to an agreement receives what was entered in the contract, and that the contractually agreed services are executed properly. Arguments for maintaining contractual justice, and thereby distributive norms, highlight how "gaming insurance" redistributes the costs of providing healthcare for an uncontracted individual onto other contract holders such that gamers are free-riding at the expense of others.

Contractual approaches form the basis for professional guidelines and codes of conduct in both human and veterinary medical healthcare providers. Yet surveys of US physicians found almost half of them are likely to sanction some form of deception to ensure that their patients receive potentially lifesaving care. Their rationale was that they believed that their primary professional responsibility was to practise as their "patient's advocate". These physicians were happy to work within the rules and restrictions imposed by third-party payers, but would willingly make an exception and deceive insurance providers to manufacture solutions once the rules began to significantly compromise the interests of their patients (Freeman, et al. 1999, Wynia, et al. 2000).

Fiduciary models
Arguments for the ethical primacy of patient advocacy find philosophical support in fiduciary approaches to professional obligations.

To overcome prohibitions on lying, fiduciary arguments assert that contractual objections to gaming and deception often rest on an idealistic fallacy. Drawing on John Rawls' (1999) theory of justice, the central premise is that the actions of moral agents (such as healthcare providers) who are caught in non-ideal circumstances (such as instances where their patients cannot pay for essential healthcare) should not be judged through normative criteria drawn from an ideal theory (a theory that assumes the external social-political contexts and institutional conditions under which agents operate are perfectly just).

If we accept that healthcare providers are obligated to pursue an overarching central internal good of using their knowledge and technical skill to do the best for their patients, it is possible that pursuit of this good, in limited circumstances, might be sufficient grounds for deception to be morally permissible (Tavaglione & Hurst 2012). If the healthcare provider considers the circumstances their patient finds themselves in to be the product of a non-ideal institutional environment, then the healthcare provider's actions should not be judged by ideal standards. Fiduciary approaches such as the one described above still accept that the restrictions placed on healthcare providers by contractual considerations can be, and most often are, warranted and justified. But when unjust, there is potentially a moral case for them to be contravened.

But what to do about Max?
It is possible to argue that Max's circumstances are unjust, but this is essentially a function of his property-status, rather than being the product of an unjust contract between the Ts and the insurance company. No contract exists, and the Ts are the unjust party should one be created based on deception. Because the attending veterinarian has a financial interest in the provision of services

to Max, we might view with some reservations any argument they make that providing the highest level of care is a moral necessity.

Indeed the vet has the means to provide this care, and if the moral imperative to do the best for their patients is overriding then they should seek a way to provide it without drawing in a third party. Rather than accepting the terms of the Ts' ultimatum the attending veterinarian should explore how their access to the funds needed to pay for the care is constrained.

It is also important to realise that a decision as to how Max will be treated for the bloody diarrhoea does not need to be made all at once. Acting as their patient's advocate the veterinarian should assist the Ts in understanding the range of different treatment approaches available, with the aim of finding a clinically and financially acceptable middle ground so important basic supportive measures can be provided. Other diagnostic procedures can be undertaken, if warranted and the owner agrees to pay, depending on how Max responds to initial treatment. If the Ts disagree to any course of action the veterinarian offers, then as David Main (2006) has argued, under the contractual model the vet has no legal or moral responsibility unless they judge that there are immediate and overriding animal welfare considerations. If the veterinarian judges that Max's current situation is unjust, and requires them as a professional to perform a rescue, then they should undertake to treat him in good faith, without expecting remuneration.

That said, there are circumstances where it is conceivable that a moral argument could be made to deceive an animal health insurance agency. But before being party to this the veterinarian must be sure that their client's dealings with the insurer have been and will continue to be manifestly unjust, and ensure that they themselves do not benefit financially or professionally from the deception.

||||||||||||||||||||||||||||||||||||||||||||||||||||||||||||||||||||||||||||

THIS is a detailed analysis drawing heavily on the literature from human healthcare, illustrating how complex the issue is. One of the major conflicts in modern healthcare is between the interests of the patient and the commercial interests of the healthcare provider. While healthcare providers may see their duty primarily as providing a high standard of care, the imperative to charge for that service quite often constrains the nature and standard of the care that is provided, or in some cases may preclude healthcare provision.

Where insurance fraud is committed, as proposed in the scenario, it is often justified by reference to a utilitarian approach. For example, one might argue that the end (treating the animal whilst ensuring the client receives the benefit of insurance) justifies the means (deception) as the majority of parties (the owners, the dog, the veterinarian and the practice) benefit whilst only a minority (the insurer) is harmed.

But is this an accurate analysis? If, as discussed, the impact of such deception is to increase premiums, we need to consider that potentially many other animals, owners and veterinarians are harmed by increased insurance premiums. The interests of easily identifiable stakeholders with immediate, direct interests in a scenario (for example Max and his owners) may be pitted against those of potential stakeholders who may be impacted in the longer term, or indirectly (for example, future insurance policy holders or future patients). One of the challenges in veterinary ethics lies in identifying and weighing the interests of these "indirect" or potential stakeholders.

CHALLENGES TO PROFESSIONALISM

What would you do?

You visit a livery yard with your senior partner to vaccinate and worm several of the horses there. One of your clients, Ms A, has bought a lovely pony for her daughter Gemma. They want you to give the pony, Toby, the "once-over" whilst you are there. As soon as Toby is brought out of the stable he dances around on the concrete, but you see he's slightly lame on his left hind. You catch the eye of your boss who's also seen it. On asking Ms A about him generally it's clear she hasn't noticed anything wrong with him. She's just in the process of sorting out her insurance for Toby – critical for her as she doesn't have the kind of money floating about to pay for unexpected vet fees. Your boss hisses to you, "Don't mention the lameness yet."

ONE_____ What should you do?

TWO_____ How would your actions differ if your boss weren't there?

△

7.32 You are asked to inspect Toby, a pony with a lame left hind.

PHOTO ANNE FAWCETT

7.7

Can professionals have a sense of humour?

It can seem as if there is an enormous weight of responsibility on professionals to maintain their professionalism at all times. Is this a realistic expectation of a human being? Does the "professional mask" slip and are there circumstances in which this is considered acceptable?

SCENARIO
SINGING CADAVER

▶   A client you do not know well rings outside of normal working hours requesting euthanasia of their small dog, Pepper. As the nurse on call you go to assist in this task. The euthanasia is straightforward and you lock up again after the clients leave. There's just the tidying up and dealing with the body to do before you and the vet can both go home. Suddenly you hear a high-pitched comedy voice singing "Look at me, I'm Sandra-Dee..." As you turn you see the head of the dead dog peeking round the corner of the consulting room, swaying and miming to the song.

Are the veterinary surgeon's antics morally permissible, and is it acceptable for the nurse to laugh at the vet's antics?

RESPONSE
STEVEN P MCCULLOCH

▶   This scenario is useful to demonstrate differences in how different ethical frameworks might judge whether there is anything morally wrong in using deceased animals to provide entertainment.

(267)

—
7.7

CAN PROFESSIONALS HAVE A SENSE OF HUMOUR?
RESPONSE
STEVEN P MCCULLOCH

△

7.33

CARTOON DR ROBERT JOHNSON

Additionally, analysis of the scenario suggests the importance of moral intuitions in veterinary ethics.

Utilitarianism prescribes that right action is dictated by the consequences of action, specifically utility, *alone* (Bentham 1962 [1789], Glover 1990). Consider if the veterinarian's antics result in the greatest overall utility for all parties impacted. In this case, utilitarian theory permits the veterinarian to perform such an act, i.e. sing "Look at me, I'm Sandra-Dee…" Indeed, in utilitarianism, if the amusement derived from such an act provides greater utility than any other possible act, it is not only morally permissible for the veterinarian to use the corpse as a prop to sing in such a way, but morally *obligatory*.

Closer inspection of the case reveals it has been constructed to make it at least possible that singing "Sandra-Dee" might produce the greatest overall utility. This is because (1) neither the veterinary surgeon nor the veterinary nurse knows the client well (they do not have "sentimental"

considerations about the client's wishes), (2) the euthanasia takes place out-of-hours, therefore no other staff are present (human utility impacts are restricted to the two agents present), and (3) the client is not present and is unaware of the event (the client's utility is not impacted). Additionally, since the canine patient is deceased, it is no longer sentient and does not have a welfare (and therefore is not impacted in terms of utility).

Utilitarian reasoning often conflicts with broadly held moral intuitions (utilitarians tend to distrust moral intuitions). Many veterinary professionals, as well as the pet-owning public, might argue that such a scene was distasteful at best, and morally repugnant at worst. In contrast to utilitarian ethics, deontological theory bases right action on laws, principles, rights and duties. Virtue-based approaches base right action on some predefined conception of virtuous professional behaviour.

The UK Royal College of Veterinary Surgeons (RCVS) Codes of Professional Conducts (for veterinary surgeons and veterinary nurses) are written in an explicitly deontological framework (McCulloch, et al. 2014, RCVS 2015a, 2015b). The five "principles of practice" include "honesty and integrity" and "client confidentiality and trust". The Codes outline responsibilities (i.e. duties) to (1) animals, (2) clients, (3) the profession, (4) the veterinary team, (5) the RCVS and (6) the public. Section 2.1 of the RCVS Code for Veterinary Surgeons states "Veterinary surgeons must be open and honest with clients and respect their needs and requirements" (RCVS 2015b). Section 6.5 states "Veterinary surgeons must not engage in any activity or behaviour that would be likely to bring the profession into disrepute or undermine public confidence in the profession" (RCVS 2015b). Section 8 of the supporting guidance of the Code states "Veterinary surgeons and veterinary nurses should be aware that these events [euthanasia] are often highly emotionally charged. In these circumstances, small actions and/or omissions can take on a disproportionate level of importance" (RCVS 2015c).

Admittedly, there is nothing in the RCVS Codes that explicitly states that veterinary professionals should not use deceased patients to provide entertainment. However, it should be clear from the above that using a euthanased animal as a prop when singing "Sandra-Dee…" is inconsistent with professional conduct, as outlined in the Codes. Such behaviour conflicts with the principles of honesty, integrity and trustworthiness. Using a euthanased patient in such a way is not consistent with the client's needs and requirements. Finally, should the scenario become public knowledge, it would risk bringing the profession into disrepute and undermining confidence in the profession. Indeed, the latter could conceivably impact on animal welfare, the overriding duty of the veterinary profession (RCVS 2015a, 2015b). If the public came to suspect that deceased pets were not always treated with due respect, they might be reluctant to seek veterinary attention for euthanasia of suffering animals.

It is instructive to mention briefly how virtue-based theory might approach the scenario. Our moral intuitions tell us there is something virtuous about professionals treating deceased animals (and humans in medical ethics) with a degree of respect. Veterinary practitioners and clients might well have different conceptions of the virtuous veterinary surgeon and the virtuous veterinary nurse. However, there is likely to be common ground in considering that using deceased animals as a prop for entertainment is not consistent with virtuous professional behaviour.

Finally, the Golden Rule is useful to provide some insight into this scenario. The Golden Rule is expressed in Christianity as "Do to others what you want them to do to you" (Matthew 7:12). The rule, which expresses reciprocity in moral behaviour, is found in almost all religious beliefs and moral theories. In this case, veterinary professionals can ask themselves how they would feel if their deceased pet were treated in such a way. It is likely that a substantial proportion of the profession would desire that their deceased pet be treated with respect.

The analysis above finds that using a deceased patient as a prop for entertainment is morally problematic. The veterinary surgeon has instigated the act, is responsible for the behaviour and should therefore bear at least the brunt of moral approbation. But what can be said of the veterinary nurse's response of laughter? According to virtue theory, it would be a vicious act for the veterinary nurse to laugh. Virtues are character traits that are deeply ingrained (Hursthouse 2013). In this scenario, a *virtuous* veterinary nurse would find the veterinary surgeon's act morally repugnant and act accordingly. Admonishment would be more appropriate than laughter. Interestingly, according to utilitarian theory, it could be argued that the veterinary nurse is morally obliged to laugh. Whether the nurse finds the behaviour humorous or not, it is likely to lead to greater utility if they laugh (consider that we often laugh at jokes of friends and acquaintances, even if we do not find them amusing, to avoid embarrassment).

The RCVS Codes provide a deontological framework for professional behaviour (McCulloch, et al. 2014, RCVS 2015a, 2015b). The RCVS Code for veterinary nurses is, in essence, the same as that for veterinary surgeons. Therefore, the principles and duties outlined in the Code would not support overt laughter from the veterinary nurse. Although it might be morally wrong for the nurse to laugh at such a scene, it is the veterinary surgeon that is most culpable. The veterinary surgeon chose consciously to perform the act, presumably, with the goal of making the veterinary nurse laugh. In contrast, laughter is a behaviour that we do not always have control over. Hence, the veterinary surgeon's antics are morally problematic in the context of professional ethics. In contrast, although it might not be morally defensible for the

7.7

CAN PROFESSIONALS HAVE A SENSE OF HUMOUR?
RESPONSE
STEVEN P MCCULLOCH

veterinary nurse to laugh, the more involuntary nature of laughter makes the nurse's behaviour far more forgivable.

||||||||||||||||||||||||||||||||||||||||||||||||||||||||||||||||||||

THE gallows humour displayed in this scenario is typical of a type of humour that "treats serious, frightening, or painful subject matter in a light or satirical way" (Watson 2011), being differentiated from cruel forms of humour by one doctor as "the difference between whistling as you go through the graveyard and kicking over the gravestones" (Wear, et al. 2009). Watson (2011) found that the use of gallows humour could be explained in a number of ways: it may help people to cope with difficult situations by maintaining a certain distance from the tragic event but also by forming a connection with their fellow joker; it may also reflect a certain incongruity, existential or otherwise, when unusual or strange events occur; it may help with managing uncertainty about diagnoses and treatment; and such humour may be exacerbated when other basic needs, such as sleep, are unmet. Patricia Morris, in her book *Blue Juice*, notes that even the title reflects an element of gallows humour she saw to be common during her ethnographic study of veterinary euthanasia. She found that in front of clients the patients and their diseases were always referred to respectfully. It was "backstage", out of earshot of clients, that gallows humour was expressed, to varying degrees by different individuals (Morris 2012).

In contrast to the hard-line professionalist approach denying an acceptability of gallows humour, Watson (2011) doubts that it is unethical when certain caveats are met, ensuring only benefits and not harms come from such jokes, ending her analysis with a patient's perspective: "So tell your jokes. Tell them somewhere I cannot hear. Then treat me well when we're together."

CAN PROFESSIONALS HAVE A SENSE OF HUMOUR?

What do you think?

ONE   Under what circumstances would gallows humour be acceptable to you?

TWO   How would you know when gallows humour would be acceptable?

THREE   Have you heard things that you feel have gone too far?

△

7.34 How would you know when gallows humour is ethically unacceptable?
PHOTO ISTOCK

Conclusion

Professionals have special obligations to act ethically in accordance with the expectations of their professional body and clients. Nevertheless it can be difficult to act professionally all the time. Clients may apply pressure to stretch the boundary of good conduct and sometimes it's not at all clear where the boundary even lies. The consequences of acting unprofessionally may be greater than for other ethical decisions as a veterinary professional's livelihood, as well as sense of identity, may be at stake. There may be guidance as to what constitutes unprofessional behaviour from the governing body, but whether or not we agree with it, or choose to abide by it, is an element of the ethical decision.

References

American Veterinary Medical Association 2016 Veterinary Dentistry. https://www.avma.org/KB/Policies/Pages/AVMA-Position-on-Veterinary-Dentistry.aspx

Anon 2010 Social media and the medical profession: a guide to online professionalism for medical practitioners and medical students. A joint initiative of the Australian Medical Association Council of Doctors-in-Training, the New Zealand Medical Association Doctors-in-Training Council, the New Zealand Medical Students' Association and the Australian Medical Students' Association.

Anon 2011 Doyle Byrnes, Student Expelled For Posting Placenta Photo, Sues Johnson County Community College. *Huffington Post.* http://www.huffingtonpost.com/2011/01/03/doyle-byrnes-student-expe_n_803839.html

Association of American Veterinary Medical Colleges 2011 Guiding Principles and Considerations: Ethical Interaction between Schools/Colleges of Veterinary Medicine and External Entities. AAVMC.

Australian Council of Professions 1997 Code of Professional Conduct: Australian Veterinary Association.

Australian Health Practitioner Regulation Agency 2013 Regulating Health Practitioners in the Public Interest Annual Report 2012/13. AHPRA.

Australian Medical Association 2010 Social media and the medical profession. A guide to online professionalism for medical practitioners and medical students.

Bentham J 1962[1789] An Introduction to the principles of morals and legislation. In: Warnock M (ed) *Utilitarianism,* 33–77. University of California: Fontana.

BMA 2011 *Using Social Media: Practical and Ethical Guidance for Doctors and Medical Students.* British Medical Association: London.

Cappola AR, and Fitzgerald GA 2015 Confluence, not conflict of interest: name change necessary. *Journal of the American Medical Association* **314:** 1791–1792.

Coe JB 2014 Reputation management: managing your risk to maximize the benefits of social media. In: *Proceedings of the Australian Veterinary Association Annual Conference.*

Coe JB, Weijs CA, Christofides E, Muise A, and Desmarais S 2011 Teaching veterinary professionalism in the Face(book) of change. *Journal of Veterinary Medical Education* **38:** 353–359.

Cousquer G 2011 Everyday ethics. *In Practice* **33:** 237–238.

D'Arcy E, and Moyniham R 2009 Can the relationship between doctors and drug companies ever be a healthy one? *PLoS Medicine* **6:** e1000075.

Dhawan S 2009 The new reality-serving legal documents through Facebook? The Court. Blog. Osgoode Hall Law School.

Fleming G 2009 Hospital staff reinstated after 'lying down game' suspensions. *Nursing Times,* 14 October.

Freeman VG, Rathore SS, Weinfurt KP, Schulman KA, and Sulmasy DP 1999 Lying for patients: physician deception of third-party payers. *Archives of Internal Medicine* **159:** 2263–2270.

Freidson E 2001 *Professionalism, the Third Logic. On the Practice of Knowledge.* University of Chicago Press: Chicago.

Fritschi L, Morrison D, Shirangi A, and Day L 2009 Psychological wellbeing of Australian veterinarians. *Australian Veterinary Journal* **87**: 76–81.

Glover J 1990 *Utilitarianism and its Critics*. Macmillan: New York.

Goldacre B 2009 *Bad Science*. Harper Perennial: London.

Goldacre B 2012 *Bad Pharma*. Fourth Estate: London.

Hampton K, Sessions Goulet L, Rainie L, and Purcell K 2011 Social networking sites and our lives. *Pew Research Center*, June.

Heyd D 2015 Supererogation. In: Zalta EN (ed) *The Stanford Encyclopedia of Philosophy (Spring 2016 Edition)*.

Hinckfuss I 1982 Morality and the resolution of conflicts. *Social Alternatives* **2**: 23–28.

Hume D 2003 [orig. 1739] *A Treatise of Human Nature*. Dover Publications: Mineola.

Hursthouse R 2013 Virtue ethics. In: Zalta EN (ed) *The Stanford Encyclopedia of Philosophy (Fall 2013 Edition)*. http://plato.stanford.edu/archives/fall2013/entries/ethics-virtue/

Kraus JS 1993 *The Limits of Hobbesian Contractarianism*. Cambridge University Press: Cambridge.

Lachance M 2016 Breaking the silence: the veterinarian's duty to report. *Animal Sentience* **1**: 1–16.

Llayton-Bennett A 2013 Honours for Harris: a vet's story. *The Veterinarian Magazine*. Sydney Magazine Publishers: Sydney.

Lundh A, Sismondo S, Lexchin J, Busuioc OA, and Bero L 2012 Industry sponsorship and research outcome. *Cochrane Database of Systematic Reviews* **12**: MR000033.

Main DC 2006 Offering the best to patients: ethical issues associated with the provision of veterinary services. *Vet Rec* **158**: 62–66.

Martinson BC, Anderson MS, and de Vries R 2005 Scientists behaving badly. *Nature* **435**: 737–738.

McCulloch S 2012 Everyday ethics. *In Practice* **34**: 494–495.

McCulloch S, Reiss M, Jinman P, and Wathes C 2014 The RCVS codes of conduct: what's in a word? *Veterinary Record* **174**: 71–72.

McHale JV 2009 Conscientious objection and the nurse: a right or a privilege? *British Journal of Nursing* **18**: 1262–1263.

Mechanic D 1998 The functions and limitations of trust in the provision of medical care. *Journal of Health Politics, Policy and Law* **23**: 661–686.

Meyer EC, Lamiani G, Foer MR, and Truog RD 2012 "What would you do if this were your child?": practitioners' responses during enacted conversation in the United States. *Pediatric Critical Care Medicine* **13**: 372–376.

Minkoff H, and Lyerly AD 2009 "Doctor, what would you do?" *Obstetrics and Gynecology* **113**: 1137–1139.

Morris P 2012 *Blue Juice: Euthanasia in Veterinary Medicine*. Temple University Press: Philadelphia.

Mullan S 2012 Comments on the dilemma in the September issue: 'What if it was your dog?' *In Practice* **34**: 551.

Nightingale D 1976 Conflict and conflict resolution. In: Strauss C, Miles RE, Snow CC, and Tannenbaum AS (eds) *Organisational Behaviour: Research and Issues*. Wadsworth: Belmont.

NUT 2010 *E-safety: Protecting School Staff. NUT Guidance and Model Policy*. National Union of Teachers: London.

Ontario Government 2011 Veterinarians Act.

Oxford Dictionaries 2016 Conflict of interest.

Press Complaints Commission 2011 Baskerville vs Daily Mail (2010).

Ramsbotham O, Woodhouse T, and Miall H 2011 *Contemporary Conflict Resolution*. Polity Press: Cambridge.

Rawls J 1999 *A Theory of Justice*. Oxford University Press: Oxford.

RCVS 2015a Code of Professional Conduct for Veterinary Nurses. http://www.rcvs.org.uk/advice-and-guidance/code-of-professional-conduct-for-veterinary-nurses/

RCVS 2015b Code of Professional Conduct for Veterinary Surgeons. http://www.rcvs.org.uk/advice-and-guidance/code-of-professional-conduct-for-veterinary-surgeons/pdf/

RCVS 2015c Euthanasia of animals. In: Code of Professional Conduct for Veterinary Surgeons, 55–57. http://www.rcvs.org.uk/advice-and-guidance/code-of-professional-conduct-for-veterinary-surgeons/pdf/

Rollin BE 2002 The use and abuse of Aesculapian authority in veterinary medicine. *JAVMA-Journal of the American Veterinary Medical Association* **220**: 1144–1149.

Shirangi A, Fritschi L, Holman CDJ, and Morrison D 2013 Mental health in female veterinarians: effects of working hours and having children. *Australian Veterinary Journal* **91**: 123–130.

Stein R 2011 New Jersey nurses charge religious discrimination over hospital abortion policy. *Washington Post.*

Tavaglione N, and Hurst SA 2012 Why physicians ought to lie for their patients. *The American Journal of Bioethics* **12**: 4–12.

Trossman S 2014 Conscientious objection. When care collides with nurses' morals, ethics. *The American Nurse* **46**: 1–11.

Watson K 2011 Gallows humor in medicine. *Hastings Center Report* **41**: 37–45.

Wear D, Aultman JM, Zarconi J, and Varley JD 2009 Derogatory and cynical humor directed towards patients: views of residents and attending doctors. *Medical Education* **43**: 34–41.

Weijs CA, Coe JB, Christofides E, Muise A, and Desmarais S 2013 Facebook use among early-career veterinarians in Ontario, Canada (March to May 2010). *JAVMA-Journal of the American Veterinary Medical Association* **242**: 1083–1090.

Weijs CA, Coe JB, Muise A, Christofides E, and Desmarais S 2014 Reputation management on Facebook: awareness is key to protecting yourself, your practice and the veterinary profession. *Journal of the American Animal Hospital Association* **50**: 227–236.

Williams V 2002 Conflicts of interest affecting the role of veterinarians in animal welfare. *ANZCCART News* **15**: 1–3.

WSAVA 2013 International Veterinary Oaths. World Small Animal Veterinary Association.

WSAVA 2014 WSAVA Global Veterinary Oath. World Small Animal Veterinary Association.

Wynia MK, Cummins DS, Van Geest JB, and Wilson IB 2000 Physician manipulation of reimbursement rules for patients: between a rock and a hard place. *Journal of the American Medical Association* **283**: 1858–1865.

Yeates JW, and Main DCJ 2010 The ethics of influencing clients. *JAVMA-Journal of the American Veterinary Medical Association* **237**: 263–267.

CHAPTER 8
ERRORS AND COMPLICATIONS

*"There are some patients that we cannot help;
there are none whom we cannot harm."*
Arthur L. Bloomfield (1888–1962)

Introduction

While there is a paucity of research about the incidence of errors in veterinary medicine, extensive research from the human medical domain suggests that error is common. It is estimated that 1 in 10 patients admitted to hospital in the developed world is a victim of medical error, and 1 in 300 admitted to hospital dies due to such an error (Reckless, et al. 2013). In the USA for example, the number of premature deaths associated with preventable harm to patients was estimated to be over 400,000 per year (James 2013).

Take errors in the administration of medication. In one study which looked at 14,041 instances of medication administration, 1271 errors were detected. Of these, 133 had the potential to cause serious or life-threatening harm to the patient and 10 actually led to patient harm. Types of errors included administering drugs too late or too soon, inappropriate route of administration, the wrong medication or wrong dose was given and documentation that was incorrect (Kale, et al. 2012).

△
8.2 This kitten received a tenfold overdose of premedication containing acepromazine and an opioid. The error was detected immediately, an opioid reversal agent administered and intravenous fluid therapy commenced to prevent hypotension. The elective surgery was postponed and the client informed.
PHOTO ANNE FAWCETT

It is difficult to determine the number of deaths due to veterinary treatment. Veterinary practices tend to be smaller, private institutions with no obligations to report error, other than perhaps to those directly impacted.

◁
8.1
PHOTO ANNE FAWCETT

△

8.3 Errors can have a profound impact not only on the patient and client, but on any and all members of the veterinary team.

CARTOON SUHADIYONO94

The morbidity and mortality conference, a confidential, thorough review of cases with adverse outcomes, has been well established in medical teaching hospitals but is rare in the veterinary clinical context (Powell, et al. 2010).

It can be challenging to distinguish genuine medical error, defined as "an act of omission or commission in planning or execution that contributes or could contribute to an unintended result" (Grober & Bohnen 2005) and unavoidable complications, defined as "a secondary disease or condition that develops in the course of a primary disease or condition and arises either as a result of it or from independent causes" (Anon 2016).

In this chapter we will consider how errors and complications can have a range of impacts on all stakeholders: the patient or patients, clients, the veterinary team and the person or persons responsible or blamed for the error.

8.1

Costs of complications

In a user-pays healthcare system, the financial cost of complications can be significant. Is it reasonable for clients to bear the costs of such complications? How do we establish who pays? The following scenario explores this issue.

SCENARIO
SUBSIDISING COMPLICATIONS

▶ You are managing a companion animal practice. The veterinary team perform an exploratory laparotomy and intestinal resection and anastomosis to remove a foreign body (peach stone) from Tiffany, a beagle puppy. The surgery goes well, and the dog remains for observation. However, 48 hours post-operatively Tiffany develops peritonitis due to leakage at the anastomosis site. The owners, Mr and Mrs S, were charged $2000 (£1200) for the original surgery.

A second surgery is performed, the original anastomosis site is resected and a second anastomosis performed. Tiffany is treated for peritonitis and heals without further incident. Mr and Mrs S demand a discount on the second surgery because the first "didn't work".

What should you do?

RESPONSE
MANUEL MAGALHÃES-SANT'ANA

▶ In a UK study, recent veterinary graduates have identified the financial aspects of practice, namely charging clients, as one of the main difficulties on entering the veterinary profession (Routly, et al. 2002). In the same vein, financial problems

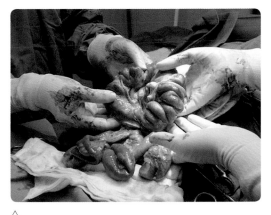

△
8.5 Codes of professional conduct offer little indication as to whether a second procedure (whether a consultation or a surgery) should cost the same as or less than the original one.
PHOTO ANNE FAWCETT

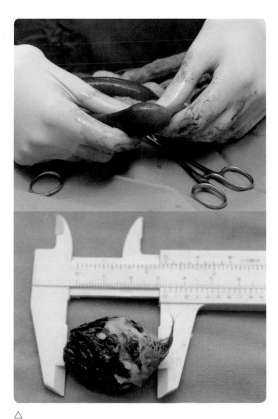

△
8.4a–b Gastrointestinal foreign bodies are associated with complications such as peritonitis. Should owners pay for revision surgery to address complications when they occur?
PHOTOS ANNE FAWCETT

are known to increase the risk of depression and suicide in veterinary medicine (Bartram & Baldwin 2010). Findings such as these might help in explaining why financial issues have been identified as a relevant teaching topic in veterinary ethics [(Magalhães-Sant'Ana 2014), see Figure 1.7].

The guiding principles of the veterinary profession, namely codes of professional conduct, offer little indication as to whether a second procedure (whether a consultation or a surgery) should cost the same as or less than the original one. Discounts

can be performed at the veterinarian's discretion but a word of caution is in order. Performing regular discounts (e.g. every time something goes wrong) can jeopardise the economic viability of the practice. In the case described above, the second surgery is not easier than the first and does not involve fewer risks (quite the contrary) or less cost. Moreover, making a discount can give the impression that you not only *feel* responsible for the poor outcome, but that you *are* indeed responsible for it.

Often in veterinary medicine, complications are not so much a result of poor patient care but a consequence of the grave clinical condition. If you firmly believe that you did everything within your power to treat the animal, it is only fair to ask for whatever is owed to you from the second surgery. However, if you consider that you were somewhat unskilled or negligent (because you remembered that the anastomosis should have been secured better, or you forgot to check for any leakages before closure), then you have a moral obligation to make a substantial discount or even to waive the payment from the second surgery.

—
8.1
COSTS OF COMPLICATIONS
RESPONSE
MANUEL MAGALHÃES-SANT'ANA

As so often happens with ethical challenges, "what appears to be a moral dilemma is the result of a breakdown in, or absence of, relevant communication between different people" (Thompson, et al. 2000). In fact, some aspects of this case seem to result from inadequate communication, suggesting that the owners' consent may not have been fully informed. A proper surgical consent form should include provisions for possible complications. In addition to signing the form (usually the same template for every surgery), clients should be made aware of the risks associated with that particular procedure and allowed to ask any questions. For example, clients should be informed that in cases of intestinal anastomosis dogs with foreign bodies are more likely to have leakage than dogs that underwent surgery for any other reason (Ralphs, et al. 2003). If, when seeking consent for the first surgery, the veterinarian clearly addresses risks of complications – including leakage, peritonitis and the need for a second surgery or prolonged post-operative care – there would be no reason to negotiate the price of the second surgery, and most of the clients would probably not ask for a discount.

As a guiding rule, pricing of veterinary services should be agreed beforehand and considered, at least in principle, independently from the outcome of the procedure. This includes, of course, agreeing on a possible discount (e.g. if the owners are financially needy or if prolonged after-care is anticipated). In addition, the principle of fairness (or justice) determines that you should consistently apply the same policy to every client under the same circumstances without discrimination. This implies that clear policies should be put into place in your practice to help prevent awkward debates and ad hoc judgements regarding the pricing of services.

||||||||||||||||||||||||||||||||||||||||||||||||||||||||||||||||||||||||||||

ONE of the guiding principles of healthcare is non-maleficence, or seeking to avoid harming the patient. Yet most owners can appreciate that all veterinary intervention carries some risk, however small, of harm to the patient.

Discounting in the context of errors and complications is a delicate issue. On the one hand, a client who perceives that the professional acted at less than their expected state may claim a right to a discount on treatment arising from the error or complication, a refund or potentially compensation. On the other hand, it may be impossible to determine if an adverse outcome was avoidable or not.

Clients may feel it is wrong to be required to pay the bill, but this raises the question of what they are paying for. The owner in this situation may refuse to pay because the outcome was not the expected outcome. However, in veterinary practice the owner pays for the service – not the outcome. While the veterinary practice must strive for the best possible outcome for the animal treated, such an outcome cannot be guaranteed.

One means of testing a particular decision is to "universalise it". No one, surely, would expect veterinarians to bear the costs of every poor outcome. It seems fairer that the client pays for the veterinarian's service – their attempt to do the best job possible for each particular animal within constraints imposed by the owner. Obtaining informed consent is aligned with respecting autonomy. There is a legitimate concern that discounting or waiving of fees in such circumstances may be perceived as an admission of liability.

On the other hand, a veterinarian can express that he or she is sorry for the outcome without admitting liability (Feinmann 2009). This is an important acknowledgement of care for the animal, the client and the bond between them.

The majority of poor outcomes in veterinary practice are due to biologic variation, low-probability risks and side effects and unrealistic

> "The most important step is to take the time to explain the outcome to the client when the client is ready to discuss the matter."

expectations rather than negligence (O'Connell & Bonvicini 2007). However, owners may not be aware of how we evaluate and weigh up risks and benefits of intervention unless we make our decision-making explicit.

Without providing this explanation, a distraught client who is simply presented with a bill may feel that the veterinarian is more interested in money than in the patient. It may take the client time to process the fact that an error or complication has occurred. The most important step is to take the time to explain the outcome to the client when the client is ready to discuss the matter.

It is helpful to explain what was done to monitor for and address errors and complications in this case, including review of case management. Quite often clients are not aware how much work goes into preventing and avoiding errors and complications, and investigating these when they do occur. This is also an opportunity for the client to ask questions and seek clarification.

Regardless of perceived fault it is important to ensure that all relevant procedures are reviewed. Advising the client of this may give them some reassurance that their animal's complication is being taken seriously, that the practice is transparent in addressing any risk factors which might have led to the outcome, and that the veterinarian is not simply concerned with charging money and moving on. It is also important for the veterinary team to feel that all potential contributing factors are identified and assessed.

In the case of severe adverse events (e.g. death), it may be reasonable to offer a "compassionate

COSTS OF COMPLICATIONS

What would you do?

A six-month-old, apparently healthy female cat presents to your practice for a routine ovario-hysterectomy. On physical examination the kitten is bright and alert, in excellent body condition, normothermic and without any audible heart murmurs or abnormalities.

Anaesthesia and surgery are performed as per routine, but the animal suffers cardiac arrest during the procedure and cannot be resuscitated despite lengthy attempts to do so. The owner is informed and is understandably distraught. He is adamant that he should not pay the bill. How should you address this?

△

8.6 There is a small, but present, risk that clinically healthy animals can suffer fatal anaesthetic complications.
PHOTO ANNE FAWCETT

discount" which is specified as such, charging costs to ensure that the practice does not lose money, but also acknowledging the client's loss. With the client's permission it may be possible to perform a post-mortem, also at cost, on the grounds that a definitive diagnosis may provide some closure and may benefit other animals (for example, littermates of the affected animal which may have the same congenital abnormality) and provide further education to the veterinarian.

In considering a way forward, it is important to note that anaesthetic deaths in healthy small animal patients are increasingly rare [0.05–0.11 per cent (Brodbelt, et al. 2008)]. Nonetheless the client may blame the veterinarian – after all, they admitted a healthy animal which subsequently died following treatment.

If we assume the veterinarian took care in pre-operative evaluation of the patient, anaesthetic administration and monitoring, it is unlikely that the anaesthetic death was the fault of the veterinarian. Similarly, in the previous case, the veterinarian may have treated the dog to the highest standard, but the underlying condition led to post-operative complications.

Where it comes down to liability, it can be difficult to apportion this based on the facts presented. In cases of anaesthetic death, clients should be offered a post-mortem, which may reveal pre-existing disease that contributed to anaesthetic death and was not detected on physical examination (Bednarski, et al. 2011). However, the client must be advised beforehand that like all other diagnostic modalities, post-mortem examination is neither 100 per cent sensitive or specific, and may not yield a definitive diagnosis.

8.2
Error disclosure

Veterinarians may be reluctant to disclose errors due to a fear of being blamed. The following scenarios explore approaches to errors and complications in practice settings.

————
SCENARIO
ASHES MIX-UP

▸ Your small animal practice offers individual cremation and return of the ashes to owners after euthanasia. The crematorium you use is very professional and you have been happy with their service over the years. Occasionally owners never come back to collect their pet's ashes and you have a small collection of these remains.

Today, the owners of Kato, a much-loved dog, have come to collect his ashes. You conducted the euthanasia and remember discussing the cremation options with them but you haven't had the ashes back from the crematorium. Now you have a sinking feeling that you ticked the wrong box on the form, requesting a "general" cremation in which ashes are not returned – after all, you were very tired. Sure enough, that seems to be what's happened, and to make matters worse the owners actually requested a special (and expensive) type of urn. You consider giving the owners one of these "spare" boxes of ashes, but you do not have one of the urns they ordered.

What should you do?

△

8.7 How would you address a mix-up with a cherished companion animal's remains?

PHOTO ANNE FAWCETT

RESPONSE

ALISON HANLON

▶ The origin of the word dilemma means "two horns" and as Rollin (1999) describes, "whichever direction you turn, you are 'impaled' on a horn". This scenario represents a true dilemma – the vet has just two options – either to be honest or to be dishonest.

There are several frameworks that can be used to evaluate this dilemma, as outlined in chapter 2. A simple step-by-step approach is advocated by Rollin (1999):

- Firstly, create a conceptual map of stakeholders and list the corresponding duties to them to identify the ethically relevant issues.
- Check to see if there is guidance in the Professional Code of Practice or legislation relevant to the case.
- If not, apply your own personal ethics – if there are two or more conflicting principles, then adopt a pluralistic approach by applying ethical theories such as utilitarianism and principalism.

The first step, stakeholder recognition and professional obligations to stakeholders, is important for increasing both moral imagination and awareness of the potential ethical issues. It's easy to identify those directly involved: the vet and the owners, but what about the veterinary practice, and other animals and clients at the veterinary practice? One action can have far-reaching effects. The veterinary profession plays an important role in society, which requires honesty and integrity and so this case also has implications for a broader range of stakeholders including all animal owners, and, for example, sectors which require veterinary inspection and certification such as the food chain, laboratory animal research, animals used in sport and so on. It may seem like an exaggeration, but the point is that vets hold a trusted position in society.

The second step in the resolution process considers professional guidelines (and legislation), which may be sufficient to support ethical decision-making. Honesty and integrity are listed in the Principles of Practice in the RCVS Code of Professional Conduct (2015) and are relevant to the current dilemma. After all, ethics stems from the Greek word *ethikos*, which focuses on the character of an individual (Tannenbaum 1995) and so applying virtue ethics is appropriate. With this in mind, you should consider being open and honest with the owners, explaining the context of the situation – it was a genuine error, made at the end of a busy day. There are risks involved with this approach it is likely to be upsetting for the owners and may damage the reputation of the vet. The alternative is to be dishonest, which may be motivated out of self-protection or to save the owners from further emotional turmoil at the loss of their pet.

There are ethical arguments for and against both approaches. A contractarian (deontologist) would contend that being honest is morally imperative, whereas being dishonest is a breach of

—
8.2
ERROR DISCLOSURE
RESPONSE
ALISON HANLON

trust and the unwritten contract between vet and owner. If you are willing to lie about this, what else might you be willing to lie about? A utilitarian, on the other hand, may advocate lying, to save the owners from further emotional harm as well as protecting the reputation of the vet, thus maximising the greatest possible good.

Taking all of the ethical arguments into account, strength of character and honesty are central to this case and other cases where mistakes are made. The cost of the cremation should be refunded to the owners and as an acknowledgement you may want to consider a token gesture, such as sending flowers or making a donation to a canine charity on their behalf. To avoid future repetition, the veterinary practice could establish a standard operating procedure, for example requiring another member of the team, such as a veterinary nurse, to verify the method of cremation.

|||||||||||||||||||||||||||||||||||||||||||||||||||||||||||||||||

IN the above scenario the animal is deceased and therefore is not considered to have any interests. Consider the following scenarios.

ERROR DISCLOSURE

What would you do?

Timmy, a Maltese terrier, is admitted for a femoral head excision after incurring a traumatic femoral neck fracture. Pre-operative radiographs are taken, but the LEFT marker is inadvertently used to indicate the fracture in the right leg. The left leg is prepped for surgery and the veterinarian asks the team to confirm the limb involved.

The procedure is performed and post-operative radiographs are taken. These confirm that a wrong-site surgery has been performed.

What would you do?

△

8.8 Timmy, a Maltese terrier, is admitted for a femoral head excision. This is performed on the wrong leg.
PHOTO ANNE FAWCETT

ERROR DISCLOSURE

What would you do?

You are working in a high-volume shelter. A cattle dog is admitted as a stray, but a microchip is detected and the owners contacted. They are relieved their lost dog has been found and they are on their way. The dog normally wears a collar with an identification tag but this has been lost since he escaped from their property.

During this time, another almost identical cattle dog, unowned and not microchipped or identified in any other way, is due to be euthanased as he has been held in the shelter for the maximum time. A kennel hand is asked to bring out the "red cattle dog". The dog is euthanased.

A staff member comments that the euthanased dog looks "exactly like that stray". The kennel hand scans the dog for a microchip and locates one. The owned dog has been euthanased.

What would you do?

△
8.9 An error as simple as mislabelling a cage in a busy shelter or hospital can have dire consequences for the patient.
PHOTO ANNE FAWCETT

SCENARIO
NEAR MISS

▶ You admit Tommy, a 12-year-old male neutered Ragdoll cat, for a dental scale and polish under anaesthetic. On pre-anaesthetic examination Tommy appears well, aside from the presence of tartar. Once the cat is anaesthetised and intubated the ET tube cuff is inflated and the pharynx packed off with swabs. The procedure is uneventful. During recovery the cat becomes dyspnoeic and cyanotic, with oxygen saturation dropping to as low as 60 per cent. A nurse alerts you that the cat has lost consciousness.

You examine the airway and find a swab obstructing the pharynx. Within seconds of removing the obstruction, the cat begins to breathe and appears to recover fully.

What should you do?

RESPONSE
SANAA ZAKI

▶ There are two aspects of this case that need to be addressed. Firstly, as the veterinarian you need to make sure you manage any complications that may arise secondary to the hypoxaemic episode that Tommy has had. Secondly, you need to make a decision about what you will communicate to the owner about this incident.

When deciding what you will communicate to the client about this incident consider the following:

(1) Episodes of reduced oxygen delivery to the brain (e.g. anaesthetic-related hypotension or hypoxaemia) have been associated with neurological deficits in cats. One common consequence is cortical blindness (Stiles, et al. 2012). In most cases this is reversible but

△

8.10 To what extent do you disclose a near miss that did not result in prolonged harm to a patient?
PHOTO ANNE FAWCETT

there are also reports of irreversible blindness in some cats (Jurk, et al. 2001). This means that although Tommy has survived this episode of dyspnoea and hypoxaemia, there may be neurological deficits evident once he fully recovers from the anaesthetic. These deficits may be significant and obvious, e.g. blindness, or they may be subtler and only really noticeable by the owner.

(2)   Not telling Tommy's owner what has happened could be construed as professional misconduct based on the veterinary practitioners' Code of Professional Conduct: "A veterinary practitioner must not mislead, deceive or behave in such a way as to have an adverse effect on the standing of any veterinary practitioner or the veterinary profession" (New South Wales Government 2013).

(3)   Telling Tommy's owner about the error that has occurred may result in enough concern about the ability of the veterinarian that the owner decides to go elsewhere for veterinary services. The vet may be resistant to telling the owner what has happened because of a fear of blame or negative consequences, rather than because he sees nothing wrong

with withholding this information from the client.

(4)   The owner of Tommy may tell others or post information about the error on social media, and this may blemish the reputation of the practice and result in significant loss of income for the owner of the practice.

(5)   The owner of Tommy may be so concerned about what has happened that they fear the risk of any future veterinary interventions for Tommy such as anaesthesia and surgery. This may mean that Tommy does not receive optimal care should he develop a serious condition in the future.

(6)   If Tommy's owner is not told about the complication and Tommy develops neurological symptoms after being discharged, he may be so concerned about the deceit that he reports the veterinarian and the practice to the Veterinary Practitioners Board.

(7)   Media coverage of this incident may tarnish the reputation of the entire veterinary profession, especially if the owner was not informed when it first happened but rather found out after the fact.

(8)   If this incident is discussed openly it may result in an improvement in the overall practices of this clinic, in areas such as record keeping, post-operative monitoring and communication between staff, and this will benefit all future patients.

(9)   The veterinarian made a mistake by forgetting to remove a swab from the cat's pharynx before extubating the cat. However, it should be acknowledged that this case also demonstrates that the vet's standard of veterinary practice is very high. The vet promotes dental prophylaxis, he secures the airway by intubating anaesthetised cats, he packs the pharynx for dental procedures to minimise the risk of aspiration, and he monitors patients as they recover from

anaesthesia following extubation so that any post-anaesthetic complications can be managed immediately. This is all part of "best practice", so it is important to not throw out the baby with the bathwater.

So who are the stakeholders in this situation?

(1) The veterinarian who has made the error.
(2) Tommy.
(3) Tommy's owner.
(4) The practice owner.
(5) Other clients of the practice.
(6) The veterinary profession.
(7) The Veterinary Practitioners Board.

Solving the ethical dilemma

If we view this case as simply a matter of honesty vs deceit, then a deontological framework would tell us that of course the vet should tell the owner the truth, the whole truth and nothing but the truth. We know it's the right thing to do. If we take a more pragmatic approach about the outcome for Tommy, the guiding principles of non-maleficence and beneficence would tell us that no harm is done by not telling the owner because Tommy survived and is OK, and certainly Tommy does not stand to gain any additional benefit from the vet telling the owner exactly what happened.

However, ethical dilemmas rarely affect only one person (or animal), and if we consider all the stakeholders that surround this case, and think about both the positive and negative impacts that a decision may have on each stakeholder, then suddenly an "easy" decision becomes very challenging.

The following approach to this complex ethical dilemma attempts to adhere to the guiding principles of non-maleficence (do no harm), beneficence (do good), autonomy (the ability to self-govern) and justice (fair distribution of benefits and consequences). It also aims to address the legal obligations that this case presents.

(1) The veterinarian needs to inform the owner of the complication that has occurred in order to not breach the veterinary practitioners Code of Professional Conduct. As a registered, practising veterinarian he must comply with this Code. By informing the owner, the veterinarian then also will be able to inform the owner of any possible delayed secondary complications that may occur as a consequence of the episode of hypoxia, and ensure that these are managed appropriately.

(2) When informing the client, the veterinarian needs to summarise exactly what happened, acknowledge the error and describe how it was managed. It will be important to highlight all the current procedures that are in place to ensure that anaesthetic complications are avoided or at the very least identified early to increase the likelihood of a good outcome. This will go a long way to reassuring the owner that although human error does sometimes occur, the risk of this occurring in the future is low. If done thoroughly, the explanation should give confidence to the owner to continue using the veterinary services of this practice, and to continue providing Tommy with optimal veterinary care should the need arise for him to have a procedure done in the future.

(3) The veterinarian should also review current procedures and make recommendations that will reduce the risk of this occurring again. For example, any animal that has packing in its pharyngeal cavity needs to be tagged (with a sign on its forehead or a special neck tag/paper collar) to indicate to all staff that the packing needs to be removed prior to turning off the anaesthetic. This will ensure this exact same complication will not happen in the future.

—

8.2

ERROR DISCLOSURE

||||||||||||||||||||||||||||||||||||||||||||||||||||||||||||||||||||

"To err is human; to cover up is unforgivable."
(Liam Donaldson, Chief Medical Officer,
England, at the World Health Organization's
World Alliance for Patient Safety launch
in 2004)

BOTH of the scenarios included support the above position. But as discussed, disclosure or open disclosure around errors or complications can be challenging.

In our society it is common to seek a source of blame, usually an individual or organisation, when things go wrong.

Such an approach is based on the "just world hypothesis" – good deeds are rewarded, bad deeds are punished. Good things happen to good people and bad things happen to bad people (Lerner 1980). When we blame someone for an error, we may argue that this is caused by character traits – forgetfulness, carelessness, recklessness or another form of moral weakness. No doubt such traits make the holder more prone to err, but this "person" approach to error does not adequately account for genuine error made by persons acting in good faith (Reason 2000).

Moreover, this approach suppresses error reporting, encourages a culture of "fear" and fails to acknowledge the circumstances or system which permitted the error in the first place (Reason 2000). By contrast, the "system approach" examines the system in which the error occurred – asking not whom to blame, but what happened, how did it happen and how could the system be altered to minimise the risk of the same error happening again? Thus:

"the basic premise in the system approach is that humans are fallible and errors are to be expected, even in the best organisations. Errors are seen as consequences rather than causes, having their origins not so much in the perversity of human nature as in 'upstream' systemic factors. These include recurrent error traps in the workplace and the organisational processes that give rise to them."
(Reason 2000)

The aim is not to absolve individuals of their responsibilities, but to foster genuine communication about and proactive prevention of errors. The big challenge is that clients and legal systems in which doctors and veterinarians practise do not always concur with such an approach (Reckless, et al. 2013). There may also be genuine differences in perception of errors and complications, with medical doctors tending to consider unexpected clinical outcomes as less serious and less in need of disclosure, than patients do. This is particularly the case when it comes to "near misses" – errors that do not result in injury or adverse effects, but which had the potential to do so (Grober & Bohnen 2005).

In addition, clinicians may be more likely to err on the side of caution when it comes to disclosure while patients expect openness and admission of responsibility (Iedema, et al. 2011). Clinicians are concerned about personal, professional and legal consequences of disclosure, as well as time and resources required for incident disclosure and management, management of relationships with and between other practitioners and patients, coping with patient emotions such as disappointment, fear and anger, and advising and justifying the response to all stakeholders (Iedema, et al. 2011). All of these function as barriers to disclosure.

Patients in turn refer to the physical, emotional and financial burden of errors and complications, the unacceptability of inadequate or denied disclosure, and the need for responses that are caring, timely, frank, accessible and frequent (Iedema, et al. 2011).

ERROR DISCLOSURE

What would you do?

You are working in a veterinary hospital that also incorporates a boarding cattery. As scheduled, two senior cats with chronic renal insufficiency are administered subcutaneous fluids from a fluid bag stationed in the treatment room. The extension set and needle are changed for individual patients but the bag is only changed once per day. One of the cats is discharged home immediately, and the owner returns reporting sudden-onset neurological signs. You realise that the bag contains ketamine. A trainee staff member has simply recycled a bag of fluids from an animal on a ketamine infusion for analgesia as he thought "this was the done thing". The other cat appears unaffected.

What should you do?

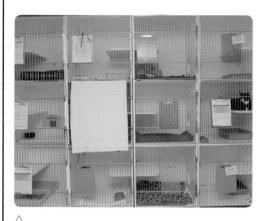

△
8.11 The wrong fluids are administered to two senior cats in a cattery.
PHOTO ANNE FAWCETT

8.3

Professional relationships and disclosure

Errors and complications can challenge relations within the team and may lead to ethical dilemmas based on conflicting duties. On the one hand we have a duty to disclose errors and complications to clients, on the other hand we have a duty not to disparage colleagues or bring the profession into disrepute. The following scenarios explore this conflict.

SCENARIO

LOYALTY VS STANDARD OF CARE

▶   You are a recent graduate and have joined a group practice of three veterinarians. You are happy to be finally putting your training and skills into caring for clients and their pets. The senior partner is well-liked in the community and was the founding partner of the group practice 20 years ago. You are on call and seeing a patient in an emergency condition that will require radical surgery. It is revealed that the senior partner had examined the patient two weeks prior, and it appears that there was a clear misdiagnosis at that time that has directly resulted in the current emergency. The client is now asking you how this could have happened.

What should you say to the client?
What should you say to your partner?

In the interest of wisdom, fairness and integrity, what if we knock a few quid off the bill?

△

8.12

CARTOON DR ROBERT JOHNSON

RESPONSE
WALTER LEE AND NIKITA CHAPURIN

▶ This scenario presents a conflict between sharing solidarity with colleagues and addressing errors of care delivered by these same colleagues. Furthermore, this situation is complicated by the fact that the error has resulted in an emergency that will now require radical surgery. The question arises as to how this scenario could be addressed and resolved among all the agents involved: the newly hired veterinarian, the animal and its owner, and the senior partner.

Approaching this dilemma with virtue ethics can provide guidance on not only how to proceed, but also a consideration of how one aspires to be a medical professional. It is an approach that has been applied in a number of medical fields and dilemmas (Barilan 2009, DuBois, et al. 2013, Gardiner 2003, Lachman 2008, Larkin, et al. 2009, Lee, et al. 2012). Virtue ethics is based on

Aristotle's work and proposes that ethical behaviour arises from character, and thus character is of fundamental importance. A virtuous character holds and aspires to act in accordance to commonly accepted virtues. There are a number of virtues that are relevant to this scenario of which we will discuss four: wisdom, integrity, fairness and kindness. What follows is a virtue ethics approach to this scenario that considers how character is expressed through the practice of these virtues.

Although the client has asked for an explanation on how this situation could have happened after being seen recently, the immediate focus should be on addressing the medical emergency. This is based on wisdom, as currently there is not enough information to make a definitive conclusion, and the senior partner should be asked about their assessment from a few weeks prior. One can respond to the client's questions by stating that the facts are still uncertain regarding how this could happen, but that you are committed to finding out the details and discussing them with the senior partner. The situation is such that focusing on what can be done now is of primary concern.

In discussing surgical options, issues of costs may arise and should be addressed. If appropriate, it may be necessary to consider discounting the cost of additional care, as there is now a shared responsibility towards the care of the animal. This could be done in a way without prematurely admitting fault, as that has yet to be ascertained.

Consideration of integrity in professional conduct is important in this discussion. This would only be available after sharing the situation with the senior partner. The intent is not a punitive one, but rather trying to understand the decision-making and to explore ways to improve delivery of care going forward. A discussion as to what may have been overlooked is important to have, and should be done in a collegial manner. Anyone who has practised long enough knows that mistakes happen. However, this should not serve as an excuse, and having the conversation with the senior partner provides an opportunity for both to become better clinicians.

Conversations on redemption of the situation can then be approached. If clear mismanagement has indeed occurred, redemption may be twofold. Firstly, how can this situation be mitigated with the client? This may include a pro bono surgery or deeply discounted one. In addition, the senior partner may need to come forward and assume the greater responsibility before the client. More importantly, the professional relationship between the client and the partner must be upheld. This should ideally entail a conversation regarding the issues at hand with the client. Some have argued for "no admission of guilt" policy when dealing with clients – but virtue ethics would support considering ways of apologising. Research regarding policies that express this approach has found them to be legally less problematic than those that "deny and defend" (Rocke & Lee 2013).

If the senior partner is unwilling to admit fault, either due to lack of understanding or for other reasons, another partner could be brought into the conversation as a neutral party. The purpose of these conversations is not to finger-point, but rather to identify lessons so that future similar medical errors do not occur. This is the concept behind mortality and morbidity conferences that are held in a peer-review fashion. Although there is a clear hierarchy and seniority at play, issues of virtue have no hierarchy. In other words, all must be held to the same ethical standard regardless of position and title.

Virtue ethics allows for consideration of not only the character of the individual, but also the "character" of the practice. We can ask – is this a practice that will strive to have a reputation of honesty and responsibility towards its clients? If there is a habit of poor care without any insight or intent to improve from the senior partners, the clinician should consider if this is a group practice that one should associate with. In other words, if this is the

case, a change in employment may unfortunately have to be considered.

In summary, virtue ethics provides an approach to this dilemma that first involves a consideration of which virtues one would wish to be exemplified by. This would not involve jumping to conclusions or staging a "cover-up" of fault. Rather, it would be one in which there would be wisdom in personal relationships, integrity to professional responsibilities and fairness to resolve issues that arise when these responsibilities are not met. Finally, there should be kindness in how these issues are approached, as the intent is to obtain personal and group improvement.

PROFESSIONAL RELATIONSHIPS AND DISCLOSURE

What do you think?

ONE   Recall a situation where you have had to discuss the error of a person senior to you (a family member, community member or senior colleague).

TWO   What made such a situation challenging?

THREE   How would you approach the situation differently if the error or complication was potentially the result of actions of a junior colleague?

△
8.13 Can you think of a time when you've had to discuss someone else's error?
PHOTO ISTOCK

SCENARIO
REFERRING VET MISMANAGEMENT

▶ You are an equine surgical specialist working in referral practice. A general practitioner has referred a two-year-old thoroughbred for a non-healing wound. Three months prior, the horse grazed the carpal extensor tendon sheath. The referring veterinarian treated this as an open wound, but failed to drain the wound, creating a nidus of infection in the tendon sheath.

The client feels the referring vet did not treat the wound appropriately.

What should you do?

△
8.14 Young horses can be lively and pick up small scrapes which occasionally become problematic.
PHOTO ISTOCK

RESPONSE
MARTIN WHITING

▶ All veterinarians have autonomy in their work and each may take different courses of action when presented with similar scenarios. All veterinarians have different skill sets and levels of experience of different cases, and this skill set difference is even more pronounced when considering veterinarians who are general practitioners versus those who are specialists. This difference is not always apparent to the client until the situation arises where two different veterinarians are treating the same animal, and achieve different results. This can cause concern for the client, but it can also become a source of tension between the two veterinarians, and potentially between two different practices. This scenario above represents one of those tense situations between the triad of the client and the two different veterinarians. It may be resolved by communication skills, but it could represent something more sinister.

This scenario will be resolved in two ways. Firstly, it will be resolved as if the primary veterinarian did everything correctly, and yet still the client is unhappy. Secondly, it will be resolved as if there was a problem with the care provided by the primary clinician.

The vast majority of veterinary systems across the world are set up so that clients attend a primary general practitioner first, and then in more complicated or serious situations, they are referred on to the specialist veterinarian. A business arrangement such as this is not just more economical for the client, but it also makes sure that the client sees the veterinarian with the most appropriate skill set and that the skill sets of veterinarians are used appropriately. It is not possible to determine from the scenario above how serious the initial injury was to the horse. An injury to a tendon sheath is usually serious enough to warrant fairly aggressive treatment, as the potential for escalation and complications is high. The primary veterinarian may not have realised quite how extensive the injury was (maybe despite a thorough investigation) and so starting initial treatment and monitoring progress may be an effective course of action. At this stage, however, the client ought to be warned that injuries near tendons and joints of horses can have serious complications and referral options should be provided. Assuming this was done, and the client and the veterinarian were happy to proceed with primary care, it turns out this was the wrong choice, even though it was a legitimate one. When the case became more complicated, it was appropriately referred on to a specialist with a more detailed skill set to deal with the problem.

If this was the case, then the primary veterinarian made the correct choice with the clinical data available, and with the client consent, proceeded with primary care, even though with hindsight it is now known that this course of action was inappropriate. It may not have been possible to know that at the time of the initial injury. It is quite justifiable that the client is upset, as their horse has a chronic injury and now has a costly referral bill on top of their original three months of treatment. However, this upset may be unjustly directed towards the primary veterinarian. The specialist ought to understand the difficult situation that the primary veterinarian was in at the onset of the problem, and communicate the dilemma that they would have been in to the client. It may have been an unfortunate, rare complication. It is the role of the specialist to put the client's mind at rest that the primary veterinarian did act appropriately. And, hopefully, with the informed consent discussion which happened at the onset of the injury, the client might remember that difficult decision that was made not to refer at that start. In this scenario, therefore, it is just an issue of communication skills of the two veterinarians to help the client understand the decisions that were made, and remind the client of all the choices that they too were involved in.

In an alternative version of events, let's consider if the primary veterinarian missed an obvious serious injury that warranted immediate referral, and the client was not informed. The client is then justified in their anger, and it is quite likely the specialist will be annoyed too, as the horse

did not receive the treatment it needed at a time when it could have resolved the problem quickly. This is where the specialist is in an awkward situation, as their duty to their client and their duty to their professional colleague are in conflict. The specialist has a duty to be objective and honest with the client, but this does not have to result in a blaming scenario on the primary veterinarian who failed in their duty of care. At this stage, the specialist might not know the full circumstances of the primary veterinarian, and so this would warrant a phone call to the primary veterinarian to discuss the management of these types of injuries in horses. After all, the primary veterinarian could have been insisting on referral, but the client was unwilling to comply. This would not be known without a direct communication with the primary veterinarian, so it is important to not make assumptions.

The primary veterinarian may have made a genuine error in their decision-making, and be incredibly remorseful over the outcome, and, once they find out from the specialist about their error, they may wish to take matters into their own hands and contact the client directly to apologise and potentially make a financial contribution to the ongoing care of the horse. A "blame game" between the specialist and the primary veterinarian, in front of the client, will rarely yield a successful outcome for anyone, and will damage the professional reputation of all involved – as well as damaging the relationship between the primary veterinarian and that particular referral practice. Good communication between the two veterinarians is likely to yield the most successful outcome and promote client confidence in the profession. Simultaneously, the primary veterinarian will likely learn from the situation and be more proactive in referring future cases, or at least considering referral and discussing it with the client.

The worst-case scenario is when the primary veterinarian was knowingly negligent and deliberately chose not to refer even when they knew the horse needed specialist treatment. Maybe the primary veterinarian wanted to keep the business for themselves, or did not like to send clients to other, potentially competing, practices. This is when the primary veterinarian is no longer acting professionally, nor holding the welfare of the animal as their primary concern. In extremely rare cases such as this, after careful communication between the specialist and the primary veterinarian, it may be warranted to report the veterinarian to the professional regulator. It is inappropriate, and damaging to the profession as a whole, to put business or personal issues above the welfare of the animal. The specialist should forewarn the primary veterinarian that they are going to report them, and provide the primary veterinarian with the letter of concern. At this stage, it is not appropriate to include the client within the letter of concern, but rather move their attention onto the continuing care and treatment of their horse. If the professional regulator would like to get in contact with the client, then that is for them to do. The specialist may wish to inform the client that they are working to help the primary veterinarian in handling these difficult injury cases, but they need not necessarily tell them they have lodged a formal concern regarding the negligence of the primary veterinarian. The specialist should continue to look after the client and their horse, as restoring the horse to its full health is their primary concern and will help the client to be more at ease with the situation.

Above all, the specialist has a duty to the animal and the client. They should not use their situation to increase the client's negative view of the primary veterinarian, but they should try to keep the focus on the horse's care. The specialist also has a duty to all future horses who may be referred by that primary veterinarian, so beginning this whole scenario with a polite and professional telephone call to the primary veterinarian to find out the circumstances of the initial injury may resolve any professional tension before it has even begun. If

there is flagrant negligence, then it is inappropriate for professions to "hide behind closed doors" and cover up problems. It is better for all involved, and especially for the animals, that the negligence, if serious, is dealt with in a way that results in no further risk of harm to animals. Effective, polite, professional communication is paramount throughout this scenario, which is a highly volatile one and can easily escalate. The primary veterinarian should be given the benefit of the doubt, and a good professional relationship maintained between the specialist and the primary veterinarian.

||||||||||||||||||||||||||||||||||||||||||||||||||||||||||||||||||||||||

THE author of the two scenarios above takes a virtue ethics or principalist approach. When communicating about error, in order to minimise harm, it is important to be aware of the limitations of the information one has immediate access to. In both cases the author recommends an approach of seeking more information, which – aside from minimising harm that might result from a false accusation of incompetence, for example – is just in allowing all parties to have their say.

In addition, in all cases the need to educate others and learn from errors and complications is stressed – this is consistent with the principle of beneficence. In providing informed consent and open disclosure, the authors are also respecting the autonomy of all parties.

Whether or not to offer an apology to the client may present a dilemma to the veterinarian. On the one hand, a genuine apology may signify care and good faith. On the other, it may be seen as an admission of liability that the veterinarian feels an employer or insurer would not authorise. Successful navigation of this dilemma requires further information.

There is extensive discussion on whether or not it is appropriate to apologise for a potential error or complication in the human medical field. In the

CHAPTER 8 ERRORS AND COMPLICATIONS

—
8.3
PROFESSIONAL RELATIONSHIPS AND DISCLOSURE
WHAT DO YOU THINK?

◁ **8.15** Veterinarians and clients impacted by errors may have very different ideas about the meaning of apologies.
CARTOON RAFAEL GALLARDO ARJONILLA

1980s and 1990s, at a time of escalating medical litigation in the USA, apologies were frowned upon. According to Quinn and Eichler:

"Physicians had been in effect encouraged, by insurers, mentors, and others, to practice a form of denial. Communication was inhibited and the relationship with the patient usually suffered. This behavior was accentuated by the legal environment and disclosure was greatly discouraged. Patients were often left with feelings of abandonment, frustration, and anger, frequently turning to the legal system for answers."

(Quinn & Eichler 2008)

The fear that an apology will lead to litigation is not borne out. Hospitals and doctors using a policy of open disclosure found that litigation was reduced (Quinn & Eichler 2008). Furthermore, patients were much more satisfied with their management and outcomes (Iedema, et al. 2011, Quinn & Eichler 2008).

PROFESSIONAL RELATIONSHIPS AND DISCLOSURE

What do you think?

ONE — How common is saying sorry in veterinary practice?

TWO — Do you think the level of apologies is appropriate?

THREE — Under what circumstances is it always advisable to apologise?

△ **8.16** When is it appropriate to say sorry?
PHOTO ISTOCK

Conclusion

Errors may have multiple victims – animals, clients, the environment and those very professionals who made the error. It is human to err, but the response to error that is important. As Reckless, et al. observe, "Doctors who make mistakes may become better at their jobs as a result. They can, and do, go on to have successful and productive careers. The key is to reflect on errors and pay heed to any lessons that can be learnt" (Reckless, et al. 2013).

As in medicine, there is scope to change culture around error disclosure in veterinary practice. For example, instituting morbidity and mortality meetings promotes the practice of reflecting on and learning from errors and complications, modifying behaviours and judgement based on previous experiences as well as review of evidence, avoiding repetition of errors which can result in complications, and educating others.

References

Anon 2016 Complication. Merriam-Webster.

Barilan YM 2009 Responsibility as a meta-virtue: truth-telling, deliberation and wisdom in medical professionalism. *Journal of Medical Ethics* **35:** 153–158.

Bartram DJ, and Baldwin DS 2010 Veterinary surgeons and suicide: a structured review of possible influences on increased risk. *Veterinary Record* **166:** 388–397.

Bednarski R, Grimm K, Harvey R, Lukasik VM, Penn S, Sargent B, and Spelts K 2011 AAHA Anaesthesia Guidelines for Dogs and Cats. *Journal of the American Animal Hospital Association* **47:** 377–385.

Brodbelt DC, Blissitt KJ, Hammond RA, Neath PJ, Young LE, Pfeiffer DU, and Wood JLN 2008 The risk of death: the confidential enquiry into perioperative small animal fatalities. *Veterinary Anaesthesia and Analgesia* **35:** 365–373.

DuBois JM, Kraus EM, Mikulec AA, Cruz-Flores S, and Bakanas E 2013 A humble task: restoring virtue in an age of conflicted interests. *Academic Medicine* **88:** 924–928.

Feinmann J 2009 Adverse events: you can say sorry. *British Medical Journal* **339:** 3057.

Gardiner P 2003 A virtue ethics approach to moral dilemmas in medicine. *Journal of Medical Ethics* **29:** 297–302.

Grober ED, and Bohnen JMA 2005 Defining medical error. *Canadian Journal of Surgery* **48:** 39–44.

Iedema R, Allen S, Britton K, Piper D, Baker A, Grbich C, Allan A, Jones L, Tuckett A, Williams A, Manias E, and Gallagher TH 2011 Patients' and family members' views on how clinicians enact and how they should enact incident disclosure: the "100 patient stories" qualitative study. *BMJ* **343:** d4423.

James JT 2013 A new, evidence-based estimate of patient harms associated with hospital care. *J Patient Saf* **9:** 122–128.

Jurk IR, Thibodeau MS, Whitney K, Gilger BC, and Davidson MG 2001 Acute vision loss after general anesthesia in a cat. *Veterinary Opthalmology* **4:** 155–158.

Kale A, Keohane CA, Maviglia S, Gandhi TK, and Poon EG 2012 Adverse drug events caused by serious medication administration errors. *BMJ Quality & Safety* **21:** 933–938.

Lachman VD 2008 Making ethical choices: weighing obligations and virtues. *Nursing* **38:** 42–46.

Larkin GL, Iserson K, Kassutto Z, Freas G, Delaney K, Krimm J, Schmidt T, Simon J, Calkins A, and Adams J 2009 Virtue in emergency medicine. *Academic Emergency Medicine: Official Journal of the Society for Academic Emergency Medicine* **16:** 51–55.

Lee WT, Schulz K, Witsell D, and Esclamado R 2012 Channelling Aristotle: virtue-based professionalism training during residency. *Medical Education* **46:** 1129–1130.

Lerner MJ 1980 *The Belief in a Just World: A Fundamental Delusion.* Plenum Press: New York.

Magalhães-Sant'Ana M 2014 Ethics teaching in European veterinary schools: a qualitative case study. *Veterinary Record* **175:** 592.

New South Wales Government 2013. Veterinary Practitioners Code of Professional Conduct. Sydney, Australia.

O'Connell D, and Bonvicini KA 2007 Addressing disappointment in veterinary practice. *Veterinary Clinics of North America: Small Animal Practice* **37:** 135–149.

Powell L, Rozanski EA, and Rush JE 2010 *Small Animal Emergency and Critical Care: Case Studies in Client Communication, Morbidity and Mortality.* Wiley-Blackwell: Oxford.

Quinn RE, and Eichler MC 2008 The 3Rs Program: the Colorado experience. *Clinical Obstetrics and Gynecology* **51:** 709–718.

Ralphs SC, Jessen CR, and Lipowitz AJ 2003 Risk factors for leakage following intestinal anastomosis in dogs and cats: 115 cases (1991-2000). *JAVMA-Journal of the American Veterinary Medical Association* **223:** 73–77.

RCVS 2015 Code of Professional Conduct for Veterinary Surgeons. http://www.rcvs.org.uk/advice-and-guidance/code-of-professional-conduct-for-veterinary-surgeons/pdf/

Reason J 2000 Human error: models and management. *Western Journal of Medicine* **172:** 393–396.

Reckless I, Reynolds DJ, Newman S, Raine JE, Williams K, and Bonser J 2013 *Avoiding Errors in Adult Medicine.* Wiley-Blackwell: Oxford.

Rocke D, and Lee WT 2013 Medical errors: teachable moments in doing the right thing. *Journal of Graduate Medical Education* **5:** 550–552.

Rollin BE 1999 *An Introduction to Veterinary Medical Ethics: Theory and Cases.* Iowa State University Press: Ames.

Routly JE, Dobson H, Taylor IR, McKernan EJ, and Turner R 2002 Support needs of veterinary surgeons during the first few years of practice: perceptions of recent graduates and senior partners. *Veterinary Record* **150:** 167–171.

Stiles J, Weil AB, Packer RA, and Lantz GC 2012 Post-anesthetic cortical blindness in cats: twenty cases. *The Veterinary Journal* **193:** 367–373.

Tannenbaum J 1995 *Veterinary Ethics: Animal Welfare, Client Relations, Collegiality.* Mosby-Year Book: St Louis.

Thompson IE, Melia KM, and Boyd KM 2000 *Nursing Ethics.* Churchill Livingstone: Edinburgh.

CHAPTER 9
CONSENT

Introduction

Informed consent has risen to the fore in medical ethics as the paternalistic "doctor knows best" approach advocated in the ancient Hippocratic oath has become superseded by a concern for patient autonomy. Veterinary usage of informed consent does not exactly mirror the medical sphere, but still, it is this respect for autonomy that is at the heart of veterinary informed consent. The requirement to obtain informed client consent before veterinary treatment is commenced is outlined in many professional codes of practice around the world. As well as providing clinical information about a range of treatment options in a landscape of private veterinary practice the associated costs are also expected to be presented. For example, in a code for all European veterinary surgeons:

> "the customer shall be informed of the benefits, risks and costs of the services proposed and the customer's informed consent should be obtained before providing any service."
>
> (FVE 2002)

Consent might be more commonly expected in written form for inpatient procedures on companion animals and equines, but for minor procedures,

such as performing an injection during a consultation, and for farm animals, verbal consent will frequently be sought.

The three tenets of informed consent in medical ethics have been proposed as (1) disclosure, the provision of sufficient information as to be able to make a decision; (2) capacity, the mental ability of someone to be able to make a decision about their treatment; and (3) voluntariness, the ability of the person to make a choice free from coercion or other forms of undue persuasion (Beauchamp & Childress 1994). That owners are acting on behalf of their animals may make the process more similar to that where proxy consent is sought from parents or others for children

△

9.2 Veterinarians should prevent and relieve suffering, disease and disability while minimising pain and fear. We must remember however that veterinary settings can cause fear in our patients.

CARTOON AILEEN DEVINE

◁

9.1

CARTOON RAFAEL GALLARDO ARJONILLA

unable to consent for themselves in the human medical arena.

The use of informed consent in veterinary practice has been described as having a particular and important role in defining the boundaries of the veterinary–client–patient relationship (Passantino, et al. 2012). In this chapter we will consider difficult situations around consent that arise through being unsure who the owner is to give consent, when owners conflict in their consent, when an owner may not have sufficient mental capacity to consent and when veterinary surgeons may be tempted to act without consent. Because animals cannot consent to veterinary intervention, the focus of this discussion will be client consent.

9.1

Difficulties in obtaining consent

We begin with an emergency situation where time is short and efforts to gain informed consent are necessarily truncated.

SCENARIO

INFORMED CONSENT IN A FIRST AID PROCEDURE

▶ You are working in a 24-hour emergency and critical care veterinary hospital. It is a busy night – you've treated one cat for severe permethrin intoxication, euthanased a moribund rabbit, stitched up three animals with lacerations, stabilised a dog with a pneumothorax and completed a caesarean.

As you are making a cup of coffee you are paged. The nursing team have admitted a Standardbred poodle, Gus, with a suspect gastric

dilatation volvulus. They have commenced intravenous fluids and attempted to place a stomach tube without success. You evaluate Gus and explain hurriedly to Mr H and his son that Gus will die without surgery. They nod tearfully.

The surgery is long and complicated but Gus survives. Everyone on the team feels fantastic. You telephone Mr H to share the good news; however, he is shocked when you update him on the costs incurred. Mr H claims he did not consent to the procedure.

What should you do?

RESPONSE
ANDREW GARDINER

▶ Communication and consent are all-important but anyone who has worked in critical care knows that, at times, it is necessary to sacrifice optimum communication, whether with clients, colleagues or others, in order to undertake immediate emergency action to save a life. Behaving in this way could be justified using several ethical theories:

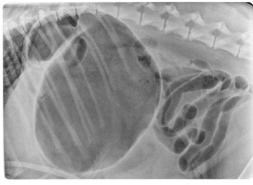

△
9.3 Lateral abdominal radiograph of a Standard poodle with gastric dilatation volvulus.
PHOTO DR JOHN CULVENOR,
NORTH SHORE VETERINARY SPECIALIST HOSPITAL

deontology (including rights), utilitarianism, virtue ethics, common morality and principalism – the latter because, in pursuing communication and discussion instead of intervening in an emergency, harm could knowingly be done, a "sin of omission".

The UK veterinary regulator, the RCVS, draws a distinction between immediate first aid and pain relief (which might involve euthanasia) and all other forms of treatment. No veterinary surgeon should unreasonably refuse to perform first aid and pain relief on any animal and this may of necessity need to be done without consent from the owner (RCVS 2015a).

The initial steps to combat shock and attempt to pass a stomach tube in this patient comprise essential first aid and pain relief. In this scenario, I assume my nursing colleagues explained that emergency first-aid action was needed, even if they did not complete an anaesthetic or treatment consent form at that point. I support their actions, although they should have liaised with me before beginning the fluids and attempting decompressive treatment. However, that is not the main issue under consideration here.

The difficulty comes when we move from immediate first aid towards definitive treatment, which in this case comprises a blurry line. The *clinical* problem was that immediate first aid was not successful, and in order to obtain the lifesaving objectives of first aid, actions which bled into definitive treatment became necessary. The crossing of this boundary also entailed moving into a different territory of consent. The RCVS states that:

"Informed consent, which is an essential part of any contract, can only be given by a client who has had the opportunity to consider a range of reasonable treatment options, with associated fee estimates, and had the significance and main risks explained to them."

(RCVS 2015b)

Consent does not always have to be written, although this is preferred. In discussing treatment, the risks and benefits of no treatment should also be considered. Discussion should also include the possibility of unexpected findings (e.g. condition worse than expected or different to that expected) and treatment options for those eventualities should be covered. In this circumstance, further discussion with the owner, for example by telephone, should take place before proceeding. The British Small Animal Veterinary Association notes that:

"It is no longer considered sufficient to add the catch-all phrase 'and any other procedures which may be considered necessary' to consent forms without some explanation as to what they might be. Such procedures will involve additional cost, and possibly additional risk, and the various options should be explained beforehand wherever possible."

(BSAVA 2015)

Note that this is not just a problem in emergency medicine. It could also feature in, for example, routine dental treatment when a vet removes many diseased and painful teeth in a patient whose mouth was much worse than expected, but the owner refuses to pay because they did not consent to such an extensive procedure, even though it was clearly and objectively the best course of action for the patient. Contacting owners by phone during a procedure is not always possible for many different reasons, phone reception and busy lines being just two. This does emphasise the need for thorough exploration of (known) scenarios before treatment is started (RCVS 2015b).

Returning to my emergency case, did the owners consent? I explained to them that their dog would die without surgery, and they nodded tearfully. I took this to mean they agreed to surgery (which, crucially, was both first aid *and* curative in intent). I considered I had obtained verbal consent for surgery. I suppose that, in nodding, the owners could just have been agreeing with me that the dog would indeed die without surgery. However, I assumed that they, like me, had their dog's best interests at heart, so I made a presumption to treat.

I did not discuss non-treatment, which in this case would mean immediate euthanasia as leaving the dog to die from gastric dilation/torsion would be unethical at two levels, personal and professional. Nor did I discuss the likely costs – there was not time. In doing this, I acknowledge this was a failing, however I feel that the actions were virtuous even in the face of lack of communication, because we acted primarily to save the life of Gus. I therefore justify these actions using a combination of virtue ethics (primarily) and deontology/rights – the rights of the dog to be treated for his severe distress when presented to a veterinary surgeon (RCVS 2015a).

I would therefore pursue the clients for payment on the grounds that they consented to the treatment of Gus and that the lack of full discussion of the details of treatment, different options/possibilities and costs, was primarily a virtuous act in favour of their dog, who could literally have died at any moment. I would hope that if this was explained in a calm and unhurried way, talking through each stage of the process, demonstrating that concern for the patient was central, and explaining the very serious nature of the problem and the details of the surgery needed, they would agree with my interpretation. If not, I would probably have to accept that the practice must bear some of the costs, as the contractual approach centred on lack of informed consent would probably win out in a court of law (which does not necessarily mean it is the most moral approach).

The scenario presented me with an uncomfortable tension between deontology and virtue ethics, but I am happy acting from within the latter. The actions were based on core focal virtues of compassion (with regard to patient suffering), discernment (with regard to making a fitting judgement given the circumstances), trustworthiness (with regard to assuming that a caring owner would agree with my actions, once they fully understood them), integrity (with regard to my interpretation of moral norms and acting on them) and conscientiousness (with regard to doing what appeared to me genuinely to be the right thing at the time).

I would feel regret at the owner's reaction if they persisted in refusing to pay for the treatment. On reflection, I would also accept that communication within the practice could develop and improve in future to try to avoid similar situations happening again. However, in virtue ethics, it is recognised that the best decision may not meet with approval from all stakeholders.

||||||||||||||||||||||||||||||||||||||||||||||||||||||||||||||||||||

IN a time-restricted and emotionally pressured situation the challenges of gaining informed consent are clear. Bateman (2007) discusses obtaining informed consent for urgent medical treatment in the emergency room, recognising that the ordinarily desirable shared decision-making between client and veterinarian may not be possible. In that instance, Bateman suggests that a practical approach to obtaining informed consent is to provide only information pertinent to the immediate decision, an estimation of the treatment outcomes, including cost and, where appropriate, the option of euthanasia. Explicit support for the client's decision in very difficult circumstances is also proposed to be important (Bateman 2007).

DIFFICULTIES IN OBTAINING CONSENT

What would you do?

You want to admit Florence, a one-year-old Hungarian Vizsla, for fluid treatment and investigation of vomiting. You suspect a foreign body and that after imaging you may need to go to surgery. Mrs C, Florence's owner, is in a great rush and leaves the practice without you being able to give a full explanation of what you expect to do, not having signed a consent form. What should you do?

△
9.4 Florence has always chewed things and now, when she needs veterinary investigation, her owner has rushed off without signing a consent form.

PHOTO ANNE FAWCETT

In this next scenario we consider a difficult situation where there are two owners, but only one who consents to euthanasia.

SCENARIO
SPLIT CONSENT

▶ For several years you've been intermittently seeing Plucky, who is now a 22-year-old gelding kept as a pet by Mr and Mrs M. They are clearly very fond of Plucky, but have never been able to afford a lot of veterinary care. Three days ago Plucky started showing signs of colic. Your initial rectal examination was inconclusive – you may have palpated some kind of impaction, but it was impossible to be sure. The Ms have repeatedly declined further diagnostics or exploratory surgery on cost grounds. You've trialled various medical therapies, but Plucky's clinical signs have progressed from mildly elevated respiratory and heart rates, to severe signs now unresponsive to strong analgesics and enteral fluids.

Earlier today, however, Mr and Mrs M consented to admit Plucky to your hospital for intravenous fluids, medical treatment and monitoring, and then went home. Plucky has now begun pacing violently in his stable and has started kicking and thrashing. Surgery is now clearly necessary.

You've informed Mr and Mrs M by phone that given that surgery is not a financial option for them, Plucky should be euthanased without further delay on humane grounds. Mrs M has reluctantly agreed, but Mr M cannot bring himself to accept this. Your colleagues are now also becoming distressed by the situation, and are heatedly discussing it, unsure what to do. They all want to euthanase Plucky, but you do not have consent from both owners.

What should you do?

9.1

DIFFICULTIES IN OBTAINING CONSENT
RESPONSE
ANDREW KNIGHT

RESPONSE
ANDREW KNIGHT

▶ The essence of this dilemma is that your duty to your patient potentially conflicts with your duty to your client. On the one hand it seems clear that without surgery, Plucky's prospects for recovery are very poor. He is clearly suffering, with clinical signs now so severe that they're unresponsive to strong analgesics and enteral fluids. To prevent further severe suffering associated with what would almost certainly be an inevitable decline toward death, Plucky should clearly be euthanased on humane grounds. On the other hand, you lack clear consent from the clients. Ms M has agreed, but Mr M has not. In such circumstances, what should you do?

△
9.5 Plucky appears in pain and surgery or euthanasia is needed.
PHOTO ISTOCK

Although as a veterinarian you do have duties to your clients, your employing practice, the wider public, the veterinary profession and yourself, it is clear within the field of veterinary professional ethics and the statements of veterinary professional associations in countries such as the UK and USA that your primary duty must always be to your patient. The AVMA (2016) Principles of Veterinary Medical Ethics, for example, state that, "Veterinarians should first consider the needs of the patient: to prevent and relieve disease, suffering, or disability while minimizing pain or fear." The RCVS (2015) Code of Professional Conduct for Veterinary Surgeons similarly states that, "Veterinary surgeons must make animal health and welfare their first consideration when attending to animals." Indeed, this is the expectation that society at large has of veterinarians, physicians and other healthcare workers. In return for entrusting these professionals with sensitive personal information and the authority to recommend treatment courses, society has a serious and reasonable expectation that they will give primacy to the interests of their patients, ahead of considerations such as commercial or self-interest, the demands of family members or animal owners who may not always be reasonable, or indeed, any other interests. Only rare exceptions are permitted, such as the duty to safeguard wider public or animal health, for example, when outbreaks of serious, transmissible animal diseases occur, whether zoonotic or otherwise.

No such exceptions being present in this case, it is clear that your primary duty is to Plucky, rather than his owners. However, Plucky does remain their legal property, which creates a problem, because if you effectively destroy your client's property without their consent, you could face a range of potentially serious consequences. These include the stressful and protracted process of attempting to defend yourself should the clients formally complain about your conduct to the veterinary licencing board, or take independent legal action to sue you or your veterinary practice for the recovery of their financial and other damages. In the worst-case scenario your veterinary licencing board could revoke your licence to practise veterinary medicine, temporarily or permanently. Even a successful civil suit could result in you or your employer having to pay the costs of replacing Plucky, and a potentially wide range of other legal and associated costs, which demonstrates the importance for veterinarians of maintaining their professional liability insurance, which was primarily created to cover such payouts. Another potentially serious consequence could be reputational damage for you and your practice, if the aggrieved clients or local journalists were to publicise the case in a one-sided manner, e.g. through social networking, or in the local newspapers. Hence, if you fulfil your primary duty and euthanase Plucky, you not only arguably fail to uphold your duty to your clients, but could also risk quite significantly damaging your own interests (your career may be at risk), the interests of your practice and even the interests of the wider veterinary profession (if this case damages public trust in it).

Unfortunately for you, at this point you are unlikely to be able to decline to have these people as your clients. Although veterinarians are free not to accept clients initially, once treatment has commenced, they may not then abrogate their responsibilities and are generally obligated to continue treatment. The AVMA (2016) Principles of Veterinary Medical Ethics, for example, state that, "Once the veterinarian and the client have agreed, and the veterinarian has begun patient care, they may not neglect their patient and must continue to provide professional services related to that injury or illness within the previously agreed limits." This demonstrates the importance of a clear prior agreement about treatment. Given the guarded nature of Plucky's prognosis without the recommended surgical intervention, this should probably have included an agreement about the course of action should medical treatment prove unsuccessful. Given that Plucky's admission was not an emergency, the agreement about treatment should have been accompanied by a signed owner consent form.

Nevertheless, in this particular case it appears that clear prior agreement has not been obtained about euthanasia, should it become necessary, and the clients currently remain unable to agree, so unfortunately your dilemma remains. Your clear overriding duty is to your patient, but fulfilling that duty by euthanasing Plucky without clear owner consent carries substantial risks. At this stage it would be wise to ensure you clearly understand the relevant laws and professional guidelines within your jurisdiction. Some jurisdictions might state, for example, that euthanasia without owner consent is illegal, although such explicit statements are rare. In others, owners or, arguably, veterinary staff, might be liable under animal protection legislation for failing to provide care considered medically necessary, to animals for which they have a responsibility, although lack of owner consent *might* provide a defence. Veterinary associations might also be able to offer advice about relevant clauses within veterinary practice Acts, licencing board regulations, or professional guidelines. This case has evolved over several days, which would hopefully have given you time to check these matters with parties likely to be knowledgeable, such as any more experienced colleagues, your practice lawyer, your veterinary association or licencing board, or your professional liability insurer. And of course you could also do your own research – most relevant material is now available online. Arguably the most difficult situation arises when the law, regulations or professional policy dictates a course of action that clearly conflicts with your primary professional duty to your patient.

If you do decide to euthanase Plucky, there are several steps it would be wise to take, to increase the defensibility of your action, if later challenged through private litigation or a complaint to your veterinary licencing board. Firstly, you should ensure the medical history of this case is clearly documented, particularly the deterioration in clinical signs to the point where it has become clear that your duty to Plucky warrants euthanasia. This should ideally be supplemented with video or photographic evidence of his clinical signs. Next, you should obtain at least one second opinion in writing from a veterinary colleague – ideally, an experienced equine veterinarian – confirming that euthanasia is warranted. To minimise suffering, you should anticipate the likely need for these and obtain them without delay. Finally, in many jurisdictions there are various legal officials, such as SPCA, humane society or animal control officers, or police constables, who do have the lawful authority to order euthanasia in the absence of owner consent. If possible a written (or at least, verbal and preferably witnessed) order should be obtained from such an official authorising euthanasia. To minimise delay, this outcome should be anticipated, with initial communications to that official as the case progresses. Veterinarians are generally permitted by their professional ethical standards to violate client confidentiality by discussing a case in this way, if they believe animal or public health may be at risk.

Depending on the level of protection afforded in the jurisdiction by animal protection legislation, and the cooperation of officials able to authorise euthanasia etc., this sort of case could provide an extremely challenging ethical dilemma. It essentially asks the veterinarian, "How far should I be prepared to go, to uphold my primary duty to the patient under my care? What price should I ultimately be willing to pay?" Given carefully documented evidence of the necessity of euthanasia

in such a case, it seems unlikely that any sanction applied by a court or veterinary licencing board would be severe, if it were even upheld. However, the element of doubt could remain. Such a dilemma poses deep questions about personal values, to which the answers will vary among individuals. Nevertheless, in most jurisdictions the primary duty of the veterinarian to the patient under their care does remain clear. And if an animal's veterinarian will not act to end their suffering, who will?

||||||||||||||||||||||||||||||||||||||||||||||||||||||||||||||||||||||

THE author is clear that the priority of the veterinarian is their duty towards their animal patient. Importantly, as well as proposing a course of action, he identifies several practical ways to minimise any potential negative consequences of the action. This form of refinement can be employed in all decisions and can help to mitigate some of the negative consequences or promote positive outcomes. For example, even if it is decided that a nervous cat needs hospitalisation, refinement could include having hiding places in the cage, being positioned in a quiet area of the ward and asking staff to keep out of the area.

DIFFICULTIES IN OBTAINING CONSENT

What do you think?

Mrs J is looking after her friend's dog Harry whilst she is away on holiday. Mrs J brings Harry to you as he seems to have been stung and has quite a swelling on his paw. You treat him with a steroid and anti-histamine injection and Mrs J pays the bill. A short while later you receive an irate phone call from the owner. "Why have you given him steroids?" she shouts. "You know I don't agree with them. I did not consent to them. You could have contacted me easily enough, I've had my phone on me all along." What should you do?

△

9.6 Harry is being looked after by a friend who consents to treatment the owner doesn't agree with.

PHOTO ANNE FAWCETT

As we've seen, it can be difficult to obtain consent when time is short or owners are conflicted in their opinions. In this next scenario we consider a case where the owner's capacity to consent is not clear.

___

SCENARIO
OWNER WITH IMPAIRED CAPACITY

▶   Mr C, a young man with learning difficulties, comes to you about his pet rat Lisa. They go everywhere together. Lisa has a large mammary tumour that is now ulcerated and affecting her ability to move around. She may be eating slightly less. You discuss the options, including surgical removal, with your client, but are not sure that he's fully understood them or even that you will see him again. However, he returns the following day with Lisa for surgery. When the nurse goes over the consent form with him he doesn't seem to understand the risks of the surgery and anaesthesia and does not give any indication that he realises that there's a chance, albeit small, that Lisa could die during the operation.

What should you do?

___

RESPONSE
JANE JOHNSON

▶   The main ethical issue in this case has to do with informed consent and whether it is possible and necessary in this instance. To begin to work through the case it is therefore useful to think through why informed consent is deemed important and how it operates.

The huge emphasis on informed consent in human medicine and research arose out of fears of paternalism by the medical profession towards patients, in part based on a failure to recognise

△
9.7 Mr C is very attached to his pet rat, Lisa.
But does he fully appreciate the risks associated
with surgery and anaesthesia?
PHOTO ANNE FAWCETT

that many decisions have a value and not just a clinical component, and in the research setting a worry that participants were being exploited by researchers. Informed consent is meant to respect autonomy by ensuring that people under-stand what will happen to them (the likely bene-fits, risks etc.) and by securing their endorsement of a course of action, ensuring it aligns with their values. If individuals have impaired capacity to consent, a third party (for instance a relative) can be invited to act in their best interests and make decisions on their behalf. Sometimes this can also involve securing the assent of the individual who is otherwise unable to give consent.

Informed consent in veterinary medicine is already at one remove as animal patients are not deemed to have the capacity to give informed con-sent, rather their owner must act as a surrogate decision-maker, protecting their best interests.

Handling this case will take time and should not be left to the vet nurse. Anticipating that Mr C

may not fully understand the ramifications of the advice you are providing, it would be helpful to flag in the first consultation that he should ring if he wishes to arrange surgery, giving you and others at the clinic a chance to prepare for his visit. You could also ask him to consider bringing along a close friend or relative to the next consultation.

Depending on the extent of his impairment it may be possible to find a way of expressing to Mr C the risks of surgery. Given you are concerned he may not understand that Lisa could die as part of the procedure it is important to attempt to convey to him the seriousness of what you are proposing, balancing this against the seriousness of Lisa's condition. Communicating with Mr C may be assisted by his having a friend or relative with him who can help express your advice in terms he will understand. If things were to go wrong and Lisa were to die or be seriously harmed, it would also be beneficial to have this support person available.

It is worth noting that in the human context much of the focus on informed consent has been motivated by a fear of litigation. To some extent, properly conducted informed consent processes are meant to protect physicians and researchers if things go wrong. This partly explains why there has been such an emphasis on ensuring persons have the capacity to make decisions. But if we take a step back, what might be more important in terms of the ethics of Mr C's case is not that he meets the formal checklist of conditions for capac-ity to give consent but that you handle his situation with sensitivity, taking the time to talk to him and elicit any values or beliefs that might influence a decision to proceed with surgery. As the vet you have the clinical knowledge relevant to assist Mr C to come to a decision about whether surgery is the right course of action, but he has knowledge of Lisa, their relationship, its importance to them both, a set of personal values and a sense of what is the right thing to do for them that all need to be taken into account.

There are underlying questions in this scenario about to whom you owe duties as a veterinarian. Contentiously in a Western setting where animals are deemed legal property, I would argue that you owe your primary ethical duty to the animal patient not the owner. So although you should try your best to help Mr C as an owner understand the implications of surgery, if on balance you are not sure that he has fully appreciated the situation but you think that it is in Lisa's best interests to have the surgery, then you are ethically justified in proceeding with surgery.

|||||||||||||||||||||||||||||||||||||||||||||||||||||||||||||||||||||||||

AGAIN, in response to this difficult scenario the author proposes that as the primary duty of the veterinarian is to the animal patient, the requirement for obtaining informed consent can be cautiously overridden. However, as Fettman and Rollin (2002) point out, the logical extension of an extreme version of this position is that there is no need for informed consent, "based on the assumption that the veterinarian knows what is best and need not explain his/her decision-making process in selecting treatment". They note that occasionally owners may bestow this form of consent for patient advocacy by demanding that the vet exercise their Aesculapian authority and do what's best for the animal, regardless of cost or inconvenience to the owner (Fettman & Rollin 2002). However, as seeking informed consent can be a useful way of providing assurance that people are not subject to deception and coercion (O'Neill 2003), it is still important. Fettman and Rollin also note that the veterinary surgeon will have some obligations to the owner, and they suggest, to the human–animal bond, proposing that considering the obligation to the bond itself will allow a balance of obligations to the patient's interests and those of the owner (Fettman & Rollin 2002).

DIFFICULTIES IN OBTAINING CONSENT

What would you do?

Mr V is a long-standing client of yours whom you've suspected of being drunk on occasions. His eight-year-old Sphynx cat, Felix, has been treated for a number of minor complaints over the years. Now Felix is inappetent and appears to have an abdominal mass. You suggest he comes back tomorrow for further investigations and, depending on the findings, possibly exploratory surgery.

The next day Mr V appears drunk when he comes to drop Felix off and pays no attention to the consent form which necessarily contains information and prices for a range of options and outcomes. How should you proceed?

△

9.8 Mr V appears drunk when you ask him to sign the consent form for Felix, his cat, to have surgery.
PHOTO ANNE FAWCETT

There are other situations where the client presenting the animal may not be the owner. For example, the presenting client may be a friend or family member of an owner who is injured, hospitalised or deceased − or a neighbour or ex-partner. Alternatively, as in the following scenario, the presenting client may be a "new owner" of an animal where ownership has not been formally transferred.

SCENARIO
OWNED "UNOWNED" ANIMAL

▶ There's great excitement when the D family bring a friendly little cat, Josie, to see you. Josie appears surprisingly relaxed in your consulting room and purrs whilst the teenage children stroke her. The Ds say that they'd seen her hanging around for some time and she just adopted them. They hadn't actually wanted another cat after you euthanased their previous one a few months ago. They are concerned that despite being hungry she doesn't seem to be able to eat well.

On examination Josie is quite thin and has a heavy flea burden and quite severe gingivitis, tartar build-up on the teeth and inflamed fauces. You take Josie out of the consulting room to give a student nurse the chance to scan her. Surprisingly, she is microchipped, and the address is a few streets away from the Ds.

You check your own records and find that the owners haven't been to you since Josie was a kitten. You had seen them intermittently with a range of pets. They appeared to live a slightly haphazard life, and have outstanding payments going back years. The last entry reads "Moved Away" and gives a new address in the next town. You suspect Josie either was left behind or has made her own way back. You really think the D family would make better owners.

What should you do?

△
9.9 Apparently unowned animals may be owned, which can lead to ethical dilemmas.
PHOTO ANNE FAWCETT

RESPONSE
ANDREW GARDINER

▶ My initial approach would be to try to explore the D family's decision to take Josie in. After their previous cat had been put to sleep, they actually told me that they did not want another cat. I would like to be sure that Josie's appearance was indeed serendipitous and that they would genuinely like to keep her as their new family cat. It is possible that they feel it is their "duty" to care for Josie (deontology), or that they see themselves as the type of people who should do this (virtue ethics), but that, really, they would prefer not to. In that case, I could suggest alternatives, for example by providing appropriate first-line medical treatment and contacting the local cat rescue, or rehoming her through the practice. I would do this not only to make the D family happier, but also to safeguard Josie's own welfare, as I believe it is in Josie's best interests to be in a home where she is valued for what she is, as the "subject of a life" (animal rights view). Either way, whether the D family decide to keep Josie or not (and I think they would like to keep her), my own actions are driven by utilitarian concerns. I think there are elements of virtue ethics in my approach, and also rights-based considerations directed at Josie with regard to her subject status and interests. I feel that Josie has expressed her own autonomy through choosing the D family, and that if possible this should figure in considerations.

However, this situation is made much more complex and difficult by the fact that Josie is microchipped, and by the history which I find in my practice records regarding her previous owners. I cannot un-know this information once I have seen it. How to act on the microchipping becomes a deontological matter which now dominates in my thinking. After all, that is why animals are microchipped in the first place – so that they can be returned to their rightful (under the law) owners. The RCVS view on microchipping is that "veterinary surgeons are *encouraged* to take appropriate steps to reunite the animal with the owner" (my italics). It seems from this a veterinary surgeon could decide not to do this; however, physical removal of any microchip would be deemed unethical and an unnecessary mutilation (RCVS 2015a). This seems a classic case of sin of omission (not reporting/ignoring the microchip) versus sin of commission (actively removing it) with the former being less evil than the latter. However, perhaps the ethical differences are less clear-cut?

The presence of the microchip problematises constructs of the "irresponsible owners" suggested by other evidence I have. If the previous owners got Josie microchipped, and it looks as though they did indeed do this because they acquired her as a kitten, then perhaps they are not so negligent after all? However, her owners do have outstanding debts and they never attended the practice regularly. They do not seem to have

made much attempt to find Josie, as phoning the practice would be an obvious first step if she went missing. Maybe they abandoned the poor cat after all. I could try to make an impartial decision with respect to the two owners to guide what action I should take, employing the theory of justice and fairness. However, that is challenging for me because, as an experienced vet, I am influenced by many narratives of similar situations which guide my gut feelings and emotions about this case (narrative ethics).

At this point, I feel a bit confused and overwhelmed and decide that I need a structured way of trying to address the different and competing dimensions of the problem. An ethical matrix approach could ensure that I include relevant stakeholder concerns and identify information gaps. One attempt at a matrix is given in Table 9.1 (overleaf).

The matrix reminds me that Josie appears to be suffering from feline chronic gingivo-stomatitis – an expensive disease to treat. I need to ensure the D family are aware of this and willing to take on the considerable cost and commitment of treatment, should they get to rehome her. The prognosis can be good with appropriate treatment, with frequent clinical cures and excellent welfare. If the D family are willing to fund and support Josie's treatment, this gives me useful information. But it does not resolve the ethical dilemma about what to do concerning Josie's "real" owners.

In the end, and after consulting my matrix, I find that deontology trumps a strong utilitarian desire to place Josie with the D family. I am not entirely happy about this, but feel it is mandated by my own ethical reasoning and because of a need not to make unjustified assumptions about the motives and behaviours of Josie's previous owners. However, the matrix also provides me with some practical suggestions as to how to handle the situation in an appropriate way which makes me feel better in terms of my own personal ethics.

9.1
DIFFICULTIES IN OBTAINING CONSENT
SCENARIO
OWNED "UNOWNED" ANIMAL

RESPECT FOR:	WELLBEING	AUTONOMY	FAIRNESS
Josie	Josie needs medical treatment for a painful mouth, and also a home.	Josie needs to be able to express normal eating behaviours. Josie herself chose the D family, so this autonomous action is relevant.	Josie should be treated appropriately regardless of her stray/owned status, i.e. her intrinsic value should be respected.

The D family	The D family's quality of life could be affected by losing Josie now they have bonded with her.
	    Or it could be adversely affected by feeling obliged to care for her, if they don't really want to do this.
		Ideally, should feel free to choose whether to adopt Josie or not.
		    However, this can be trumped by other concerns (legal ownership of Josie).
			Should be informed as to what they are embarking on in treating Josie's problems: need appropriate information to guide choices.

Josie's original owners	Quality of life: could be mourning loss of Josie; or may be happier without her.	May have valid reasons for not attending practice regularly or not engaging in certain preventives/ neutering. Or may have abandoned Josie knowing she had a significant health problem.	They should be treated "symmetrically" without undue assumptions until sufficient facts are known.
Veterinarian/ practice	To do the right thing for Josie in line with personal, practice and professional ethics.	My veterinary skill and judgement should be used to decide how best to manage Josie.	The practice and staff should be fairly paid for past and future treatment of Josie.
Biota (local neighbourhood)	Effects of stray cats on local wildlife and environment affect the wellbeing of the biota.	Local wildlife biodiversity needs protection.	Sustainability of the local ecosystems/ green spaces needs to be protected by measures to control strays and protect birds, plants and animals (and possibly buildings in urban landscapes).

△
   Table 9.1 An ethical matrix for Josie.

Once I have explained the whole situation to the D family, and obtained their consent, I will prepare an estimate to cover the likely costs of treating Josie's gingivo-stomatitis, including full mouth extraction and interferon, as published in veterinary clinical literature (Gorrel 2013). I will then write (by registered mail) to Josie's original owners explaining the situation, enclosing the fee estimate and stating the need for immediate treatment due to her hyporexia and weight loss. I will emphasise that treating Josie is an urgent animal welfare concern which must be addressed.

I will explain that she was brought in by the D family, who are willing to take on the costs and after-care that the treatment might require. They will also bring Josie's basic care (vaccines, etc.) up to date, which according to our records appears to have lapsed. Josie will need to be signed over to the D family.

The decision will have to remain with Josie's original and legal owners, but I am hopeful they would relinquish Josie to what I envisage would be a happy life with the D family. A priority would be addressing basic, urgent welfare needs (removal of significant pain and distress caused by her gingivo-stomatitis) and setting her up for a good life as a companion cat.

If Josie's original owners do not agree to sign her over, then my only option is actively to advocate for Josie's medical treatment, and take what action I can if that is not happening on the grounds of owner obligations to animal welfare as stated in law, for example in the UK the Animal Health and Welfare (Scotland) Act (HMSO 2006a) and the Animal Welfare Act (HMSO 2006b) and the Welfare of Animals Act (Northern Ireland) 2011. http://www.legislation.gov.uk/nia/2011/16/enacted.

DIFFICULTIES IN OBTAINING CONSENT

What do you think?

ONE _____ Develop a practice policy for the treatment of animals whose ownership is not clear.

TWO _____ In cases where ownership is disputed, some have suggested that the animal should be able to choose. How might you facilitate this and how might you justify any challenge to the animal's choice?

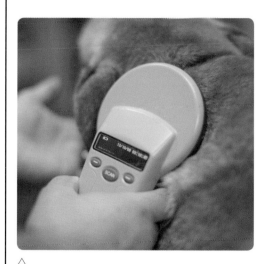

△

9.10 Microchips can help establish the owner and are now compulsory in some countries.
PHOTO ISTOCK

9.2
Acting without consent

The scenarios we have looked at so far have involved cases where obtaining consent has been difficult. But, acting when consent is not fully obtained is not the same as deliberately going against the wishes of a client. Here we consider a situation where a vet is tempted to do just that.

SCENARIO
REHOME WITHOUT CONSENT

▶ You're a small animal veterinarian with several years of experience in a multi-vet small animal practice in central London. A newly graduated colleague has just asked your advice. She was presented with a healthy and well-socialised two-year-old intact male West Highland White Terrier called Archie. The dog's owner, Mr G, is moving to a different apartment that does not allow dogs, and has requested euthanasia. When your colleague initially expressed discomfort at the request, Mr G apparently told her that he would let the dog loose on the street. When she suggested that he think about adoption he said he could not bear to have someone else own him. And so your colleague accepted the client's request and fee for euthanasia. After Mr G left, one of the nurses asked if she could give the dog to her aunt, who lives several hundred miles away. Apparently she would provide the dog with a wonderful home, and it would be extremely unlikely the client would ever find out.

What should you do?

△
9.11 Can you rehome a dog without owner consent?
PHOTO ANNE FAWCETT

RESPONSE
ANDREW KNIGHT

▶ Obviously the best option for the patient is not to be euthanased, which is contrary to the client's initial request. Accordingly, this case should have been handled carefully and sensitively from the beginning.

Your colleague should have more thoroughly explored the possible options with this client.

She should have explained that while euthanasia might initially appear to be an option, as a veterinarian her primary duty is to her patients, and that it is clearly not in the interests of a young, healthy dog to be euthanased and denied the many years of life he would otherwise be expected to enjoy. Hence, unfortunately she would be in violation of veterinary professional ethics if she performed euthanasia in this case. This might also be an appropriate time to check that the client understands what euthanasia really means. There have been rare cases in which, after a dispute has arisen, clients have alleged that they did not understand that words such as "putting to sleep" actually implied death. Accordingly, the veterinarian should delicately confirm that the client is indeed requesting euthanasia, and that this is irreversible.

She should also have more thoroughly explored the owner's concerns. Through open-ended questioning and good communication techniques she should have tried to determine the reason for Mr G's reluctance to have the dog live out his life in another caring home. His assertion that "he could not bear to have someone else own him" suggests strong attachment. It may be that his reluctance to have someone else own his dog stems from an insufficient understanding of the rehoming process and options. These might have been explored with him further. And your colleague might have noted that he obviously cares a great deal for his dog, and might have sensitively asked him to weigh any distress he might feel against the potential for his dog to enjoy the remainder of his life.

Given the client's threat to let the dog loose, the ramifications of abandonment should also have been discussed. In some jurisdictions this is actually illegal, which the owner would probably not have been aware of. Additionally, this could be likely to result in the dog being collected by animal control officers, placed in a shelter and possibly adopted contrary to the owner's stated wishes – without him even having any control over the ultimate choice of new home.

The owner may be initially reluctant to discuss some of these matters. However, by trying to empathise with the owner about his difficulties, handling the case sensitively and explaining that any veterinarian nevertheless has a duty to discuss

these matters when considering a case of euthanasia, the owner may consent, and indeed may benefit from having thought through the case more thoroughly. It could also be emphasised that such a discussion may also help the owner to be as certain as possible about their choice, which should serve to minimise any later regrets.

If, after such a discussion, the owner remained adamant about his request for euthanasia, the veterinarian could explain that euthanasia in the absence of a sound medical reason would contravene her personal and professional ethics, and politely offer to refer the client to another veterinarian for a second opinion. As stated by the RCVS (2015a) Code of Professional Conduct for Veterinary Surgeons, "No veterinary surgeon is obliged to kill a healthy animal unless required to do so under statutory powers as part of their conditions of employment." And further, "Where, in all conscience, a veterinary surgeon cannot accede to a client's request for euthanasia, he or she should recognise the extreme sensitivity of the situation and make sympathetic efforts to direct the client to alternative sources of advice." However, the owner should be warned that another veterinarian might also decline the procedure. Indeed, in the UK for example, most veterinarians would decline the euthanasia of a healthy animal unless wider animal or public health was endangered.

Unfortunately, however, in this particular case such discussion was much briefer, and your colleague did agree to the request. She also accepted the client's fee for euthanasia. In a legal sense, she entered into a contract with the client to provide a service. A range of consequences could now occur if she failed to fulfil her contractual obligation to complete the euthanasia.

Particularly given the lack of in-depth discussion previously about alternatives, it would be appropriate for your colleague to contact the client again before proceeding, and explain that a new

> "There have been numerous cases when clients have later learnt clinical truths that were previously denied to them."

option has arisen, namely, rehoming in a caring home so far removed from the client that it would be unlikely he would ever need to see his dog again, if that experience might distress him. Your colleague could offer this as a new alternative, along with a refund of the euthanasia fees. If the client accepted this new option, then the dilemma would be resolved.

If however, despite your colleague's best efforts, the client continued to request euthanasia – that is, declined to release your colleague from her contractual obligation – then her dilemma would remain. On the one hand, if she declined the euthanasia, a range of consequences could accrue, particularly if the client found out. As well as being very unethical, attempting to deceive clients in this way can be unsuccessful in the long term. There have been numerous cases when clients have later learnt clinical truths that were previously denied to them. This has occurred when clinical notes are transferred to a new veterinarian, or when a falling out occurs among certain staff members, who then choose to report unethical activity.

If the client were to later discover that his dog had been rehomed contrary to his expressed wishes, he might well choose to pursue the case further, e.g. by complaining to the veterinary licencing board that malpractice had been committed, or by launching independent legal action (e.g. alleging theft of the dog, or claiming other damages). He might even attempt to publicise the case, to damage the reputation of the veterinarian or practice within the community. Public

trust is extremely important for the successful functioning of the veterinary profession, so it is unsurprising that veterinary licencing boards take a very dim view of professional dishonesty by veterinarians, and sanctions in such cases are likely to be significant. This could mean the temporary or even permanent (albeit less likely) loss of your colleague's licence to practice, as well as financial damages, if the client launched independent legal action.

On the other hand, it is clear that your colleague's first duty is to her patient. This axiom of veterinary professional ethics is made clear in the statements of veterinary associations. The AVMA (2016) Principles of Veterinary Medical Ethics, for example, state that, "Veterinarians should first consider the needs of the patient: to prevent and relieve disease, suffering, or disability while minimizing pain or fear." The RCVS (2015a) Code of Professional Conduct for Veterinary Surgeons similarly states that, "Veterinary surgeons must make animal health and welfare their first consideration when attending to animals." Euthanasing a young, healthy and well-socialised dog is clearly not in its interests. As well as constituting a clear violation of the veterinarian's primary duty to their patients, such actions may well contribute to an inability to live with oneself, burnout and stress. These factors may be more important than are first apparent, given the relatively high levels of depression, anxiety, stress and burnout within the veterinary profession (Hatch, et al. 2011).

Hence, if the client cannot be dissuaded from his request for euthanasia, your colleague clearly faces a very serious dilemma. She is essentially challenged by the question, "How far should I be prepared to go, to uphold my primary duty to the patient under my care?" It could even be that choosing to save the life of this patient through deception ultimately results in the loss of her veterinary licence, impacting her ability to help other animals in the future. This outcome is not certain, but a temporary suspension of licensure, at least, could be a real risk, should her veterinary licencing board find out about her dishonesty.

The values placed on various competing factors, and hence choices made in response to this dilemma, will vary among individuals. Clearly, however, such cases should be handled more thoroughly from the outset, to minimise the occurrence of such dilemmas. And if presented with a dilemma already extant from which one cannot escape, the various outcomes and their probabilities should be very carefully weighed.

|||||||||||||||||||||||||||||||||||||||||||||||||||||||||||||||||||||||||

WHATEVER course of action you choose when faced with euthanasia of a healthy animal you can be sure that you are not alone in having to face this problem. In one small survey of 58 veterinarians in the UK most had faced being asked to euthanase animals they did not want to, and for a few this occurred as regularly as monthly (Yeates & Main 2011). There are also media reports of dogs that had supposedly been euthanased appearing alive and well elsewhere. In one case, a dog was rehomed after the owners presented the dog for and paid for its destruction due to aggressive behaviour. The dog subsequently attacked other animals and was traced back to the original family via a microchip, 18 months after it was supposed to have been destroyed by a veterinarian (Armitstead 2013).

ACTING WITHOUT CONSENT

What would you do?

You are working in a busy veterinary clinic where it is not unusual for clients to "drop off" animals for a consultation. Your nurse brings in the next patient – an apparently healthy six-year-old cat. The cat, apparently named Lily, is booked in for euthanasia. The client, a Mrs S, signed and dated the consent form at reception but stated that she did not wish to be present for the procedure and could not wait.

You perform a physical examination. The cat appears to be in excellent health. What do you do?

△
9.12 A healthy cat is left at the clinic for euthanasia, but no history is provided.
PHOTO ANNE FAWCETT

ACTING WITHOUT CONSENT

What would you do?

Mr D comes to you in quite an agitated state. Having had periods of unsatisfactory housing he has now decided to move in with his girlfriend in her flat. She hates his dog Stan, and will not have him in the flat, and anyway dogs are not allowed by the landlord. Mr D insists that euthanasia is the only answer, he will not consider rehoming as he doesn't think Stan will be happy. Mr D is quite angry and upset and just wants to leave Stan with you to "do the deed". Your mum has just lost her dog and an idea starts to form – Stan would be great for her. She lives in another part of the country.

   What should you do?

△

9.13 Mr D's living arrangements make it challenging to keep his dog Stan. Is euthanasia the answer?

PHOTO ANNE FAWCETT

9.3

How much disclosure is required for consent?

Part of the process of obtaining proper consent requires disclosure of the appropriate amount of information so that someone can make an informed decision. Here we consider two common scenarios where vets have to decide how much information is relevant for client consent.

SCENARIO 1
INFORMED CONSENT

▶   You are a new graduate in your second week of work in a three-vet mixed practice. During your first week you were well supported in both consultations and small amounts of surgery. Today you are due to be doing more operating alongside the practice principal. Bella, an overweight two-year-old golden retriever, has been booked in for a spay and you feel a little nervous. Just through

△

9.14 Will you attempt a routine procedure that you have little experience performing without supervision? If so, will you inform the client?

PHOTO ANNE FAWCETT

chance, you ended up only doing one bitch spay during your neutering rotation in final year. Before the clinic is open your practice principal is called to an emergency away from the practice and she will be gone all morning.

When you admit Bella, how much information about your neutering experience and likely supervision should you give?

SCENARIO 2
INFORMED CONSENT

▶ You admit Henry, a six-year-old Labrador, for observation overnight following vomiting and diarrhoea, assumed to be a case of gastroenteritis. Mrs O and her children are fussing over Henry. "You will look after him won't you?" she asks. You reply that "Of course we will" but you are a little concerned that Mrs O thinks you will be sitting up watching Henry, when actually you will sleep in the practice flat and check him last thing at night and first thing in the morning.

Should you tell the owners that you will not be watching Henry all night?

RESPONSE
STEVEN P MCCULLOCH

▶ Both of these scenarios are interesting in that although they do not describe dramatic ethical conflicts, they are both important and common in general veterinary practice. For the purposes of discussion, the first scenario is called "bitch spay" and the second "overnight care". The author would argue that the right action in both cases is relatively uncontroversial. In "bitch spay" the veterinary surgeon should defer the surgery until assistance is available. In "overnight care"

△
9.15 When you promise to "look after" Henry the Labrador, what do you mean?
PHOTO ANNE FAWCETT

the veterinary surgeon is duty-bound to inform the owner about the level of care the patient will receive during the night. Despite the relatively uncontroversial nature of these cases, which I will discuss below, the veterinary surgeon may feel a degree of indecision in each case.

In "bitch spay" the inexperienced veterinary surgeon may feel the urge to perform the spay without the availability of assistance. The major motivating factor here is reputational, where the admitting veterinary surgeon is concerned about what the practice principal, their colleagues or the client thinks about their professional skills (a second motivational factor might be the desire to prove to themselves that they can perform the surgery in a sole-charge context). In "overnight care", the veterinary surgeon may be motivated to disclose limited information to the owner for financial/business reasons. Additionally, and perhaps equally importantly, the veterinary surgeon in "overnight care" might consider it more burdensome to discuss other options, and perhaps arrange such options, rather than simply keep the patient hospitalised in-house.

9.3

HOW MUCH DISCLOSURE IS REQUIRED FOR CONSENT?
RESPONSE
STEVEN P MCCULLOCH

The motivational factors described above in "bitch spay" and "overnight care" are themselves suggestive of the right course of action in both cases. Prima facie, reasons for action motivated by reputation and self-esteem ("bitch spay") and financial gain and expedience ("overnight care") do not appear to be strong ones. But what of analysis of these scenarios from first principles? Consider first the case of the inexperienced veterinarian performing a bitch spay in a sole-charge context. It is instructive to highlight how the *nature* of a bitch spay makes this an interesting ethical decision. A bitch spay is the common term for an ovariohysterectomy in a female dog. At least in the UK, the veterinary profession almost universally recommends spaying female dogs at an early age. The major reasons are to prevent or reduce the incidence of pathological disease such as pyometra and mammary gland carcinoma. Additionally, spaying female dogs is performed as a method of population control and to prevent undesirable behaviours in the bitch associated with being in season that are considered stressful to her.

Since bitch spays are almost universally recommended, they are very common operations and are labelled as "routine" in veterinary discourse. The "routine" label of the bitch spay, however, is potentially deceptive. It would be mistaken to confuse the routine nature of the bitch spay, with its more intrinsic nature as a major operation with potentially serious, even life-threatening consequences. A bitch spay involves the surgical ligation of the ovarian and uterine vessels. These vessels are direct tributaries of the aorta, the major artery in the body. There is therefore a significant risk of intra- and post-operative haemorrhage in a bitch spay. In the worst-case scenario, bleeding from the ovarian or uterine vessels, for instance due to slippage of ligatures, can result in the death of a patient.

It is likely the combination of the "routine" nature – the frequency with which the operation is performed – and the intrinsic nature of a bitch spay – as major abdominal surgery – contribute to uncertainty in the newly qualified veterinary surgeon. The veterinary surgeon may feel that his/her boss – the principal veterinarian – would expect her new employee to perform such a "routine" operation. On the other hand, the newly qualified veterinary surgeon will be all too aware of the potential complications of the operation. These potential complications are not academic. The author, like many veterinary surgeons, has experienced bitch spays bleeding on the theatre table. Surgically, there are two factors which speak for the availability of assistance, should it be needed. Firstly, some bitch spays are technically difficult, normally because it can be difficult to exteriorise the ovaries and there may be significant intra-abdominal fat impeding visibility. Secondly, in the event of a serious bleed, it is technically difficult – and requires a certain degree of surgical experience – to locate the source of bleeding and (re-)ligate the offending vessel. In the event of a bleed, even for experienced surgeons, an assistant can facilitate exposure of the surgical field by swabbing/suctioning blood and displacing abdominal viscera away from the vessel.

The picture painted in the "bitch spay" scenario suggests the operation is to be performed on a healthy patient for preventative reasons. The patient is two years old and the scenario does not mention that she is suffering from pyometra or any other disease. Therefore, the surgery is elective, i.e. it can, and should, be performed at a time and place to *minimise anaesthetic and surgical risks*. To date, the new graduate has performed one bitch spay, under the supervision of competent surgeons. It would not be minimising anaesthetic and surgical risks if the newly qualified veterinary surgeon were to perform the bitch spay in a sole-charge context. Hence, the veterinary surgeon has two reasonable options.

Firstly, s/he could admit the bitch as normal, and simply wait for the principal to return to the clinic prior to taking the patient to theatre. Secondly, s/he could advise the client of the circumstances and reschedule the (elective) operation for a more suitable time. Arguably, the first option is most favourable if the principal is expected to return and can be present, at least in the building, for the spay. The second option, although it inconveniences the client, is more appropriate if the principal is not expected to return, because despite the inconvenience, the second option is far more defensible in the event that the principal cannot be expected to return. The primary and overriding duty of the veterinary surgeon, at least in the UK, is to the welfare of the animal under his/her care (McCulloch, et al. 2014, RCVS 2015a). This author would argue that, even in the event that the owner, fully informed, requests that the newly qualified veterinary surgeon perform the operation in a sole-charge context, it is problematic for him/her to acquiesce. Such acquiescence would prioritise the owner's wishes over the veterinary surgeon's duty to animals. This position might seem extreme, but it follows both from the RCVS Code (RCVS 2015a) and from argument by analogy. Consider if a mother/father or legal guardian in human medicine were to insist elective surgery was performed on their child/legal dependant against the advice of the healthcare provider, based on availability of consultant surgeons to oversee the operation. For an elective procedure at least, it would be absurd to claim that the patient's interests ought to be overridden by the parent's/guardian's wishes, based on convenience. This is because the health of a child is far more important than the convenience of parents/legal guardians, no matter how busy their lives might be. If we have different intuitions about veterinary medicine (the author does not, but perhaps some do), it is only because (1) we value the welfare of animals less than the convenience of clients (or business considerations), or (2) we are not wanting to cause a stir/ we are following the status quo.

The "overnight care" scenario is assessed here only briefly. In the scenario, the client is unaware that the patient will not be checked overnight. The scenario poses the question whether the veterinary surgeon should inform the client of the level of care. "Overnight care" is best assessed in the context of treatment options and informed consent. In many cases that warrant hospitalisation, there are three broad management options ("management" here means treatment considered very broadly). These are (1) treat the case at home as an outpatient (recheck tomorrow), (2) treat the case as an inpatient in-house and (3) referral to some other form of overnight care. Consider, for example, a six-year-old Labrador with a two-day history of vomiting and diarrhoea, assumed to be suffering from gastroenteritis. In such a case, the veterinarian could offer (1) treatment and recheck the following day, (2) admit, hospitalise and administer intravenous fluids, with fairly minimal overnight attention, or (3) referral to a dedicated overnight veterinary hospital, with staff present throughout the night.

The supporting guidance of the RCVS Code of Professional Conduct states that informed consent "can *only* be given by a client who has had the opportunity to consider a *range of reasonable treatment options*, with associated fee estimates, and had the significance and main risks *explained to them*" (RCVS 2015b, italics mine). In the context of the RCVS Code, the question posed by the scenario "Should you tell the owners?" must, in the professional context, be answered in the affirmative. To rephrase the above quotation, a range of reasonable treatment options is a necessary condition of informed consent. Clients must have these reasonable treatment options, including their significance (e.g. efficacy – what use is the fluid therapy if the drip stops in the middle of the night?) and risks (what if the patient deteriorates without the presence of veterinary attention?), explained to them.

||||||||||||||||||||||||||||||||||||||||||||||||||||||||||||||||||||||||||

JUST how much information is the right amount to provide to clients? Some people will want to know far more than others about the various options available. It has been suggested that in the medical context patients should be able to have control over the amount of information they receive as part of the consent procedure. This can be achieved by layering the information available so that patients who want to can "dig deeper" to find out more detail (O'Neill 2003). It has been argued that the veterinarian has a particular responsibility to owners in being able to provide information that owners cannot ordinarily be expected to have, and that this at least has the potential to influence a client's decision about treatment (Flemming & Scott 2004).

Veterinarians shall earn the trust of their customers through full communication and by providing appropriate information.

HOW MUCH DISCLOSURE IS REQUIRED FOR CONSENT?

What would you do?

You diagnose an anterior cruciate ligament rupture in a five-year-old mixed-breed dog. There are multiple surgical approaches that may be taken but as you begin to describe these the owner claps his hands over his ears and says, "Stop! – Sorry… I can't hear this. I have a thing about surgery, I just can't stand to think about it. Just do what you think is best and I'll take your word for it."

What should you do?

△
9.16 How do you obtain informed consent from a client for a procedure that the client does not wish to discuss?
PHOTO ANNE FAWCETT

9.4

Consent for novel procedures

A novel therapy or procedure is one that has had "a limited historical use for a (potential) treatment for the particular condition in that species, or which combines or modifies an accepted therapy in a way that has had limited previous use and where the modification has a potential for altering clinical outcomes" (Yeates 2016).

Novel procedures require particular consideration for consent. Because they are novel there may be a very small evidence base, if any, to predict the outcome. Similarly there may be limited knowledge about potential risks. Where there is a dearth of information, it can be difficult for a client to give "informed" consent. Nonetheless, novel procedures test the boundaries of the veterinarian–client–patient relationship and therefore informed consent is particularly important to obtain in a free and fair way. We have considered so far two of the three tenets of informed consent: disclosure and

△

9.17 Would you use a sample of a new product on this pony?

PHOTO ANNE FAWCETT

capacity. In this next scenario, focussed around the use of novel surgical procedures, we consider the "voluntariness" of informed consent when there may be a conflict of interest.

SCENARIO 1
SURGICAL INNOVATION

▶ You are working in equine practice. A friend and client works for a biomedical company that manufactures patches that can be used to reduce the risk of leakage at enterotomy sites. The material has been tested in dogs only. Nonetheless, your friend says if you need such a product, samples can be procured at no cost if you are happy to share the results. Later you admit Wintona, a pony, for colic due to enteroliths. You are concerned about leakage at the site and suspect a patch may reduce the risk.

SCENARIO 2
SURGICAL INNOVATION

▶ Treacle, a guinea pig, has presented with a mass, diagnosed as a sarcoma. Removal of the mass will leave a large skin defect. You consider the use of a skin flap to repair the defect, but cannot find any reports of such a procedure in this species.

What should you do?

RESPONSE
JANE JOHNSON

▶ Both these scenarios involve a form of innovation – taking knowledge of something that works in one context and attempting to transpose it to another. This knowledge could involve

9.18 How might you approach novel treatment in a guinea pig?

PHOTO ANNE FAWCETT

a technique, a drug or in this instance something tested in one species being tried in another. Other examples of novel therapies include new pharmaceuticals, surgical procedures, clinical protocols and novel applications of existing pharmaceuticals (Yeates 2016).

Though innovation is important to progress, it involves risks and uncertainty. For this reason if you intend to suggest these treatments, it is essential that the owners of both animals be informed of the innovative nature of what you propose. Owners should be made aware of the lack of species-specific evidence regarding the safety and efficacy of the patches and of the surgery respectively, and of the alternatives to trying something new, including not doing anything at all. In the case of the guinea pig, it is important to consider whether or not a large skin defect is of sufficient concern to warrant a novel treatment. Just as in human medicine, owners will have different risk profiles; some may be happy for their animal to be a pioneer and others not so. You should be aware, however, that many people fall prey to the misconception that new means better and may not hear your warnings about a lack of species-specific data.

If the owners agree and you do proceed with these novel approaches you should strongly consider publishing the results regardless of the outcomes, since a major issue around innovation, particularly in surgery, is that data is not collected. This means flawed strategies are reattempted and harms and benefits are not tracked. When the company suggests supplying the patches at no charge if you agree to share the results, you need to clarify what they will do with these results. For instance, will they publish, or will the results remain in-house and be effectively buried if they are negative?

A further and significant ethical issue raised by the pony case has to do with conflicts of interest. The conflicts in this case involve your professional role as a vet which may run up against what you feel you owe your friend and client from the company, and if you accept the "free" patches, then what you owe the biomedical company. These relationships establish a feeling of reciprocal obligation which can compromise your clinical decision-making.

In advising Wintona's owner you should be focused on what is best for the pony irrespective of how this may impact on your friendship and client relationship with the company representative or on the relationship with the biomedical company. Having said this, the unconscious nature of the bias introduced by conflicts of interest makes it difficult to establish what your advice might be in the absence of the conflict. And this is part of the reason why conflicts of interest are so problematic, because the conflicted person is generally unaware of the actual impact and influence of the conflict.

Disclosure is one of the strategies often proposed to manage and disarm conflicts of interest; would revealing your conflict to the pony's owner deal with the conflict in this situation? In short — no, disclosure would be inadequate in this case. Given the unconscious bias introduced by conflicts of interest, neither you nor Wintona's owner are in a position to assess how your relationships impact on your recommendations, so disclosure of the conflict does not work to address the situation. Empirical research in the social sciences has also pointed to the perverse impacts of disclosure. Having disclosed a conflict, a professional appears liberated to exaggerate evidence related to the conflict, and the person receiving the advice tends to assume that the professional is being candid and relies on the accuracy of the advice.

Another strategy for handling conflicts is recusal; removing oneself from decision-making that might be impacted by the conflict. Depending on other factors (such as the availability of another vet in the practice to handle Wintona's case) this might be an appropriate course of action. However, arguably in not taking up the opportunity presented by the patches you are acting unethically and impeding the development of what might turn out to be a promising way of dealing with leakage at enterotomy sites in ponies. To assuage this worry you could work with the company to design a proper research trial for these patches to ensure the scientific integrity of the data. After all, in addition to securing brand loyalty, part of the rationale of the company in dispensing free samples was to recruit your patients into what amounts to de facto research.

|||||||||||||||||||||||||||||||||||||||||||||||||||||||||||||||||||||||||||||||

THE use of novel treatments within veterinary practice is far less regulated than for medical procedures. It is therefore all the more important to ensure that adequate consent is obtained. Informed consent for novel therapies has been suggested as needing to include an understanding by the owner of:

- the subjective assessment of the risks and benefits of the novel therapy,
- all relevant information regarding the risks and benefits of accepted treatments,

- that the novel treatment is unproven and not established,
- and that there is "adequate advance consent for emergency procedures, such as euthanasia" (Yeates, et al. 2013).

Importantly, it has been proposed that novel therapies should only be undertaken where "no undue influence, including that of a financial nature, is exerted on owners to participate" (Yeates 2016).

Novel therapies are often evaluated on a cost:benefit or utilitarian analysis; however, such an approach falls short if it does not incorporate aspects of deontology or principalism, notably respect for the patient's own interests. As with any other veterinary therapy, "treatments should involve as little pain, discomfort, fear and any other foreseeable risks as possible and should safeguard each patient's physical and mental integrity" (Yeates 2016).

The 3Rs framework can be useful in considering novel therapeutic interventions:

(1) Can this novel therapy be *replaced* in whole or part with conventional, proven therapies? Is there a need to delay use in this patient in favour of *in vitro* and limited *in vivo* testing?
(2) Can the patient's risk exposure be *reduced*?
(3) Can use of the therapy be *refined* by pain-scoring and the judicious use of analgesia, discussion with specialists and colleagues and publication of results to the wider scientific community?

Before performing a novel procedure, for example, it is important to accurately describe the procedure, and document anticipated risks and benefits in light of the clinical objectives (Yeates 2016). The veterinarian should pre-determine the threshold of harm which, if exceeded, would prompt reversion to conventional treatment or euthanasia (Yeates 2016). Risk exposure may be more readily reduced in cases of non-urgent novel therapies where conventional treatment is available, but may be more difficult if the novel therapy appears to be in the immediate interests of an individual patient (Yeates 2016). In the above scenarios the author stresses the need to publish negative results, so that such therapies are not widely applied and so that future patients at the very least may not be harmed by them. In some settings it may be possible to involve at least two veterinarians: one who treats the animal with the novel therapy, and another one to act as advocate and decision-maker for the animal (for example, deciding when to withdraw the novel therapy and/or revert to conventional treatment) (Yeates 2016).

Conclusion

In this chapter we have explored a range of scenarios around client consent, including barriers to informed consent, disclosure, acting against client wishes (without consent) and clients consenting to novel therapies. A common theme in these scenarios is the need for shared decision-making, which is facilitated by clear communication about our own thinking processes. It may be helpful to explicitly discuss the ethical framework you are using with clients. For example, if you are weighing up different treatments or approaches based on a cost:benefit analysis, making explicit reference to this may improve communication and allow you to determine points of agreement and disagreement.

CONSENT FOR NOVEL PROCEDURES

What would you do?

You have heard "on the grapevine" that a new coccidiostat on the market for sheep may be effective and safe for rabbits. You have a large rabbit breeder as one of your clients that has outbreaks of coccidial disease from time to time which you suspect is due to partial resistance to the current treatments. What should you do?

9.19 Would you use a sheep medication on rabbits?
PHOTO ANNE FAWCETT

CONSENT FOR NOVEL PROCEDURES

What would you do?

Your friend is a great human orthopaedic surgeon and has offered to show you how to use one of their latest implants. How might you limit the harms associated with trying a new surgical technique? What would be the ethical issues around using a cadaver for training in the first instance?

9.20 Would you try out a new human orthopaedic implant on a cadaver?
PHOTO ISTOCK

References

Armitstead J 2013 Sasha the dog horrifies Townsville owners after she is found alive 18 months after being euthanised. *Townsville Bulletin, October 1 Edition.* News Ltd.

AVMA 2016 Principles of Veterinary Medical Ethics of the AVMA. American Veterinary Medical Association.

Bateman SW 2007 Communication in the veterinary emergency setting. *Veterinary Clinics of North America-Small Animal Practice* 37: 109–121.

Beauchamp TL, and Childress JF 1994 *Principles of Biomedical Ethics.* Oxford University Press: New York.

BSAVA 2015 Informed consent. British Small Animal Veterinary Association. https://www.bsava.com/Resources/BSAVAMedicinesGuide/Consent.aspx

Fettman MJ, and Rollin BE 2002 Modern elements of informed consent for general veterinary practitioners. *JAVMA-Journal of the American Veterinary Medical Association* 221: 1386–1393.

Flemming DD, and Scott JF 2004 The informed consent doctrine: what veterinarians should tell their clients. *JAVMA-Journal of the American Veterinary Medical Association* 224: 1436–1439.

FVE 2002 Code of Good Veterinary Practice. Federation of Veterinarians of Europe: Brussels, Belgium.

Gorrel C 2013 *Veterinary Dentistry for the General Practitioner.* Saunders, Elsevier: Edinburgh.

Hatch PH, Winefield HR, Christiec BA, and Lievaart JJ 2011 Workplace stress, mental health, and burnout of veterinarians in Australia. *Australian Veterinary Journal* 89: 460–468.

HMSO 2006a Animal Health and Welfare (Scotland) Act.

HMSO 2006b Animal Welfare Act (c.45).

McCulloch S, Reiss M, Jinman P, and Wathes C 2014 The RCVS codes of conduct: what's in a word? *Veterinary Record* 174: 71–72.

O'Neill O 2003 Some limits of informed consent. *Journal of Medical Ethics* 29: 4–7.

Passantino A, Quartarone V, and Russo M 2012 Informed consent in Italy: its ethical and legal viewpoints and its applications in veterinary medicine. *ARBS Annual Review of Biomedical Sciences* 14: 16–26.

RCVS 2015a Code of Professional Conduct for Veterinary Surgeons. Royal College of Veterinary Surgeons. http://www.rcvs.org.uk/advice-and-guidance/code-of-professional-conduct-for-veterinary-surgeons/pdf/

RCVS 2015b Supporting guidance: Communication and consent. Royal College of Veterinary Surgeons. http://www.rcvs.org.uk/advice-and-guidance/code-of-professional-conduct-for-veterinary-surgeons/supporting-guidance/communication-and-consent/

Yeates J, Everitt S, Innes JF, and Day MJ 2013 Ethical and evidential considerations on the use of novel therapies in veterinary practice. *Journal of Small Animal Practice* 54: 119–123.

Yeates JW 2016 Ethical principles for novel therapies in veterinary practice. *Journal of Small Animal Practice* 57: 67–73.

Yeates JW, and Main DCJ 2011 Veterinary opinions on refusing euthanasia: justifications and philosophical frameworks. *Veterinary Record* 168: 263.

CHAPTER 10

EDUCATION AND TRAINING

Veterinarians must undertake extensive tertiary education, and in some cases must sit external examinations before they can register to practise. Similarly, many nurses and technicians undertake training to equip them with skills and qualifications to perform their jobs.

Veterinary training must equip students with a knowledge and skills base that allows them to achieve "Day 1" competencies and be eligible for registration with the appropriate veterinary board or authority.

In addition, veterinary professionals may take extensive post-qualification or post-graduate training. In some cases, this is a requirement of ongoing registration or workplace requirements, but additional education and training may be taken as a means of further specialisation, transitioning into a new field or returning to practice after a hiatus.

Education and training may be formal, through courses, workshops or graduate programmes, or informal, for example, on-the-job training to improve skills in a particular area, such as ultrasound.

In addition to direct impacts on animals used, education may have indirect impacts on the welfare of animals. According to James Yeates, for example, "Educators can… cause wider iatrogenic welfare harms through their recommendations, for example,

where lectures or textbooks promote overtreatment or suggest overly inflexible protocols that do not allow practitioners to adequately consider each individual patient's welfare" (Yeates 2013).

In this chapter we explore ethical dilemmas arising through education and training, including those associated with animal use, and the potential impacts of education and training on patients, clients and communities.

10.1

Animal use in education and training

Veterinarians and associated professionals are required to be competent at performing certain tasks on their first day of work (in some countries these are formally called "Day 1 Competencies"). Developing competence requires exposure to situations to learn and practise those skills, but what sort of exposure and how much? Veterinary curricula, for example, incorporate variable amounts of animal use, from the dissection of cadavers to learn anatomy, through the use of live animals to practise surface anatomy and clinical examination, to utilising live patients on which to practise surgical and other clinical skills.

◁

10.1

PHOTO ANNE FAWCETT

The harm of animals in education and training is traditionally justified on utilitarian grounds, accepting that the use of animals in training may incur some degree of harm, but that this is a "lesser evil" when compared with harming future patients (and clients). Increasingly, the 3Rs – replacement of animals where possible, reduction of the number of animals used and refinement of technique to minimise harm – originally designed as guiding principles for experiments involving animals, are applied to teaching.

The use of animals in education and training is becoming more closely regulated in many countries (Lairmore & Ilkiw 2015). It may be a requirement that those proposing animal use in teaching must now demonstrate the absence of alternatives. Nonetheless, for some people the use of animals remains objectionable and goes against their values. These students may raise a conscientious objection to some or all animal use in a teaching programme. The following scenario explores conscientious objection.

△

10.2 Use of cadavers and live animals in teaching can be controversial.
PHOTO ANNE FAWCETT

SCENARIO
STUDENT CONSCIENTIOUS OBJECTION

▶ You are a technician working in a university veterinary faculty. In the preclinical years, students learn anatomy by undertaking lectures as well as dissections of cadavers sourced from a pound. The animals are not euthanased for the purpose of teaching.

As part of your role you are on the faculty's teaching committee. At the beginning of the semester, two students inform the committee that they cannot participate in the classes as they conscientiously object to the use of cadavers in teaching.

How should you respond?

RESPONSE
MANUEL MAGALHÃES SANT'ANA

▶ The ethical quandary posed by this scenario can be summarised by the following question: can the duty of providing anatomy hands-on training to veterinary students be subdued by their right to object in taking part in those procedures? The use of live animals in surgical training can suggest the same apprehension, therefore it is important to state clearly that, contrary to live animals, cadavers do not have *interests* (let alone *rights*) and cannot be considered as relevant stakeholders (cf. chapter 2). Altogether, the rationale used for analysing conscientious objection to using live animals for teaching purposes may not be adequate for cadavers.

The first step for analysing the scenario is to investigate the legitimacy of the students' objection. Repugnancy and distress at handling animals' remains (organs, tissues, blood) can be claimed; students may also believe that it is against their *ethos* to dissect what was once a sentient being.

These claims can be perceived as valid personal vindications but they cannot justify conscientious objection since they do not necessarily reflect an ethical concern for the sourcing of the cadaver. Nonetheless, the stated reasons are still relevant since they should promote a reflection on the required skill set for future veterinarians.

Being able to recognise anatomic structures from all common domestic species is amongst the most important competences for a veterinarian. This applies for virtually every veterinary role, including small and large animal practitioners, inspectors and researchers (amongst many others). Accrediting organisations such as the American Veterinary Medical Association and the

△
10.3 Being able to recognise anatomic structures from all common domestic species is amongst the most important competences for a veterinarian.
PHOTO ANNE FAWCETT

European Association of Establishments for Veterinary Education establish that graduation can only be awarded after exposing students to relevant hands-on training with (both living and dead) domestic animal species. Although alternatives to the use of animals such as computer-based resources, mannequins and haptic simulators can work as adjuncts to anatomy teaching, dissecting (live) tissues from (dead) animals is still a fundamental veterinary procedure and a prerequisite for qualifying as a veterinarian all across the world. In sum, the main ethical question is not if cadavers should be used in anatomy teaching but whether the use of those specific cadavers is justified (including the numbers involved).

This brings us to the second issue to consider: is the pound an ethical source of cadavers? The fact that animals were not killed for the purpose of teaching may not be enough reason for considering the source as ethical. A recent report on greyhound racing in Great Britain (GREY2K USA 2014) showed that hundreds of unwanted greyhounds from the UK dog racing industry had been bought by British and Irish veterinary schools and used in anatomy teaching. Cases of euthanasia of healthy greyhounds in private veterinary clinics and subsequent trade with veterinary schools have also been reported. Even though veterinary schools rejected the allegation that these animals were euthanased *specifically* for teaching purposes (since the dogs would have been euthanased anyway, had the schools not purchased them), this is still a case where more could have been done in terms of ethical sourcing. In effect, this kind of exposition impacts the reputation of the veterinary profession by suggesting (to students, and also to the public) that veterinary schools support the status quo regarding the use of animals in sport.

A similar rationale can be applied to pound dogs; using surrendered or unclaimed pound animals (that are otherwise destined for euthanasia)

for their anatomy classes, might give the impression that veterinary schools support the systematic killing of stray companion animals. The belief that using stray dogs for anatomy teaching is an unacceptable instrumentation of vulnerable animals that are in most need of care and compassion could provide a valid argument for conscientious objection. Nevertheless, it might be the case that the pound professes a no-kill policy, which may not prevent performing euthanasia in extreme circumstances, such as terminal or incurable illnesses and in the case of dangerously aggressive dogs. These animals would have been killed for their own benefit and/or for the benefit of society and not purposively for teaching, thus making the sourcing ethical and conscientious objection probably inadmissible. The same would apply to farm animals, which might have been humanely slaughtered within the food chain, and some of their organs or body parts redirected to anatomy teaching. Matters such as these should be explained to the students, in order to help them make a more informed decision.

In order to claim conscientious objection, veterinary students must present their arguments addressing the ethics of animal sourcing before the teaching session, and preferably at the beginning of the semester. This should be done in a safe and supportive environment. Some schools have a student support scheme or students' advisors. If such support does not exist, a tutor of their confidence should work as a facilitator between the students and the teaching committee. In turn, the teaching committee should be able to track back the sourcing of the cadavers and provide evidence of the steps taken to ensure that it is in accordance with the principles of the 3Rs (Russell & Burch 1959). This dialogical process will hopefully support students' decision to participate in anatomy classes, make them reflect on whether they have the skill set needed to pursue veterinary training, or else promote a change in the sourcing of cadavers for teaching purposes, if and where applicable.

In conclusion, if students are allowed to conscientiously object, that should not mean they can progress in the course without performing anatomical dissections; it means that the school must work around the objection and provide an alternative teaching session with ethically sourced cadavers or body parts. On the other hand, even if the teaching committee rejects the legitimacy of the objection, students should still be encouraged to suggest alternatives, in case they believe that more can be done in terms of ethical sourcing [e.g. donated cadavers (Tiplady, et al. 2011)] and in reducing the number of cadavers used. To inform the future decisions, clear guidelines as to the use of animals in teaching (grounded on the principles of the 3Rs) should be developed and presented to prospective students before they are allowed to enrol in veterinary training (Knight 2014, Whittaker & Anderson 2013).

As a way of assisting future ethical decision-making, an algorithm for conscientious objection to the use of cadavers in teaching is suggested (Figure 10.4).

||||||||||||||||||||||||||||||||||||||||||||||||||||||||||||||||||

THIS scenario illustrates the application of the "3Rs" in reviewing animal use in teaching. The 3Rs framework is utilitarian, essentially prompting a cost:benefit analysis and aimed at maximising knowledge ("good") and minimising harm (Graham & Prescott 2015). As the author points out, the key ethical issue here is the sourcing of these animals, as this could potentially impact on the interests of animals while they are living. For example, if animals are euthanased for the sole purpose of use in teaching, we may consider them to be harmed directly by that teaching if death does in fact harm the interests of an animal. However, if – as in many contexts – there is no shortage of animals killed for other purposes, it may be seen as "good" to maximise potential benefits by using these animals for teaching.

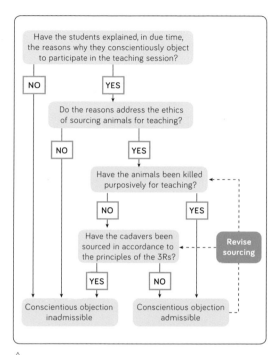

Have the students explained, in due time, the reasons why they conscientiously object to participate in the teaching session?

NO | **YES**

Do the reasons address the ethics of sourcing animals for teaching?

NO | **YES**

Have the animals been killed purposively for teaching?

NO | **YES**

Have the cadavers been sourced in accordance to the principles of the 3Rs?

Revise sourcing

YES | **NO**

Conscientious objection inadmissible | Conscientious objection admissible

△

10.4 Algorithm for conscientious objection to the use of cadavers in veterinary teaching.
SOURCE MANUEL MAGALHÃES SANT'ANA

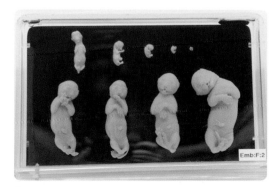

Emb:F:2

△

10.5 The purpose for which an animal is killed is ethically relevant.
PHOTO ANNE FAWCETT

In designing curricula, education and training institutions must consider requirements for course accreditation, compliance with animal welfare legislation and codes and also anti-discrimination legislation (Knight 2014).

Far more contentious than the use of (ethically sourced) cadavers in teaching is the use of live animals. Such use may be minimally invasive (for example, use of cats to teach principles of feline physical examination, the use of horses to teach equine surface anatomy), moderately invasive (for example, the use of cattle to teach pregnancy detection via rectal palpation) or majorly invasive (performing surgery on animals).

Students may experience moral stress when forced to choose between harming or killing an animal to further their own careers, or potentially forfeiting their career (Knight 2014). According to Knight:

"although when put to the test a conscientiously held belief should ordinarily be combined with a willingness to incur personal discomfort or suffering or material loss, the essence of non-discriminatory principles is that no student should be required to incur such losses, as a result of conscientiously objecting to participation in any nonessential educational activity. For the reasons provided previously, this clearly includes teaching or assessment activities involving harmful animal use."

(Knight 2014)

Increasingly sophisticated models and alternatives to live animals are available, with studies showing that non-harmful teaching methods were as effective as, if not more effective than, the use of live animals (Lairmore & Ilkiw 2015). Benefits of using humane teaching methods include time and cost savings, the ability to customise the learning experience, repeatability of learning, increased student confidence and satisfaction, compliance with animal use legislation, elimination of objections to use of purpose-killed animals and integration of clinical studies with ethics in the curriculum (Lairmore & Ilkiw 2015).

Certainly the use of humane alternatives in education and training is aligned with the values professed in many veterinary oaths. If the 3Rs are applied strictly, the use of animals should be completely replaced in education and training if humane alternatives provide the same or better educational outcomes.

ANIMAL USE IN EDUCATION AND TRAINING

What would you do?

Each year your university buys in a few weaner piglets for animal handling practice, keeping them for a few months and selling them for slaughter at the end. This year, one of your students wants to rescue the piglets, saying they would never have signed up for the course if they knew animals would be used like this. Somehow this storm has made it to the national newspapers. What should you do?

△
10.6 One of your students wants to rescue the pigs that have been used for animal handling practice.
PHOTO ANNE FAWCETT

10.2

Volunteering

Many students and graduates learn by volunteering. On the face of it, volunteering is a win-win scenario – others benefit from time given pro bono, and the volunteer benefits by gaining knowledge, practical skills and life experience. But if communication is not clear and care is not taken, volunteering may not be entirely benign.

SCENARIO

VOLUNTOURISM

▶ The veterinary school where you work is keen to facilitate students in gaining useful "extra-mural experience" abroad. It is anticipated that foreign placements will broaden the students' experiences and will allow them to see cases that are almost unheard of in developed countries as well as gaining valuable surgical experience.

An overseas animal charity which actively promotes volunteering opportunities for veterinary students has offered places on their programme to two of your final-year veterinary students. You become aware of this when reviewing the students' plans for seeing practice. You discuss these with the students and can see that they are very keen to develop and use their skills to help the stray animals in a developing world country. They have been told that they will be working on a programme that will involve the mass vaccination and neutering of owned and stray animals in a field surgery.

You establish that they will receive some supervision from the charity's in-country staff and satisfy yourself that the technical learning to be gained from the experience appears to have been considered. You are, however, concerned that there has been little or no thought given to the ethics of the project or the ethical learning for the students involved.

RESPONSE
GLEN COUSQUER

▶ Veterinary voluntourism is a little-researched and poorly understood phenomenon that can be compared with other forms of voluntourism, including medical voluntourism. As these phenomena have grown, they have been subject to increasing critical scrutiny (McLennan 2014). This has, in turn, drawn attention to a number of serious consequences, including unsafe practices, inadequate training and lack of sustainability, not to mention poor communication with, and understanding of, local communities.

Emboldened by a desire to "do good" and equipped with new knowledge and skills that need to be put into practice and developed, it is all too

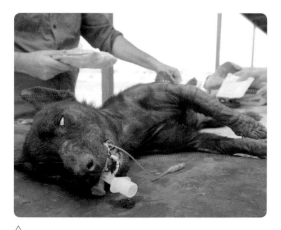

△
10.7 Some desexing programs involve field surgery. This dog with severe mange is being prepped for ovariohysterectomy and administration of anti-parasitic agents.
PHOTO ANNE FAWCETT

easy for the veterinary student on the threshold of their professional career to rush out into the field and get down to work. Faculty staff who fail to recognise and address ethical aspects of the curriculum are complicit in this oversight. As McLennan (2014) points out, "many volunteers come with limited experience of working in developing world environments and with little knowledge of medicine outside the context in which they have been trained." A well-meaning, but ill-considered, rush into the field is thus inherently problematic.

It is all too easy to overlook the power and privilege that allow Western students to impose interventions on those (animals and communities) whose views, concerns and fears may go unheeded. According to Simpson (2004):

"The processes that allow young Westerners to access the financial resources and moral imperatives necessary to travel and volunteer in a third world country are the same ones that make the reverse process almost impossible."

Failure to recognise this power imbalance leaves volunteer programmes open to claims that they are built on the structures of colonialism. McLennan (2014) identifies the themes of neo-colonialism and paternalism as recurring ones within her study of medical voluntourism in Honduras. Paternalism in this case is defined as "the process by which providers intentionally confer a treatment or service upon a person without their consent".

It is worth noting that there are many concepts in the scenario described above that should be afforded closer scrutiny. These include:

(1) The nature of the "good", which all too often appears to be parachuted in from outside. How has this good been determined? For whom, how, where and when is it a good?

(2) What is understood by the terms "owned and stray animals" in these communities?

10.2
VOLUNTEERING
RESPONSE
GLEN COUSQUER

These concepts are often quite different in other cultures where companion animals have a quite different profile and status.

We should not forget either that mass sterilisation of humans was, within living memory and with the support of NGOs, imposed on rural communities in countries such as India with little consideration of the ethics involved. The hazy line between want and need and between coercion and consent was never addressed and, in some cases, people were literally rounded up and subjected to procedures and treatments that they were not in a position to consent to (Citrin 2010, Menon 2003, Rao 2004, Tarlo 2003).

The need for critical theory and a pedagogy of social justice

An unquestioning presumption that one knows best led certain people to undertake these human sterilisation programmes. The same may be true of animal sterilisation campaigns. That this is so highlights the need for a pedagogy of social justice within all volunteer programmes (Simpson 2004).

In the case of veterinary voluntourism, a valuable opportunity exists for the students involved to gain exposure to critical literature and undertake valuable ethical, as opposed to technical, learning.

In particular, they should consider how health becomes medicalised and how vets can abuse their power and status.

According to Whyte, et al. (2003), the medicalisation of health occurs when medicine is used to solve problems that should be addressed in other ways, when the conditions that foster poverty, scarcity, sickness and suffering are understood in the wider contexts in which they occur.

A useful starting point

In an attempt to address these complex issues and in recognition of the finding that "positive outcomes are not a natural consequence of

△
10.8 Many free-roaming dogs in communities are owned.
PHOTO ANNE FAWCETT

voluntourism, but must be nurtured" (McLennan 2014, p. 164), it is suggested that students:

(1) carefully consider their own motivations.
(2) undertake a detailed ethical study prior to undertaking the placement.
(3) undertake a review of the work undertaken upon their return.
(4) share and discuss this learning with past and future participants in the programme.

This approach will allow some of the consequences of "short-term voluntourism" to be addressed. In particular, it addresses the lack of time and space set aside for critical reflection and learning and deliberately sets out to create an opportunity for volunteers to truly engage with and learn about the place, the people and the animals they are there to serve (McLennan 2014).

In order to frame the moral dimension of the students' learning, it is suggested that they engage with the framework for evaluating the effects of a voluntourism project proposed by Scheyvens (2011).

Scheyvens' framework for voluntourism

In this framework, six perspectives are proposed. These are:

(1)   Voluntourism as harmful.
(2)   Voluntourism as egocentric.
(3)   Voluntourism as harmless.
(4)   Voluntourism as helpful.
(5)   Voluntourism as educational.
(6)   Voluntourism as social action.

Students should be challenged, at the outset, to reflect on their motivations. This provides an opportunity for them to recognise that, in yielding to their desire to travel and to have an authentic experience, in undertaking a "self-serving quest for career and personal development" (Devereux 2008), their motivations are primarily egotistical.

This, in turn, provides students with an opportunity to recognise that what they may consider to be a life-changing experience is likely to be "based on an emotional response to a situation they do not really understand". Recognising this can challenge them to develop a more detailed insight into the causes of poverty and deprivation and what can be done to alleviate them (McLennan 2014).

Where voluntourism becomes educational, it has the potential to promote a better understanding of the causes of poverty, deprivation and poor health – both human and animal – and can ultimately lead to social action that seeks to address some of these causes.

This scenario highlights the need to create time and space for the ethical learning that can arise from a placement in the developing world. Students should be encouraged to develop their own understanding of the many socio-economic causes of poor health and welfare and to share this with the communities wherever possible. In doing so, veterinary voluntourists can come to understand the limitations of "band-aid medicine" in which the same activities are repeated again

and again. This, in turn, opens up possibilities for more sustainable interventions that address some of the root causes of the problems requiring veterinary treatment. This represents a move away from egocentric voluntourism and towards voluntourism as education and social action.

|||||||||||||||||||||||||||||||||||||||||||||||||||||||||||||||||||||||||||||||

VOLUNTEER tourism, or voluntourism, typically combines travel and volunteering, in social, economic development or conservation oriented projects (Sin, et al. 2015). In a veterinary context, voluntourism programmes are often based around conservation, provision of emergency aid or neutering and vaccinating animals in an effort to improve animal welfare as well as public health. Veterinary voluntourism is based on the assumption that efforts of volunteers will benefit humans and animals in host communities – but a good outcome is not guaranteed.

The literature around voluntourism is particularly interesting from an ethical point of view because it highlights that good intentions are not enough:

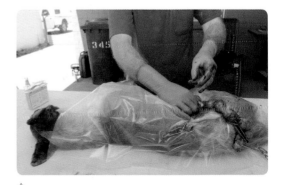

△
10.9 Field surgery brings its own challenges, such as achieving asepsis in field conditions. This bitch is pregnant which makes the surgery more difficult.
PHOTO ANNE FAWCETT

sometimes by doing "the right thing", harm may be done to individuals, or a community.

There are four key themes addressed in current academic research on voluntourism:

(1)   The pre-trip motivations of volunteers compared with those of mainstream tourists;
(2)   Whether voluntourism is motivated by altruism or self-interest;
(3)   Impacts and outcomes of voluntourism on host communities;
(4)   Impacts and outcomes of voluntourism on the volunteers themselves (Sin, et al. 2015).

For example, one criticism of voluntourism is that the key beneficiaries are in fact the volunteers who achieve a sense of accomplishment, self-worth and something to add to their résumé, rather than members of the host community.

A principalist framework is well suited to discussing the ethics of voluntourism. According to the principal of non-maleficence, volunteers should seek to minimise or mitigate any potential harm. In a veterinary clinical context this involves minimising harm to animals during capture, anaesthesia and surgery and understanding the ecology of the dogs in the region.

For example, because many of the animals in such contexts are free-roaming, they can be challenging to capture. The capture itself may be distressing, impacting the animal's welfare (its interaction with humans may be forever changed; also injury can occur when animals are overly restrained). This can also be very risky for volunteers and as such, dog handlers must be experienced.

One medical author suggests steps to minimise harm include focusing on one country or region (rather than simply flying into different destinations then flying out again before any meaningful or sustainable contribution is made), learning the local language, learning about local health problems, learning about traditional and introduced health-care systems and respecting local cultural norms (Bezruchka 2000).

In a veterinary context, one must consider that clinical inexperience may increase anaesthetic risk and surgical trauma, in a field setting where there are typically few resources to manage complications (for example, intravenous fluids to treat hypovolaemia or antibiotics to treat post-operative infections). It is essential to consider the supervision and resources available and – if students are to perform surgery – community members should be informed, consulted and their understanding and support solicited.

Beneficence, or doing good, is ensured by working with the community to address areas of actual rather than perceived need, ensuring that all treatments are performed to the best standard permitted by circumstances, and that details of all patients and (where possible and applicable) owners are recorded so that this information can be used to measure outcomes. In implementing the principle of autonomy it is important that consent is obtained. Volunteers should accept refusal of their services. Where there is a large stray population it is not possible to obtain owner consent, but community consultation is essential to ensure that such a programme is indeed welcome. Increasingly, veterinary organisations that take volunteers have established relationships with communities.

Programmes that have a negligible or negative impact (for example, rebound of local dog population, too many volunteers injured, too many post-surgical complications) should be reviewed and placed on hold until such issues can be addressed. One-off programmes are rarely successful. There needs to be a commitment by the organisation to revisit the same community to ensure that a programme is sustainable and not simply tokenistic.

It is important to acknowledge that volunteering itself may be very beneficial to volunteers, in ways that impact host communities, current and future patients and colleagues. For example, through participation in the Community Veterinary Outreach program in Canada, "students have challenged and redefined preconceptions of those who are homeless, and their respective pet ownership. Students have a self-identified need to exercise more compassion and empathy towards people regardless of circumstance" (Jordan & Lem 2014).

The promotion of autonomy of clients of voluntouring programmes centres around respect for their values and providing sufficient information that they can agree to partake or not without coercion. This may be more problematic than it first seems in situations where the only realistic availability of veterinary care is through a voluntourism programme. Here, clients do not have the luxury of being able to operate an autonomous choice of veterinary providers that accord with their own values.

Finally, in terms of justice, it is important – as discussed – that the reasons for underlying animal overpopulation, such as poverty, are identified and explored. Although this may not be feasible within the remit of a voluntourism session, organisations involved in voluntourism should also be actively seeking more just solutions for the animals and owners they seek to help.

VOLUNTEERING
What would you do?

A veterinary charity has a programme of providing free equine healthcare for low-income owners in South America manned mostly by volunteers from the USA. What ethical issues would you consider before deciding whether to go? What questions would you need to ask the charity?

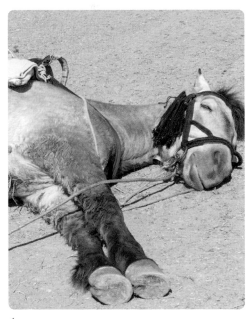

10.10 What factors would you take into account when deciding whether to volunteer to provide free equine treatment for low-income owners abroad?
PHOTO ISTOCK

10.3

Moral development and professional identity formation

—
SCENARIO
STUDENT OMISSION

▸ You are a senior student on your clinical rotation paired with another classmate to work with a famous faculty member. It is your first day, and the teacher sends you both to do a history and physical exam on a patient. When you return, the attending asks you to present the history and your classmate to present the exam. During the physical exam presentation, the attending asks about the neurological exam, which you both forgot to do. As you feel embarrassment developing from within your gut when you realise your oversight, your classmate states it was "normal" and continues on with the physical exam presentation.

What, if anything, should you say or do?

—
RESPONSE
WALTER LEE AND NIKITA CHAPURIN

▸ This situation involves the relationship between four entities – namely the student and their classmate, the team's patient, and the famous faculty member. The ethical dilemma that the student is facing comes from a number of issues that include: having good interpersonal relations and trust between classmates, being subjectively evaluated by a faculty member, finding themselves complicit in a false statement of performing something that was not done, and the

△

10.11 What happens when a neurological examination is omitted?
PHOTO ANNE FAWCETT

potential harm to the patient from inaccurate clinical evaluation. We can utilise a variety of ethical frameworks in approaching this dilemma.

We propose that a virtue ethics approach provides the framework that can bring about a meaningful resolution that addresses the relationships involved. Unlike the consequentialist ethical models that focus on the end result (e.g. the impact of our actions on the patient's care, consequences of bringing up the fact a neurology exam was not performed), a virtue-based approach emphasises the effects on an individual's character rather than consequences of one's actions. In other words, virtue ethics reckons the nature of our character and how we exhibit those virtues as of fundamental importance when making decisions (Gardiner 2003). Responsibility, discernment, trustworthiness and compassion are four virtues that are relevant for consideration in this case and may be used in resolving similar dilemmas.

Finding the virtuous way forward

Firstly, the student should act with responsibility towards the patient, which implies doing everything necessary to provide the best care for the patient. This should involve a serious consideration on whether failing to perform a neurological exam and stating that it was "normal" directly compromised the animal's care. For example, for a patient presenting with neuromuscular symptoms, neurological examination is essential in proper diagnosis and management. If it were likely that care was compromised, a responsible clinician would go and perform the neurological exam and report it to the faculty mentor, who could then act on the information regardless of the consequences.

However, in making this judgement, the student should strive to show discernment, which involves our ability to make sound decisions. This would involve formulating judgements without undue influences such as fear of harming personal relationships, or secondary gains such as attempting to appear "better" than the student who made the mistake. In this instance, the need to point out that a neurological exam was not done on the patient may come at the expense of upsetting a classmate or a lower evaluation grade from the faculty member. However, while personal relationships are crucial in medicine and we must make an effort to foster them, they should never come between us and doing what is right for the patient's welfare.

If the neurological exam was not pertinent for the patient's evaluation, or in fact normal, the student may be tempted to write this off as a harmless "white lie" in an attempt to save face in front of the famous faculty member. However, if we utilise the virtue ethics approach, we must place a much greater emphasis on character and the type of clinician one aims to be. Here, the virtue of trustworthiness plays a role. For example, a good clinician will strive to be trustworthy and report history and physical exam findings as they are. While one

action will not make one "not virtuous", one must strive to do our best in accordance to our moral values. In this scenario, showing trustworthiness should also entail approaching the classmate to determine if they recognise the mistake, as well as being upfront with them in that their misrepresentation of the exam as "normal" made you feel uncomfortable.

The intent behind the actions, rather than the actions themselves, is central to virtue ethics. The intent to approach the student is not meant to be punitive, but rather to provide an opportunity to grow professionally. The conversation with the other student should be done in a manner that provides an opportunity for both an understanding as to how this incident unfolded, as well as discussion on how their actions reflect on the student's and classmate's character. It may be decided after this discussion that it would be best to approach the famous faculty member to explain what had happened.

More specifically, reflection and discussion of the incident are important for resolution of this dilemma and preparation for similar future situations. By understanding the factors that led to the false response, and understanding how this behaviour reflects on one's character, the other student may self-correct professional habits early in training. This is especially important since mistakes like these in students may predict future behaviour in them as practising clinicians. Indeed, research with medical students has shown that problematic behaviour as early as in medical school is associated with subsequent disciplinary actions by state medical boards (Papadakis, et al. 2004).

One must not overlook the importance of the teacher and how they can support or suppress an environment conducive to character development and professional identity formation. This is because one of the goals of professional education is promotion of professional identity formation.

This is especially true in the medical field, where we are held to a high moral standard and expected to act in the best interest of our patients. For this reason, virtue ethics is oftentimes taught and applied throughout training (Gardiner 2003, Lee, et al. 2012).

Serving as an educator, the famous faculty member should reflect on how one provides a safe and an open environment for professional growth. For example, if both students agree to approach the faculty member to explain what happened and ask for forgiveness, how the faculty member responds is critical to the professional identity formation of the students. Consider if the faculty member responded by recommending suspension for lying. How would this response impact building a healthy environment for meaningful professional identity development? It may suppress any further actions by students to confess mistakes and any opportunity for formative character building. In contrast, if the faculty expressed disappointment about the lie, but also recognised how brave it was to admit a mistake, and subsequently granted forgiveness, the impact on the character of the student and classmate would likely be substantial. This would be an opportunity for a "teaching moment" that could have a tremendous impact on the students' professional identity formation. Clearly, repeated offences must be addressed, since such habitual issues reflect poorly on one's character. All educators should seek discernment on what is the appropriate response to make in specific situations, especially in the context of how it impacts professional identity formation.

To conclude, the virtue ethics approach focuses on one's moral character. Striving to show responsibility to the patient, one must think first about delivering appropriate care and showing discernment in judgement. Furthermore, as a future clinician that aspires to be trustworthy, the student should approach the classmate about the misrepresentation of the exam, but in a manner that focuses on how these actions reflect on one's character. Furthermore, those in teaching roles should seek to provide a learning environment that supports meaningful professional identity formation.

|||||||||||||||||||||||||||||||||||||||||||||||||||||||||||||||||||||||||

THE authors mention that if the neurological examination was not essential to the case, the student may be tempted to write off its omission as a harmless "white lie". But even a consequentialist approach might challenge this. In relation to errors in human healthcare, and indeed in the field of airline safety, there has been an increasing trend in documenting "near misses" on the grounds that addressing factors that lead to a "near miss" in one case may avert iatrogenic morbidity and mortality in future cases (Powell, et al. 2010).

The authors also consider the intention behind the decisions in this case. The decision to lie is an attempt to save face – it prioritises ego over patient welfare. Admission of an error or omission can be confronting, and it may be tempting to lie or cover up errors to avoid adverse consequences or judgement by a colleague or client. However, preferencing one's ego or interests over those of a client really contradicts what professionalism stands for. As Freidson noted, "the ideology of professionalism asserts above all else devotion to the use of disciplined knowledge and skill for the public good" (Freidson 2001) – not private advancement.

In terms of the role of teachers, in the veterinary practice setting, veterinarians may be in a leadership position and can consider how they can support "an environment conducive to character development and professional identity formation" suggested by the authors of the above scenario.

MORAL DEVELOPMENT

AND PROFESSIONAL IDENTITY FORMATION

What would you do?

As a final-year veterinary student you are respon-sible for checking and medicating certain inpa-tients. On one occasion you administer 10 times the stated dose of penicillin to a cat. You do not realise this until the next day, at which point the cat is clearly fine. What do you do?

△

10.12 What should a student do who later discovers they gave a penicillin overdose to a cat with no ill effect?

PHOTO ISTOCK

10.4

Learning from patients

It is commonly stated that a veterinarian is only as good as his or her caseload. We cannot help but learn from our patients, whether it's gaining experience in treating common conditions, learn-ing the hard way that uncommon things occur more commonly than we might expect, or that some patients and diseases contradict or chal-lenge the textbooks. It is important that veterinar-ians and animal health professionals learn from their patients, but even here we need to exercise caution. The following scenario explores some of the ethical challenges involved.

SCENARIO

BEYOND SKILL SET

▶ You see a dog, Pepper, weighing 30 kg, with a suspected ruptured cruciate ligament of the right hind leg. The other vet in your practice has only ever performed an extra-capsular stabilising

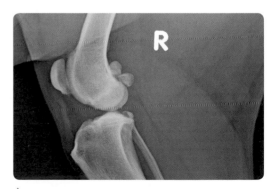

△

10.13 In what conditions is it appropriate to attempt a procedure you have not performed before on a patient?

PHOTO ANNE FAWCETT

technique for over 20 years with what she considers to be reasonable success, although you haven't conducted a clinical audit. In your year since graduating you have become quite a keen surgeon and have performed some orthopaedic work, including fracture repair. You remember that during your final-year rotation in orthopaedics some of the clinicians strongly recommended the Tibeial Plateau Levelling Osteotomy (TPLO) as the surgical treatment of choice for such cases, and indeed you were able to observe one and suture up the skin afterwards. You are considering orthopaedic surgery as a route for specialisation and are keen to try out a TPLO on Pepper. You have discussed referral for the case, but the client has limited finances to afford the referral costs and you are considering undertaking the procedure for the first time.

What should you do?

RESPONSE

MARTIN WHITING

▸   The case involves a common condition for which there are several well-established treatment options; the patient has an injury that requires an intervention to restore them but which intervention will be selected? There are many facets to such a question. Firstly, which intervention is in the patient's *best* interest – this will require an examination of the evidence-based data for the different procedures, their chances of a successful outcome or failure and an analysis of clinical indications for each. This is largely an empirical question and the data for this can be presented factually to the client. The pros and cons of a TPLO vs the extra-capsular technique are discussed elsewhere (Bergh, et al. 2014, Conzemius, et al. 2005, Taylor-Brown, et al. 2015): this scenario uses cruciate repair as an

example to highlight the ethical dilemma of when it is acceptable to attempt to expand one's skill set far beyond our current competency, it is not meant as a review of the two techniques, which are highly debated. Secondly, client choice is paramount. Ensuring the client is part of the decision-making process, and is offered all treatment options and their associated risks, harms and benefits, is a critical part of informed consent. This must include the costs of treatment, ongoing care and the costs associated with failure. The process of informed consent may mean the client elects for a procedure that isn't the "gold standard". Selection of an option that is in Pepper's interest, but not best interest, is perfectly acceptable. Clients will often have limitations on them that prohibit their selection of the best option, but this will form part of the informed consent discussion. The topic of informed consent is dealt with in chapter 9.

Let's assume at this stage, the harms, risks and benefits of each procedure have been explained to the client. The client is knowledgeable and readily understands each intervention and its merits. Some surgical interventions require a great deal of surgical experience and skill to be able to achieve the published success rate. The statement that "procedure A has a 90 per cent chance of success and procedure B has only a 60 per cent chance of success" needs to be qualified by who undertakes the procedure. It could be that even the most novice surgeon could achieve a 60 per cent success with procedure B, while only the elite surgeons will achieve a 90 per cent success with procedure A (these numbers are for illustrative purposes only to highlight how surgical outcome can be dependent on surgical skill and experience). Those of us without elite skills and attempting procedure A may have a success rate well below this. Pepper's chance of a successful outcome following surgery is then highly dependent on not just which procedure is selected, but also the competency of the surgeon undertaking the procedure. The client is now trying to decide between the procedures. One option will be the extra-capsular technique and a second option will be to refer to a specialist for TPLO. The heart of this ethical dilemma is perhaps, should new graduate veterinarians even offer to attempt a complex procedure they have not performed themselves?

It does not come as too much of a surprise that most veterinary professional regulators require their members to "stay within their own area of competency". However, this seems in direct conflict with the principle of continuing professional development! If we take the first requirement, it seems unequivocal that the veterinarian should refer Pepper to someone who is more competent at undertaking the desired procedure (or that an extra-capsular technique should be offered if finances are problematic). If we take the second requirement in isolation, one could argue that it is a necessity for the new graduate to undertake the procedure so as to improve their skills. A purely deontological approach, which follows the "rules" of professional regulation, can lead us into difficulty where these rules seemingly conflict.

This scenario is a really nice example that highlights the differences between the ethical theories explored in chapter 2. For example, if we are to take a utilitarian view, we would need to calculate a cost:benefit analysis of all the affected parties to determine if the greatest good for the greatest number could be achieved by the new graduate undertaking the procedure. The most cursory analysis shows this as positive for the veterinarian who will learn new skills, the practice who will earn the money from the surgery, the client may also benefit if it is successful as it will likely be cheaper than a referral surgery and the dog will probably be the major party who does not benefit. Even if the procedure is successful there is likely to be an increased morbidity associated with a less experienced surgeon undertaking the procedure; if it is an unsuccessful procedure then the dog may suffer a great deal. This analysis does not take into account the requirement to stay within the veterinarian's area of competency.

Perhaps the scenario could be considered under virtue theory; where the virtue of being a good veterinarian should correspond to societal expectations. But this is also confusing as a virtuous veterinarian is one who would refer complex cases beyond their competency to a more experienced veterinarian and they would *also* seek to expand their skill set and increase their practical knowledge.

Staying with virtue theory, there could be a way forward that is a "win-win" resolution for all affected parties, including the referral centre. It is an appropriate, and noble, endeavour to increase your skill set; not only will this improve your ability to treat similar animals, but it will provide wider, more generalised, associated skills (e.g. generally improved surgical technique). Also, by learning and gaining competency in new and interesting procedures the veterinarian can flourish as a "lifelong learner" and induce a greater personal reward from the vocation. The veterinarian in this scenario has a great interest in orthopaedic surgery and wishes to develop their skills further; such desire should be encouraged and is often strongly promoted by professional regulators. The problem of how to develop new skills without essentially "experimenting" on the patient and putting their welfare at risk is hard to resolve.

As a novice veterinarian there are going to be a great many interventions that have not yet been attempted, and these skills need to be learnt. Sometimes, these interventions are simple variations on existing skills and may require little extra guidance from "experts". It would be completely appropriate for a novice veterinarian to attempt something that is new to them, if it is just a minor extrapolation of existing skills and the risk to the patient is minimal. However, patient welfare must always be held as our paramount concern. How can a novice veterinarian learn a complex novel procedure without putting patient welfare at risk? They may have read all there is to read on the procedure, watched videos and discussed it with other colleagues, but doing the procedure for the first time can be incredibly daunting, especially for interventions as complex as a TPLO. When is it ethical to gain this experience for the first time?

The simple answer is supervision. The more distant a novel skill set is from your existing skill set, or the more welfare compromise that could result from an error in the novel procedure, then the closer the level of supervision of your work by an experienced colleague should be. If you are very competent at castrations but have limited skill at an ovariohysterectomy, then it is wise to have an experienced colleague nearby to assist and offer advice as you proceed. As surgeries become more complex then it may require the novice to assist a more experienced surgeon the first few times, before the novice themselves becomes the primary surgeon under the expert guidance. Returning to the scenario above, it is unlikely that the skills to undertake a TPLO exist within the practice, so how can this supervision be achieved?

Some referral centres and general practices have established excellent working relationships. The specialists at the referral centres offer discount continuing professional development courses to those veterinarians who refer cases to them. This relationship could be extended further. Perhaps CPD days could be arranged where the general practitioner spends the day at the referral centre with an element of formal teaching and then assists with the surgery as a practical element of the training. This way the novice

veterinarian gets the benefit of the experience and knowledge of the specialist and the specialist is able to financially gain from the experience. However, Pepper's owner cannot afford a referral level of care, which may seem to thwart this plan. This is where the financial arrangement between primary and referral practice comes into play. The novice veterinarian can use their CPD allowance from their practice to subsidise the referral surgery, as they are receiving specialist and individualised training. Thus, the client is able to get the surgery they desire undertaken by a specialist – this also ensures that Pepper's welfare is put as our first consideration. The novice veterinarian gets a day of CPD and learns a novel set of skills. The referral practice receives the financial benefit of the TPLO surgery and benefits from improved relations with the general practice. Both veterinarians then also flourish in the professional sense as one learns and the other gets the rewards from imparting knowledge.

‖‖

HEALTHCARE professionals learn from patients. No matter what their training background, the realities of practice, the variety of skills required, technological advances and new techniques mean that at some point we will be attempting a procedure on a patient for the first time. As medical writer Atul Gawande noted with regards to medical practice, "there has long been a conflict between the imperative to give patients the best possible care and the need to provide novices with experience" (Gawande 2002).

Residences are structured to mitigate harm in this process by providing supervision and graduated responsibilities. Such an approach may also satisfy the principle of beneficence by providing additional benefit from teaching. Gawande claims that teaching hospitals enjoy better outcomes than non-teaching hospitals: "Residents may be amateurs, but having them around checking on patients, asking questions, and keeping faculty on their toes seems to help" (Gawande 2002).

However, he notes that when given a choice, for example if the patient happens to be a doctor and knows how things work, patients will elect for experienced surgeons to perform procedures – even though they appreciate that novices need the experience. Although this has not been formally studied in veterinary hospitals, it is the author's observation that veterinary nurses and veterinarians will – when given a choice – elect for the most experienced team member to perform more complex procedures.

This is paradoxical, as Gawande notes:

"and this is the uncomfortable truth about teaching. By traditional ethics and public insistence (not to mention court rulings), a patient's right to the best care possible must trump the objective of training novices. We want perfection without practice. Yet everyone is harmed if no one is trained for the future."

(Gawande 2002)

LEARNING FROM PATIENTS

What would you do?

You adopted a two-year-old stray female dog and would like to have her neutered. Your employer asks if you are comfortable to let M, a new graduate colleague, perform the procedure. What would you do?

△
10.14 Would you let a new graduate colleague spay your dog?
PHOTO ANNE FAWCETT

LEARNING FROM PATIENTS

What would you do?

A client presents a rabbit which you diagnose with gastrointestinal stasis. You offer to admit the animal for treatment. "I'm very worried about her, she is my baby," he says. "Have you treated this before?" You have only treated one rabbit for gut stasis and it died. How might you address his concerns?

△

10.15 You've only ever treated gut stasis in a rabbit once before, and the animal died.
PHOTO ANNE FAWCETT

LEARNING FROM PATIENTS

What do you think?

In his article "The Learning Curve", Gawande (2002) observed that it seemed unjust that some people, for example doctors, could elect to have a more experienced surgeon, whereas ordinary patients were not given this option.

ONE _____ What features of a policy for a veterinary teaching hospital would help to ensure it is just for everyone: student veterinarians, staff, clients and patients?

△

10.16 How could a policy help ensure fair treatment for student veterinarians, staff, clients and patients?
PHOTO ISTOCK

10.5

Evidence-based veterinary medicine

The term evidence-based veterinary medicine (EBVM) refers to evidence-based decisions which "combine clinical expertise, the most relevant and best available scientific evidence, patient circumstances and owner's values" (RCVS Knowledge 2016).

Or, as the Evidence-Based Veterinary Medicine Association puts it:

"Evidence-based veterinary medicine is the formal strategy to integrate the best research evidence available combined with clinical expertise as well as the unique needs or wishes of each client in clinical practice. Much of this is based on results from research studies that have been critically-designed and statistically evaluated."

(Evidence-Based Veterinary Medicine Association 2016)

△
10.17 Should EBVM take centre stage in all of our decision-making?
PHOTO ISTOCK

Meaningful EBVM has been made possible by the ability to analyse clinical data. The following scenarios explore the use of EBVM.

———
SCENARIO
EBVM

▶ Evidence-based medicine is the "conscientious, explicit and judicious use of current best evidence, primarily from clinical trials, in making decisions about the care of individual patients" (Hasnain-Wynia 2006).

EBVM seems to be the phrase of the moment, increasingly being taught in its own right in veterinary schools and promoted widely.

Should EBVM take centre stage in all of our ethical decision-making?

———
RESPONSE
MARTIN WHITING

▶ EBVM certainly has become a key phrase. Research groups are focused on the concept (e.g. Centre for Evidence-based Veterinary Medicine at Nottingham University: http://www.nottingham.ac.uk/cevm/index.aspx) and big data analysis of the veterinary case records of over 4 million unique animals is under way (VetCompass 2014). This is providing a startling amount of new information for use by practitioners of veterinary medicine. Professional regulators are also citing EBVM as an important responsibility for veterinary practitioners (Jorge & Pfeiffer 2012). In fact, in the UK the RCVS have stated that the welfare of animals is dependent upon EBVM – a very strong claim. They explain that "in order to be considered fit-to-practice, veterinary practitioners hold the responsibility to ground their decisions on sound, objective and up-to-date evidence, when available" (Jorge & Pfeiffer 2012). The final point of reference from the RCVS states that practising EBVM is in the public interest, making EBVM worthy of ethical consideration.

There are problems with translating population medicine into individual healthcare which are endemic within EBVM, but these will be addressed separately in the next scenario regarding EBVM outliers. The current scenario examines whether EBVM should take a priority in our ethical approach to patient care. Is there a time when ethics and EBVM are in conflict? There have been angry letters from many practitioners of medicine and veterinary medicine expressing that EBVM is not necessary or that it represents a "dumbing down" of professional judgement; some even state it is unethical to undertake "protocol-driven medicine". These largely represent a misunderstanding of the concept of EBVM and it is probably best explained by rephrasing the question to "is it ever ethical to make and advise on clinical decision-making using fewer and less evidence than is practicably available?" When the idea is rearranged in this format, it is hard to see why EBVM should not take centre stage. The idea of fewer and less evidence needs to be explored.

In 2010, 18,000 veterinary papers were published in various scientific journals (Jorge & Pfeiffer 2012). This is a staggering quantity of new data to be consumed, processed and implemented by general practitioners. 2010 was not a unique year for research; this number of articles is representative and increasing each year. Any one of these papers could substantially change the way in which the practitioner manages a case. A newly discovered contraindication or drug reaction could be lifesaving for many animals. To electively choose to not engage with new data, new discoveries or CPD can have a substantially negative impact on animal welfare and, subsequently, the reputation of the veterinary profession.

Professionals who ignore new data yearly, cumulatively degrade their own clinical practice. This was most publicly seen in a RCVS disciplinary case in 2011. Mr H had not kept up to date with modern advances in veterinary medicine through peer-reviewed literature or CPD and consequently he was using outmoded techniques and medications which were substantially detrimental to the welfare of the animals under his care (RCVS 2011). The Disciplinary Committee of the RCVS found Mr H guilty of serious misconduct for a number of reasons, including for recommending and undertaking surgical procedures without sufficient clinical grounds and without considering alternative options, and undertaking procedures outside his area of competence (RCVS 2011). As a consequence, the Committee directed that Mr H's name be removed from the register.

Mr H's case is an extreme example, but the principles of ignoring new data are omnipresent. The advantages of EBVM are to keep practitioners up to date with not just the latest, but the most comprehensive data regarding clinical practice. Peer-reviewed review articles, like systemic reviews and meta-analysis, take all supporting and conflicting data and report these in a digestible comparative fashion to allow the busy practitioner to assimilate the data without the need to read all 18,000 articles every year.

As veterinarians we do not leave veterinary school fully trained, fully educated and fully competent for the remainder of our practising life. The need to continually develop and modify practice depending on new data is inherent to the role of a professional, and indeed a commitment to lifelong learning is written into some oaths that veterinarians may be required to swear on graduation. This is what is meant by EBVM being in the public interest. It is in the public's interest that the professional before them is fully appraised of the latest information and is able to advise the client using robust evidence for their decisions rather than anecdotal or assumed knowledge.

EBVM is not about protocol-driven medicine, it is not about following flow charts saying "if patient has condition x then treat with drug y". It is complex interaction that brings the best available evidence to the client and the clinician to help them both make the decision that is best for the situation before them. EBVM is "the conscientious, explicit and judicious use of current best evidence in making decisions about the care of individual patients. This means integrating individual clinical expertise and the best available external clinical evidence from systematic research" (Cockcroft & Holmes 2003). EBVM therefore integrates the knowledge from scientific research with the clinical experience of the individual veterinarian and presents it in a format suitable to the particular client regarding the animal before them with its own special and individualised needs.

In order to get informed consent from a client to proceed with an intervention, the veterinarian must first explain the risks, the benefits and the harms of each appropriate intervention in a way that the client understands. This information can only be imparted to the client if it is first known by the veterinarian. Without the evidence base then informed consent is meaningless and the client will not have the necessary information to be able to choose between the different intervention options. This does not mean that the only treatment options are those that have the strongest evidence base, but that the certainty clinicians have for the outcome should be shared with the client. EBVM provides the tool to present this information to the client. The presence of a systematic review does not make any one treatment more effective than another, but it gives us the information we need to make the decision about the efficacy of treatments. It is hard to assert an efficacy if there is no evidence provided, but EBVM is in its early stages and studies and knowledge are always increasing.

The primary interest of a veterinarian, and hopefully the client, is patient welfare and EBVM serves to provide the information tools that are needed to discern part of what is needed to ensure welfare is maximised.

EBVM does not provide all the information though, so care needs to be taken. Combining the data from EBVM about the outcome (both positive and negative) of any intervention needs to be integrated with the client's narrative of the patient's life to determine what is best for the individual patient and client in any given scenario.

On the negative side of patient care are complications and contraindications. These also form part of EBVM. As an ever increasing number of drugs and interventions are being developed and big data analysis of their impact is undertaken, new complications and contraindications are being discovered. Knowledge of this part of EBVM is essential for ethical patient care. Little explanation is needed to understand that staying abreast of the negative effects of certain treatments is in the patient's interest. To instigate a course of treatment that has known and established risks which have not been explained to the client during the process of informed consent, and where this could have been avoided if the risks had been known, can lead to serious claims of negligence.

EBVM is not everything! Ethical practice should place the shared decision-making process at centre stage, with the primary focus of doing what is best for the patient given the limitations of the circumstances (such as client finances). EBVM provides the information and a tool to do that. There will always be a qualifier needed in reading EBVM data – what outcome is desired? Reviewing data for any intervention should report both morbidity and mortality data, as well as likely costs. These will provide essential information for the shared decision-making process. Some clients may prioritise length of life within a given quality, while others may set the standard of the quality of life much higher over quantity. For others, it will be how to achieve the best quality/quantity within a given cost. So, EBVM comes at the stage of shared-decision making and informed consent. It does not stretch beyond that to a dictatorial practice of medicine that eliminates clinician choice.

In summary, EBVM does not dictate to the veterinarian how they should practise; it should not lead to protocol-driven medicine and remove freedom of choice. EBVM should be one of a veterinarian's most powerful tools to ensure that public interest is preserved, animal welfare is served and the profession continuously improves. EBVM enhances clinical practice, client informed consent and ultimately provides the power behind shared decision-making. EBVM is essential in informed ethical practice, but it does not dictate it.

SCENARIO

EBVM – 10 PER CENT CHANCE OF SURVIVAL
OR 90 PER CENT CHANCE OF DYING

▶ You are called to a much-loved pony, Calvin, that has a stiff-limbed stance and difficulty eating – typical signs of tetanus. Calvin is sweating slightly but can still walk slowly. Over the next 24 hours he deteriorates further, becoming recumbent and now unable to eat. You know from reviews that tetanus cases have a reported survival rate of 25 per cent (Green, et al. 1994) to 41 per cent (Kay & Knottenbelt 2007) but that this pony has a particularly poor prognosis, with probably no more than a 13 per cent chance of survival (Reichmann, et al. 2008). However, you remember a very seriously affected case a few years back that your colleague treated (against your judgement), which recovered despite very poor prognostic indicators.

How relevant are these different sources of information to decisions about Calvin?

RESPONSE

MARTIN WHITING

▶ Tetanus is a condition that is well recognised in veterinary medicine in many species. It has serious welfare implications for the animals concerned and particularly for horses because of the difficulties in providing nursing care, and special concerns associated with recumbency in such large animals. There have been many reviews of tetanus treatments and outcomes (Green, et al. 1994, Kay & Knottenbelt 2007, Reichmann, et al. 2008, South 2014) but the dilemma in this case is about how we manage available data in opting for treatment choices and advising clients on success. This can have a major impact on informed consent.

Regarding the treatment of tetanus, different papers cite different levels of success from 13 per cent to 41 per cent in horses. There is also the case that the colleague treated, which was severely affected but recovered; an n = 1 study. How should these different elements of evidence affect our approach to such a case? The figures cited, on face value, show that it is unlikely the

△
10.18 Tetanus carries a poor prognosis in horses.
CARTOON SUHADIYONO94

horse will recover. Therefore, it will likely undergo a substantial welfare compromise, slow deterioration and ultimately death in a manner that is distressing to the horse, the owners and the veterinary staff. However, the papers do all also refer to animals that survived. How do we use the data provided to determine the best course of action? Take, for example, the Reichmann study. When these data are presented to clients, they are likely to want to know if their horse is going to be one of the lucky 13 per cent, in which case it would be worth pursuing treatment, or if it will be one of the unlucky 87 per cent, in which case it might be better to end the suffering with euthanasia sooner rather than later.

This case represents the apparent conflict that exists between population medicine and individual healthcare. Evidence-based medicine focuses on providing data on the average outcome for the average patient, so it essentially tries to "ignore" the outliers of a population, while personalised medicine places equal consideration on the outliers of an EBVM approach to those who form its mode (de Leon 2012). Let us assume that one study is based on a completely homogenous population of horses, all with tetanus. All of these horses are clinically identical, with identical treatment plans and nursing care. The outcome of these identical horses was that 10 per cent survive, and 90 per cent did not. Let's also assume that Calvin is identical to the horses in this study. How does the clinician determine if Calvin is one of the 10 per cent or one of the 90 per cent when advising the client? The hardened approach to EBVM might suggest that with such a painfully debilitating condition, the 90 per cent chance of not surviving is so poor that treatment should not be attempted and the horse should be euthanased. The negative welfare of a horse with severe tetanus is *so great* that a 10 per cent chance of recovery is not "good enough" to pursue. But the client is going to want to know more about the chance of their individual

horse, Calvin, surviving or not, before they make a decision to euthanase him.

EBVM in this case is very useful at presenting the data to the client about the chances of Calvin surviving. The different studies give us different data about the likelihood of Calvin surviving. EBVM should not necessarily lead to formulaic or protocol-driven medicine. Veterinarians are not required to deduce that "because condition x results in death 90 per cent of the time, all patients should be euthanased". The complicated part of EBVM is relating the data about past studies to the case of Calvin presented to the clinician, and to Calvin's owner. Perhaps each of the three studies listed instigated different treatments, leading to different conclusions about the chances of successful treatment for Calvin. Perhaps each study represents a slightly different population of horses and therefore it needs to be determined which population most closely resembles Calvin.

Ultimately, EBVM and the data it provides us with become the final stage in the conversation about informed consent and progression to treatment options. All of these studies represent different chances of outcome, and helping the client to understand these differences will help them to come to terms with the treatment options available. Each study may have examined different treatments but elected for the same outcome of euthanasia once a horse in the study reached a certain level of clinical severity or of welfare compromise. These data can help the clinician and the client to pre-decide an end point of care before embarking on treatment. Sometimes this "limit" might be pain-scoring, consciousness level or ability to feed themselves. Determining an end point can be easier if there is an agreed or commonly established end point in the literature.

It is human nature to respond to statistical facts in some unusual ways. A gambler may think it is far more likely they will win the lottery rather than be hit by a meteor, even though the statistical chance of either may be equal. A dieter may be happier to consume a biscuit that is 90 per cent fat free, than one which is advertised as consisting of 10 per cent fat. Calvin may have a 13 per cent chance of survival according to Reichmann, but this same data may induce a very different reaction in the client if it is stated as an 87 per cent chance of dying.

It is not just the data of EBVM itself which help us construct the narrative in which we discuss prognosis or treatment options with clients. But veterinarians must also take great care in being balanced in that narrative and not unduly biasing clients or unwittingly manipulating them into one course of action or another. Stating the data in both the positive and negative forms will help counter this bias.

> "Aligning the particular patient and the wishes of the client with the data available through EBVM can help the client understand how their choices will influence patient outcome."

In conclusion, EBVM provides us with data to assist the client in their decision-making process. It should not lead to formulaic medicine that removes clinician or client choice. Aligning the particular patient and the wishes of the client with the data available through EBVM can help the client understand how their choices will influence patient outcome. Great care needs to be taken when explaining these facts to prevent unintended biases and influence.

THERE are different possible ethical justifications for employing EBVM. From a utilitarian perspective, a good outcome depends on an accurate prediction of the consequences. The accuracy of prediction is surely improved (albeit not guaranteed) when based on the best, most comprehensive evidence. Similarly when it comes to a deontological analysis, surely the client or the consumer has the right to the highest standard of care – with which comes an expectation that the clinician will be held to across current evidence regarding a condition. As stated by one proponent of EBVM:

> "Some years ago, the RCVS tried to define the minimum skills the public had a right to expect from a new graduate. These became known as 'day one competencies'. The public has a similar right to expect the profession as a whole to be competently aware of the effects of its interventions in animal health using the best data available. We may have been able to exist without our present capability of widespread analysis for 225 years, but now we have it there is a legitimate public expectation that we should use it."
>
> (Lanyon 2016)

Finally, assessment of an intervention applied to a large number of cases, when compared to a control cohort, is the only way to ascertain that the intervention is minimally harmful whilst providing an actual benefit. Thus according to the RCVS, "When rigorous research underpins medical decisions, adverse events can be minimised (i.e. unintended injuries caused by medical management rather than the disease process) and patient outcomes can be improved" (Jorge & Pfeiffer 2012).

The use of this data in decision-making is the basis of just and fair practice, thereby satisfying the requirements of principalism.

EBVM is an offshoot of evidence-based medicine (EBM). In medicine, EBM is associated with treatment guidelines, algorithms and protocols with the ultimate aim of improving and standardising patient care. One of the advantages is that treatment decisions aren't simply based on the whims or limitations of a single practitioner.

However, EBM or EBVM can only be as good as its evidence base, which may be minimal. In the words of David Mills:

"the upper echelons of the EBVM pyramid are sparsely populated – there are probably fewer than 10 well-conducted, adequately powered veterinary randomised controlled trials in existence (and most of these are in cardiology). Most studies are retrospective, with low numbers, conducted in highly selective referral populations: the first-opinion animals which don't quite fit the criteria, being different breeds, ages, with co-morbidities, in clinics with clinicians less experienced or competent in a field, may be very poorly served."

(Mills 2015)

In some cases the only evidence available is a small case series, a single case report or extrapolation from first principles. The application of the findings of underpowered studies to individual patients may even cause harm. For this reason Mills queries whether "the translation of EBM to EBVM" is logically flawed and unjustifiable (Mills 2015).

The gold standard of evidence is the randomised controlled trial (RCT), but these can be an ethical minefield. For example, subjecting a patient with a particular condition to sham surgery to compare the outcome with those who receive surgical treatment may have an unacceptable welfare cost (for example, progression of the underlying condition, or unnecessary pain). For this reason there are some interventions for which observational studies may be acceptable evidence. This point is made rather colourfully in a paper suggesting that "the most radical protagonists of evidence based medicine organised and participated in a double blind, randomised, placebo controlled, crossover trial of the parachute" (Smith & Pell 2006).

Another criticism of EBM is that the emphasis on standardisation of care conflicts with another movement in medicine – cultural competence – which emphasises individualisation of care. According to Dr Hasnain-Wynia, cultural competence in medicine (CCM) is "the delivery of health services that acknowledges and understands cultural diversity in the clinical setting and respects individuals' health beliefs, values and behaviours" (Hasnain-Wynia 2006). The emphasis of CCM has shifted from the provision of specific knowledge about specific cultural groups and minorities to promoting humility, communication and understanding the patient narrative.

Both EBM and CCM base recommendations on information from studies of populations or subgroups of patients – so critics can argue that their findings either don't apply broadly enough, or marginalise some individuals. As addressed by Dr Whiting, the classic criticism of EBM is that it promotes "cookbook" medicine or a one-size-fits-all approach, and removes the art of practice. The classic criticism of CCM – though less relevant in a veterinary context – is that it promotes cultural stereotyping of patients or clients.

The application of standardised population probability-defined data to an individual patient may not result in holistic improvement (Mills 2015).

A practical issue raised here is how one practitioner can be expected to process the volume of information required to provide absolutely current information about a single condition, let alone multiple conditions. This is a real challenge and may require a revolution in the way veterinarians currently access information. Whilst review articles are helpful they may be biased (for example, review articles published in a surgery journal may have an inherent bias towards surgical treatment of a condition), and are not performed regularly enough to ensure that every practitioner has access to current information at the time they need it.

Currently, veterinarians are responsible for developing their own postgraduate CPD plans, and choosing which material they read. There is a real danger that key information may be lost in the white noise of information overload. For example, it is unrealistic to expect a general practitioner to read 18,000 papers, yet if we recommend that it is reasonable to read just 365 papers (one paper for every day of the year) there is a high risk that many papers that would have provided current information relevant to case management would be overlooked.

The "cost" of processing this information has been pointed out in relation to the study of medicine:

"In the 1940s there were three major medical journals in the United States: all three were monthly subscriptions, two of which were newsletters. Keeping up with the literature meant pouring yourself a cup of coffee on a Sunday, sitting in your comfy chair, and reading for an hour. Since that time, the volume of medical literature has grown exponentially. More research is being done worldwide, and it is more accessible than in the past. In addition, the average physician receives multiple journals, including throwaway journals, weekly. It is impossible to keep up with the literature anymore, yet when we see the overwhelming pile next to the bed, we feel incompetent as physicians and scientists. After all, we were told 'if you don't keep up with the literature, you are not enough'. One colleague of mine was told, 'if you don't keep up with the literature,

people will die'. What a ridiculous guilt trip. This creates anxiety and frustration for us all. It hits one of our shared personality traits smack in the face – perfectionism."

(Lipsenthal 2007)

The volume of information available, coupled with the fact that veterinarians (even those working in academia and those who produce scientific studies) are time-poor, leads to a lag between the release of new evidence and its adoption into everyday practice. One of the aims of EBM is to reduce this lag (Jorge & Pfeiffer 2012).

But EBVM also requires veterinarians in all fields to take responsibility in collecting data, so that it can be collated and analysed (Lanyon 2016).

EVIDENCE-BASED VETERINARY MEDICINE

What do you think?

ONE _____ To what extent should a single practitioner be responsible for selecting and processing all relevant information?

TWO _____ Is there scope for professional organisations to take some responsibility in the presentation of information in a format that is highly accessible to practitioners?

THREE _____ How should data collection be regulated and overseen, and who should pay for data analysis?

FOUR _____ Is it acceptable for practices to sell clinical data?

⚠ **10.19** Where does the responsibility lie for assimilating evidence about veterinary practice?
PHOTO ISTOCK

SCENARIO
HOMEOPATHY

▶ You are keen to take on an interesting position you have been offered in a new veterinary practice. You have visited the practice and are impressed by the staff and facilities. However, when you find that the practice runs complementary medicine clinics and that one of the veterinary surgeons uses homeopathy alongside other treatments, you begin to have some doubts.

In an era of EBM, can the use of homeopathy by a registered veterinarian ever be considered ethical?

RESPONSE
ANDREW GARDINER

▶ Homeopathy comes under the category of alternative or complementary medicine. A search using the Advanced Search facility of the Royal College of Veterinary Surgeons' "Find a Vet" search engine returned 695 UK practices

⚠ **10.20** In the era of EBVM, is the use of homeopathy ethical?
CARTOON RAFAEL GALLARDO ARJONILLA

EVIDENCE-BASED VETERINARY MEDICINE
RESPONSE
ANDREW GARDINER

which state an interest in complementary medicine (RCVS 2015). Complementary medicine is usually used alongside "regular" (or allopathic) medicine; alternative medicine, as the name suggests, may be used instead of it. This can be an important distinction ethically, as we shall see. Other terms encountered are holistic medicine and integrated/integrative medicine. However, all the terms are often used quite loosely and, in its widest sense, complementary medicine can include many facets of general nursing and patient care, as well as rehabilitative treatments such as physiotherapy and hydrotherapy.

Historically, homeopathy emerged in the late eighteenth and early nineteenth centuries as a result of ideas promoted by the German physician Samuel Hahnemann, and homeopathy was used regularly in mainstream German medicine until well into the twentieth century, and still retains significant popularity. Homeopathy emerged as a radical alternative to the practices of the "heroic medicine" era. The latter featured bleeding, purging, blistering and the administration of potent drug recipes, the ingredients of which were often kept secret by practitioners working in competition with each other. Hahnemann's system of highly diluted remedies aimed to potentiate the body's innate healing mechanisms and to move away from empiricism and trade secrets towards a system of medicine based on drug "provings", which are considered central to classical homeopathy. These general thoughts were in keeping with a system of medicine that satisfied the ethical theory of principalism and its core directive of, "First, do no harm" – i.e. whatever you do, do not make the patient's condition worse. In contrast, many "heroic" treatments probably did a great deal of harm. The dramatic effects of purging, bleeding and blistering indicated that the administered treatment was potent and was definitely having some sort of effect on the patient, thus justifying a fee, but whether this actually helped the patient or not is open to debate.

A quote from a standard veterinary textbook shows how the movement away from empiricism took some time and was still being translated into veterinary education as late as 1933:

> "The irrational treatment adopted in former times must be attributed to lack of appreciation of the natural powers of recovery, to a firm belief in the virtues of certain potent drugs, and to the erroneous idea that practitioners possessed the means of directly overcoming the effects of disease. Although new medicinal agents are constantly being introduced, drugs are now less frequently prescribed than formerly. We now endeavour to assist the natural powers of recovery by attending to hygienic and dietetic details, and to the careful nursing of the patient. As has been indicated, animals may recover spontaneously from many affections, and the indiscriminate administration of drugs frequently tends to impede recovery."
>
> (Greig 1933)

It is important to situate homeopathy in this historical context, however briefly, before considering the ethics and controversies of the subject today. This is not to say that the homeopathy was non-problematic when it first appeared – its radical reframing of health, disease and treatment has always stirred up considerable controversy. However, in its position as an alternative to heroics, homeopathy was, in some ways, in keeping with a more rational and less interventionist form of medicine compared with what went before.

So much for then, what about now? There are few veterinary homeopathy papers published in peer-reviewed journals (Mathie & Clousen 2014). They would not satisfy standard EBM criteria as proof of efficacy and, whilst the same can be said for many conventional veterinary studies, the nature of homeopathy and its putative action tends to put it in a different category. Homeopathy is not currently deemed professionally unethical per se in Britain – there is no prohibition coming from the RCVS. However, homeopathy does draw on a very different theoretical underpinning which many argue is impossible to reconcile with scientific positivism/rationalism. This has led to calls for an ethical ban on the grounds that veterinary medicine is a science-based profession.

Homeopathy is currently featured in professional veterinary textbooks on alternative/complementary medicine published by reputable publishers (Schoen & Wynn 1998), and occasionally gets a mention in other mainstream texts, e.g. on pain management (Gaynor & Muir 2015), as part of complementary or adjunctive approaches. The subject is not taught in veterinary schools and it is often confused with herbalism. Homeopathy appears quite frequently in the correspondence pages of journals such as *Veterinary Times* where it tends to cause heated debate. It remains the most controversial of the alternative or complementary treatments. A frequent criticism is that homoeopathic preparations, because of the sequential dilutions that they are subject to, contain no measurable quantity of the putative therapeutic substance detectable by current technology. Such massive dilution means they will apparently do no harm (and so satisfy the first ethical criterion of principalism); the question is, can they do any active good?

The most relevant recent veterinary paper in connection with EBM approaches appeared in 2009 (Hill, et al. 2009). This paper described a collaborative investigation between a practising homoeopath and a veterinary clinical specialist working in a veterinary school. The study was rigorously designed and involved 20 dogs. As reported on the University of Bristol's news web pages:

"The dogs were prescribed individualised homeopathic medicines by vet John Hoare.

Two months after starting the treatment, the owners of 15 of the dogs reported no improvement. However, owners of the other five dogs reported pruritus scores that were at least 50 per cent improved compared to their pets' score at recruitment. One of the five dogs improved by 100 per cent and needed no further treatment. The other four dogs that responded well in this first phase were then put forward into a blinded randomised trial in which they received their homeopathic prescription at some times and placebo at other times. The three dogs that completed this phase of the study improved more with the active remedy than with placebo, and owners were able to distinguish correctly which pill was which."

(University of Bristol 2009)

The conclusion was that a larger, multi-centre study was needed to explore the preliminary results, which could not in themselves be taken as proof that homeopathy was working because positive results in a small sample could be explained as "outliers". Unfortunately, that larger multi-centre study has not yet gone ahead, and homeopathy is so controversial that socio-cultural pressure from within the scientific community could make conducting such a multi-centre investigation difficult.

Homeopaths may cite difficulty with the gold standard of EBM, the double-blind, placebo-controlled RCT, because their prescriptions are individualised to each patient. However, in Hill, et al.'s (2009) case, individualised prescribing was incorporated, making the study, and a larger variant of it, of particular value whatever the outcome since optimal homeopathic prescribing was allowed.

But is EBM the ultimate test? In standard medical trials, patient difference deemed relevant to homeopathic prescribing tends to be smoothed out. Indeed, one of the problems with EBM is this tendency to homogenise the study group. There are other related philosophical and ethical issues

10.5

EVIDENCE-BASED VETERINARY MEDICINE
RESPONSE
ANDREW GARDINER

with EBM which are increasingly discussed as the movement becomes more prominent, including a risk that the practice of medicine becomes "evidence tyrannised" (Lambert 2006). Medical historians also rightly ask, if EBM is so new, what was medicine based on before? The answer is evidence – just a different kind of evidence, such as accumulated clinical experience and reasoning from basic science and pathophysiology. It is likely that in a future of highly individualised personal medicine based on genetic predisposition, EBM will be seen by historians as another era of medicine which had its flowering phase, a phase of relative prominence, and was then largely superseded by something else, because this is how science invariably progresses (Goldenberg 2006). Nevertheless, EBM is currently viewed as the gold standard for assessing treatment and, as long as adherents are not blind to its weaknesses, would seem to be the most ethical protocol currently available.

A historical-ethical view of homeopathy, as given here, might therefore place it in the context of a reaction against heroic medicine, which did too much to the patient to the extent of causing actual harm. The response, in homeopathy, was a movement in the other direction, by doing far less (or, critics might argue, nothing at all, at least in terms of drug therapy). Homopathic cures may say as much about epistemological framings of disease and cure as about tissue-level explanations of drug action. A present-day reading of "heroic medicine" could be technological advances and an atomistic view of disease which runs the risk of losing sight of whole-patient considerations, leading to unnecessarily invasive tests and treatments. Such considerations are debated in both veterinary and human medicine but do not help us with deciding about homeopathy specifically (Jarvis 2010).

Whether the use of homeopathy by a registered veterinarian is unethical therefore remains open to debate. Any veterinarian should be able to be called to account for their treatment decisions by their professional body (peers). However, clinical freedom is a valued commodity and restriction and over-policing of practitioners' treatment choices (an inherent risk of the EBM "tyranny") may not be in patients' best interests, especially when freedom to treat is securely situated within a generally robust and functioning overall ethical framework (Greenhalgh, et al. 2014). Homeopathy therefore presents a difficult problem for professional regulatory bodies, which possibly explains their reluctance, in some cases, to impose an outright ban.

As mentioned already, homeopathy satisfies a core requirement of principalism ("Do no harm"). Thus it could be argued that it is ethically unproblematic, given current knowledge, to use it as a *complementary* therapy alongside conventional approaches. If used as an *alternative* treatment by a registered veterinary surgeon, then it must be judged on the same ethical basis as any other treatment. If a patient is deemed harmed because a conventional treatment should have been used instead, then use of homeopathy as a sole therapy could be unethical: no direct harm by the therapy may have taken place, but there could be a sin of omission by withholding a generally accepted and proven therapy. In this sense, homeopathy could be considered symmetrically with other forms of treatment and a clear distinction made between complementary and alternative usage, pending further research which is urgently needed.

So should you be worried about taking the job? It depends on whether you feel you would somehow be "homeopathically contaminated" by working in the practice. Assuming not, issues could still potentially arise with case transfer, client communication etc. These would need to be resolved on a case-by-case basis or via some general understanding or protocol within the practice team. It may be useful to discuss these issues prior to accepting the post.

||||||||||||||||||||||||||||||||||||||||||||||||||||||||||||||||

FOR another perspective on homeopathy read-ers are directed to section 5.7 in the chapter on veterinary treatment.

EVIDENCE-BASED VETERINARY MEDICINE

What do you think?

ONE _____ What features of an "alternative" treatment would be important to con-sider when coming to a view about whether it was ethical or unethical to allow it to be available for veterinary use?

TWO _____ What ethical principles are important to you in this case?

THREE _____ How would it be best for governing bodies to regulate such treatments?

△
10.21 How would you decide if an "alternative" treatment should be available for veterinary use?
PHOTO ISTOCK

10.6

Oversupply of veterinarians

One major point of discussion in veterinary educa-tion and training is an actual, perceived or poten-tial oversupply of graduates. Many students and veterinarians are concerned about oversupply due to the risk of reduced employment opportuni-ties, increased demand for graduate jobs driving down salaries, increased competition between practices, increased demand for non-traditional veterinary roles, and ability of employers to select overqualified applicants without remunerating them for their additional skills (American Veter-inary Medical Association 2013, Fish & Griffith 2014, Heath 2008, McCormick 2013).

The veterinary profession is not the only profes-sion with an oversupply of graduates. At varying

△
10.22 In some countries there is an oversupply of graduating vets.
PHOTO ISTOCK

times and in various contexts, doctors, dentists, lawyers and even the clergy have suffered from an oversupply. But how should concerns about oversupply alter our behaviour and in what ways? The following scenario examines concerns about oversupply of veterinarians.

SCENARIO
OVERSUPPLY OF VETS

▶ You are a veterinary surgeon working in a mixed animal practice in a semi-rural location. Hannah, the daughter of one of your close friends, a medical doctor, contacts you seeking advice about pursuing a career in veterinary science. She tells you that she is very passionate about working with animals, and believes this is "her calling". Over the past two years your practice has received dozens of job applications from veterinarians looking for work, despite the fact that no position has been advertised. A new veterinary school has opened in the nearest town.

How would you advise Hannah?

RESPONSE
JOHN BAGULEY

▶ Beauchamp and Childress (1979) introduced an ethical model widely adopted in medical clinical practice and bioethics. This model suggests four equally important principles: autonomy, beneficence, non-maleficence and justice. These principles however can also be applied to ethical dilemmas such as the one presented in this scenario.

An autonomous individual is free to choose and importantly possesses an adequate understanding to facilitate meaningful choice. This principle is aligned with that of informed consent; as

veterinarians we ensure our clients are able to make informed decisions about the care of their animals.

We can apply this same principle to this scenario. Hannah has a passionate desire to become a vet and your ethical duty is to ensure she is able to make an informed decision. This means discussing your role in veterinary practice, possibly other potential veterinary roles, the study required to become a veterinarian, working conditions and remuneration, moral stressors in the role and of course current issues with respect to the supply and demand for veterinarians and veterinary graduate debt. You could also assist her with places where she could find appropriate resources and possibly introduce her to some of your colleagues. It may even be possible for her to spend time at your veterinary practice although I would expect given her level of commitment she has already completed some relevant work experience.

Your assistance means that Hannah is now free to make an informed choice regarding her future.

Veterinary schools similarly have an ethical responsibility to provide potential veterinary students and veterinary students in the first year of their study with information about the current issues, challenges and benefits associated with a career in veterinary science. The time from beginning studies to the time when you become a veterinarian is at least five years and may be seven years or longer and factors such as supply and demand could change significantly during this time.

As mentioned already, there are the other principles in this ethical model of beneficence, non-maleficence and justice. By providing information and assistance to Hannah you are "doing good" and certainly not "doing harm". It is important to respect her autonomy and not try to persuade her either way; crushing her dreams could "do harm". Your role is simply to provide her with the best available information in an objective manner (or with assistance on how to find this information).

Justice requires that you would similarly provide this information and assistance to other potential veterinary students and veterinarians.

||||||||||||||||||||||||||||||||||||||||||||||||||||||||||||||||

THE oversupply of veterinary graduates impacts a number of stakeholders: actual and potential veterinarians; actual and potential veterinary students; veterinary schools themselves; veterinary employers and employees; veterinarians seeking employment; veterinary professional associations; accrediting bodies; and clients and patients.

Many objections to increasing numbers of graduates are made on consequentialist grounds. For example, increasing graduate numbers by introducing a new veterinary school in the area will increase competition for jobs, drive down salaries and mean that those veterinarians who do work must work longer hours for lower remuneration. Certainly there is evidence to show that these consequences are borne out in some regions (Heath 2008). But that is not always the case.

For example, there may remain desperate shortages of veterinarians in some (typically rural or remote) areas. An increase in overall graduate numbers may reduce this shortage marginally, yet still yield an overall increase in "good". Let's say, hypothetically, that the new veterinary school produces 150 graduates, and 2 of those graduates work in a rural area that has not had a permanent veterinarian for 10 years. On a utilitarian analysis, it may be that those 2 veterinarians improve the lives of thousands of patients and hundreds of clients significantly, possibly to such an extent that it outweighs the fact that 148 graduates work for lower-than-ideal pay closer to major cities and towns.

But is such an outcome just?

The veterinarian in the above scenario may feel obligated to talk Hannah out of her proposed career path on the grounds that it may be financially catastrophic for her to pursue her dream. Yet, as the author points out, such an approach seems paternalistic and flies in the face of Hannah's autonomy.

Could there be a conflict of interest? Does the veterinarian see Hannah as potential competition? If so, placing undue influence on a single student seems an inefficient means of addressing future competition. Instead, there may be scope for the veterinarian providing helpful information to Hannah, as well as getting involved with their professional association or engaging directly with the university about their concerns.

Another issue raised in the scenario is the potential gulf between expectations about working in the veterinary profession and the reality. These may be financial, or otherwise.

Veterinary specialist Dr Joanne Intile wrote:

"When I decided to change careers and become a veterinarian, like so many of my peers, the concept of taking on triple digit student loan debt was negated by my pure and noble intentions. This was my calling. This was my aspiration. And there simply was *no price* to be placed on my ability to follow my dream… I now wish for such things as owning a home, taking a vacation, raising a family, and (gasp) retiring one day."

(Intile 2014)

But does educating future veterinarians about potential stressors risk demonising their profession, ignoring the fact that many known stressors such as long working hours and low remuneration are found in other professions (Cake, et al. 2015)?

OVERSUPPLY OF VETERINARIANS

What do you think?

ONE — Who could and should take responsibility for the number of veterinarians that graduate?

TWO — In contexts where there is a proven oversupply of veterinarians, whose responsibility is it to address this issue?

THREE — What ethical framework would you recommend?

△

10.23 Veterinary training is very long and intensive with both academic and practical elements.
PHOTO ISTOCK

Conclusion

In this chapter we have seen that there are ethical issues that arise through the education and training processes for veterinary surgeons, but also opportunities for equipping students to better face the challenging ethical landscape ahead of them. Universities and other training providers should be aware of the importance of ethical issues to students and have practices in place that promote both good welfare of the animals used in teaching and excellent ethical reasoning skills in their students.

References

American Veterinary Medical Association 2013 2013 U.S. Veterinary Workforce Study: Modeling Capacity Utilization.

Beauchamp TL, and Childress JF 1979 *Principles of Biomedical Ethics*. Oxford University Press: New York.

Bergh MS, Sullivan C, Ferrell CL, Troy J, and Budsberg SC 2014 Systematic review of surgical treatments for cranial cruciate ligament disease in dogs. *Journal of the American Animal Hospital Association* **50:** 315–321.

Bezruchka S 2000 Medical tourism as medical harm to the Third World: Why? For whom? *Wilderness & Environmental Medicine* **11:** 77–78.

Cake MA, Bell MA, Bickley N, and Bartram DJ 2015 The life of meaning: a model of the positive contributions to wellbeing from veterinary work. *Journal of Veterinary Medical Education* **42:** 184–193.

Citrin DM 2010 The anatomy of ephemeral health care: "health camps" and short-term voluntourism in remote Nepal. *Studies in Nepali History and Society* **15:** 27–72.

Cockcroft P, and Holmes M 2003 *The Handbook of Evidence-Based Veterinary Medicine*. Wiley-Blackwell: Oxford.

Conzemius MG, Evans RB, Besancon MF, Gordon WJ, Horstman CL, Hoefle WD, Nieves MA, and Wagner SD 2005 Effect of surgical technique on limb function after surgery for rupture of the cranial cruciate ligament in dogs. *JAVMA-Journal of the American Veterinary Medical Association* **226**: 232–236.

de Leon J 2012 Evidence-based medicine versus personalized medicine. Are they enemies? *Journal of Clinical Psychopharmacology* **32**: 153–164.

Devereux P 2008 International volunteering for development and sustainability: outdated paternalism or a radical response to globalisation? *Development in Practice* **18**: 357–370.

Evidence-Based Veterinary Medicine Association 2016 About us. www.ebvma.org/

Fish RE, and Griffith EH 2014 Career attitudes of first-year veterinary students before and after a required course on veterinary careers. *Journal of Veterinary Medical Education* **41**: 243–252.

Freidson E 2001 *Professionalism, The Third Logic. On the Practice of Knowledge.* University of Chicago Press: Chicago.

Gardiner P 2003 A virtue ethics approach to moral dilemmas in medicine. *Journal of Medical Ethics* **29**: 297–302.

Gawande A 2002 The learning curve. *New Yorker*. Bell & Howell Information and Learning Company.

Gaynor JS, and Muir WW 2015 *Handbook of Veterinary Pain Management.* Elsevier: St Louis. pp372–373.

Goldenberg MJ 2006 On evidence and evidence-based medicine: lessons from the philosophy of science. *Social Science and Medicine* **62**: 2621–2632.

Graham ML, and Prescott MJ 2015 The multifactorial role of the 3Rs in shifting the harm-benefit analysis in animal models of disease. *European Journal of Pharmacology* **759**: 19–29.

Green SL, Little CB, Baird JD, Tremblay RRM, and Smith-Maxie LL 1994 Tetanus in the horse – a review of 20 cases (1970 to 1990). *Journal of Veterinary Internal Medicine* **8**: 128–132.

Greenhalgh P, Howick J, and Maskrey N 2014 Evidence based medicine: a movement in crisis? *BMJ* **348**: g3725.

Greig JR 1933 *Hoare's Veterinary Materia Medica and Therapeutics, 5th Edition.* Baillière, Tindall & Cox: London.

GREY2K USA LACS 2014 Greyhound Racing in Great Britain – a Mandate for Change. League Against Cruel Sports: Godalming, UK.

Hasnain-Wynia R 2006 Is evidence-based medicine patient-centered and is patient-centered care evidence-based? *Health Services Research* **41**: 1–8.

Heath TJ 2008 Number, distribution and concentration of Australian veterinarians in 2006, compared with 1981, 1991 and 2001. *Australian Veterinary Journal* **86**: 283–289.

Hill PB, Hoare J, Lau-Gillard P, Rybnicek J, and Mathie RT 2009 Pilot study of the effect of individualised homeopathy on the pruritis associated with atopic dermatitis in dogs. *Veterinary Record* **164**: 364–370.

Intile J 2014 The rising cost of becoming a veterinarian. http://www.petmd.com/blogs/thedailyvet/drjintile/2014/october/rising-cost-becoming-veterinarian-32082

Jarvis S 2010 Where do you draw the line on treatment? *Veterinary Record* **167**: 636–637.

Jordan T, and Lem M 2014 One health, one welfare: education in practice veterinary students' experiences with community veterinary outreach. *The Canadian Veterinary Journal* **55**: 1203–1206.

Jorge R, and Pfeiffer D 2012 RCVS Position Paper on Evidence Based Veterinary Medicine (EBVM). https://knowledge.rcvs.org.uk/document-library/rcvs-position-on-ebvm/

Kay G, and Knottenbelt DC 2007 Tetanus in equids: a report of 56 cases. *Equine Veterinary Education* **9**: 107–112.

Knight A 2014 Conscientious objection to harmful animal use within veterinary and other biomedical education. *Animals* **4**: 16–34.

Lairmore MD, and Ilkiw J 2015 Animals used in research and education, 1966–2016: evolving attitudes, policies, and relationships. *Journal of Veterinary Medical Education* **42**: 425–440.

Lambert H 2006 Accounting for EBM: notions of evidence in medicine. *Social Science and Medicine* **62**: 2633–2645.

Lanyon L 2016 Collecting the evidence for EBVM: who pays? *The Veterinary Record* **178**: 120–121.

Lee WT, Schulz K, Witsell D, and Esclamado R 2012 Channelling Aristotle: virtue-based professionalism training during residency. *Medical Education* **46**: 1129–1130.

Lipsenthal L 2007 *Finding Balance in Medical Life.* Finding Balance Inc.: San Anselmo.

Mathie RT, and Clousen J 2014 Veterinary homeopathy: systematic review of medical conditions studied by randomised placebo-controlled trials. *Veterinary Record* **175**: 373–381.

McCormick RH 2013 Veterinarian oversupply in small animal clinical practice. *JAVMA-Journal of the American Veterinary Medical Association* **242**: 308.

McLennan S 2014 Medical voluntourism in Honduras: 'helping' the poor? *Progress in Development Studies* **14**: 163–179.

Menon S 2003 A bad day in Usayini: the sinister targets of Indian health camps. *Third World Resurgence.*

Mills D 2015 Is EBVM ethical? *Veterinary Record* **177**: 181–182.

Papadakis MA, Hodgson CS, Teherani A, and Kohatsu ND 2004 Unprofessional behavior in medical school is associated with subsequent disciplinary action by a state medical board. *Academic Medicine* **79**: 244–249.

Powell L, Rozanski EA, and Rush JE 2010 *Small Animal Emergency and Critical Care: Case Studies in Client Communication, Morbidity and Mortality.* Wiley-Blackwell: Oxford.

Rao M 2004 *The Unheard Scream: Reproductive Health and Women's Lives in India.* Panos Institute: New Delhi.

RCVS 2011 Grimsby vet struck off for serious professional misconduct. http://www.rcvs.org.uk/news-and-events/news/grimsby-vet-struck-off-for-serious-professional-misconduct/

RCVS 2015 Find a Vet. http://www.rcvs.org.uk/find-a-vet/

RCVS Knowledge 2016 What is EBVM? RCVS Knowledge: https://knowledge.rcvs.org.uk/evidence-based-veterinary-medicine/what-is-ebvm/

Reichmann P, Lisboa JAN, and Araujo RG 2008 Tetanus in equids: a review of 76 cases. *Journal of Equine Veterinary Science* **28**: 518–523.

Russell WMS, and Burch RL 1959 *The Principles of Humane Experimental Technique.* Methuen: London.

Scheyvens R 2011 *Tourism and Poverty.* Routledge: London.

Schoen AM, and Wynn SG 1998 *Complementary and Alternative Veterinary Medicine: Principles and Practice.* Mosby Elsevier: Oxford.

Simpson K 2004 'Doing development': the gap year, volunteer-tourists and a popular practice of development. *Journal of International Development* **16**: 681–692.

Sin HL, Oakes T, and Mostafanezhad M 2015 Traveling for a cause: critical examinations of volunteer tourism and social justice. *Tourist Studies* **15**: 119–131.

Smith GCS, and Pell JP 2006 Parachute use to prevent death and major trauma related to gravitational challenge: systematic review of randomised controlled trials. *International Journal of Prosthodontics* **19**: 126–128.

South V 2014 Clostridial diseases of the horse. *In Practice* **36**: 27–33.

Tarlo E 2003 *Unsettling Memories: Narratives of the Emergency in Delhi.* University of California Press: Berkeley.

Taylor-Brown FE, Meeson RL, Brodbelt DC, Church DB, McGreevy PD, Thomson PC, and O'Neill DG 2015 Epidemiology of cranial cruciate ligament disease in dogs attending primary-care veterinary practices in England. *Veterinary Surgery* **44**: 777–783.

Tiplady C, Lloyd S, and Morton J 2011 Veterinary science student preferences for the source of dog cadavers used in anatomy teaching. *Alternatives to Laboratory Animals* **39**: 461–469.

University of Bristol 2009 Homeopathy in dogs pilot indicates need for larger clinical trial. Bristol. http://www.bristol.ac.uk/news/2009/6274.html (accessed June 2015).

VetCompass 2014 Health surveillance for UK companion animals. Royal Veterinary College.

Whittaker AL, and Anderson GI 2013 A policy at the University of Adelaide for student objection to the use of animals in teaching. *Journal of Veterinary Medical Education* **40**: 52–57.

Whyte SR, Van der Geest S, and Hardoe A 2003 *The Social Lives of Medicines.* Cambridge University Press: Cambridge.

Yeates J 2013 *Animal Welfare in Veterinary Practice.* Wiley-Blackwell: Oxford.

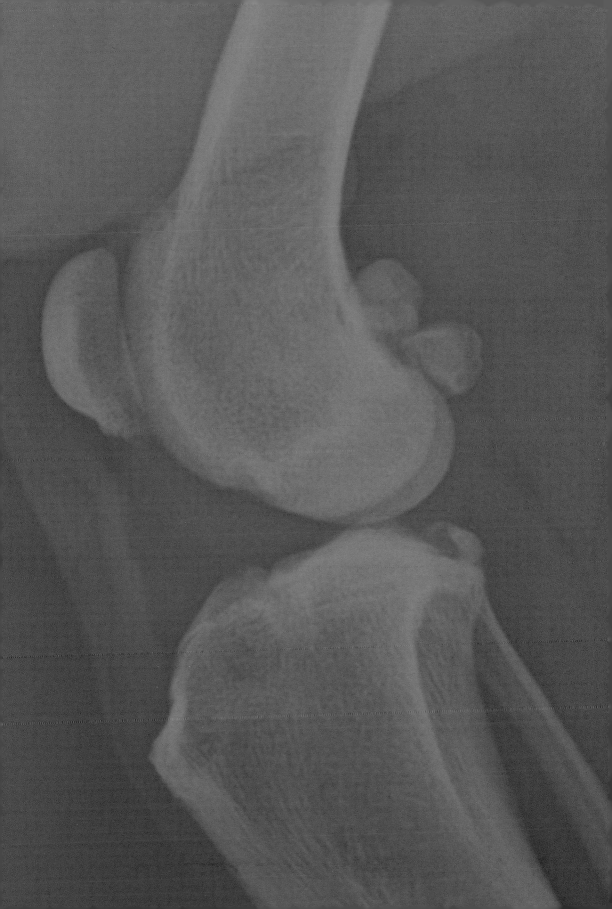

CHAPTER 11
TEAM RELATIONS

Introduction

Working as part of a team is integral to almost all elements of veterinary practice. Veterinary technicians, nurses, receptionists, administrative staff and others all contribute to the successful care of animals. Furthermore, with factors such as a general shift from smaller, single-vet practices to large veterinary hospitals and chains, referral and specialty hospitals, increased specialisation within the profession, expanded teaching opportunities and pet insurance, there is an increased need for collaboration and communication within and between practices and other organisations (Armitage-Chan, et al. 2016). In different spheres of work veterinarians may be working with carers of experimental animals, abattoir staff or other government colleagues.

At its best, a veterinary team is a slick operation, injects fun into the daily routine and has been shown to be an important source of support at difficult times (Ballantyne & Buller 2015, Hartnack, et al. 2016). The role of the team was seen as particularly valuable in discussing clinical decision-making, and high-intensity situations such as management of an emergency (Armitage-Chan, et al. 2016). For example, in cases where one

◁

11.1

PHOTO ANNE FAWCETT

individual's reasoning is impaired or biased (by factors such as previous involvement with a case or owner pressure) a colleague with no prior involvement of the case was able to provide a useful, more rational perspective (Armitage-Chan, et al. 2016).

Relationships with colleagues can have a huge impact on clinical outcomes and job satisfaction. In one study veterinarians reported that favourable clinical outcomes and good relationships with colleagues were the most important sources of satisfaction for them (Bartram, et al. 2009b). But that isn't always the case. Sometimes relationships between colleagues can be strained and actually stressful for one or more parties (Gardner & Hini 2006). This can occur for many reasons.

For some, the concept of veterinarian as team player is viewed as a challenge to their professional role, with potential for miscommunication cited as a source of stress. Aspects of modern veterinary practice that challenge team communication include situations where multiple vets see an individual patient, shift work, handover for overnight care and part-time work (Armitage-Chan, et al. 2016).

Similarly, when groups of individuals set high personal expectations, they may judge others to the same standard. If expectations aren't met, some individuals within the group may become critical of others, not only souring team relations but also damaging collaboration and communication within the team, potentially jeopardising patient safety. Importantly this may inhibit provision of support within the team or the wider profession, and can lead to a loss of trust within and of the profession (Armitage-Chan, et al. 2016).

In this chapter we will consider situations relating to relationships between colleagues, both on and off duty, where our moral obligations lie and how to resolve them.

11.1

Social media and the veterinary team

The advent of social media has changed the way we communicate. Not only do many of those working with animals maintain personal social media profiles, but veterinary organisations and practices increasingly have a social media presence. This can be an excellent way to promote the practice brand and communicate with clients, but communication on social media is not always positive. For example, clients are able to post complaints on a public forum. It can be challenging to delineate boundaries between one's personal and professional life on social media. As discussed in this next scenario, our behaviour online can have widespread impacts on our professional reputation, collegial relations and career opportunities.

SCENARIO

VENTING ON SOCIAL MEDIA

▶ You work for a companion animal veterinary practice. As you head to the front of the clinic to check the list of morning appointments, one

of the other staff members of the clinic calls you over to their computer terminal and draws your attention to a public post made on Facebook by another employee of the clinic. The post reads:

△

11.2

PHOTO ANNE FAWCETT

What should you do?

RESPONSE

JASON B COE

▶ Social media is rapidly, and profoundly, changing the way people communicate in society today. Facebook specifically has been developed to share information broadly among individuals, and research suggests people, including veterinary personnel, are likely to share more thoughts or ideas on Facebook than they would otherwise (Christofides, et al. 2009, 2012, Weijs, et al. 2014). As the uptake and use of social media continue to grow, identifying an approach to manage the potential risks and repercussions resulting from scenarios such as that described above is important to safeguard the veterinary profession and its members.

To start off, it is important to identify the stakeholders affected in this situation. In this scenario, at minimum, consideration must be given to the direct and indirect effects that the post could have on the staff member who posted the comment, his or her peers including his or her boss and other staff members, the veterinary practice as a whole, and the broader veterinary profession that all veterinary personnel represent. The ethical principles of non-maleficence, beneficence and autonomy can be used to help develop a response. Weighing each principle in relation to each stakeholder affected by the post can assist in navigating this type of situation.

Firstly, the employee of the clinic who made the public post on their personal Facebook profile needs to be considered. As long as the post does not infringe upon societal laws or professional standards within the employee's jurisdiction, the employee is entitled to choose the content they post to their own Facebook profile (Coe, et al. 2011). This includes the person also choosing the harms (e.g. reputational damage, job loss) and benefits (e.g. popularity, cathartic release) they are willing to accept. In the current scenario, the staff member's right to choose what they disclose online and the resulting harm or benefits they are willing to accept (i.e. autonomy) need to be weighed against the potential for harm (i.e. maleficence) to their peers, the veterinary practice in which they work and the broader veterinary profession.

In this scenario the staff member has publicly berated their boss using Facebook. The resultant harm brought to their boss is likely to be both personal (e.g. embarrassment) and professional (e.g. reputation loss). In addition, demeaning or derogatory workplace comments contribute to the creation of a toxic workplace environment, which can in turn lead to reduced job satisfaction and increased burnout among all veterinary staff (Moore, et al. 2014). Further, the implications for the veterinary practice as a whole need to be considered. In a recent study exploring veterinarians' use of and attitudes toward Facebook, participants identified several personal experiences where venting on Facebook by other staff about work or colleagues had implications not only for

the person making the post but also the veterinary practice as a whole (Coe 2014). As shared by one veterinarian participating in the study:

"A fellow associate veterinarian made negative posts about our clinic that were sent anonymously back to the clinic. Although they were just 'venting', she didn't see that clients could see that post, and just how negatively it may have affected the clinic and other veterinarians."

Therefore, the potential harms brought to the veterinary practice, other members of the veterinary staff and the broader veterinary profession need to be accounted for in developing a response to this scenario.

Unfortunately, venting online is not absent from the veterinary profession (Coe, et al. 2011, Weijs, et al. 2013), and although there may be individual benefits perceived by the staff member who decides to vent their feelings and frustrations publicly on Facebook, greater consideration needs to be given to the potential harm the post could bring to their peers, the veterinary practice in which they are employed and the broader veterinary profession. In this scenario the prevention of potential harm to the staff member, their peers, the veterinary practice and the broader veterinary profession outweighs any benefit the individual posting the comment would receive. Furthermore, within at least some jurisdictions, this post may violate the employee's common-law duty to act in the best interest of their employer (Douglas C Jack, personal communication, 18 June 2015), negating the autonomy an individual holds to post whatever they choose to social media such as Facebook. In addition, the venting language of the post could also be considered libellous in that it may be perceived to defame the reputation of the employer by suggesting that the employer is lazy and lacks professional competency (Douglas C Jack, personal communication, 18 June 2015). In the end,

11.1

SOCIAL MEDIA AND THE VETERINARY TEAM
RESPONSE
JASON B COE

the staff members observing the post need to take action (e.g. discuss the potential consequences with the employee that made the post, notify management) to have the post removed or modified to eliminate the potential harm involved.

Arising from this scenario is also an opportunity to promote good (i.e. beneficence) for both the veterinary staff and the practice. Disclosure of questionable content through Facebook exists within the veterinary profession (Coe, et al. 2011, Weijs, et al. 2013) and this medium increases the potential for information to reach a broader audience (Coe, et al. 2012, Weijs, et al. 2014). As a result veterinary practices need to consider steps that can be taken to safeguard themselves, their staff and the veterinary profession against the inherent risks of social media. It would behove the veterinary practice in this scenario to take the opportunity to pursue a discussion and educate staff members about their personal use of social media and the associated benefits and risks for all stakeholders. Research within various populations has shown that raising an individual's awareness of the consequences of publicly posting questionable content to Facebook reduces online disclosure (Christofides, et al. 2012, Coe, et al. 2012, Weijs, et al. 2014). Understanding the risks associated with one's personal use of social media will allow all veterinary staff to maximise the benefits by avoiding the potential pitfalls.

In addition, proactively promoting a practice culture in which staff informally regulate one another's use of social media offers an environment in which indiscretions are managed quickly from within the veterinary team. This in itself affects staff learning on the issue while limiting the potential harms that could be incurred. In this scenario, educating and discussing the use of social media with staff members would better position them to immediately address their colleague in this type of situation by asking them to reflect on whether they

had thought through the intentions of the post and whether they had considered the harm it could bring to themselves, other staff members, their boss or the veterinary practice as a whole (Weijs, et al. 2013). Further, staff members are likely to feel more empowered within this practice culture to request that their colleague remove or modify the content of the post in the interest of reducing the potential harm it could cause.

|||||||||||||||||||||||||||||||||||||||||||||||||||||||||||||||||||||||||||||||||

COE'S discussion raises a number of important points around venting, privacy and social media. Whilst he has taken a principalist approach to analyse the scenario, one could emphasise virtue ethics here. The individual responsible for the post may be honest; however, they fail to demonstrate virtues we may expect, including compassion (for their employer and colleagues), trustworthiness (the person is venting to others, likely without their employer's knowledge, instead of addressing their concerns directly with their employer) and prudence in their judgement about what to post. The timing of the post may also be an issue. If the individual concerned is posting this during work hours, it is clearly interfering with their ability to perform their duties.

Increasingly, organisations draft social media policies which employees are required to abide by as part of their working conditions. This is, in part, to ensure that employees are aware of potential consequences of social media use in relation to the workplace; however, such policy may function to gag or censor employees.

For example, such guidelines may forbid posting that could damage the reputation of the organisation; could lead to criminal or civil liability; breaches confidentiality of employers, employees or contractors; may be construed to discriminate against, bully, harass or otherwise harm others; or may be seen to be making a comment on behalf of an organisation when the employee is not author-ised to do so.

More generic guidance may be available both within (RCVS 2015) and outside the profession. For example, in 2010 the Australian Medical Asso-ciation published *Social Media and the Medical Profession: A Guide to Online Professionalism for Medical Practitioners and Medical Students* (Australian Medical Association 2010).

These guidelines explain potential conse-quences of posting material on social media, even if it may not be construed as inappropriate at the time of posting. For example, it explains how seemingly innocent posts may breach client confidentiality. The guidelines also discuss the importance of professional boundaries, the need to maintain professionalism online, challenges in maintaining privacy and the destiny of data. As the guidelines state:

"although doctors and medical students are increasingly participating in online social media, evidence is emerging from studies, legal cases, and media reports that the use of these media can pose risks for medical profession-als. Inappropriate online behaviour can poten-tially damage personal integrity, doctor-patient and doctor-colleague relationships, and future employment opportunities. Our perceptions and regulations regarding professional behav-iour must evolve to encompass these new forms of media."

(Australian Medical Association 2010)

Veterinarians, nurses and animal health profes-sionals must appreciate – rightly or wrongly – that their online behaviour may be judged by the same standards by which their behaviour in the consul-tation room or workplace may be judged (Fawcett & Baguley 2011).

Although whistle-blowing is an important aspect of professional life if we see negligent

conduct by others that threatens animal welfare, we should look to rectify this in an appropriate way. Posting vitriol online is easy but unprofes-sional and likely to cause more problems for you, as well as your intended target. In one case, based on a real post, cited in the guidance provided by the Australian and New Zealand Medical Associ-ations (Australian Medical Association 2010), you can feel the frustration of the attending doctor, but also the inappropriateness of such comments to a wide audience:

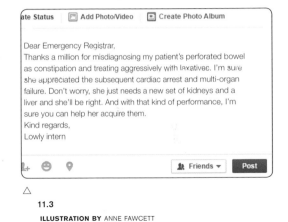

Dear Emergency Registrar,
Thanks a million for misdiagnosing my patient's perforated bowel as constipation and treating aggressively with laxatives. I'm sure she appreciated the subsequent cardiac arrest and multi-organ failure. Don't worry, she just needs a new set of kidneys and a liver and she'll be right. And with that kind of performance, I'm sure you can help her acquire them.
Kind regards,
Lowly intern

△
11.3
ILLUSTRATION BY ANNE FAWCETT

Allegations of negligence can also occur inad-vertently. Fawcett and Baguley (2011) discussed the potential ethical ramifications of a student mis-takenly posting that a dog's operation had been "botched" as an aside to their disappointment at missing a party. Not only would the student have to answer for their behaviour, which could include having to defend themselves in a court of law, but the vet concerned may find it difficult to appease the owner of the dog.

CHAPTER 11 TEAM RELATIONS

——
11.1
SOCIAL MEDIA AND THE VETERINARY TEAM
WHAT WOULD YOU DO?

SOCIAL MEDIA AND THE VETERINARY TEAM

What would you do?

A petition calling for the closure of a veterinary practice on animal welfare grounds appears in your social media feed. The preamble to the petition includes claims that the veterinarian who owns the practice has engaged in neglect and at times acts of cruelty towards hospitalised patients, and bullied employees. You note that the petition, instigated by a former employee of the practice that you studied with, has been signed by a number of your colleagues. What would you do?

△

11.4 A former employee of another practice circulates an online petition that claims, among other things, their former boss is neglectful and cruel to animals.

PHOTO ANNE FAWCETT

SOCIAL MEDIA AND THE VETERINARY TEAM

What would you do?

A client posts a complaint on your practice website about a colleague. The complaint states that the client's dog died because your colleague (named in full in the post) misdiagnosed the dog, treated it for a condition it did not have and failed to recognise the underlying condition from which the dog ultimately died. You know for a fact that this is not true as while diagnostic work-up was offered at your clinic, the client declined tests and opted for very limited treatment (antibiotics), even though extensive supportive care was recommended. What would you do?

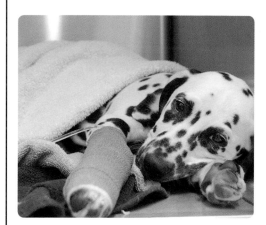

△

11.5 A client publishes claims their dog died as a result of mistreatment at your practice on social media, but you know this is not the case.

PHOTO ANNE FAWCETT

11.2

Substandard practice

Practising to the highest possible standard is a laudable aim and yet, on occasion, circumstances may conspire against practitioners and gold standard care is not possible. When substandard care becomes routine for an individual or practice, team members may feel overwhelmed or incapable of challenging the status quo. Locum practitioners brought in to cover absent staff will see a wide range of ways of doing things as they go from practice to practice. This may allow them to glean the best methods, ensuring a high level of competency. However, locums or other staff may also see situations that fall below what they consider acceptable, or they themselves may not meet the required standard. We look at both of these situations in the following two cases.

△
11.6 You are concerned about the practice standards and conditions.
CARTOON SUHADIYONO94

SCENARIO

GETTING YOUR HANDS DIRTY
– SUBSTANDARD CONDITIONS

▶   Your friend owns a practice and has asked you to locum for a month while she travels. Within hours of arriving you are concerned about the practice standards and conditions.

The anaesthetic machine leaks and thus staff use total-intravenous anaesthesia. An eye enucleation (normally a sterile procedure) is performed in the treatment room – the same room in which dentistry, abscess lancing, wound dressing and so on are performed. Radiographs are performed in an adjoining building. Instead of placing an intravenous catheter to maintain venous access for total intravenous anaesthesia, staff simply tape the needle with the syringe attached onto the forelimb as they carry the animal, typically resulting in the vein being blown. The attitude to sterility seems at best cavalier, with one vet removing a urinary catheter from the sterile packet, holding it in her mouth as she repositions the patient, then proceeding to place the catheter.

What should you do?

RESPONSE

ALISON HANLON

▶   Temple Grandin talks about "when bad becomes normal" and this is the situation faced by the locum. This scenario is particularly relevant to new veterinary graduates – what should you do if your new workplace has poor clinical standards, putting the health and welfare of the animals at risk?

There are a number of options to support ethical decision-making such as a resolution process advocated by Rollin (1999):

- Firstly, create a conceptual map of stakeholders and list the corresponding duties to them to identify the ethically relevant issues.
- Check to see if there is guidance in the Professional Code of Practice or legislation relevant to the case.
- If not, apply your own personal ethics – if there are two or more conflicting principles, then adopt a pluralistic approach by applying ethical theories such as utilitarianism.

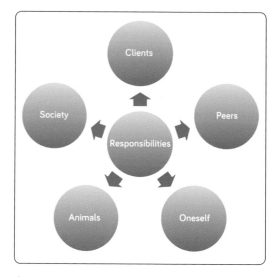

△

Conceptual Map: Professional responsibilities.

Poor professional standards have implications for a number of stakeholders such as the locum (oneself), the veterinary team, all patients and clients of the practice, and extend to other animals, the animal-owning public, other veterinary clinics, the veterinary profession in general, veterinary suppliers and society. What professional obligations does the locum have towards each of these stakeholders? Examples of ethically relevant issues relating to the stakeholders include:

- The practices are substandard, conflicting with the locum's personal ethics.
- Clients are paying for a professional service and expect that their animals are treated according to recognised professional guidelines.
- Substandard practices put the reputation of individual veterinarians and the wider profession at risk.
- Animals have an interest in being well and avoiding suffering.
- Vets have a trusted role in society and substandard practices are a breach of trust.

Once the stakeholders and ethically relevant issues have been identified, the next step in the process is to check professional guidelines, codes of practice or legislation that can be applied to the circumstances. The RCVS Code of Professional Conduct (2015) contains professional responsibilities relevant to this scenario:

- Veterinarians "must maintain minimum practice standards equivalent to the Core Standards of the RCVS Practice Standards Scheme".
- Veterinarians "must not impede professional colleagues seeking to comply with legislation and the RCVS Code of Professional Conduct".
- "Veterinarians must not engage in any activity or behaviour that would be likely to bring the profession into disrepute or undermine public confidence."

Furthermore, in the UK, the RCVS Practice Standards Scheme (www.rcvs.org.uk/practice-standards-scheme) provides guidance such as the number of consulting rooms, hygiene standards, routine maintenance of anaesthetic equipment and a system for recording the clinical outcomes of cases.

Despite clear guidelines, the outcome will often depend on the strength of character of the locum (see virtue ethics). The locum is in a difficult position with conflicting interests. For example, acting on concerns about the substandard practices could jeopardise your friendship with the practice owner and gain you the reputation of being a trouble-maker, thus affecting future employment opportu-nities. However, placing animals at unnecessary risk could equally damage your professional repu-tation and your future employment opportunities.

An inexperienced clinician may be concerned about rocking the boat, or feel intimidated and therefore do nothing. This can cause moral stress (Rollin 1999) and be detrimental to the health and wellbeing of the locum. Rollin contends, "the only way to alleviate moral stress is by way of moral action that is aimed at eliminating the practice giving rise to the stress." A "look before you leap" strategy is one approach, for example, arranging to meet your friend at the practice before agreeing to locum. This will help you to see the practice standards and discuss potential concerns with your friend in advance of any locum agreement. Checklists are available for selecting locums, which could easily be adapted by the locum to "vet" prospective workplaces (Brookfield 2011).

A more experienced locum may choose direct action. For example, having one-to-one discus-sions with members of the veterinary team in the practice, to raise awareness of the ethical issues, and providing the locum with an opportunity to voice their concerns. By so doing, it will enable the locum to adopt a different approach to proce-dures, even if veterinary colleagues are unrespon-sive or unsupportive. Ultimately, the locum should discuss their concerns with the practice owner, on her return. Good communication skills as well as having an understanding of ethical theories and decision-making frameworks are key competen-cies relevant to this case.

Acknowledgements
I would like to thank my colleague, Dr Alan Wolfe (Univer-sity College Dublin, School of Veterinary Medicine), for helping to unravel the clinical implications of this scenario.

|||||||||||||||||||||||||||||||||||||||||||||||||||||||||||||||||||||||||||||||||

THE difficulties faced by locums are not unique to veterinary practice. To cope with these it has been suggested that particular qualities are required for locum community doctors too:

> "They [locums] must be able to live with risk. Locum work is not for those who suffer sleep-less nights about the day's consultations or worry about periods of unemployment... Locums have to develop the ability to size up a practice, to slot into its team, and to protect themselves from risks in practices with poor standards and systems."
> (Fieldhouse & Harvey 2010)

The risks of working in substandard conditions may go beyond the immediate situation as such circumstances may actually predispose one to poor ethical decision-making. One study of doc-toral students found that greater prior exposure to unethical events was related to higher levels

△
11.7 "Locum work is not for those who suffer sleepless nights."
PHOTO ANNE FAWCETT

of unethical responses reported for hypothetical scenarios. The authors stated:

"exposure to unethical events led people to incorporate these events into their body of knowledge about how work is conducted. This knowledge is, in turn, used as people make decisions about how to conduct their work, with exposure to unethical events giving rise to unethical decisions."

(Mumford, et al. 2009)

Less experienced subjects exposed to unethical events were more likely to be affected negatively in their ethical decision-making than experienced subjects, and were negatively affected by a greater range of unethical events (Mumford, et al. 2009). There is good reason therefore to take care to protect particularly younger staff and students from such unethical practices during training and their early career.

As discussed by the author, virtue ethics can prove a useful lens for looking at such a scenario. Just as virtues can be role-modelled, so can vices. The virtuous team member would recognise that one cannot instantly rid a practice of vices, but will appreciate the need to role-model best practice as they see it. In using discernment, the virtuous team member would focus on identifying the most pressing problems and addressing these. For example, if the team is generally cavalier about sterility, spending time discussing its importance, demonstrating best practice and supporting team members who make positive changes are all possible, even within an imperfect environment.

Of course if a practice is below standard then it is likely in breach of legislation and related professional codes. A deontologist may report the practice to the relevant veterinary authorities. This is likely to sour future relations with the team, but it may effect change more rapidly.

SUBSTANDARD PRACTICE

What do you think?

In some countries there are schemes that accredit veterinary practices, for example using a combination of self-certification and annual inspection.

ONE — What elements of a practice do you think would be most suitable for accreditation?

TWO — What elements would be very difficult to certify?

THREE — Would you be in favour of such a scheme?

△
11.8 Feedback from formal practice inspections can help improve standards.
PHOTO ISTOCK

This next scenario considers what to do when the tables are reversed and it is the locum who is causing concern.

SCENARIO
SUBSTANDARD LOCUM

▶ You are a nurse in a companion animal practice. Your boss hires Ben, a locum, to cover for him while he is away at an overseas conference. As you give Ben a tour of the practice, he reveals that he graduated two years ago and is very keen to dive into consults and surgeries.

But things do not run smoothly. The first three clients who have consults appear flustered and ask when your boss is returning. The surgery nurse notes that Ben appears very uncertain, and in fact has to get a senior colleague to scrub in on all procedures.

Over lunch, you learn that Ben took a gap year after graduating and has only been working as a vet for two months. He will be working sole-charge every evening.

That evening, a 10-year-old dog is presented for acute collapse just before close of business.

△
11.9 What do you do if you feel a patient is receiving substandard care?
PHOTO ANNE FAWCETT

After performing an examination Ben collects blood, a procedure that is challenging due to the low blood pressure of the patient.

The nurse on duty suggests that he contacts a senior colleague, but Ben says he "can handle this".

He sends the dog home with the owners and promises to call them when the blood results are returned in the morning. The dog dies overnight.

What should you do?

RESPONSE
JOHN BAGULEY

▶ As a veterinary nurse you are faced with two options: do nothing or based on your suspicions confront the issue.

It is difficult for a recent graduate veterinarian to gain the respect and trust of clients, particularly when compared to an experienced veterinary practice owner whom many clients would have been seeing for many years. Recent veterinary graduates must also gain the respect and trust of paraprofessional staff, which is arguably even more important for their early career survival. Clearly, for various reasons, Ben has not been able to gain your respect or the respect of some clients. Recent graduates are in a very vulnerable position and positive feedback, support and patience can quite literally save a life. That does not prevent you or others from providing criticism but it does stress the need to create an environment where such feedback can be more readily accepted so that it will lead to positive change.

Whilst the first option presented was do nothing, the above response, whilst not directly confronting your suspicions, does provide a possible way forward with potentially the greatest good for the greatest number. To this you could add approaching Ben to tell him what a great job he did to get that blood sample and recall a story when you or another veterinarian struggled to get a sample or place an intravenous catheter and required the assistance of a colleague. When the results come back ask him why he thinks the dog died. It is vital to ensure you approach this task with respect and empathy.

An ideal workplace would provide an open, supportive environment and we would have sufficient respect, empathy and trust for each other, confidence in ourselves, and the skills to provide and receive feedback which would lead to further improvements in our individual and team performance. Clinical rounds, morbidity and mortality meetings, team involvement in the development of policies and procedures, and clinical audits would be some of the more veterinary-specific features of such a workplace.

It is important to note that even if we have developed this ideal workplace, it is highly unlikely that a new employee would readily fit into such an environment and easily fit into such a team. Once again, a starting position of respect and empathy will allow trust to develop and similarly a lack of respect and empathy may prevent trust from developing and ensure suboptimal team performance.

How could you more directly confront the issue? In this scenario a senior vet has already been involved in assisting Ben with surgeries. The most effective way to deal with your suspicions and confront the issue is to privately consult with this senior veterinarian and present your specific concerns. Your comments must be specific to be effective. If your suspicions are confirmed the responsibility for corrective action (from management, legal and ethical perspectives) lies with this senior veterinarian. If this veterinarian fails to act under these circumstances then you will be faced with those original options: do nothing or confront the issue.

Ultimately we are all responsible for creating an open, supportive environment which optimises

> "Starting from a position of respect and empathy for others and openly sharing your stories will enable you to help lead your colleagues into creating such an environment."

team performance and animal welfare outcomes. Starting from a position of respect and empathy for others and openly sharing your stories will enable you to help lead your colleagues into creating such an environment. Through our choices we can create the greatest good for the greatest number.

IN his response the author focuses on the importance of building an excellent team and the workplace qualities that could foster it. Great relationships with team members can be one of the most rewarding aspects of professional veterinary life (Bartram, et al. 2009b) as well as providing support during difficult times (Gardner & Hini 2006). But more than individual relationships within a veterinary team the whole team effectiveness can significantly impact on the job satisfaction and likelihood of burnout of individual team members (Moore, et al. 2014). In an analysis of a large number of studies of healthcare teams it was found that "collaboration, conflict resolution, participation, and cohesion are most likely to influence staff satisfaction and perceived team effectiveness" (Lemieux-Charles & McGuire 2006).

One way of approaching interpersonal relationship-building is to deal with difficult situations by aiming to embody the "virtuous" veterinary team member. Here we might strive to replace scornfulness with respect, harshly judgemental attributes with more forgiving ones and so on. From these virtuous states the correct actions will follow.

SUBSTANDARD PRACTICE

What would you do?

An experienced team member has inadvertently administered a tenfold anaesthetic overdose, resulting in a life-threatening complication for a patient. The patient recovers due to rapid and coordinated action by the team. Afterwards, the team member approaches you, distraught.

"I cannot believe I made such a stupid mistake," he says. "I should not be in this job."

What would you do?

△

11.10 A patient receives a tenfold anaesthetic overdose, but survives due to rapid, coordinated action from the team. How do you address the team member responsible?

PHOTO ANNE FAWCETT

SUBSTANDARD PRACTICE

What do you think?

You are concerned that a colleague is handling animals in a manner that causes animals to become more, rather than less, stressed and fearful.

ONE_____What factors would you take into account when deciding what to do?

TWO_____Is there a point at which you would simply report such handling to a superior, rather than raising your concerns with the colleague themselves?

THREE_____Where would this come?

△

11.11 Poor animal handling can lead animals to become more fearful, stressed and unsafe to handle. How do you address concerns about animal handling within the team?

PHOTO ANNE FAWCETT

11.3

Off duty

One of the nice things about veterinary-related jobs is that they are, on the whole, respected by society and can enable a close relationship with a particular community. But sometimes, as in this next scenario, that closeness may present its own problems.

———
SCENARIO
HEN'S NIGHT

▶ You attend a colleague's hen party at a local pub. Over the course of the evening, much alcohol is consumed and the group becomes rowdy, to the point that the publican requests that you leave for another venue. Everyone in the group, which includes vets and vet nurses, is wearing costume. As you walk down the street, you run into a client who stops you to say hello.

△
11.12 Hen nights are meant to be fun, but would you like to run into a client during the evening?
PHOTO ISTOCK

"What's the occasion?" she asks, surveying the group.

You point to the bride and say, "Its Julia's hen night."

At that very moment, Julia, who has been the recipient of numerous drinks bought by well-wishers, vomits on the street and bursts into tears.

"Doesn't she work in your practice?" the client queries.

What should you say?

———
RESPONSE
JOHN BAGULEY

▶ By getting drunk and vomiting in the street Julia has created a situation which is potentially embarrassing for her, for her colleagues and her profession but does that mean veterinarians are not allowed to get drunk?

Firstly, it is important to remember that veterinarians are responsible for providing care to animals and veterinarians must not be affected by alcohol whilst at work. Whether this is covered by veterinary law, employment law, contract law or not covered by any law in different jurisdictions is irrelevant. No one would accept their surgeon operating whilst under the influence of alcohol, their pilot flying a plane under the influence of alcohol or their taxi driver driving under the influence of alcohol. Alcohol consumption leads to impairment of reaction time, reasoning, judgement, memory and coordination and a big night out can lead to a hangover and fatigue. If you choose to get drunk you should ensure that you are not under the influence of or affected by alcohol consumption when you next go to work.

Secondly, a fundamental human right is that we are free to make choices within the boundaries created by legislation as well as administrative (legislated) and official codes of conduct and our personal values but with this freedom comes responsibility; ethically we must take responsibility for the choices we make.

In most jurisdictions the law does not prevent individuals from getting drunk (although there may be a technical legal issue with the responsible service of alcohol in this scenario). Administrative codes of conduct are legislative codes that are generally confined to situations when you are acting as a veterinarian; getting drunk when not working as a veterinarian and not subsequently going to work affected by alcohol would most likely not be a breach of an administrative code. However, conduct not during the course of your professional practice may be considered in breach of an administrative code if it is considered to be sufficiently connected to practice or sufficiently serious to be considered incompatible with practice and therefore test the notion of the veterinarian being of good character.

Official codes of conduct generally apply within the confines of your "membership" but can again apply in personal situations. For example, there have been examples of sportspeople found guilty by their sporting clubs of acting in a way that brings their club or their sport into disrepute despite the behaviour being neither on the field, within the club nor when representing the club. In most cases a sanction would be applied before removing someone as a member due to breach of an official code. Such decisions appear to be rare in veterinary professional organisations but we are moving into a brave new world of social media; what if the client takes a video of the bride-to-be vomiting in the street and posts it online with the caption "that's my vet"? Social media allows an indiscretion which may previously have only been known to a small few to be viewed by millions. Does that increase the likelihood that such an indiscretion will have an adverse effect on the standing of the profession?

So, what do you say to your client?

In addition to deontological and consequentialist approaches to ethical dilemmas we could also consider virtue ethics in this scenario. If you read this scenario and answered "well of course the answer is yes" then you most likely possess the virtue of honesty. Your honesty defines who you are, you could not be dishonest, and you have the practical wisdom (phronesis) to deal with this situation honestly and tactfully, aware of the consequences but, importantly, you are not acting honestly simply because of the consequences of being dishonest. Your honesty and wisdom provide you with a happy, fulfilling life (eudaimonia) which others also recognise.

If, like me, you still need just a couple more years to reach these heights, you can still apply either a deontological or consequentialist approach to derive the answer "I must be honest with my client". To lie is wrong (deontological) or if you lie to your client and this lie is discovered your client will believe you are dishonest (consequentialist). Similar consequences result from an attempt to avoid the question or somehow provide a response which is "economical with the truth". Honesty is considered a personal characteristic and as such transferrable between your private life and your professional life.

Fundamentally all professions rely on trust. The public trusts that you have the knowledge and skill to perform your work as a veterinarian and society affords professions the privilege of self-regulation based on this trust. Any deception will erode that trust in you and may erode some trust in the profession. Most people will forgive a veterinarian for over-indulging but most people will not easily forgive a veterinarian who lies. For many, the fundamental reasons for honesty in this situation will be derived from either "doing the right thing" or the possible consequences of not being honest, prompting an honest response but strictly speaking not necessarily a virtuous response.

11.3
OFF DUTY
WHAT WOULD YOU DO?

||||||||||||||||||||||||||||||||||||||||||||||||||||||||||||||

OFF-DUTY obligations of professional veterinarians are hard to define, and in contrast to guidance about working practices professional guidance may be limited to avoiding bringing the profession into disrepute. Usually the threshold for disrepute is high and associated with substantial criminal activity. But it's clear that the consequences of off-duty mild "misbehaviour" may be greater for professionals than others because undesirable character traits or actions may be thought to negatively impact on one's ability to carry out one's professional activity. But, people are flawed, and less boring for it, and when indiscretions do occur they can be easily compounded by trying to cover them up.

OFF DUTY

What would you do?

A junior vet you employ has produced a calendar of nude photographs of herself which she intends to sell for charity. The calendar was shot in your clinic. This involved the staff member bringing two photographers in on weekends.

The images are sexually suggestive and some feature her covering her private parts with equipment or animals. In one case she used a wildlife patient who subsequently died due to unrelated causes, but you are concerned that she may have held the animal roughly. In another case she has used the practice cat to cover herself.

How should you respond?

△

11.13 An employee poses for photos for a charity calendar in the practice.
CARTOON RAFAEL GALLARDO ARJONILLA

11.4

Colleagues in distress

Veterinary mental health has long been under the spotlight. Veterinarians and nurses are at increased risk of psychological distress and suicide compared with the general population and other healthcare professionals (Bartram, et al. 2009a, Milner, et al. 2015, Nett, et al. 2015, Platt, et al. 2010, 2012b). Recent evidence has suggested that such predispositions may be present even before university (Cardwell, et al. 2013), raising important questions about the characteristics that are associated with being drawn to veterinary professions and the selection process for training. In one study, between 49 per cent and 69 per cent of veterinary students reported levels of depression at or above the clinical cut-off during early training (Reisbig, et al. 2012) and in another study 54 per cent of students had experienced mental ill-health (Cardwell, et al. 2013). However, an intervention focused on improving mental health of veterinary students through five weekly hour-long group sessions was shown to be effective at reducing depression and the stress associated with homesickness and unclear professional expectations (Drake, et al. 2014).

Amongst graduated veterinary professionals surveys investigating mental health have shown, for example, depression affecting 31 per cent of respondents (Nett, et al. 2015), signs of minor psychological distress affecting 37 per cent of female veterinarians (Shirangi, et al. 2013) and suicide ideation occurring in 17–21 per cent of respondents (Bartram, et al. 2009a, Nett, et al. 2015). Some risk factors for poor mental health and increased suicide rates have been identified as long working hours, heavy workload, managerial duties, poor work–life balance, difficult client interactions and performing euthanasia (Platt, et al. 2012b, Shirangi, et al. 2013) although difficulties

outside work were important contributory factors for many (Platt, et al. 2012a).

Being regularly exposed to or inducing death has been hypothesised as contributing to poor mental health and suicide risk. Amongst veterinary technicians and animal care staff those directly involved in euthanasia reported higher levels of stress (Scotney, et al. 2015). Ready access to means of suicide, e.g. self-poisoning or firearms (Milner, et al. 2015, Platt, et al. 2010, 2012a), coupled with poor mental health, has proven to be a toxic and sometimes lethal combination for the profession.

In this next scenario we consider how to deal with a colleague exhibiting signs of distress at work.

SCENARIO

COLLEAGUE TAKING MEDICATION

▶ You are working in an equine practice in a regional area. You are suspicious that a colleague who has recently returned to work following surgery for intervertebral disc disease has been accessing methadone from the practice supply.

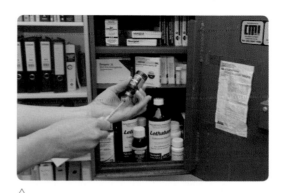

△

11.14 You suspect that a colleague who has been experiencing personal upheaval has been accessing restricted drugs.
PHOTO ANNE FAWCETT

Over the past few months he has been increasingly moody and challenging to speak to, appearing at times to be visibly in pain. He has recently separated from his wife of nine years and is negotiating custody of their children.

What should you do?

—
RESPONSE
SANAA ZAKI

▶ There are a number of issues to consider before a decision can be made about how to best manage this situation:

(1) In most jurisdictions methadone is a restricted substance and so there are legal requirements regarding its acquisition, storage and dispensation. By taking pharmaceutical products from the practice without informing the practice manager or paying for the goods, the veterinarian is in effect stealing. This veterinarian therefore may be breaking the law on a number of counts. This veterinarian's actions also constitute professional misconduct. This places his eligibility for continued registration as a veterinary practitioner into question.

In Australia, the Veterinary Practice and the Poisons and Therapeutic Goods Regulation are under the control of the State Government. The Veterinary Practitioners Code of Professional Conduct: located in the Veterinary Practice Regulation 2013 (schedule 2), states:

"A veterinary practitioner must not mislead, deceive or behave in such a way as to have an adverse effect on the standing of any veterinary practitioner or the veterinary profession."

"A veterinary practitioner must not obtain any restricted substance medications in order to take that substance himself or herself."

(2) Methadone is a synthetic opioid that has a number of actions on the nervous system. Not only is it a potent analgesic, but it also causes dose-dependent sedation and in humans produces a euphoric state of wellbeing where a person's perception of reality becomes distorted, their motor function compromised and their ability to make rational decisions is diminished. Chronic use results in the development of tolerance and addiction. Tolerance means a person requires increasingly higher doses of a drug to achieve an equivalent effect. Addiction is both physical and psychological, and can lead to irrational behaviour where the acquisition of the drug becomes the person's primary focus regardless of any negative consequences. Physical addiction manifests as severe withdrawal symptoms when there is sudden cessation of the drug following its chronic use. All of this means that this veterinarian may unknowingly be compromising the health and welfare of animals in his care and placing their lives at risk.

(3) This veterinarian is already under significant stress, both physically and emotionally. He suffers from chronic pain due to intervertebral disc (IVD) disease, and has recently had surgery on his back. According to the Global Burden of Disease study by the WHO, back pain is a major contributor to years lived with disability, ranking sixth in the world (Murray, et al. 2012). The veterinarian is also going through a marital separation, and is currently trying to negotiate custody of his children. Accusing him of taking a restricted substance for personal use may worsen

his mental and emotional state. Falsely accusing the veterinarian could tarnish his reputation and add unwanted stress to his already stressful life. It may also jeopardise his ability to gain custody of his children. The veterinarian is a colleague and staff member who needs to be supported if he is having personal and/or physical health problems. http://www.aihw.gov.au/back-problems/health-burden/

(4) Currently there is no evidence proving that the veterinarian is taking methadone from the practice for personal use. This is the conclusion you have drawn based on the veterinarian's unusual behaviour recently and repeated incidences of unaccounted-for reductions in the volume of methadone currently listed in the drug register.

(5) The practice owner is responsible for ensuring that the storage, reporting and dispensation of controlled substances meet government department regulations.

In Australia, Therapeutic Goods legislation for restricted substances (S8: drugs of addiction) requires a veterinary practitioner to make a record in the hospital patient database each time he/she prescribes one. This record must include: date of prescription, name and address of animal owners, species, name of drug, strength, quantity, number of repeats, repeat intervals, and directions for use shown on the prescription.

Restricted drugs must be stored separately in a room, safe, cupboard or drawer that is securely fixed and kept locked when not in use, and only accessible by a veterinary practitioner. A drug register must be kept where the drugs are stored and must detail receipt, supply and use. Details required are: date of entry, name and address of supplier or name and address of animal's owner, species, quantity received, supplied,

used and the balance, name of veterinarian authorising supply/use, and signature.

Poor record keeping at the practice makes it hard to trace whether drugs are going missing. The restricted drug register is not filled in properly. The patient medical records are also incomplete and very little detail is included in the patient record when pain relief is given or when animals are sedated.

If your concerns are reported to the regulatory bodies the practice owner will be held accountable for potential breaches of the Therapeutic Goods legislation for restricted substances. You know that an audit of the practice would identify several breaches in procedure when it comes to handling and reporting of restricted drugs.

(6) There are significant legal implications associated with this situation: unauthorised possession or use of a drug of addiction is an offence; it is an offence to make false or misleading entries in a drug register; and if a restricted drug is lost or stolen, it must be reported immediately by notifying Government Pharmaceutical Services.

So who are the stakeholders in this situation?

(1) The veterinarian suspected of taking the drugs.
(2) The family of the veterinarian (especially his children).
(3) The owner of the practice.
(4) The patients and clients of the practice.
(5) The government bodies that regulate controlled substances.
(6) The Veterinary Practitioners Board that licences veterinarians in NSW.
(7) The veterinary profession as a whole.
(8) You as the veterinarian who notices that this is happening.

Ethical decision-making process

In this situation, utilising one single ethical framework such as utilitarianism or deontology may not yield an outcome that protects all the stakeholders that are at risk of harm, or ensure compliance with the laws and regulations that govern how veterinarians practise.

The following approach to this complex ethical dilemma attempts to adhere to the guiding principles of non-maleficence (do no harm), beneficence (do good), autonomy (the ability to self-govern), and justice (fair distribution of benefits and consequences). It also aims to address the legal obligations that this case presents.

(1) Discuss your concerns about the vet's unusual behaviour with the other veterinarians in the practice.

Regardless of whether or not he is accessing methadone for his personal use, he is a colleague who obviously needs support and help while he deals with a very serious health issue and his current personal challenges.

(2) Nominate one of you to speak with the veterinarian. Voice your concerns about his unusual behaviour; offer support (including professional counselling) and give him the opportunity to explain his current behaviour.

(3) Work with the veterinarian to develop some strategies for reducing his stress levels. This may include taking some time off, reducing his work hours and/or the number of consultations he performs each day, reviewing the workflow in the clinic and modifying the set-up to minimise any added strain on his back.

(4) Review the practice's handling of restricted drugs, their storage and the associated record keeping. Is the register up to date and accurate? Do patient records reflect administration of the drugs that have been signed out to a particular patient? Are stocktakes done regularly and do current stock levels match what is recorded in the register?

(5) Remind all staff of the procedures that need to be followed when restricted drugs are prescribed. This may require staff training.

(6) Is there any evidence of missing restricted drugs? If so, this needs to be reported immediately to the Poisons and Therapeutic Goods Regulatory Body.

(7) Seek advice from your professional organisation about your suspicions that the veterinarian is self-prescribing methadone. They may be able to advise you on how to begin a discussion with the veterinarian.

|||||||||||||||||||||||||||||||||||||||||||||||||||||||||||||||||||||||||||

BY taking a principalist approach the author has incorporated a number of different ethical perspectives with minimising harm being central. Although we usually value autonomy highly, legislation restricting the sale and use of drugs, including those that have a professional use, has been invoked by competing deontological ideological principles surrounding the immorality of self-harm, neglecting social responsibilities, harming others and so on. A utilitarian approach would focus on means to reduce the negative impacts on the user and wider community. Finally, virtue ethics may be used to invoke support for compassionate harm-reduction programmes (Christie, et al. 2008).

The fact that there are seven steps proposed to deal with this difficult situation illustrates how there is rarely a single action that should be invoked in response. Most ethical dilemmas are multifaceted, and dealing with these elements can be time-consuming and require different skills and potentially help from within or outside the team.

COLLEAGUES IN DISTRESS
What do you think?

ONE   **Do you think you know signs of mental ill-health to look out for in colleagues?**

TWO   **What signs in particular would encourage you to seek advice about a colleague?**

THREE   **Whom would you turn to for help for yourself or a colleague?**

△
11.15 Would you know the signs of mental ill-health to look out for?
PHOTO ISTOCK

Conclusion

Working within a team brings its own rewards as well as being a potential source of tension. The support offered by a team is beneficial to both team members and patients. Preventing and handling difficult situations requires careful communication, particularly as new technologies may make it difficult to convey meaning accurately and social media can broadcast problems far and wide. Public displays of unwise behaviour may be considered unprofessional and have negative consequences within the local community, for example through loss of trust by clients, or more widely, for example, by eroding trust in the profession as a whole. High levels of mental ill-health amongst veterinary professionals are a cause for concern and all team members should aim to ameliorate rather than exacerbate stress in themselves and others.

△
11.16 At its best, a functioning veterinary team works together to achieve the best outcomes.
PHOTO ANNE FAWCETT

References

Armitage-Chan E, Maddison J, and May SA 2016 What is the veterinary professional identity? Preliminary findings from web-based continuing professional development in veterinary professionalism. *Veterinary Record* **178:** 318.

Australian Medical Association 2010 Social Media and the Medical Profession: A Guide to Online Professionalism for Medical Practitioners and Medical Students.

Ballantyne KC, and Buller K 2015 Experiences of veterinarians in clinical behavior practice: a mixed-methods study. *Journal of Veterinary Behavior – Clinical Applications and Research* **10:** 376–383.

Bartram DJ, Yadegarfar G, and Baldwin DS 2009a A cross-sectional study of mental health and wellbeing and their associations in the UK veterinary profession. *Social Psychiatry and Psychiatric Epidemiology* **44:** 1075–1085.

Bartram DJ, Yadegarfar G, and Baldwin DS 2009b Psychosocial working conditions and work-related stressors among UK veterinary surgeons. *Occupational Medicine-Oxford* **59:** 334–341.

Brookfield C 2011 Practice tips: checklist for hiring a great locum. *Canadian Veterinary Journal* **52:** 439–440.

Cardwell JM, Lewis EG, Smith KC, Holt ER, Baillie S, Allister R, and Adams VJ 2013 A cross-sectional study of mental health in UK veterinary undergraduates. *Veterinary Record* **173:** 266.

Christie T, Groarke L, and Sweet W 2008 Virtue ethics as an alternative to deontological and consequential reasoning in the harm reduction debate. *International Journal of Drug Policy* **19:** 52–58.

Christofides E, Muise A, and Desmarais S 2009 Information disclosure and control on Facebook: are they two sides of the same coin or two different processes? *Cyberpsychology and Behavior* **12:** 341–345.

Christofides E, Muise A, and Desmarais S 2012 Hey Mom, what's on your Facebook? Comparing Facebook disclosure and privacy in adolescents and adults. *Social Psychological and Personality Science* **3:** 48–54.

Coe JB 2014 Reputation management: managing your risk to maximize the benefits of social media. In: *Proceedings of the Australian Veterinary Association Annual Conference.*

Coe JB, Weijs CA, Christofides E, Muise A, and Desmarais S 2011 Teaching veterinary professionalism in the Face(book) of change. *Journal of Veterinary Medical Education* **38:** 353–359.

Coe JB, Weijs CA, Muise A, Christofides E, and Desmarais S 2012 Understanding veterinary students' use of and attitudes toward the social networking site, Facebook, to assist in developing curricula to address online professionalism. *Journal of Veterinary Medical Education* **39:** 297–303.

Drake AS, Hafen MJ, and Rush BR 2014 Promoting wellbeing among veterinary medical students: protocol and preliminary findings. *Journal of Veterinary Medical Education* **41:** 294–300.

Fawcett A, and Baguley J 2011 Social media menace? *In Practice* **33:** 190–191.

Fieldhouse R, and Harvey J 2010 Locums – The challenges of being a GP locum. *GP Online*, 15 April. http://www.gponline.com/locums-challenges-gp-locum/article/995946

Gardner DH, and Hini D 2006 Work-related stress in the veterinary profession in New Zealand. *New Zealand Veterinary Journal* **54:** 119–124.

Hartnack S, Springer S, Pittavino M, and Grimm H 2016 Attitudes of Austrian veterinarians towards euthanasia in small animal practice: impacts of age and gender on views on euthanasia. *BMC Veterinary Research* **12:** 26.

Lemieux-Charles L, and McGuire WL 2006 What do we know about health care team effectiveness? A review of the literature. *Medical Care Research and Review* **63:** 263–300.

Milner AJ, Niven H, Page K, and LaMontagne AD 2015 Suicide in veterinarians and veterinary nurses in Australia: 2001–2012. *Australian Veterinary Journal* **93:** 308–310.

Moore I, Coe JB, Adams CL, Conlon PD, and Sargeant JM 2014 The role of veterinary team effectiveness in job satisfaction and burnout within companion animal veterinary clinics. *JAVMA-Journal of the American Veterinary Medical Association* **245:** 513–524.

Mumford MD, Waples EP, Antes AL, Murphy ST, Connelly S, Brown RP, and Devenport LD 2009 Exposure to unethical career events: effects on decision making, climate, and socialization. *Ethics and Behavior* **19**: 351–378.

Murray CJL, Vos T, Lozano R, Naghavi M, Flaxman AD, Michaud C, Ezzati M, Shibuya K, Salomon JA, Abdalla S, Aboyans V, Abraham J, Ackerman I, Aggarwal R, Ahn SY, Ali MK, Alvarado M, Anderson HR, Anderson LM, Andrews KG, Atkinson C, Baddour LM, Bahalim AN, Barker-Collo S, Barrero LH, Bartels DH, Basanez MG, Baxter A, Bell ML, Benjamin EJ, Bennett D, Bernabe E, Bhalla K, Bhandari B, Bikbov B, Bin Abdulhak A, Birbeck G, Black JA, Blencowe H, Blore JD, Blyth F, Bolliger I, Bonaventure A, Boufous SA, Bourne R, Boussinesq M, Braithwaite T, Brayne C, Bridgett L, Brooker S, Brooks P, Brugha TS, Bryan-Hancock C, Bucello C, Buchbinder R, Buckle G, Budke CM, Burch M, Burney P, Burstein R, Calabria B, Campbell B, Canter CE, Carabin H, Carapetis J, Carmona L, Cella C, Charlson F, Chen HL, Cheng ATA, Chou D, Chugh SS, Coffeng LE, Colan SD, Colquhoun S, Colson KE, Condon J, Connor MD, Cooper LT, Corriere M, Cortinovis M, Courville de Vaccaro K, Couser W, Cowie BC, Criqui MH, Cross M, Dabhadkar KC, Dahiya M, Dahodwala N, Damsere-Derry J, Danaei G, Davis A, De Leo D, Degenhardt L, Dellavalle R, Delossantos A, Denenberg J, Derrett S, Des Jarlais DC, Dharmaratne SD, et al. 2012 Disability-adjusted life years (DALYs) for 291 diseases and injuries in 21 regions, 1990–2010: a systematic analysis for the Global Burden of Disease Study 2010. *Lancet* **380**: 2197–2223.

Nett RJ, Witte TK, Holzbauer SM, Elchos BL, Campagnolo ER, Musgrave KJ, Carter KK, Kurkjian KM, Vanicek CF, O'Leary DR, Pride KR, and Funk RH 2015 Risk factors for suicide, attitudes toward mental illness, and practice-related stressors among US veterinarians. *JAVMA-Journal of the American Veterinary Medical Association* **247**: 945–955.

Platt B, Hawton K, Simkin S, Dean R, and Mellanby RJ 2012a Suicidality in the veterinary profession: interview study of veterinarians with a history of suicidal ideation or behavior. *Crisis – the Journal of Crisis Intervention and Suicide Prevention* **33**: 280–289.

Platt B, Hawton K, Simkin S, and Mellanby RJ 2010 Systematic review of the prevalence of suicide in veterinary surgeons. *Occupational Medicine-Oxford* **60**: 436–446.

Platt B, Hawton K, Simkin S, and Mellanby RJ 2012b Suicidal behaviour and psychosocial problems in veterinary surgeons: a systematic review. *Social Psychiatry and Psychiatric Epidemiology* **47**: 223–240.

RCVS 2015 Code of Professional Conduct for Veterinary Surgeons. http://www.rcvs.org.uk/advice-and-guidance/code-of-professional-conduct-for-veterinary-surgeons/pdf/

Reisbig AMJ, Danielson JA, Wu T-F, Hafen MJ, Krienert A, Girard D, and Garlock J 2012 A study of depression and anxiety, general health, and academic performance in three cohorts of veterinary medical students across the first three semesters of veterinary school. *Journal of Veterinary Medical Education* **39**: 341–358.

Rollin BE 1999 *An Introduction to Veterinary Medical Ethics: Theory and Cases.* Iowa State University Press: Ames.

Scotney RL, McLaughlin D, and Keates HL 2015 A systematic review of the effects of euthanasia and occupational stress in personnel working with animals in animal shelters, veterinary clinics, and biomedical research facilities. *JAVMA-Journal of the American Veterinary Medical Association* **247**: 1121–1130.

Shirangi A, Fritschi L, Holman CDJ, and Morrison D 2013 Mental health in female veterinarians: effects of working hours and having children. *Australian Veterinary Journal* **91**: 123–130.

Weijs CA, Coe JB, Christofides E, Muise A, and Desmarais S 2013 Facebook use among early-career veterinarians in Ontario, Canada (March to May 2010). *JAVMA-Journal of the American Veterinary Medical Association* **242**: 1083–1090.

Weijs CA, Coe JB, Muise A, Christofides E, and Desmarais S 2014 Reputation management on Facebook: awareness is key to protecting yourself, your practice and the veterinary profession. *Journal of the American Animal Hospital Association* **50**: 227–236.

CHAPTER 12
WORKING
WITH THE LAW

Legal frameworks aim to reflect the ethical values of the society they serve. Usually legislative requirements are deontological in nature and correspond with core baseline duties placed on a citizen. These legal "bottom lines" may be exceeded by other ethical values that are not enshrined in law. For example, it may be a legal imperative to tell the truth in a court of law, but it is only an ethical principle to tell the truth more generally.

How each of us interacts with our legal duties is an ethical decision in itself. Some people consider it more important to be law-abiding than others, even when they think that, as in the words of Mr Bumble, the fictional Dickens character, "the law is an ass – an idiot"! Legislation changes slowly, such that the law may be slow to reflect the values of society. There are many instances, for example, where the law disadvantages particular groups of people, such as laws permitting slavery, and legislation that discriminates against women,

non-heterosexuals, and people of different ethnic or cultural background including Indigenous peoples to name but a few. Indeed, it is argued that the law – by treating non-human animals as a type of property – disadvantages non-human animals (Cao 2015).

Veterinary surgeons, just as other citizens, have to consider how and whether they will act in accordance with the law, bearing in mind that unlawful action can result in professional as well as legal consequences. Nonetheless there may arise a situation where one's professional duty, for example as expressed in an oath, conflicts with the law. In such situations veterinarians may "act according to that which is right, not simply that which is regulated" (Webster, personal communication, December 2015).

Unlike many other citizens, veterinary surgeons are often operating in roles where they are required to "police" the legislative requirements in a given jurisdiction. Finally, veterinary professionals may be asked to give advice on the legislative compliance of their clients' proposed actions.

A review of relevant legislation in different countries is beyond the scope of this book. In this chapter we will explore ethical dilemmas pertaining to both obeying and policing laws.

◁

12.1 Practices that are legal in some countries or jurisdictions may be prohibited in others. Cockfighting is illegal in many countries but is permitted in some.
PHOTO ANNE FAWCETT

12.1

Animal welfare legislation

Legislation to protect animal welfare is common around the world but not yet universal. The Global charity World Animal Protection have ranked countries based on a range of indicators on their "Animal Welfare Index". The index includes a theme called "Recognising Animal Protection" that has a range of indicators covering the degree of legal protection for animal welfare. Of the top 50 countries producing beef, poultry, pork, sheep and goats, milk and eggs, World Animal Protection gave its highest rating (A) for "laws against causing animal suffering" to Austria, Denmark, New Zealand, Sweden, Switzerland, The Netherlands and the UK, and the lowest ranking (G) to just four countries: Azerbaijan, Belarus, Iran and Vietnam (WAP 2014). The OIE Terrestrial Animal Health Code states that "legislation should contain, as a minimum, a legal definition of cruelty as an offence, and provisions for direct intervention of the Competent Authority in the case of neglect by animal keepers" (OIE 2015).

In the following scenarios we will explore whether farm animals are legally fit to travel, from two different angles.

△

12.2 Upon gathering the sheep it's clear some are in poor condition.
PHOTO ISTOCK

severely lame and one has injured her leg badly, possibly during the gathering, but has kept up with the flock coming in by walking on three legs. Mrs W will make less money than usual from the lambs as they have not grown as well as in previous years. She is keen to make as much money as she can from her cull ewes by sending them to the abattoir. She wants you to state in writing that the animals are fit to travel.

What should you do?

RESPONSE

PETE GODDARD

▶ The scenario describes a not uncommon dilemma and one where views may differ so it is important to consider relevant legal and ethical aspects. While the author is steeped in UK legislation, an attempt will be made to ensure the response is applicable more generically.

Of course, being presented with such a scenario as part of clinical veterinary work is

SCENARIO

FIT TO TRAVEL?

▶ Mrs W, a sheep farmer, has gathered her sheep from the extensive grazing area they've been on all summer. It has been a difficult year, with bad weather and poor grass growth. The lambs will be sent to slaughter and Mrs W is considering which ewes to retain for breeding. Several of the ewes are in poor condition, some are disappointing for both Mrs W, the owner, who is forced to confront shortcomings in her stock-person skills, and the attending veterinarian as there were likely to have been some preventative or advisory measures that could have been put in place which would have eliminated or reduced the severity of the problems. Also the vet will undoubtedly have a compassionate view towards Mrs W's difficult financial situation. It is, however, not uncommon for enterprises where profits are marginal or non-existent to fail to properly feed and tend to their sheep. Unfortunately, the situation as described does not allow the clock to be turned back and action is required.

There are two main issues described: the poor condition of the ewes and the lameness in the ewes (in at least one case with a ewe walking on three legs). Lameness is one of the most common signs of ill-health and discomfort in sheep and clearly there are adverse welfare effects. The level of lameness in a flock may also indicate poor over-all welfare.

In the UK, the Royal College of Veterinary Surgeons (and veterinary jurisdictions in many countries) is rightly very strict about certification so the vet must adhere to this high standard. Animals which are unfit to transport must not be transported (unless to veterinary premises for veterinary care providing the condition will not be worsened or cause unnecessary pain during transport). Thus none of the sheep which are described as severely lame are likely to be fit to travel as transport would result in additional suffering and the veterinarian cannot sign to allow them to do so. (Were Mrs W to transport them against veterinary advice they would be identified at ante-mortem inspection in the abattoir lairage and a prosecution against her might be taken). As some of the cull ewes were described as being in poor general condition, this may also impact on their fitness to travel and if they are in a severely undernourished state (that may influence their ability to cope with the journey) the

veterinarian likewise cannot sign to say that they can do so – though in this case there may be a greater element of subjectivity involved. However, animals suffering from unintended weight loss, particularly chronic weight loss, could be considered unfit. The fact that the ewes are going for slaughter rather than to market does not affect the decision about fitness to be transported. (It is worth recognising that the ewes in poor condition may realise very little value once transport and abattoir costs are deducted).

Fitness to travel is a potentially difficult issue especially if there is not a marked degree of lameness or poor condition (and other reasons not included in this scenario). While it has been the subject of much debate, the overriding requirement is that no animal should be transported unless it is fit before the intended journey starts and is likely to remain sufficiently fit throughout the journey and unless the journey conditions will not cause it unnecessary suffering. In the current case, the hardship of Mrs W may also be troubling the veterinarian for humanitarian reasons but this should not affect her/his clinical judgement (for which s/he would be liable). The severely lame sheep are not fit to transport; they have a painful condition which would result in additional suffering and they would likely arrive in a worse condition than that when they set out. In some cases sick or injured animals may be considered fit for transport if they are only slightly ill or injured and transport would not cause additional suffering. However, any painful condition obviously affecting the normal walking of the animal would render it unfit for transport. Animals must be able to walk freely and unassisted on all limbs without the need for more than the normal encouragement. So an animal showing lameness is likely, in the majority of circumstances, to be in pain and must not be transported (unless for veterinary diagnosis or treatment) as this would be likely to exacerbate the injury.

The best course of action in the current scenario may be for the vet to diagnose the cause(s) of lameness and suggest appropriate treatment and care to allow the ewes to become sound and be sent for slaughter at a later date. This should include a time frame for the sheep to respond to recommended treatment. During this time there is an opportunity for their condition to improve (feed and housing resources permitting) and treatment considerations would need to include any drug withdrawal time advice. In the case of the ewe unable to support its weight on one leg, a treatment or an immediate on-farm euthanasia decision would need to be taken, the latter to be followed by collection through a fallen stock scheme (or on-farm burial where permitted), with Mrs W fully engaged with the decision process. Accurate diagnosis as to the cause of lameness across all the affected ewes would also allow a recommendation for more long-term flock foot care to help avoid problems for other sheep in the future.

Since body condition is also poor, it is quite possible that Mrs W's flock has a parasite problem or she may simply need advice about appropriate nutrition. However, in both these situations, the marginal nature of the enterprise may mean that there is no money to be spent on additional feed or anthelmintics. The attending vet could at least advise on best use of pasture resources and on management actions to reduce parasite exposure in the future. It may be that the number of sheep on the farm exceeds the carrying capacity and reducing stock numbers may, in following years, lead to a lesser number of lambs in better condition and with fewer problems. Part of the advice could result in a switch from involuntary culling of lame or otherwise ill sheep to a more selective voluntary culling of older ewes that would allow a targeted flock improvement programme. Thus there is an excellent opportunity for the attending vet to add value to this enterprise through basic flock care

and nutritional advice detailed as part of the flock's health plan, thereby establishing a proactive relationship with the owner. Information could be as straightforward as introducing or enhancing the use of body condition scoring and the setting of target scores for different classes of sheep at different points in the production cycle – again to be detailed in the flock health plan.

Finally, an ethical difficulty is presented to the attending vet if the poor condition and lameness are such as to suggest a present or ongoing welfare concern but the owner fails to act. In this circumstance, asking a colleague to inspect the sheep to provide a second opinion and hopefully reiterate the initial view would be a sensible step before initiating any action involving the relevant authorities, though this may be needed to protect the welfare of the sheep as a last resort, as is the veterinarian's ethical duty.

||||||||||||||||||||||||||||||||||||||||||||||||||||||||||||||||||

COMING under pressure to act unlawfully will likely happen from time to time to many veterinarians, particularly where there is a conflict between the interests of a client and relevant legislation. Veterinarians may use a utilitarian approach in weighing up costs and benefits (the costs may be losing the client's respect, trust and income, while the benefit may be avoiding a charge that one is complicit in animal cruelty or neglect). But is a utilitarian analysis alone appropriate in such cases?

ANIMAL WELFARE LEGISLATION

What do you think?

Feeding backyard hens kitchen scraps is illegal in many countries to prevent disease spread but may improve the welfare through increasing pleasure derived from feeding. In what way would the outcome for the animal influence your actions regarding unlawful activity of a client?

△
12.3 Hens may enjoy kitchen scraps.
PHOTO ANNE FAWCETT

This next scenario considers what might happen when a veterinary surgeon makes a different decision to that presented in the preceding case.

SCENARIO
DIFFERING VETERINARY OPINION
ON FITNESS TO TRAVEL

▶ You are the Official Veterinarian at a rural abattoir. A small-time farmer brings his trailer with two large fattening pigs in it for slaughter. He says they are "a bit off their legs" but he's already been to his vet who suggested he drive them straight for slaughter without delay, a journey of approximately eight minutes, and he has a note from the practice to this effect. On opening the trailer it appears that each pig is severely lame, and you suspect one has a broken leg. In your opinion these pigs were not fit to travel to your abattoir.

How should you proceed?

RESPONSE
ED VAN KLINK AND PIA PRESTMO

▶ The law, in Europe in this case the Council Regulation on welfare of animals during transport (European Council 2005), is quite clear: no animal is allowed to be transported unless it is fit for the intended journey. Animals are not considered to be fit for transport if they are unable to move independently without pain or to walk unassisted. Based on this, the Official Veterinarian has every reason to enforce their opinion.

The farmer can be considered to have dealt with the problem in good faith. If the farmer is in doubt as to whether animals can be transported, the legislation recommends to seek advice from his vet. As he has done this, and the practising vet

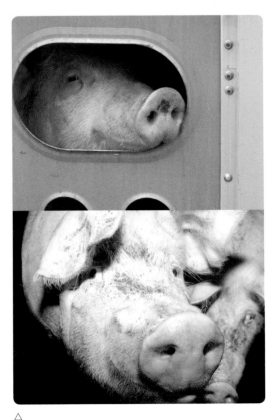

△
12.4a–b When the trailer is opened you find both pigs are severely lame.
PHOTOS ISTOCK

case the Official Veterinarian needs to collect all relevant evidence and write a witness statement that needs to contain the words that the animal suffered unnecessarily.

However, the reality of life means that it is impossible to act upon each and every case of animal suffering that the Official Veterinarian is witness to. It is a fact that there will be animals transported that do suffer. Also, there are probably quite a number of vets in practice, who do not exactly know what is required and acceptable, and who give advice like this in good faith, thinking that recommending to take the animals to the nearest place to be killed, is in the animals' best interest. At the same time, both the farmer and the practitioner may have thought that this was the best of the options, if, for example, the animals would have had to wait considerably longer to be killed on-farm by either the practitioner or a third person.

In cases like this, the animals would probably have been fit for emergency slaughter, which means that they can be killed on farm, after an ante-mortem inspection that can be done by the practising vet. While the practitioner would hopefully have been on the farm to see the animals when writing the note about the animals being fit enough to travel the short journey, he/she could have written an emergency slaughter certificate for the same or a lower fee. The process here would be that the practitioner assesses whether there are public health risks associated with the slaughter of the animal, and if that is not the case, the animal can be stunned and bled on farm and transported dead, accompanied by an emergency slaughter certificate, to the nearest abattoir. The emergency slaughter certificate must contain a proper declaration of the reason of emergency slaughter.

The second option, not to report the colleague, would have to result in the Official Veterinarian getting in touch with the practitioner to explain what should have happened, and what could have

has given him the advice that the animals can be moved, as quickly as possible and over as short a distance as possible, he is no longer liable.

The Official Veterinarian at the abattoir therefore has two options: to report or not report the colleague. If the Official Veterinarian chooses to make a case and report the practitioner who gave the faulty advice to the relevant authority they will then take appropriate action. In actual fact, the Official Veterinarian is obliged to report, because the non-compliance has taken place and action needs to be taken. In order to make the been the consequences should it have come to a case. Educating the practitioner in this way can be expected to be much more effective in preventing this kind of problem from happening again. Also, it is the much cheaper option as to make a case will involve further investigation, legal action and possible convictions, resulting in the practitioner concerned being struck off or suspended for a certain period.

||||||||||||||||||||||||||||||||||||||||||||||||||||||||||||||

THE implication of this utilitarian analysis is that greater good will be served through having a "quiet word" with the practitioner and, as such, this should trump the deontological and legal duty to report. Whilst this might seem a dereliction of duty on the one hand it is possible to incorporate this type of softer approach to animal welfare law enforcement into legal processes. The potential to use written warnings to insist upon changes to improve welfare before a prosecution is considered can be enshrined in law (for example in the UK under the Animal Welfare Act 2006 amongst others). In addition, guidance on how to deal with infringements may be issued to those involved in law enforcement that, depending on the severity or the certainty of the infringement, could range from verbal advice, through warning notices insisting on improvement and finally to prosecution in law courts (Havant Borough Council 2013).

ANIMAL WELFARE LEGISLATION

What do you think?

How might you define the threshold at which you would abandon the softer, collaborative approach to legislation breach and report an owner?

△

12.5 Softer approaches to law enforcement always require dialogue.
PHOTO ISTOCK

This next scenario (overleaf) explores when and how to decide what level of law enforcement is required when faced with farmers with personal difficulties.

SCENARIO

A DAY IN THE LIFE OF A
FARM ANIMAL WELFARE INSPECTOR

▸   You are a local authority animal health inspector and attend Mr G's farm in follow-up to animals being presented at market in poor condition. There are around 60 dairy cows, their calves and most of last season's young stock. You are aware that nine months earlier the owner had been served an improvement notice to ensure prompt

△
12.6a–b On a follow-up inspection you find the condition of some of the cows and calves is still poor.
PHOTOS ISTOCK

treatment of lame cattle and to improve poor body condition through additional feeding. This had been complied with and, after a number of visits with advice and support, at the last visit four months ago the condition of the stock had markedly improved. However, on this revisit, you find again that some of the housed young stock are in poor condition, there is no dry lying area for them and some are extremely lame. You call the local animal health office and the attending government veterinary inspector confirms some of the cattle are in a worse condition than the previous year, exacerbated by the poor housing conditions. In consideration of the worsened situation and the similar welfare problems identified you caution Mr G and explain that evidence will be collected in relation to the non-compliances found and that he will be invited to provide a statement. You supply Mr G with your local authority guide on enforcement policy. Further improvement notices are served by the veterinary inspector with one week's notice given to attend to the housing conditions and the lameness and two months to improve the poor body condition scores.

Mr G, who looks physically exhausted, then explains he has been caring for his elderly mother recently and has been unable to do everything as he should have. He cannot afford to get staff help and community nursing support for his mother has been limited. You provide Mr G with details for contacting the Farm Community Network to see if they can offer her any support at this time. The following day, having read through the guidance on enforcement action he telephones the office and breaks down in tears, saying he cannot afford a lawyer to go to court and thinks this will affect his mother badly as well as her ability to care for the stock.

What should you do?

RESPONSE

SOPHIA HEPPLE

▶ Whilst your immediate response may be not to take this any further on hearing Mr G's mitigating situation, as a local authority inspector you have acted correctly in cautioning Mr G under the circumstances; welfare non-compliances have been identified, the same as before and in a worsening situation, also confirmed by the veterinary inspector. Many local government authorities have specific guidance with respect to whether to take investigations forward for presentation to the court, as guided by the Regulators' Code (Department for Business Innovation & Skills 2014) in the UK. However, this does not mean that they should not be properly investigated in the first instance.

The Regulators' Code is produced to support the statutory principles of good regulation (under the Legislative and Regulatory Reform Act 2006). This Act requires local authorities to have regard to the statutory Principles of Good Regulation when exercising a specified regulatory function (as defined by the Legislative and Regulatory Reform (Regulatory Functions) Order 2007). For local authorities, these include those activities enforcing animal health and welfare. This means the welfare enforcement activity you carry out must be performed in a way that is:

(1) Proportionate – your activities reflect the level of public risk and enforcement action taken will relate to the seriousness of the offence;

(2) Accountable – your activities will be open to public scrutiny, with clear and accessible policies, and fair and efficient complaints procedures;

(3) Consistent – your advice will be robust and reliable and will respect advice provided by others. Where circumstances are similar, you will endeavour to act in similar ways to other local authorities;

(4) Transparent – you will ensure that those you regulate can understand what is expected of them and what they can anticipate in return; and

(5) Targeted – you will focus your resources on higher-risk enterprises and activities, reflecting local need and national priorities.

There are two further influences on the local authority approach to actions including potential sanctions and penalties.

Firstly, when deciding whether to prosecute, the local authority must also have regard to the provisions of The Code for Crown Prosecutors (Director of Public Prosecutors 2013), a public document that sets out the general principles to follow for the Crown Prosecution Service (CPS) when decisions are made in respect of prosecuting cases. The Code sets out two tests that must be satisfied, commonly referred to as the "Evidential Test" and the "Public Interest Test":

(A) Evidential Test – is there enough evidence against the defendant?

When deciding whether there is enough evidence to prosecute, you need to consider what evidence can be used in court and is reliable. The Crown Prosecutors must be satisfied there is enough evidence to provide a "realistic prospect of conviction" against each alleged offender.

(B) Public Interest Test – is it in the public interest for the case to be brought to court?

The Crown Prosecutors will balance factors for and against prosecution carefully and fairly, considering each case on its merits. The public interest factors that are taken into account are: seriousness of offence, culpability of the offender, circumstances of and harm caused to victim, age of offender, impact on community, is prosecution proportionate (specifically in relation

12.1

ANIMAL WELFARE LEGISLATION
RESPONSE
SOPHIA HEPPLE

to court costs) and whether sources of information need protecting.

Secondly, action taken should reflect the principles set out in the Macrory Review (2006) on making sanctions effective. Sanctions should:

(A) aim to change the behaviour of the offender;
(B) aim to eliminate any financial gain or benefit from non-compliance;
(C) be responsive and consider what is appropriate for the particular offender and regulatory issue, which can include punishment and the public stigma that should be associated with a criminal conviction;
(D) be proportionate to the nature of the offence and the harm caused;
(E) aim to restore the harm caused by regulatory non-compliance, where appropriate; and,
(F) aim to deter future non-compliance.

So as a local authority inspector you already have the Regulators' Code (2014), the Macrory Review (2006) and the Code for Crown Prosecution (2013) influencing decision-making regarding further action after the evidence collection stage. Now these are all in essence ethical codes of conduct with respect to treating offenders or suspects in a similar manner, although not necessarily equally. Further, the term "proportionate" is stated in all three guides. All codes appear to follow the principles of "retributive justice" such that the "punishment fits the crime". This is where the punishment deserved depends on the magnitude H of the wrongness of the act and the person's degrees of responsibility r or degree of flouting of the wrongness, and is equal in magnitude to their product: $r \times H$, where r ranges from 0 (zero responsibility of flouting of the law) to 1 (full responsibility / flouting of the law).

Nozick (1981) argues that retribution is not the same as revenge, which requires a personal element of pleasure in exacting punishment for a wrongdoing. This is why the various codes refer to elements of fairness, transparency, independence and consistency in approach to investigating and taking action on wrongdoing which reflect a more structured, dispassionate approach, with specified limits associated with retributive justice. The "public interest test" essentially provides for interpreting the r element – personal responsibility and/or degree of flouting of the law.

This is now where judgement on further action comes into play. It could be argued that irrespective of what is going on in Mr G's life, he has full responsibility for the cattle under his care and should be meeting their basic needs and that he should face full prosecution, particularly as he has previously acted wrongly towards them in the past. However, Mr G's responses made under caution and his actions in the intervening time between the inspection and decision-making can certainly influence the decision.

For example, it could be that under caution, Mr G readily admits to wrongdoing and recognises that his behaviour has not been that of a reasonable stockperson. Since the inspectors' visit he has been in touch with the Farm Community Network (FCN) and has had some help from the local members in sorting out the young stock, bedding up and supplying feed when he has had to take his mother to hospital. The veterinary inspector subsequently reports that lame animals have been appropriately treated, the cattle have dry lying areas and separation of young stock into appropriate groups has reduced aggression and bullying at feeding time. In this case then, in line with the Macrory principles, the offender has already modified his behaviour. If the evidential test under the CPS code is sufficient then the public interest test would argue that a "simple caution" may suffice.

In contrast, it could be that Mr G takes the attitude that he has done nothing wrong, as responsibility for his mother takes clear priority over the cattle care. The young stock remain in a poor state with only one visit by the private vet to attend to the lameness and Mr G further justifies his lack of action because of a failure of the community health care team to support him in caring for his mother. In this instance, if the evidential test is sufficient then it may be in the public interest to take this forward as a prosecution.

These two examples reflect Nozick's position on the non-teleological approach to retributive justice in that punishment should link the wrongdoer with correct values, and is a vehicle whereby the nature and magnitude of the offender's wrongness have a corresponding significant effect in his or her life. If the threat of prosecution alone has resulted in a behavioural change to correct wrongs, this may suggest further formal action is not necessary.

||||||||||||||||||||||||||||||||||||||||||||||||||||||||||||||||||||||||

THE detailed guidance about when to prosecute that is available for UK inspectors may not be mirrored in all jurisdictions. Then, it will be up to each individual to make the decision for themselves, taking into account the circumstances of both the animals and the owner. Factors commonly associated with poor farm animal welfare were identified by government and private vets in one study in Ireland as "physical and social isolation among herd owners, addiction and mental health problems, including depression" (Devitt, et al. 2014). In that study the authors explored dilemmas experienced in the course of enforcement duties undertaken by vets, noting that these often arose when it was difficult for vets to determine whether their primary responsibility was to the animals or the person. In addition, they found that "empathy and feelings of attachment and proximity to the human situation in turn influences the decision-making process… of the government veterinarian" but recognised that "empathy towards the animal owner can encourage greater compliance with veterinary recommendations" (Devitt, et al. 2014).

ANIMAL WELFARE LEGISLATION

What would you do?

You attend the farm of Ms D following a complaint by a neighbour about the welfare of a small group of sheep on the premises. On inspection the sheep are in poor condition, with some bordering on emaciation. One of these is collapsed, dehydrated and unable to stand. The ewe's temperature is subnormal and she is generally unresponsive. You determine that this ewe requires emergency euthanasia. When you explain that the animal needs to be euthanased, that you are willing to do this now or that Ms D can call her private vet out, she screams that these sheep are her only family and that you cannot kill them. She insists that her animal just needs some treatment and that she would rather die than have her 10-year-old pet sheep killed. Ms D has no routine private vet on enquiry. She provides a list of the three vets she has used in the past and two have outstanding unpaid bills.

△
12.7 Your suggestion of euthanasia of one ewe in very poor condition is met with horror.
PHOTO ISTOCK

Finally, we turn to a case of persistently poor farm welfare standards.

SCENARIO
LOW FARM WELFARE STANDARDS

▸   Your practice has been providing veterinary services to a small mixed farm for over 50 years. The senior partner in the practice remembers the farm when it was run by old Mr J. Even back then

△

12.8a–b In contrast to his neighbours Mr J fails to maintain his fields and buildings, compromising the welfare of his animals.
PHOTOS SIOBHAN MULLAN

it was ramshackle, with animals kept here and there, and everything done on a shoestring. Mr J's entrepreneurial spirit always meant there was something unexpected – a barn stuffed with turkeys fattening for Christmas, or an extra load of calves bought cheaply at the market.

Now Mr J's son John has let standards on the farm slip further. You have told him often enough to provide what seem to you to be the most basic requirements – good-quality food, shelter and care – to animals he appears to take pride in. You are sure these would be cost effective in the long run as you are often there to treat animals that have become run-down and have been left too long before calling you. Sometimes you see some improvements but somehow things always seem to slip back to how they were – or worse.

Today you are there to attend to three sick calves, one of which you don't think would have lived out the day. You spot a cow that's looking very thin but are told, "She's just put it all into her calf, that's all."

What should you do?

RESPONSE
PATRICIA V TURNER

▸   This scenario gets at a common ethical issue in veterinary medicine – whether the needs of the client trump those of their animals. While the client has a right to individual freedom of expression and they should be able to live their life in the manner they desire, this does not mean that continuing poor animal management or care practices can be tolerated indefinitely, particularly when this results in animal neglect, suffering and death. Just because a client is pleasant and is well-intentioned towards the animals they care for does not excuse ignorance or animal neglect. Similarly, while the production of a ready

"Just because a client is pleasant and is well-intentioned towards the animals they care for does not excuse ignorance or animal neglect."

source of nutritious food through farming may be considered a public good, Western society as a whole has been very vocal in recent years in indicating that the end result (the production of protein from an animal source, in this case) is not justified by any and all means. There is a public expectation that food animals will be well cared for during their lives and that they will be treated with respect. The veterinary professional is entrusted by the public as an expert who is expected to safeguard the health and welfare of client animals through attending to their needs. Thus, the veterinarian is morally obliged and expected by the public to try to rectify the problem in this scenario. In fact, in some jurisdictions, the veterinarian could be charged with professional misconduct for failing to recognise and directly deal with a situation of animal neglect or abuse.

Obviously, the veterinarian does not want to act as a whistle-blower and contact the authorities every time they deal with a sick animal on any farm as they risk losing the client in so doing, and thus the opportunity to act in a positive way to benefit animal welfare, in addition to lost practice revenue. In general, clients should be given an opportunity to rectify a situation through education, training and acquisition of appropriate skills. The fact that things have continued in the same way for over five decades on this particular farm suggests that the client has not been amenable to subtle or not-so-subtle suggestions for improving their practices.

The client needs to be informed clearly by the veterinary clinic that their animal husbandry practices are unacceptable according to current performance standards and that they must change if they are going to continue to farm animals in the future. This communication must also be documented in the medical records for this farm. The seriousness of the situation should be conveyed to the farmer. The possibility of having animals seized by an animal protection society can come as a shock to the client, particularly when they genuinely care for the animals that they are managing, and can help them to understand how serious the situation is. A written plan of action can be used to help the producer define goals and an appropriate course of action. Often the greater community in which these individuals live is also aware of ongoing problems with animal care on the farm and there may be peer support networks available that can be tapped into. For example, clients in these circumstances may be encouraged to work with local or regional animal producer groups on a voluntary basis to improve their practices and knowledge. Regular visits to the farm during the period of transition by these peer support producers as well as periodic visits by the veterinarian can help to ensure that the level of care for the animals is improving over time. Outright defiance, an unwillingness to change their methods or reverting to past practices are all signals that the action plan is not working. If the veterinarian believes this to be the case, they have a duty to notify the local animal protection authorities to conduct more formal on-farm animal welfare assessments.

HERE again a "softer" approach is advocated as a possible solution, at least initially. Private veterinarians may be unused to setting and monitoring such an action plan, and may feel conflicted in doing so. After all, shouldn't their practice have already solved this farm's problems over the last 50 years? The veterinarian may be influenced by an "escalation of commitment" bias that could easily come with serving the client for so long. This apparently irrational decision-making response, where the more involvement individuals have had in a dilemma, the more they appear blinded to poor outcomes, may explain why new, successful courses of action were not pursued earlier by the practice. This bias has been well studied for economic and business decisions but also in social and health care situations (Sleesman, et al. 2012). Amongst many possible explanations for commitment bias the continuing of a course of action as a justification of one's previous behaviour is one that may apply readily to the veterinary context. Indeed, the authors of a large meta-analytical review found "the desire to 'save face' and maintain one's reputation appears to be a strong situational force affecting the tendency to escalate" (Sleesman, et al. 2012). Other work has suggested that an individual's preference for a particular course of action is influential and they will therefore prefer to choose it even in the face of continued negative outcomes (Schulz-Hardt, et al. 2009).

De-escalating commitment bias has been shown to be possible in a number of ways, including through tapping into desires for personal growth and promotion (Molden & Hui 2011) and using meditation to focus on the present rather than the past and future (Hafenbrack, et al. 2014). In addition, focusing on one's core values that don't relate to decision-making, so-called "self-affirmation", was also found to be effective in reducing commitment bias (Sivanathan, et al. 2008). Any such mechanisms could be employed within the veterinary setting, where there is naturally a high likelihood of escalation of commitment biasing decisions about future courses of action.

ANIMAL WELFARE LEGISLATION

What would you do?

Mr S has been farming ducks for over 10 years and has mostly just about reached minimal legal standards in your opinion. Now he has joined with some other nearby producers to market his birds as "happy and local" with an assurance they are "inspected by vets". How willing are you to certify Mr S's birds? What are your main concerns? How could you bring about an improvement in standards?

△

12.9 Farm assurance schemes usually have clear standards and inspection processes to provide assurance to customers.
PHOTO SIOBHAN MULLAN

12.2

Legislation with animal welfare implications

Veterinarians and animal health professionals may at times engage with legislation or proposed legislation which, while not directly relating to animal welfare or the role of the veterinarian, may impact on one or both of these. A current example is so-called ag-gag legislation, as outlined in the following scenario.

△
12.10 "Ag-gag" legislation has been proposed to ensure the integrity of agricultural industries – but is this something the veterinary profession should support?
PHOTO SIOBHAN MULLAN

SCENARIO
"AG GAG" LEGISLATION

▶ You are a member of a veterinary association that is required to submit a response, on behalf of its members, to the government about legislation which has been proposed to ensure the integrity of agricultural industries.

The proposed legislation prohibits the taking of photos or video footage on or in an agricultural facility or property without permission and the publication of such material; requires that any material is immediately handed over to authorities if it is gathered; prohibits seeking employment with an agricultural business without disclosing ties to animal rights organisations; and imposes tougher penalties for trespassing on agricultural properties.

The legislation has been proposed in the months following the exposure of a major poultry producer for animal welfare violations thanks to footage obtained from hidden cameras planted by animal rights activists. As a result of an investigation sparked by the publication of disturbing images in the media, the producer was fined a large amount of money. The share price of the company dropped, but consumer confidence in other poultry operations also dropped to an all-time-low.

The veterinary organisation represents veterinarians who work in the poultry and other agricultural industries, as well as those who work in companion animal practice, public health and education.

Should the organisation support or reject the proposed legislation?

RESPONSE
JED GOODFELLOW

▶ Over the past decade there has been a marked increase in the use of covert surveillance by animal activists to detect and expose cases of animal cruelty, particularly within agricultural industries. The increasing prevalence of these private investigations can, in part, be attributed to their effectiveness in influencing major reforms in animal welfare practices within certain industries. In the USA and Australia for example, such investigations have led to criminal prosecutions, animal trade suspensions, multiple parliamentary inquiries, the largest meat recall in US history and the introduction of new regulatory requirements for livestock and slaughter facilities. When footage of cruelty to farm animals is published it often receives extensive media coverage and stimulates public debate about society's broader obligations to the welfare of animals used to produce food and fibre. The activists involved believe that these ends justify the law-breaking means that are often used to gather the necessary evidence and bring cruelty offences to public attention.

In response, livestock industries and their political representatives have sought to introduce laws designed to curb such investigations. The proposals often contain one or more of the prohibitions described in the above scenario, including prohibitions on taking footage of agricultural practices without permission, offences for publishing the resultant images and special trespass provisions that go beyond those already provided for under the criminal law. These proposals have been dubbed "ag-gag" laws by the US media as they are seen as an attempt to keep the public in the dark about animal cruelty and poor welfare practices that may be occurring within certain livestock industries.

The veterinary profession is often called upon to provide comment on such legislative proposals as it is regarded by the community as a trusted authority on all matters relating to animal health and welfare. The position taken by national veterinary associations can therefore wield significant influence in the policy and political debates surrounding such laws. The task for the veterinary association in formulating a considered position is not an easy one. There are many complex and competing issues to consider and the debate will invariably take place within a heated political environment due to the various interests at stake.

Those representing the agricultural sector will emphasise the risks to biosecurity, and to animal health and welfare, posed by activists entering farming facilities unlawfully. They will raise concerns about the property and privacy rights of farmers who may be covertly filmed and have their identities exposed.

On the other side of the debate, animal welfare groups will raise concerns over the lack of transparency within certain livestock industries and the limited government oversight to ensure welfare standards are complied with. In addition, consumer, media and free speech organisations may raise concerns over the potentially stifling effect such laws will have on the right to free speech and political communication regarding questions of animal welfare. In 2015, the US Federal Court considered these issues when it was asked to rule on whether a law containing ag-gag provisions in the state of Idaho infringed the First Amendment right to free speech. The Court held that it did and the law was struck down accordingly [see, *Animal Legal Defense Fund et al. v the Governor of Idaho* (2015)].

Balancing these competing interests will be a challenging exercise. As the representative body of a respected profession, a veterinary association cannot condone law-breaking behaviour. But questions must be asked as to whether ag-gag laws are the appropriate means of addressing such behaviour. Are they likely to achieve their desired ends and could there be unintended consequences of going down this particular legislative path? The association should consider how the broader public may perceive such laws and what may be the impact on consumer confidence if livestock industries are seen to be targeting activists and closing their doors at a time when demand for openness and transparency in food production has never been greater. An editorial by the *New York Times* (Editorial Board 2013) provides one example of how ag-gag laws may be received by the general public:

"The ag-gag laws guarantee one thing for certain: increased distrust of American farmers and our food supply in general. They are exactly the wrong solution to a problem entirely of big agriculture's own making."

Another helpful exercise for the veterinary association to undertake when considering its position may be to review the extent to which existing law already covers the "mischief" sought to be addressed by the proposed legislation. Do laws of criminal trespass, private surveillance regulation, and state and federal biosecurity legislation already provide legal mechanisms with which to address the conduct in question? If they do, questions must be asked about the true intent of the proposed law – is it really designed to protect animal health and welfare, and the legitimate property rights of farmers, or is it really directed towards silencing critics of certain livestock industry practices? These questions played a decisive role in the US Federal Court case mentioned above. The judge found that existing law already served the purposes that the ag-gag law purported to address:

"Laws against trespass, fraud, theft, and defamation already exist. These types of laws serve the property and privacy interests that the State [of Idaho] professes to protect.

The existence of these laws necessarily casts considerable doubt upon the proposition that [the ag-gag law] could have rationally been intended to prevent those very same abuses.

The overwhelming evidence gleaned from the legislative history indicates that [the ag-gag law] was intended to silence animal welfare activists, or other whistleblowers, who seek to publish speech critical of the agricultural production industry."

So as this brief overview has sought to demonstrate, ag-gag laws raise many significant public policy questions, and these are but a snapshot of the many vexed issues the veterinary association will have to contend with in determining its position on such laws. Ultimately, guidance for the veterinary association will be found in coming back to first principles of the profession, its fundamental values and core duties. Accordingly, potential impacts on animal health and welfare, both immediate and long term, direct and indirect, should be of core concern. It will be incumbent upon the veterinary association to consider all sides of the debate, to navigate its way through the political rhetoric and to consider each legislative proposal on its merits without fear or favour.

||||||||||||||||||||||||||||||||||||||||||||||||||||||||||||||||

THE above scenario is topical because so-called ag-gag legislation has been proposed in various states in the USA and Australia. In considering trespass in order to take footage of animal abuse, a utilitarian may argue that the means (trespass) is justified by the end (for example, prosecution of an individual, farmer, producer or organisation for offences of animal cruelty and neglect – if such is successful). A deontologist might argue the case that property rights and rights to privacy should be respected. Neither approach will solve the impasse between animal activists and representatives of industry – the former condones illegal behaviour, some of which may have a detrimental effect on welfare. The latter minimises the risk of exposure and prosecution of unscrupulous producers and eliminates transparency which may erode public trust.

LEGISLATION WITH ANIMAL WELFARE IMPLICATIONS

What do you think?

The above discussion points out numerous risks to animals, including the risk to animal welfare and biosecurity due to trespass by animal "activists", as well as risks to animal welfare due to lack of transparency and limited regulatory oversight.

ONE   **How might you address these competing concerns?**

⚠ **12.11** The most intensive farming systems tend to be targeted by activists.
PHOTO ISTOCK

12.3

Exotic and illegal animals

Veterinarians have the ability to treat all animals, yet the owning and treatment of some animals may be restricted or outlawed in some jurisdictions. This presents a dilemma on a number of fronts. In addition, there may be genuine confusion about the veterinarian's role.

For example, in Queensland, Australia, it is illegal to keep rabbits and ferrets as pets – but legal for veterinarians to treat them. Some rabbit owners attempt to evade the law by identifying rabbits as "long-eared guinea pigs" (Williams 2016). Additionally, there may be genuine safety considerations in treating dangerous breeds or species, for example, venomous snakes.

The veterinary team must consider the legal implications, the interests of the animal as well as other animals that may be impacted, biosecurity and public health aspects, the interests of the client and their role as health provider but also in some cases an obligation to police legislation. The following scenario explores issues around breed identification in a context where a particular breed is banned.

_____

SCENARIO
BREED IDENTIFICATION

▸ You are working at an animal shelter in a region where pit bull terriers are banned. You are presented with a friendly dog that looks like it could be a pit bull. You are required to perform a breed assessment.

How should you approach the assessment of this dog?

△

12.12a–b Photographs from Sophie Gamand's *Flower Power: Pit Bulls of the Revolution* series. As upward of 1 million pit bull-type dogs are destroyed in the USA per annum, Gamand sought to challenge the way we look at and treat these dogs. All dogs photographed in the series are from shelters. Gamand is concerned that pit bulls "have the false reputation of being more dangerous than other dogs, hence attracting irresponsible primary owners who are looking for a 'scary dog'".
FIGURES REPRODUCED WITH PERMISSION WWW.SOPHIEGAMAND.COM/FLOWERPOWER @SOPHIEGAMAND

—

RESPONSE
MARTIN WHITING

▶ In England the Dangerous Dogs Act 1991 applies, which classifies some types and breeds of dog as being potentially so harmful to the public that a special license is needed in order to own these dogs. Criticism of the legislation has existed since its enactment; the opposing view was that the control of dogs should relate only to their aggressive actions, rather than the breed of which they are part. The legislation does cover aggressive dogs of all breeds also, but it is the condemnation of entire breeds that has caused some people to object to this legislation. Similar legislation exists in several other countries, or it may be enacted in a different way, for example some shelters in other parts of the world will employ a policy of destruction of dogs belonging to breeds that are listed as dangerous or fighting breeds, or certain types of dog such as the pit bull type.

This creates a serious moral dilemma for the veterinarians involved. The dog, although potentially of a pit bull-type breed, may be otherwise healthy and non-aggressive, it could be a suitable family pet. However, the veterinarian is asked to comply with policy or legislation and certify this dog as either being a pit bull type, or not, the former of which will ultimately lead to its destruction. The importance of correct identification cannot be underestimated. A false positive could result in a dog which is not a pit bull type being euthanased, while a false negative could result in a pit bull type being saved and rehomed, when the policy or legislation is for its destruction. A false positive has obvious ethical concerns, which will be discussed below, but a false negative carries with it the potential for greater human harm. If a pit bull type was not identified as such, and then was rehomed with a family with small children, and then it went on to attack the children,

would the original veterinarian be liable for their misidentification? Could the harm to the children have been prevented?

The question of how to accurately detect a pit bull-type dog has perplexed courts and shelters since the legislation was first enacted. This is partly due to the fact that included animals in the UK legislation are not just pit bull terriers (of which there is a recognised conformational breed standard guideline) but it also includes pit bull-*type* dogs, those that closely resemble pit bulls. The most common method for determining a pit bull-type dog relies on a conformational analysis. This may be necessary as it might not be possible to determine the dog's heritage as this data may not be available. There is no conformational standard available from many of the kennel clubs and so, in the UK, the government produced a 15-point guide to determine pit bull-type dogs (DEFRA 2009). Some courts have permitted the use of genetic testing to determine the heritage of the dog. However, using this method to determine if a dog is a pit bull type, rather than just a pit bull by the American standard, is both costly and potentially inaccurate (Barnett 2011, Wisdom Panel 2013). The difficulty comes from determining the dogs on the edge of the pit bull definition, as genetic testing may only determine if they are of a pit bull heritage. The use of conformational measurements and ratios of conformational points tends to form the basis of much of the identification process. However, two studies found that visual identification is also an inaccurate process, highlighting a large inconsistency amongst staff in shelters regarding being able to determine if a dog was a pit bull or not (Hoffman, et al. 2014, Olson, et al. 2015). This difficulty in determining if a dog is a pit bull-type dog has promoted continuing research in the field of veterinary forensic science, for example, using hair samples from the dog (Wharton, et al. 2015). Wharton, et al. (2015) used specialised light microscopy to measure aspects of the dogs' hair such as thickness and medullary ratios. From these data, it may be possible to determine, against breed reference intervals, if a dog is of a pit bull type, as it can reference across many associated and similar breeds. There is a wealth of measurement tools available to the veterinarian, each with their own confidence intervals and areas of inaccuracy, which can be used to determine if a dog is a pit bull-type breed. The next question is to determine how this data is used by the courts.

It is an interesting quirk of UK courts that the burden of proof changes when they consider determining a dog to be a dangerous breed or not. Normally, in courts, people are considered innocent of the crime of which they are accused, until it is proven against them that they are guilty. The burden of proof shifts to the converse burden when dangerous dogs are considered. Any dog may be accused of being a pit bull type, and the burden of proof resides with the owner (or person responsible for the dog) to prove that it is not a pit bull. So, it is necessary to know in the courtroom how much the dog does not appear to be a pit bull type, rather than how much it does appear to be one. Each of the tests specified above can provide some rebuttable evidence to determine the likelihood of a dog being a pit bull type or not. Ultimately, it is then for the court to decide upon the likelihood. It is the role of the veterinarian to provide as accurate as possible determination using the available methods, and then to enact the verdict of the court.

The veterinarian is being asked to confirm the breed so that euthanasia of a (probably healthy) dog may be undertaken to comply with legislation or policy. This has its own ethical dilemma. In some countries or cultures, it is not permissible to euthanase healthy companion animals. Pit bull dogs can make very good companion animals, and can be induced to be aggressive like any other

dog. Veterinarians may feel ethically compromised to certify for destruction a healthy dog. There ought not to be a requirement for veterinarians to do this, as provisions for conscientious objections from certain domains of veterinary work are permitted in most national professional regulatory frameworks. It is an interesting complexity to veterinary professional ethics that in some domains of work, such as in meat hygiene, it is essential for a veterinarian to certify an animal is healthy for slaughter, while in other domains that involve companion animals, if they are it is often a cause of moral distress. This ultimately resides in the purposelessness of the destruction of a healthy dog that may never go on to be dangerous, while farm slaughter has the purpose of providing food or preservation of biosecurity for the safety of other animals.

The moral dissonance that is produced in certifying a dog for destruction, especially when fit and healthy, can cause some veterinarians to, legitimately, opt out of this work. But while it remains a necessity for some jurisdictions, there will always have to be veterinarians available to do this work, and ensuring that it is done with utmost accuracy will help to preserve the lives of those dogs who do not need to be euthanased and will uphold the legislation designed to protect people against dangerous dogs, no matter how illegitimate the premise of that legislation may be. It may be of comfort to some that the role of the veterinarian is to be objective and as professionally accurate as possible in providing evidence to the court. When the verdict is returned, the veterinarian is again required to undertake their duties professionally. The actual decision regarding the fate of an individual dog, however, is determined outwith the veterinarian's role.

|||||||||||||||||||||||||||||||||||||||||||||||||||||||||||||||||||||

BREED-SPECIFIC legislation (BSL) and policy exists in different countries, although it is controversial. As discussed, critics argue that banning a breed or type unfairly discriminates against a whole group of dogs, when in fact a minority of these may display aggressive or dangerous behaviour. Hence the catch-cry of the anti-BSL groups is, "blame the deed, not the breed."

BSL is opposed by some veterinary organisations on the grounds that breed alone is not an effective predictor of canine aggression; it is impossible to definitively identify some "dangerous" breeds using appearance or DNA technology; and BSL ignores the responsibility of humans (Australian Veterinary Association 2014). Another criticism of BSL is that it may give members of the public a false sense of security about breeds not listed as dangerous.

There may be some advantages. That veterinarians are not ultimately responsible for the fate of pit bull-type dogs that come before the law may be of some comfort to them. However, shelter staff and veterinary surgeons working in rehoming centres may find themselves directly responsible for the life or death of pit bull-type animals that cannot be rehomed if labelled as an illegal breed. One study found that 41 per cent of shelter staff admitted they may be tempted to overlook pit bull-type features of a dog and knowingly mislabel it as a non-restricted breed type, presumably to save the dog and give it a chance of being adopted (Hoffman, et al. 2014). The effect of this type of breed-specific legislation on the actions of shelter staff, who choose that vocation because of their love of animals, may not be what legislators were aiming for, but is perhaps unsurprising. This suggests that these people may be experiencing moral stress about the responsibility of labelling a dog when the label is attached to a particular fate.

In the above case, the author argues that perhaps the greatest good is achieved by upholding policy and ensuring that the rate of false positive identification is minimised, thereby mitigating harm. In practice this may not occur. From a consequentialist analysis, if 4 out of 10 dogs are mislabelled this suggests the system is failing. Moreover it does not correspond well with a principalist approach: staff mislabelling dogs are perhaps doing so because they feel that to label dogs means those dogs will be harmed, and the perception of the risk posed by these dogs may be so low that mitigating such potential harm may not count as beneficence. Shelter staff may feel that in being ordered to label dogs, rather than judge each dog's suitability for rehoming, they have no autonomy and that the only way to exercise this is to lie. Finally, if animals with good temperaments are destroyed because of an association with individuals that have bad temperaments, it seems unjust.

A consequentialist might argue that the harming of a single child justifies drastic action to eliminate the risk, but again does this really do so? From a virtue ethics approach it would seem another system is needed as there remains risk, a huge number of potentially rehomeable dogs are destroyed and we have published evidence that those charged with identifying breeds are not comfortable doing so.

LEGISLATION WITH ANIMAL WELFARE IMPLICATIONS

What do you think?

The purpose of breed-restricting legislation is to actively prevent harm through the requirements. Legislators weighed up the risk of harm by certain animals, the likely level of any harm (the hazard) and the estimated degree of enforcement of the legislation.

ONE   What non-breed characteristics would you consider to be more important than breed for reducing harm to humans by dogs?

TWO   How could you use regulatory approaches, including legislation and advisory guidance, to reduce the harm?

△

12.13 Regardless of the breed there are measures that can be taken by owners to reduce the likelihood of their dog causing harm to people.
PHOTO ISTOCK

Exotic species may be banned or restricted in some jurisdictions, again presenting a dilemma for the veterinary team: should they ensure that owners comply with legislation, or should they treat the animal and leave the policing to other authorities?

SCENARIO
ANACONDA

▶   You are a veterinarian with a keen interest in exotic and unusual pets. You have joined a prominent exotic practice to increase your exotics caseload. You learn that the boss is treating an anaconda belonging to a private client. This species is exotic to the country you are working in and is in fact illegal to keep (with the exception of some zoos that can apply for permits).

The animal is in poor body condition with generalised weakness. It is also suffering from stomatitis.

You ask your boss in conversation if the owner has a licence.

"No," he says. "But I won't turn them away and I won't report them. That will just push the owners underground. Someone has to help these animals."

You feel uneasy about this.

What should you do?

RESPONSE
MANUEL MAGALHÃES-SANT'ANA

▶   This scenario describes a genuine moral dilemma. On the one hand, you have a duty of loyalty and commitment to your workplace and veterinary colleagues, and your boss indeed expects you to act accordingly. On the other hand, you have the right to follow your conscience and abide by the law. Failing to do so could result in

12.14a–b A client presents a prohibited exotic species for treatment.

PHOTOS ANNE FAWCETT

use of three chief principles of normative philosophical traditions (respect for: wellbeing, autonomy, and justice) and combines them with several interest groups in a matrix. In this particular case, some of the most relevant interest groups include the anaconda (and other exotic animals), the client, the veterinary team (including your boss) and society at large (including the natural environment). Table 12.1 illustrates the application of the matrix to the veterinary treatment of protected exotic species.

It is not possible here to go into detail for all aspects of the matrix, but some points should be highlighted. The utilitarian principle of wellbeing aims to maximise happiness and minimise suffering. Neurological disease is a particular concern in reptiles and may be present here as evidenced by weakness and poor body condition. In this regard, treating the animal and educating the client regarding appropriate husbandry could, in the short run, improve the anaconda's health and welfare and that of other animals under the client's care, whereas seizing the animal and reporting the case to the authorities could prevent future harm to the anaconda but could also, in the long run, put other animals from the same client at increased risk of not receiving veterinary care or even. of being culled (1). The best interest of society and the environment should also be acknowledged. This animal may put other animals in the owner's collection at risk but can also represent a biosecurity hazard and a public safety risk, if it escapes (10). Conservation efforts also require that traded exotic animals and trafficking routes are identified, and exotic veterinarians have a particular responsibility in this respect (10).

From a deontological point of view, some principles must be respected regardless of their consequences, namely the Kantian principle of autonomy and the Rawlsian principle of justice. The freedom of exotic animals is at stake, and a protected species should be left in the

moral stress as well as legal liability. As is often the case with moral dilemmas, one can only expect to reach the *best possible* solution, since a *right* solution might not be available.

A useful tool for exploring this case is Ben Mepham's Ethical Matrix, which has been presented in chapter 2. The Ethical Matrix is a conceptual tool that aims to facilitate ethical reasoning and decision-making, which will help not only yourself, but also your boss (and the team) to deal with similar cases in the future. It makes

RESPECT FOR:	WELLBEING	AUTONOMY	JUSTICE
Exotic species	(1) Animal health and welfare	(2) Freedom	(3) Duty of care
Client	(4) Human–animal bond	(5) Confidentiality and informed choice	(6) Responsible ownership
Veterinary team	(7) Reputation	(8) Autonomous professional judgements	(9) Fairness in practice
Society and environment	(10) Public health and safety; conservation efforts	(11) Biodiversity	(12) Sustainability

△

Table 12.1 The ethical matrix applied to the veterinary treatment of protected exotic species. Boxes have been numbered for descriptive purposes.

wild or, if already in captivity such as this anaconda, moved into a sanctuary instead of living in a private house (2). As leading advocates for animals, veterinarians have a duty of care which might include providing appropriate treatment, performing humane euthanasia, if required, and reporting the case to the authorities (3). The veterinary team, on the other hand, is expected to act with professionalism. This involves framing their autonomous judgements in accordance with best professional conduct (8). Fairness in practice could include informing the owner of his options and responsibilities, and the legal duty to cooperate with authorities (9). Finally, from an environmental perspective, biodiversity and the intrinsic value of living creatures (11), as well as sustainability concerns, such as a fair exotic pet trade (12), should also be considered.

With all these concepts in mind, how should you proceed? Should you conscientiously object to treat this particular animal on environmental grounds? Probably not, unless there are reasons to believe that the owner is an illegal trader. In my view, a sound approach to this scenario would involve treating the anaconda and ensuring that its husbandry is improved. Animal health and welfare are compromised and, as an exotic veterinarian, it is your primary responsibility to relieve the anaconda's suffering and restore its health (1). This measure would also reinforce the human–animal bond (4). But your responsibilities are far from ending there. Your duty of care involves preventing harm to future animals, in addition to the anaconda (3). You also have the duty of raising the awareness of the veterinary team as to the welfare, ethical and legal issues that emerge from such a case, maybe by holding a work meeting on the subject (and maybe using the ethical matrix as an aid).

Amongst other possible solutions, you could try to convince your boss that, while his attitude

12.3
EXOTIC AND ILLEGAL ANIMALS
RESPONSE
MANUEL MAGALHÃES-SANT'ANA

can work for the first few animals, it fails to prevent the future captivity of protected species, and so a more structured approach is needed. This can include the development of practical ethical guidelines to be consulted in similar circumstances, and that may help prevent systematically treating illegal exotic species. Such a measure would have a positive impact on the reputation of the veterinary practice (7). As a way to raise awareness on responsible ownership (6) and promote informed client choices (5), your boss could try, in subsequent consultations, to convince the client to donate the animal to an animal sanctuary or zoo. If this is done via the practice, client confidentiality could be maintained (5) while fulfilling wider legal responsibilities (9).

|||||||||||||||||||||||||||||||||||||||||||||||||||||||||||||||||||||||||

WHAT if you didn't know that anacondas could not be kept without a licence? Some might argue that ignorance is bliss, and on some occasions that might be so. On the other hand, ignorance could be a blissful precursor to a sticky situation – ignorance is rarely an acceptable excuse for illegality. It may be prudent to employ a short delaying tactic with a client to buy some time to find out the exact regulatory position of the situation. Only then can an appropriate ethical decision be made.

Other factors to consider are the lifetime welfare of the animal. Wild capture may involve injurious and stressful handling, stressful transport and poor husbandry and sanitation during "storage", all of which may involve significant suffering to the animal (Warwick 2014). It has been argued by some authors that the majority of captive reptiles suffer prolonged stress, morbidity and premature mortality because they are non-domesticated and frequently subjected to poor husbandry (Warwick 2014).

In clinical contexts, veterinarians may be oriented towards taking steps for immediate risk reduction (that is, treating the patient and working with the client to improve husbandry) rather than adhering to more abstract guidelines, rules and legislation. This phenomenon has been documented in healthcare settings where – despite being aware of the longer-term risks of antimicrobial resistance – doctors would flout prescribing guidelines in order to manage immediate risks, their reputation and to ensure consistency with one's peers (Broom, et al. 2014). This bias in decision-making operated even where the prescribing doctors were aware of it. Thus, for example, at the moment of treatment one's relationship with the client and animal can be stronger than one's relationship with other stakeholders that may not be directly impacted by the actions taken.

LEGISLATION WITH ANIMAL WELFARE IMPLICATIONS

What do you think?

ONE _____ If an illegal animal escapes once it has been seen by the veterinary team, how complicit is the veterinarian?

TWO _____ What are your reasons for your answer?

△
12.15 Some bird of prey enthusiasts may be tempted to keep animals that haven't been properly licenced.
PHOTO ISTOCK

Conclusion

In this chapter we have considered a number of different scenarios where veterinary ethics and the law interact. It is only possible to frame any possible courses of action within any regulatory requirements if one is actually aware of the relevant legislation. There are many ethical dilemmas that involve legal considerations and some of these are covered in other chapters. For example, we discuss reporting of animal abuse in section 13.3, and insurance fraud in section 6.3.

References

Animal Legal Defense Fund et al. v the Governor of Idaho 2015 *USFC District of Idaho 1-14-cv-00104-BLW.*

Australian Veterinary Association 2014 Breed specific legislation policy. http://www.ava.com.au/policy/614-breed-specific-legislation

Barnett KB 2011 Breed discriminatory legislation: how DNA will remedy the unfairness. *Journal of Animal Law and Ethics* **4**: 161–205.

Broom A, Broom J, and Kirby E 2014 Cultures of resistance? A Bourdieusian analysis of doctors' antibiotic prescribing. *Social Science & Medicine* **110**: 81–88.

Cao D 2015 *Animal Law In Australia*. Lawbook Co. Thomson Reuters: Sydney.

DEFRA 2009 Dangerous Dogs Law: Guidance for Enforcers. Department for Environment, Food and Rural Affairs: London.

Department for Business Innovation & Skills 2014 Better Regulation Delivery Office: Regulators' Code.

Devitt C, Kelly P, Blake M, Hanlon A, and More SJ 2014 Dilemmas experienced by government veterinarians when responding professionally to farm animal welfare incidents in Ireland. *Veterinary Record Open* **1:e000003**.

Director of Public Prosecutors 2013 The Code for Crown Prosecutors. https://www.cps.gov.uk/publications/docs/code_2013_accessible_english.pdf

Editorial Board 2013 Eating with our eyes closed. *New York Times*, 9 April.

European Council 2005 Council Regulation (EC) No 1/2005 on the protection of animals during transport and related operations and amending Directives 64/432/EEC and 93/119/EC and Regulation (EC) No 1255/97. Official Journal of the European Union L 3/1-44.

Hafenbrack AC, Kinias Z, and Barsade SG 2014 Debiasing the mind through meditation: mindfulness and the sunk-cost bias. *Psychological Science* **25:** 369–376.

Havant Borough Council 2013 Animal Welfare Enforcement & Prosecution Guidelines. http://www.havant.gov.uk/animals/animal-welfare-enforcement-prosecution-guidelines

Hoffman CL, Harrison N, Wolff L, and Westgarth C 2014 Is that dog a pit bull? A cross-country comparison of perceptions of shelter workers regarding breed identification. *Journal of Applied Animal Welfare Science* **17:** 322–339.

Macrory RB 2006 *Regulatory Justice: Making Sanctions Effective.* Better Regulation Executive: London.

Molden DC, and Hui CM 2011 Promoting de-escalation of commitment: a regulatory-focus perspective on sunk costs. *Psychological Science* **22:** 8–12.

Nozick R 1981 *Philosophical Explanations.* Harvard University Press: Cambridge.

OIE 2015 Terrestrial Animal Health Code. OIE: Paris, France.

Olson KR, Levy JK, Norby B, Crandall MM, Broadhurst JE, Jacks S, Barton RC, and Zimmerman MS 2015 Inconsistent identification of pit bull-type dogs by shelter staff. *Veterinary Journal* **206:** 197–202.

Schulz-Hardt S, Thurow-Kroening B, and Frey D 2009 Preference-based escalation: a new interpretation for the responsibility effect in escalating commitment and entrapment. *Organizational Behavior and Human Decision Processes* **108:** 175–186.

Sivanathan N, Molden DC, Galinsky AD, and Ku G 2008 The promise and peril of self-affirmation in de-escalation of commitment. *Organizational Behavior and Human Decision Processes* **107:** 1–14.

Sleesman DJ, Conlon DE, McNamara G, and Miles JE 2012 Cleaning up the big muddy: a meta-analytic review of the determinants of escalation of commitment. *Academy of Management Journal* **55:** 541–562.

WAP 2014 Animal Protection Index. World Animal Protection: London.

Warwick C 2014 The morality of the reptile "pet" trade. *Journal of Animal Ethics* **4:** 74–94.

Wharton A, Bailey D, and Gwinnett C 2015 Pit bull type aid to identification protocol. In: *AWSELVA-ECAWBM-ESVCE Congress*, 2. Animal Welfare Science, Ethics and Law Veterinary Association: Bristol.

Williams P 2016 Rabbit owners try to convince police it is a guinea pig after illegal pet found. *ABC News.* Australian Broadcasting Corporation.

Wisdom Panel 2013 Wisdom Panel FAQ. http://www.wisdompanel.co.uk/why_test_your_dog/faqs/

CHAPTER 13
ONE HEALTH

Introduction

One Health is defined as "a worldwide strategy for expanding interdisciplinary collaborations and communications in all aspects of health and care for humans, animals and the environment" (One Health Initiative 2016).

In this model, human health, animal health and environmental health are inextricably linked. The One Health movement "seeks to promote, improve and defend the health and wellbeing of all species by enhancing cooperation and collaboration between physicians, veterinarians, other scientific health and environmental professionals and by promoting strengths in leadership and management to achieve these goals" (One Health Initiative 2016).

According to the One Health Initiative, therefore, proponents of One Health are dedicated to improving the lives of all species. One Health is best seen as an umbrella under which many disciplines and specialisations contribute to the overall health and wellbeing of humans, animals and the environment.

The One Health concept has been embraced by the Food and Agriculture Organisation (FAO),

the World Health Organization (WHO) and the World Organisation for Animal Health (OIE) in addition to many veterinary schools and professional organisations (Gibbs 2014).

One of the key drivers of the One Health movement over the past decade has been zoonotic disease, particularly severe acute respiratory syndrome (SARS), H5N1 influenza and rabies. In addition, up to 75 per cent of emerging and re-emerging diseases are zoonotic or vector-borne, highlighting the need to characterise disease pathophysiology in one species in order to develop management strategies for another (Mellanby 2015). Other factors that have prompted the One Health approach include, but are not limited to:

- Implications of an increasing human population;
- Geographic expansion of human populations increasing contact of humans with animals and vectors;
- The impact of climate change on vector and disease distribution, as well as livestock and food production;
- Increased awareness of the human–animal bond;
- Increased understanding that food security requires an awareness and management of the interaction between humans, animals and the environment;
- Concerns about antimicrobial resistance;
- Concerns about foodborne illnesses;

13.0
INTRODUCTION

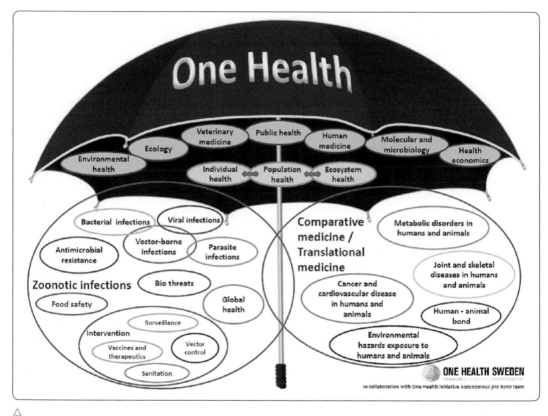

△

13.2

CREATED BY ONE HEALTH SWEDEN IN COLLABORATION WITH THE ONE HEALTH INITIATIVE AUTONOMOUS PRO BONO TEAM

- Increased interest in using naturally occurring models of human and animal disease (Mellanby 2015, One Health Initiative 2016).

Example

Experimental animal models are often used to study diseases that occur in human patients, with the knowledge gained used to benefit human patients. There have been public concerns expressed about the use of animals as experimental models of human disease, and scientific concerns that models such as mouse models do not accurately mimic human disease. However, companion animals may develop spontaneous diseases (such as some tumours) that are similar to those induced in experimental models. Companion animal veterinarians, in diagnosing and treating these patients, gather valuable data in phenotyping and treating disease, some of which may be useful in managing human patients with the same disease(s). By sharing data on spontaneous clinical cases, veterinarians, physicians and scientists can develop management strategies that benefit both humans and animals with

△

13.3 Companion animals may develop spontaneous diseases (such as this melanoma) that are similar to those induced in experimental models.

PHOTO ANNE FAWCETT

In *Human-Animal Medicine: Clinical Approaches to Zoonoses, Toxicants and Other Shared Health Risks*, the three basic areas of relevance of non-human animals to human health are explained in terms of benefits to human patients:

"First, a number of health risks are related to animal contact, including zoonotic infectious diseases transmitted from animals to human beings; and animal bites, stings and other direct trauma. Second, important psychosocial effects of the human-animal bond may have physical benefits as well. Third, animals may serve as 'sentinels' for toxic or infectious health hazards in the environment that are also a risk for human beings."

(Rabinowitz & Conti 2010b)

While the shared health risks of humans and animals are emphasised throughout, the implication is that animal health is of interest insofar as it impacts human health and wellbeing. Veterinarians must come to their own conclusions about how they serve the interests of their patients within this framework, and must fulfil public health duties without working beyond their professional scope of practice (Rabinowitz & Conti 2010a).

A newer, aligned concept is that of One Welfare. The One Welfare approach aims to highlight the direct and indirect benefits of animal protection to the welfare of humans (Pinillos, et al. 2015). In a similar approach to One Health, One Welfare encompasses multiple disciplines to achieve both human and animal wellbeing. Examples include reduction in abuse in humans and animals, improved welfare of livestock and farmers, improved animal welfare and food safety and improved conservation and human wellbeing (Pinillos, et al. 2015). There is a large degree of crossover between One Health and One Welfare, so we have elected to use the term One Health in this chapter.

these diseases – relying less on experimental models (Mellanby 2015).

While there are efforts to integrate a One Health approach into veterinary curricula, this approach has been criticised for a number of reasons, including:

- The definition is too broad, making it particularly hard to define goals and measure outcomes;
- Some members of the medical profession see One Health as a veterinary "land grab" for funding and resources (Gibbs 2014);
- The term has not been used widely in the media, possibly because there are multiple definitions used (Gibbs 2014);
- Professions remain relatively isolated by institutional and disciplinary barriers, hindering collaboration (Speare, et al. 2015);
- Many veterinary professionals work in the private sector with little to no scope for collaboration with medical professionals (Speare, et al. 2015).

13.1

Medicine for all?

SCENARIO

ADVANCES FOR ANIMALS VERSUS PEOPLE

▶ You have been asked to be interviewed for a local radio programme about the advances (and associated advancing cost) of veterinary medicine. It makes you think of one of your wealthy clients, Mrs A, who over the years has spent a small fortune on her pets. She has already spent £4000 on hip replacement for her Labrador, Amelia, and now the other hip appears painful and a likely candidate for surgery. She joked to you: "Oh, well, it's a third of the price of my husband's hip replacement!" You also think of another client, Mrs J, who has been waiting for nine months for her own hip replacement on the National Health Service, although her treatment will eventually be free. You know, of course, that in other countries there are plenty of people who cannot afford healthcare or insurance and would never be able to benefit from a hip replacement.

How should you approach this interview?

RESPONSE

ALISON HANLON

▶ This scenario focuses on the intrinsic and extrinsic value that we place on animals. For companion animals in particular, the human–animal bond is a key determinant of value. For a beloved dog or cat, if they can afford to, the owner may be willing to spend vast sums of money on veterinary treatments to ensure the health and wellbeing of their companion.

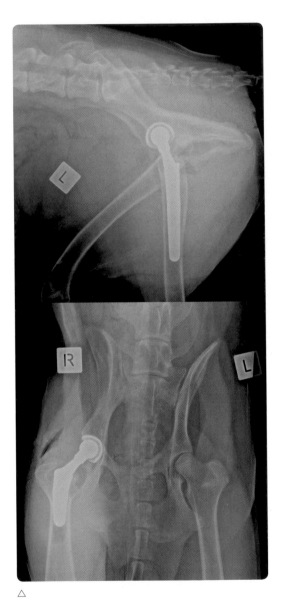

△ **13.4a–b** Amelia the Labrador has had a hip replacement and will likely need a second.
RADIOGRAPHS DR DAVID LIDBETTER

Chapter 2 presents several frameworks that can be used to evaluate this scenario. A simple step-by-step approach is advocated by Rollin (1999):

- Firstly, create a conceptual map of stakeholders and list the corresponding duties to them to identify the ethically relevant issues.
- Check to see if there is guidance in the Professional Code of Practice or legislation relevant to the case.
- If not, apply your own personal ethics – if there are two or more conflicting principles, then adopt a pluralistic approach by applying ethical theories such as utilitarianism etc.

The stakeholders involved in advanced veterinary treatments include the client, animal, veterinary practitioner, veterinary practice, other clients and animals at the practice, referral veterinary practices in general, veterinarians undertaking postgraduate training, the veterinary profession, veterinary suppliers (e.g. facilities and equipment), the veterinary pharmaceutical industry, pet insurance companies and society.

Ethically relevant issues associated with the stakeholder community include the quality and quantity of life of the animal (in other words, where to "draw the line"); efficacy of the veterinary treatment, especially if it is experimental or exploratory (e.g. consider the statistical concepts of the "numbers needed to treat" versus the "numbers needed to harm"); and the financial costs associated with the treatment and the owner's ability to pay. Financial aspects are likely to be addressed in the interview and may include profit-seeking by the veterinary practitioner or practice, or the owner putting themselves into financial difficulty to pay for treatment of pets not covered by insurance. Furthermore, as the scenario describes, a key concern for society relates to access to treatments for people and pets.

The next step involves referring to Professional Codes of Practice and legislation. In this case, except for renal transplantation in cats ("currently suspended pending a review"), there is no specific guidance on advanced treatments in the RCVS Code (2015). However, veterinary norms will apply, for example, ensuring the health and welfare of the animals.

To prepare for the radio interview, step three involves exploring different ethical perspectives. In this case it may be beneficial to focus on conflicts of interest. For example:

- A contractarian has an anthropocentric approach and as long as the client or society accepts advanced veterinary treatments, the contractarian vet will be willing to provide the service. This perspective is based on a contract between the vet and the owner. A situation may arise where the owner requests a course of action that goes against societal norms (for example, pursuing treatment of an animal with a progressive, terminal illness with no hope of improvement or recovery). In this case the contractarian vet will have a dilemma – to decide on which is more important: their relationship with this individual client or their reputation in the wider community.
- In contrast, a utilitarian will consider the interests of the owner and animal. Information is a key driver for this approach – so advice provided by the utilitarian vet should be based on published data and experience of the effectiveness of the treatments, to help the owner to make an informed decision. Thus the harms and benefits of the treatment can be fully explained by the vet. If the treatment has a poor success rate or will impede the quality of life of the animal or the owner (due to the degree of post-treatment care required and/or the emotional burden), then a utilitarian is likely to advise against advanced treatments. A

utilitarian will also be concerned with the wider societal implications of advanced treatments such as any conflicts of using medicines or other resources on animals, which could otherwise be made available for human patients. Where conflicts arise, a utilitarian will look for a solution or a compromise to offset the harms and thus enhance the benefits for the greatest number of stakeholders.

- The relational perspective considers the human–animal bond. The ethical evaluation will depend on the context, for example, will the advanced treatment alter the animal and thus impair the human–animal bond? In contrast, advanced treatments may be advocated in cases where the animal has a similar clinical condition to the owner such as a cancer, adding a further dimension to the ethical analysis.

- Another perspective, which can be described as "Respect for Nature", may be against advanced treatments. It is a biocentric viewpoint, which is particularly interested in safeguarding species-integrity. This perspective often argues against interfering with the genetic make-up or natural processes of a species. Whilst an individual animal may not be of great concern according to this perspective, veterinary or animal husbandry practices which alter natural processes may be considered ethically problematic.

Adopting a pluralistic approach to the radio interview by presenting a range of perspectives will help to demonstrate the complexity of the issue.

This topic encompasses socio-economics, areas that veterinary professionals are not typically qualified to talk about. In preparation for the interview and during the interview, you should set your boundaries and acknowledge your area of competency. Good communication skills and an understanding of ethical theories and applied ethics are key competencies to help the vet to enter an informed debate about the advances in veterinary medicine.

||||||||||||||||||||||||||||||||||||||||||||||||||||||||||||||||||||||||

HUMAN and animal healthcare share many similarities including extensive undergraduate and postgraduate training (i.e. basic sciences and clinical studies) of healthcare professionals, as well as primary and referral treatment facilities. It is natural to compare these systems and indeed comparisons can provide important insights, for example about the management of patients and standard of care in the respective healthcare systems.

Comparison between human and veterinary healthcare systems is likely to become more common due to One Health initiatives and the need for cross-disciplinary collaboration to address challenges such as antimicrobial resistance (Garcia-Alvarez, et al. 2012). In addition, patients and clients of medical and veterinary services may be more likely to compare the standards of care and relative expense as the costs of healthcare increase.

Another cause for comparison is personal. For example, on occasion clients suffer from the same or similar conditions as their pet, such as diabetes mellitus, hyperthyroidism or neoplasia. Such a situation invites comparison in regard to respective management (including costs) of disease(s) (Boston 2014).

One concern commonly raised, as implied in the above scenario, is that animal patients receiving advanced veterinary care are receiving valuable resources that human patients in need might receive, and that this is sometimes perceived as being inherently unjust and unfair.

For example, Hadley and O'Sullivan argue that constraints should be placed on the amount of money spent on veterinary treatment which extends an animal's life (Hadley & O'Sullivan 2009). This, they argue, is a "luxury", albeit a less frivolous luxury than some non-sentient, luxury items such as expensive cars or jewellery. They go on to argue that instead such animals should be painlessly euthanased and the money given instead "to an organisation willing to alleviate [human] suffering and death from preventable diseases" (Hadley & O'Sullivan 2009). The utilitarian argument rests on the contentious assumption that animals are not self-conscious.

It invites a "slippery slope" argument. As the authors note, if spending money on treating companion animals diverts funds from more important causes, one could regard all companion animal relationships as ethically suspect. After all, owners invest time, effort and resources in providing healthcare such as vaccinations, parasite treatment, food, water, shelter and exercise. Surely this could be similarly devoted to "greater" causes? The authors claim that this sets the bar too high, requiring too much of people. However, this raises the troubling question of where one draws the line between acceptable treatment of animals and over-investment. Perhaps such an assessment pertains more to the justice of the economy in which those healthcare systems operate. According to Andrew Gardiner, it is important to recognise that "Veterinary medicine is, for the most part, a form of private medicine. Money those clients choose to spend on animals is not necessarily money that would be otherwise spent on human healthcare, either at home or abroad" (Gardiner 2007).

He argues that the marketplace in which veterinarians operate dictates what constitutes acceptable and appropriate intervention: "Poor medical facilities in less developed nations are more to do with global politics and macroeconomics than with advanced veterinary treatments in rich countries" (Gardiner 2007).

MEDICINE FOR ALL?

What do you think?

ONE _____ How can the discrepancy between the availability of advanced veterinary care in some areas and lack of availability of human healthcare in others be addressed by veterinarians?

TWO _____ How might veterinarians ensure that the distribution of healthcare resources is just?

THREE _____ How can veterinarians work to alter the marketplace in which they operate to ensure it is just?

△
13.5 How can veterinarians positively impact on the lives of people living in inadequate conditions with limited healthcare?
PHOTO ISTOCK

△
13.6 A two-year-old pet rat exhibiting neurological signs was referred to a specialist centre for a CT scan. A pituitary adenoma was diagnosed, which was managed medically for several months before the animal died.
PHOTO ANNE FAWCETT

13.2

Antimicrobial use

Antimicrobial therapy has been used in veterinary medicine since antimicrobials became available in the twentieth century. However, antimicrobial use can and does lead to antimicrobial resistance. Use of antimicrobials exposes bacterial pathogens and commensal organisms to varying concentrations of antimicrobials for variable time periods, creating a selection pressure which can lead to the emergence of resistance or an increase in numbers of resistant bacteria (Weese, et al. 2015).

In its Consensus Statement on Therapeutic Antimicrobial Use in Animals and Antimicrobial Resistance, the American College of Veterinary Internal Medicine states that:

> "While the overall role of therapeutic antimicrobial use in animals in the development of AMR [antimicrobial resistance] in animal and human pathogens is poorly defined, veterinarians must consider the impacts of antimicrobial use in animals and take steps to optimise antimicrobial use, so as to maximise the health benefits to animals while minimising the likelihood of antimicrobial resistance and other adverse effects."
> (Weese, et al. 2015)

Also see https://www.avma.org/KB/Resources/Reference/Pages/Antimicrobial-Use-and-Antimicrobial-Resistance.aspx.

Concerns about use of antimicrobials have led, among other things, to a debate about the relative impact of animal versus human use of antimicrobials on antimicrobial resistance. Such arguments may be couched in terms of pitting the interests of an animal or a producer against the interests of the general public and users of human healthcare services.

SCENARIO
ANTIBIOTIC RESISTANCE

▶ Henry, a four-year-old Persian cat, is brought to you by the M family. He has been unwell for a couple of days, not eating and lethargic. On examination you find nothing remarkable except that he has a slight pyrexia. You would previously have suggested an anti-inflammatory injection and, to be on the safe side, a long-acting antibiotic, but you are aware of increasing concern about over-prescribing of antibiotics leading to antibiotic resistance, affecting your ability to treat cats like Henry in the future.

What should you do?

RESPONSE
ALISON HANLON

▶ The responsible use of antibiotics in veterinary medicine is essential to both human and animal health. Antimicrobials have been used in veterinary medicine for over 60 years, and whilst most of the debate has focused on the risk to human health, the overuse of antibiotics also jeopardises animal health.

There are several frameworks that can be used to evaluate this dilemma, as outlined in chapter 2. A simple step-by-step approach is advocated by Rollin (1999):

"Antimicrobials have been used in veterinary medicine for over 60 years, and whilst most of the debate has focused on the risk to human health, the overuse of antibiotics also jeopardises animal health."

of antibiotic resistance in all animal sectors such as food production and companions. The stakeholder community also extends for example to veterinary educators, the pharmaceutical industry and research organisations such as those involved in drug discovery.

The next step in the ethical evaluation is to consider the ethically relevant issues including conflicts of interests between the stakeholders. For example:

- The clients are concerned for Henry's health and want him to be treated appropriately by the vet.
- The vet wants to keep the clients happy by providing an effective treatment but further tests are required to establish a diagnosis, which take time and money and Henry's condition may continue to deteriorate.
- Use of antibiotics on Henry today may increase the risk of antibiotic resistance and thus impair veterinary care options for Henry in the future.
- There is a professional obligation for veterinarians (and medical doctors) to use antibiotics prudently.

△

13.7 Henry is lethargic, inappetent and suffering from mild pyrexia.

PHOTO ANNE FAWCETT

- Firstly, create a conceptual map of stakeholders and list the corresponding duties to them to identify the ethically relevant issues.
- Check to see if there is guidance in the Professional Code of Practice or legislation relevant to the case.
- If not, apply your own personal ethics – if there are two or more conflicting principles, then adopt a pluralistic approach by applying ethical theories such as utilitarianism etc.

Using Rollin's (1999) conceptual map to identify the stakeholders and responsibilities, it is evident that administering an antibiotic to Henry (without further investigating the clinical signs) has repercussions for Henry, other animal patients, clients of the practice and other veterinary practices, the veterinary team in the practice and other practices and the wider veterinary profession such as the veterinary regulatory authority (for example the RCVS in the UK, AVMA in the USA and AVA in Australia). It also impacts society as a whole, through the increased risk

Seeking guidance from the Code of Professional Conduct and legislation forms the second step of the ethical resolution process (Rollin 1999). A provision on antimicrobial resistance is included in the UK's RCVS Code (2015), placing responsibility for the decision to use antibiotics on the veterinary practitioner:

"The development and spread of antimicrobial resistance is a global public health problem that is affected by use of these medicinal products in both humans and animals. Veterinary surgeons must be seen to ensure that when using antimicrobials they do so responsibly, and be accountable for the choices made in such use."

13.2

ANTIMICROBIAL USE
RESPONSE
ALISON HANLON

To support ethical decision-making, it is appropriate to further evaluate the situation by moving on to the next step of the resolution process, to take account of both personal ethics and other perspectives. Whilst it may seem that prescribing an antibiotic to Henry is harmless, we could ask – well what if every vet did that? We know that the widespread use of antibiotics in humans and animals has resulted in growing antimicrobial resistance. There may be a number of reasons for the clinical signs described for Henry, which may or may not be due to a bacterial infection. In cases such as this, good practice is to conduct a full clinical examination, and collect a blood sample for haematological analysis to establish a diagnosis. The decision to delay treatment or not prescribe antibiotics may be at odds with the expectations of Henry's owners and thus communication will play a key role, explaining the rationale for the decision, to resolve this conflict of interest.

As a long-term measure, joining initiatives such as "Antibiotic Guardian" (Public Health England 2015) and educating clients by putting up posters in the clinic's reception area will help to address client expectations.

Acknowledgements
I would like to thank my colleague, Alan Wolfe, BSc (hons), PhD, BVMS, MVS, ECVP (University College Dublin, School of Veterinary Medicine), for helping to unravel the clinical implications of this scenario.

THIS scenario requires the veterinarian to base their decision on up-to-date evidence about antimicrobial resistance, as well as current guidelines, which will be discussed below. It also raises a conflict between the interests of an individual or minority (companion animal patients and their owners) versus the interests of the majority.

The "Tragedy of the Commons" is a concept introduced in 1968 by Garrett Hardin (1915–2003), an American ecologist and philosopher who was concerned with the dangers of overpopulation. His famous thought experiment is based on a commons or common pasture, on which self-interested farmers will aim to keep as many cattle as possible.

"What is the utility of adding one more animal?... Since the herdsman receives all the proceeds from the sale of the additional animal, the positive utility [to that herdsman] is nearly +1... Since, however, the effects of overgrazing are shared by all the herdsmen, the negative utility for any particular decision-making herdsman is only a fraction of -1. Adding together the... partial utilities, the rational herdsman concludes that the only sensible course for him to pursue is to add another animal to [the] herd. And another; and another... Therein is the tragedy. Each [person] is locked into a system that [causes] him to increase his herd without limit... in a world that is limited. Freedom in a commons brings ruin to all."

(Hardin 1998)

Hardin argues that individual wellbeing is valued and individualism gives freedom, but the more the population exceeds the carrying capacity of the environment, the more freedoms must be given up in order to ensure that the commons is not depleted (Hardin 1998).

In the case of antimicrobial resistance, antimicrobials represent the common resource that cannot cope with limitless exploitation (Baquero & Campos 2003). If they are used purely to benefit individuals, without consideration for others, antimicrobial resistance is likely to develop more rapidly.

However, restriction on antimicrobial prescribing has been seen to be interfering with prescriber autonomy.

Furthermore, it is difficult to determine when excessive or increased use of antimicrobials by individuals becomes a problem. In a utilitarian analysis, use of antimicrobials may be positive even in the event of amplification of drug-resistant strains or organisms. Alternatively, in some situations increased use of antimicrobials does lead to less favourable outcomes for society. A case-by-case assessment of appropriate use of antimicrobials is recommended (Porco, et al. 2012).

ANTIMICROBIAL USE

What do you think?

ONE   How would you use an ethical framework to develop guidelines for the use of a common resource such as antimicrobials?

TWO   How much responsibility do we have for the general health of future generations of the general population, versus that of our individual patients?

△

13.8 How should vets balance any responsibility towards patients and future generations of animals?
PHOTO ANNE FAWCETT

SCENARIO
ANTIBIOTIC RESISTANCE ON FARM

▶   As a pig vet you often prescribe antibiotics for large numbers of animals at once, and the drugs are incorporated into feed. There's been growing concern about antibiotic resistance reducing our ability to treat human infections, and the finger has been firmly pointed by some medics at the widespread use of antibiotics in farm animals. Low-dose antibiotics are used as growth promoters in many countries, but in theory at least, not in the EU since 2006. Every day you see the suffering and poor growth of pigs with respiratory and other infections. It has been suggested by the Chief Medical Officer of the UK that sick farm animals should be killed rather than treated with antibiotics.

What would ethical use of antibiotics involve for you?

△

13.9 Antimicrobial resistance is a major threat to the health and wellbeing of humans and animals. Use of antimicrobials in production animals is one part of the global problem.
PHOTO ANNE FAWCETT

RESPONSE

HENRY BULLER

▶ Although here in the UK and in the EU countries, we have banned the use of antimicrobial medicines for growth promotion (though this still goes on in other large-scale producer countries like the USA and China), I think many people are still very much in that same mindset, that unless they are used regularly and systematically, even when the disease being targeted is not actually present, then production levels and profitability will suffer. We might not call it growth promotion but that, in effect, is what it seems to be. We are improving growth (at the herd level) by looking to remove a possible disease threat. It's risk management if you like. And because the risk is thought to be always there, and because it can have a big effect on farms and their economic performance, we have come to rely on what is easiest – medicating food with increasingly powerful antimicrobial medicines at pre-determined moments during the animal's time on the farm.

It's population health management, rather than an individual pig approach (which would be to make all pigs better) and I think we have got so used to it that it has become the norm, and the animals are getting used to it, becoming less tolerant or resistant when infection does take hold. The healthiest herds may be the ones that have no disease but ironically they have less capacity to fight disease were they to get it.

In the past, we have not had to worry so much about it, after all, the pigs are not around for that long, the medicines are cheap and with the cephalosporins and the flouroquinolones the withdrawal times are getting shorter. But now, with global concerns about antimicrobial resistance (AMR), I don't think that ethically we can simply go on using these medicines metaphylactically for a new population or prophylactically, as an advance guard against possible infection, at the levels we currently do. If the AMR threat is as significant as they say, and the use of antimicrobials in pig farming is a significant enough contributor to AMR, then we need to do a number of things to reduce reliance on antimicrobials: firstly, increase the adoption of vaccines; secondly, move to more targeted or selective use of antimicrobial medicines; thirdly, where possible engage in more disease eradication programmes within defined areas; fourthly, improve our delivery systems for getting the medicines into the pigs in a more effective way; and fifthly, adopt management changes in our production units to reduce the need for medicines. We are always going to need some antibiotics because animals are always going to get ill but I am all for reducing antimicrobial drug use where it is not needed. The issue is, when is it needed and how can we reduce that need through other measures?

There are many misconceptions about animal farming and pig production is a good case in point. Veterinarians are not dealing with individual animals like a doctor [physician] deals with individual humans. What they do is really population medicine, controlling the dynamics of a disease or infection within a group, herd or population of farm animals. When pigs get ill, it can spread fast. There's a lot of disease there. It's not clean; you're dealing with lots of animals, many of which might have a number of different disease potentials. That is the reality of production diseases. In one form or another, they are always present. Our role is to keep them in check, to stop them getting out of control.

If we can reduce the threat or risk of pig disease through housing and post-weaning group management, through reducing farrowing frequency, through having less mixed age groups together, by using batch systems and synchronising the heats in the gilts, delayed weaning, having better ventilation, even having less temperature difference between day and night, then I am all for that and these things may well mitigate welfare losses due to reduced medicine use. But there is a general impression that we should simply stop using antibiotics. That's just naive. There are situations where we have to use antibiotics to prevent pigs dying.

Like so many ethical questions in veterinary medicine, it all for me comes down to who (or what) is the subject of our primary obligations: is it the individual animal or is it the population (or herd), is it the farmer, whose profitability and livelihood depend on healthy stock, or is it the wider human population who benefit nutritionally from animal farming? The trouble is, the ethical subject changes with the circumstance and the context. Instead of feeding humans, farm animals may lead to human deaths through zoonotic infections like SARS, H5N1 or H1N1 or by contributing to medicine resistance in human pathogens. To have a healthy stock, you might need to cull individuals. So when human populations are threatened by the transmission of disease from farm animals, the latter are culled and the former are vaccinated (or treated). We do have that ultimate sanction. We don't like it but we can cull all the animals and start again. Once you accept the function of livestock production is for human consumption, the culling of individual animals to protect a herd, or the culling of a herd to protect humans, though regrettable, can sometimes become necessary, but it must only be used when necessary, not as an alternative.

|||||||||||||||||||||||||||||||||||||||||||||||||||||||||||||||||||||||||||

ANTIMICROBIAL stewardship pertains to the way we look after the resource(s). Page, et al. (2014) propose "5Rs" of antimicrobial stewardship: these include Russell and Burch's well-known 3Rs (where possible, use of antimicrobials should be replaced, reduced and refined) in addition to responsibility of the prescriber, and review of stewardship outcomes. In terms of responsibility, the prescribing veterinarian must accept responsibility for the decision to use an antimicrobial agent, recognising that such use may have adverse consequences beyond the recipient. The prescribing veterinarian must also perform a risk assessment of the particular circumstances, and determine that the benefits of such antimicrobial use together with any risk management measures recommended will minimise the likelihood of any immediate or longer-term adverse impacts on the individual patient, other patients or public health.

In addition, the fifth "R" stipulates that any antimicrobial stewardship initiatives are reviewed regularly to ensure compliance with initiatives and ensure that antimicrobial use practices act or reflect contemporary best practice.

In the above scenario, the Chief Medical Officer calls for the abolition of antimicrobial use in farm animals, with culling posed as the only alternative. Such a policy would be against the interests of animals culled and consumers and producers impacted, without knowing the extent of (if any) benefits to other parties. The respondent calls for prudent use of antimicrobials, an approach more in line with principalism as it seeks to reduce harm, promote good, provide justice for all parties and respect autonomy.

The relative contribution of the use of antimicrobials in animals to AMR in humans is inadequately understood (Weese, et al. 2015). There is a danger therefore that risks may be over- or understated. Furthermore, the relative impact of strategies to reduce resistance remains to be characterised.

Refining, reducing and replacing antimicrobials requires veterinarians to ensure that these agents are only used when indicated. This may in turn require more diagnostic testing, improved animal husbandry and early intervention in the event of disease.

13.3
ZOONOSES

ANTIMICROBIAL USE

What do you think?

ONE — Who should bear the costs of increased diagnostic testing to ensure antimicrobials are indicated?

TWO — Should the use of antimicrobials be restricted? If so, in what circumstances and to whom should restrictions apply?

THREE — How would you ethically justify restriction of antimicrobials to a veterinarian, a farmer or a companion animal owner?

13.3

Zoonoses

Zoonoses are diseases that can be transmitted between animals and humans (some authors reserve the term zoonoses for diseases that can be transmitted from animals to humans, and

△

13.10 Antimicrobials being used on a quail farm.

PHOTO ISTOCK

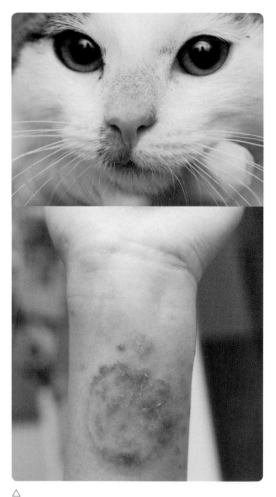

△

13.11a–b Dermatophytes are a common cause of zoonoses in veterinary staff.

PHOTO ANNE FAWCETT

anthropozoonoses for diseases that can be transmitted from humans to animals – we will use it to refer to both). As already mentioned, the majority of emerging and re-emerging infectious diseases are zoonotic. Zoonoses may result in mild, self-limiting disease, such as dermatophytosis which is typically a minor skin ailment in immunocompetent individuals, or life-threatening diseases, such as Hendra virus, rabies, highly pathogenic avian influenza or bovine spongiform encephalopathy.

Zoonoses may present ethical dilemmas as they require veterinarians to make decisions that directly impact on human health, including their own (Dvorak, et al. 2013). For example, in Australia at present some veterinarians are refusing to treat horses that are not vaccinated against Hendra virus, a disease with a 50 per cent mortality rate in humans, due to their risk of occupational exposure and potential liability (Spillman 2015). At least two veterinarians and one veterinary nurse have been infected with Hendra virus in the course of their work, with one of the veterinarians dying.

||||||||||||||||||||||||||||||||||||||||||||||||||||||||||||||||||||||||||||||

ZOONOSES require veterinarians to know how to manage the risks of human infection. Some authors argue that a One Health approach necessitates collaboration in the clinical management of zoonoses. For example, veterinarians could advise on risk reduction strategies for a human patient and close contacts in the event of salmonella acquired from working with sheep and spread to family members on work clothes and other fomites. Or, in another example, veterinarians could advise on management strategies for leptospirosis in source animals, where a farmer has been diagnosed with the disease and must weigh up the costs and benefits of vaccinating the herd against leptospirosis (Speare, et al. 2015). In the next scenario we look at how veterinarians may approach serious zoonotic disease.

ZOONOSES

What do you think?

ONE — To what extent should veterinarians prioritise the health of humans (including themselves) over that of animals?

TWO — On what ethical grounds might you justify refusal to treat an animal that is sick or injured?

△
13.12 A veterinarian wearing full personal protective equipment when examining a horse which may have Hendra virus.

PHOTO AUSTRALIAN VETERINARY ASSOCIATION

SCENARIO

ZOONOSIS

▶ Mr and Mrs S run a small caravan park. They have a little "petting" area with some rabbits and guinea pigs in a pen that people can lean over and stroke. They bring you the body of the second rabbit that has died in recent days for a post-mortem. You find multiple granulomatous lesions that, after culturing, are found to contain *Yersinia pseudotuberculosis*. This bacteria can cause disease in humans, which occasionally can be very serious, particularly in immunosuppressed people. Direct zoonotic transmission from rabbits has been reported, but there is a lack of information about the zoonotic potential in this type of setting. The other animals appear healthy at the moment.

What should you do?

△

13.13 Mr and Mrs S keep and breed multiple animals for the petting zoo they run. You have just diagnosed a notifiable disease in one of their animals.

PHOTO ANNE FAWCETT

RESPONSE

ANDREW KNIGHT

▶ Once alerted to the diagnosis by the laboratory results, you should research this unusual disease in your textbooks and online. You would discover that although *Yersinia pseudotuberculosis* does not appear to be a common pathogen in domestic rabbits (Meredith & Redrobe 2002), it can nevertheless occur. This enteropathogenic bacterium normally causes signs such as weight loss, lethargy, abdominal masses (presumably the granulomatous lesions in this case), hyperthermia and leukocytosis, and is spread by faecal contamination of the environment. However, the disease is normally self-limiting and the prognosis generally good (Cousquer & Meredith 2014). The fact that the disease has already proven fatal to two rabbits in this case should therefore stimulate you to question this diagnosis. You should first check with the laboratory about the specificity or false positive rate for this test. Although few if any tests have a 100 per cent specificity/ zero per cent false positive rate, if the test parameters are close to these levels the results may be considered highly reliable. If not, you should ask the laboratory whether any more accurate tests might be available, following this initial screening test. And although the remaining animals appear healthy, they should be rechecked for clinical signs of this disease. The guinea pigs are also at risk (Quesenberry & Boschert 2011). If in doubt a blood test for leukocytosis could be considered. If the diagnosis remains probable or even reasonably possible, then a precautionary approach would be wise, given the potentially serious consequences of this disease for some affected animals, and its zoonotic risk.

Additionally, the fact that two rabbits have probably already died from a disease not normally fatal should trigger a search for concurrent illnesses or pathogens, including a check of general husbandry, preventative healthcare measures including parasiticides, and the general health status of the remaining animals.

The owner should be advised about treatment of the remaining animals, which is normally by supportive care. Particular attention should be paid to diet and feeding. The owner should be advised that loss of appetite and consequent loss of normal bowel movements can be very serious in rabbits, and requires early veterinary intervention should it occur. Until the disease has demonstrably resolved, and the environment has been disinfected, sensible precautions should be implemented, including gloves, hand-washing, and the provision of care by people who are not otherwise suffering from concurrent illnesses or immunocompromise.

Although uncommonly transmitted to humans from domestic rabbits, such transmission is possible, most commonly through foodborne routes. Far East scarlet-like fever usually manifests. Signs include fever and right-sided abdominal pain, which can mimic appendicitis. Signs typically resolve without treatment, but disease can be more serious and treatment necessary in complex cases or immunocompromised patients (Jani 2013).

Accordingly, you should alert the owners to the risk of zoonotic transmission, and this information. Any owners or family members who experience possible symptoms, or are at risk of developing more serious disease due to concurrent illness or immunocompromise, should be advised to consult their doctor. The health status of members of the public who might come into contact through petting is unknown, so to prevent ongoing risk the surviving animals should be isolated from the public until the disease has demonstrably resolved.

Given that this disease does not normally cause serious signs in humans it might seem tempting to skip this advice. However, given the potential for serious human disease, this would be unethical. Proffering this advice, and recording in the clinical notes that it has been provided, would also provide a legal defence, in the unlikely event that the client or another affected person attempted to initiate action against you for failing to warn of the risks. Such "defensible practice" is a necessity of contemporary veterinary practice, and should be instinctive for experienced veterinarians.

Technically speaking, other options are available in this case, including euthanasia and no treatment. While euthanasia of all surviving animals and decontamination of cages would indeed eliminate the risks to humans, it would be contrary to the interests of the surviving animals, who appear to be healthy and presumably have the potential to enjoy a good life. Should closer examination of the remaining animals reveal a level of clinical disease, the owner should be strongly encouraged to isolate and treat the rabbits as advised previously, given the clear risks to animal welfare and, potentially, to human health.

|||||||||||||||||||||||||||||||||||||||||||||||||||||||||||||||||||||||

IN this scenario the author advises the veterinarian to gather as much information as possible by:

• Researching the disease.
• Confirming the diagnosis.
• Advising of the risk and hazard of human transmission and recording any advice given.
• Considering the welfare implications for animals.

One potential approach to this dilemma is to weigh up the interests of all stakeholders (the owners of the petting zoo, the animals, the veterinarian, the clients of the caravan park and the wider public, for example). In the case of a serious, notifiable disease, culling animals humanely may be justified on utilitarian grounds. Should the owners become infected in the course of treating their animals, the consequences may be devastating. Similarly, should other animals become infected the consequences may be devastating for them.

Isolating the animals in a hospital setting may reduce this risk, as long as risks to veterinary hospital staff can be managed.

Another interesting element to managing zoonotic diseases like *Yersinia pestis* is that these are notifiable diseases in some regions. Reporting a disease to the relevant health authorities is a requirement which legally overrides the veterinarian's obligation to maintain client confidentiality. Client confidentiality is discussed at greater length in sections 7.1 and 13.5.

ZOONOSES

What would you do?

How would you approach a feline patient that you diagnose with dermatophytosis in a shelter? The cat has mild, scaly skin lesions on its ears and chin, but is not at all bothered by these and is otherwise healthy on examination. In this case, the shelter is full to capacity, and has a policy of only rehoming healthy animals.

ONE   **Consider what information you will need to make your decision.**

TWO   **What is your ethical justification for this decision?**

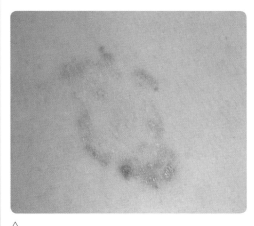

△

13.14 Ringworm in humans is highly contagious but rarely serious.
PHOTO ISTOCK

ZOONOSES

What would you do?

How would you approach a canine patient that has tested positive to exposure to Hendra virus? The dog – a family pet – has no clinical signs, but the disease is known to be transmitted from horses to humans and has a 50 per cent mortality rate in human patients. The dog lives on a property where three horses have died due to Hendra virus.

ONE — Consider what information you will need to make your decision.

TWO — What is your ethical justification for this decision?

△
13.15 Close contact has probably allowed the spread of Hendra virus between horses and a farm dog.
PHOTO ISTOCK

13.4

Human–animal bond

The term "human–animal bond" was coined in the 1980s by Dr Leo Bustad (Holcombe, et al. 2016), spawning a large amount of literature about the health benefits of the bond (largely to humans). There is some debate about the benefits, and potential negative impacts, of the human–animal bond, particularly with regards to human health and wellbeing. Most of the literature focuses on the bond we have with companion animals, particularly dogs and cats.

Benefits of the human–animal bond include:

- Fewer visits to the family physician, with significant savings to the individual and the overall healthcare budget.
- Decreased cardiovascular disease.
- Reduction in obesity in dog walkers.
- Psychological benefits from companionship, as well as animal-assisted therapy (Clower & Neaves 2015, Johnson 2010).

△
13.16 Benefits of the human–animal bond to the health of humans are now well recognised in published literature.
PHOTO ANNE FAWCETT

Potential detrimental effects include:

- Zoonoses.
- Refusal to evacuate premises in the case of an emergency.
- Refusal to move into care or a nursing home.
- Detrimental effects to the animal (for example, animal hoarding, obesity, neglect or animal abuse) (Joffe, et al. 2014).

There is less information available about the impact of the human–animal bond on the animal. Attachment of a human to an animal does not automatically mean that the animal's best interests will be taken into account. This can present a challenge for veterinary staff when assessing quality of life and making decisions around euthanasia.

SCENARIO
OWNER AND PET HAVE SIMILAR CONDITION

▶ You diagnose a nine-year-old dog, Jack, with a humeral fracture secondary to osteosarcoma. Thoracic radiographs reveal severe metastatic disease. As Jack has severe pre-existing arthritis on the contralateral limb, he is a poor candidate for amputation. According to a veterinary oncologist you consult with, Jack has a very poor prognosis even with chemotherapy.

As you reveal the findings of your investigations to Jack's owners, a couple, the mother bursts into tears.

△
13.17 There is less information about the impact of the human–animal bond on the animal.
PHOTO ANNE FAWCETT

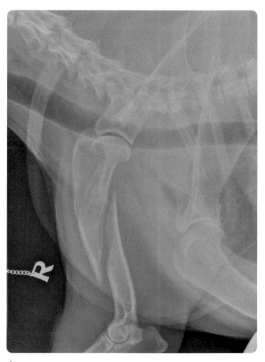

△
13.18 Jack is diagnosed with osteosarcoma at the same time his owner is undergoing treatment for osteosarcoma.
RADIOGRAPH DR JOHN CULVENOR,
NORTH SHORE VETERINARY SPECIALIST CENTRE

"Our son Adrian has just been diagnosed with a bone tumour in his leg," she cries. "It's like Jack has gone out in sympathy. The doctors tell us it is very bad but we're holding onto hope that we can keep him comfortable."

You believe the most humane option for Jack is euthanasia.

What should you do?

RESPONSE

MANUEL MAGALHÃES-SANT'ANA

▶ This case illustrates the difficult dilemma of how far a veterinarian should go in influencing clients' decisions. Between a weak suggestion and a strong persuasion, there is no consensus as to what degree of influence a veterinarian may exert over clients (Yeates & Main 2010). Moreover, few cases are more challenging for veterinary practitioners than those dealing with euthanasia and end-of-life care (Shaw & Lagoni 2007). In everyday practice virtually every veterinarian will be faced with such cases, with very little support from ethical guidelines or regulatory frameworks. This is in contrast with human medicine, where end-of-life decisions are more strictly regulated and fewer decisions are left to the physician's discretion.

Decision-making in veterinary practice can be seen as a shared responsibility, involving "a joint venture between the veterinarian and client" (Shaw, et al. 2008). Even so, veterinarians should be able to take the leading role in the decision-making process, when and if the situation requires it (Rollin 2002). That seems to be the case with Jack's owners, since their emotional involvement may prevent them from fully assessing the available options and all their consequences. A satisfactory resolution of this case will probably require a multiple-step approach, one that allows

for elements of animal welfare (e.g. quality of life), professionalism (e.g. communication, human–animal bond) and theoretical concepts (e.g. moral values) to emerge (cf. Figure 1.2).

As for Jack, his main interest is to avoid suffering and the veterinarian has a prima facie duty to ensure this goal; a straightforward, unpassionate approach to the case would demand Jack's euthanasia, regardless of the circumstances. However, Jack's life cannot be seen in isolation. As a companion animal, Jack's life should be considered in relation to those of his owners. Jack is not only a dog with a tumour. Jack is someone's companion (and may even be considered a son, or at least a surrogate of a son) with a terminal condition. To ignore the owners' needs would be uncompassionate and disrespectful.

Jack may have a relatively poor quality of life (QoL), which will inevitably deteriorate due to progression of this condition. The decision to implement chemotherapy would only worsen QoL while bringing very few benefits or none at all. Nonetheless, there are other forms of palliative care which may increase Jack's QoL, even if only for a short period of time. On that note, it would be possible – and even desirable – to adjourn decisions regarding euthanasia, for the sake of raising the awareness of the clients as to Jack's critical condition and extremely poor prognosis. Such an approach would prepare them for a more conscious and informed decision to euthanase, without further compromising Jack's welfare.

In order for this approach to work effectively, quality end-of-life communication is paramount (Shaw & Lagoni 2007). The veterinarian should demonstrate empathy and respect for the owners' feelings and reassure them regarding their decisions. Probably one of the most important aspects of this case is to highlight the differences between human and animal palliative care. The owners should understand that while humans can surpass the side effects of chemotherapy with the hope of recovery, that is not the case for Jack. His present suffering is all he has, and he will not be able to assess hypothetical future benefits of treatment. In addition, they should be aware that a poor outcome of Jack's chemotherapy could harm, rather than help, their son's treatment. In order to promote awareness and responsibility, clients should be provided with a QoL checklist [e.g. HHHH-HMM QoL Scale (Villalobos 2002)] and trained to assess Jack's QoL. Another important element of this case is to maintain rapport, allowing the owners to call you whenever they need help (including out-of-hours). This will certainly bring them comfort and increase their confidence in the level of veterinary care being provided.

But what if Jack's QoL deteriorates to a level below a life worth living before his owners are ready to euthanase? A conflict between the best interest of the animal and those of the clients will require revising the decision-making process. This may involve a different approach, moving from suggestion to persuasion. The first thing to do is to invite the owners to the clinic (in case they didn't take the initiative) and take the time to reassess the case together, going through the QoL scale and evaluating the decision-making steps. Probe the owners with questions that might direct them to euthanasia, such as "Jack seems to be fading away, have you noticed that?" or "The most important thing is to avoid suffering and making Jack comfortable. Do you still think we are achieving it?" From my experience, the owners sooner, rather than later, will ask for the euthanasia.

As a final note, it is probably useful to draw the reader's attention to at least two alternative approaches that must be avoided. One consists of (ab)using veterinary authority by exerting coercion over clients (through resentment, threatening or litigation). The other approach involves removing yourself from the decision-making process, whether by refusing to provide your informed opinion (e.g. "you know best" or "this is not my dog") or

by withholding invaluable information that may help in reaching an informed decision (e.g. side effects of chemotherapy). Although these two approaches fall beyond opposite ends of the acceptable spectrum of influence, they are closer to one another than they may appear at first glance. Both forcing and hampering decision-making violate clients' right to self-determination by preventing autonomous judgements and informed choices. In fact, withholding options to clients has been described as "the strongest form of coercion" in veterinary decision-making (Yeates & Main 2010).

IN this scenario, there is an implicit request to weigh up the interests of the family in the dog's continued existence against the interests of the dog (which are presumed to be geared towards not suffering). A utilitarian might argue that the family consists of more stakeholders, and therefore if the dog is euthanased before they have come to terms with this, more stakeholders will be suffering.

This assumes that the nature of their suffering is equal. However, some authors argue that pain can be worse for animals than for some people. The basis for this argument is that our cognitive sophistication may mitigate the intensity of pain, through our expectations, memories and ability to shift our focus.

For example, I can endure the pain of a local anaesthetic at the dentist because I know that this is transient, controlled, non-harmful and will eliminate further pain. I can communicate with the dentist and I can withdraw my consent at any time. I know that the pain I am experiencing is a *means to an end*. We cannot say the same for animals.

Secondly, cognitively sophisticated beings can engage in "inter-temporal calculations" with respect to our interests, and we can discount pain to achieve other, higher-order interests. So if we are undergoing major surgery to remove a tumour, we can focus on the fact that the intent is curative or palliative. All going well, we may wake up sore but cancer-free, or at least with a sense that we will have improved QoL beyond the immediate post-operative period. A dog in a comparable situation could not discount pain in the same way. According to Sahar Akhtar (PhD(s), University of Virginia, USA), animals without this ability to engage in inter-temporal calculations:

"…are not able to choose to endure pain for the sake of satisfying long-term interests, and there is no global or higher order perspective to consider. Unlike us, an animal that cannot see itself as existing over time cannot reflect on the value or meaning of its life taken entirely, cannot form interests for life taken as a whole, and cannot formulate lifelong objectives. Thus for animals, it makes far less sense to think of pain as something that can be discounted for the sake of other long-term or complex interests. We thus have reason to think that a given measure of pain can be a larger detriment to their welfare than a comparable measure of pain is to ours."

(Akhtar 2011)

As the author points out, this scenario is particularly challenging because Jack is suffering from neoplasia with a poor prognosis, as is a family member. Because these states are readily comparable, justification of euthanasia as the most appropriate treatment may be challenging.

In this scenario the veterinarian has prioritised the welfare and interests of Jack but recognised the need to sensitively explain some of the differences in human and animal suffering by way of justifying his or her recommendation. Where palliation or hospice are viable options, these can be discussed. If they are not suitable, it is appropriate to explain why.

One of the challenges of QoL assessment is that there is no universally agreed standard QoL measurement below which euthanasia should be performed. Nonetheless, QoL scoring systems can aid in mutual decision-making by providing structure for discussion.

ZOONOSES

What would you do?

Translational medical research involves the use of animals, especially larger animals such as pigs and sheep, prior to clinical trials in humans. This is often cited as an example of One Health in action. You have been asked to provide veterinary support to a translational medical research trial testing the long-term effect of surgical implants to treat kidney disease. The trial will use the pig as a model for the second most common cause of children requiring kidney dialysis.

ONE   What factors would you consider in deciding whether to assist the trial?

TWO   Would you speak at a One Health conference about this work?

△
13.19 Pigs are used in translational medical research before human clinical trials.
PHOTO ANNE FAWCETT

13.5
Animal abuse

Animal abuse and animal cruelty are serious forms of antisocial behaviour which have been linked to interpersonal violence, including domestic violence. Perpetrators of abuse, particularly children, may be victims of violence themselves (Flynn 2011). For this reason, animal abuse and cruelty are viewed not solely as animal welfare issues, but also as a matter that impacts the well-being of humans.

Veterinarians, technicians and nurses are in a unique position to identify cases of animal mistreatment and abuse. Animal abuse may be categorised as physical, sexual, neglect and emotional abuse (Tong 2014). Witnessing animal abuse can have a significant detrimental impact on those present. Children may become perpetrators of abuse (McDonald, et al. 2015).

As animal abuse may be an indicator of domestic or intimate-partner violence, the question of whether veterinarians should be legally mandated to report suspected abuse is often raised.

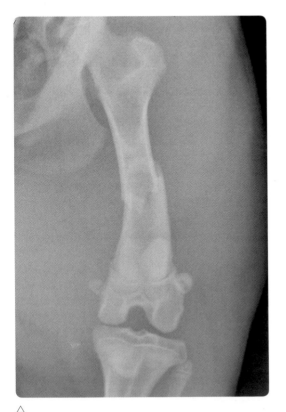

△
13.20 The dog presented with a fractured femur.
PHOTO JOHN CULVENOR

SCENARIO
REPORTING ANIMAL ABUSE

▶ Mrs V and her teenage children accompany their dog Alfie into your consulting room. Alfie is non-weight bearing and you suspect he has a broken right femur. You pinned the same leg a year ago. Aside from extensive bruising around the leg and abdomen, Alfie appears otherwise well, if timid.

You discuss treatment options and associated costs. On hearing the costs, the daughter looks at her brother and says, "HE should have to pay! HE kicked him!"

Suddenly the mother looks very uncomfortable.

"Leave him alone," she says. "We're just going to fix it."

Alfie's injuries are consistent with blunt trauma.

What should you do?

RESPONSE

SANAA ZAKI

▸ Issues to consider:

(1) The Veterinary Practitioners Code of Professional Conduct (New South Wales Government 2013) states: "A veterinary practitioner must at all times consider the welfare of animals when practising veterinary science." On this basis, failure to report a case of suspected animal abuse could be viewed as professional misconduct, if failure to report the incident results in repeated or ongoing abuse, and therefore compromises the safety and welfare of the animal.

(2) The Australian Veterinary Association (www.ava.com.au/policy/12-animal-abuse) has developed an Animal Abuse policy document. In this document it recommends that veterinarians report all suspected cases of animal abuse to the relevant authorities. However, it does not support the introduction of a legislative requirement for veterinarians to report such cases. It argues that such laws would only serve to discourage owners from seeking veterinary treatment for their injured pet.

(3) There is a strong link between animal abuse and human interpersonal violence – in particular, abuse of women by their male partner (Febres, et al. 2014) and abuse of children by a close family member. One study found that in 88 per cent of families where physical child abuse was occurring, pets were also abused (Deviney, et al. 1983). In addition, persistent childhood cruelty to animals is a marker for maltreatment and abuse in children (McEwen, et al. 2014).

(4) In response to this growing body of evidence that links animal abuse and domestic violence, some countries such as the USA have introduced laws that extend domestic violence orders of protection to the pets of households where domestic violence has occurred.

(5) This close link between domestic violence and animal abuse means that a veterinarian who suspects that animal abuse has occurred must consider the effect of discussing the matter with the client or owner of the dog, on other family members, especially in cases where children are the instigators of the animal abuse. It would be a difficult conversation for a veterinarian to have if it had the potential to expose a child or spouse to further domestic violence.

(6) Animal abuse reporting is mandatory in some regions. In the USA, where they do occur, mandatory reporting laws extend civil immunity for good faith reporting to protect veterinarians who report cases of suspected animal abuse. In Australia, reporting of animal abuse is not mandatory, but in its animal-abuse policy document, the AVA states that veterinarians have an ethical obligation to report abuse.

(7) Cruelty to animals is an offence in many countries, including all states in Australia and the USA. The Prevention of Cruelty to Animals Act (NSW) gives veterinarians the power to take possession of an animal that is so severely injured, so diseased or in such a physical condition that it is cruel to keep it alive, and humanely euthanase it.

(8) There are a number of organisations that can provide information and advice to veterinarians who suspect that an injured patient presented to them is a victim of animal abuse. These include animal welfare organisations such as the Royal Society for the Prevention of Cruelty to Animals (RSPCA) or American Society for the Prevention of Cruelty to Animals (ASPCA), the

Veterinary Registration Board and professional associations.

(9) A veterinarian that reports a suspected case of animal abuse may risk significant reprisal in the form of threats to their own personal safety or litigation. This may act as a deterrent for some veterinarians that are faced with a case of animal abuse where the evidence appears circumstantial.

(10) A veterinarian must maintain confidentiality of information, and so may decide to not report a suspected case of animal abuse because this breaches that confidentiality. However, the Veterinary Practitioners Board code of conduct states that confidentiality must be maintained except where otherwise required by the Code, such as a case where the welfare of an animal is at risk.

So who are the stakeholders in this situation?

(1) Alfie.
(2) Mrs V and her children.
(3) Animal welfare bodies such as the RSPCA and ASPCA.
(4) Veterinary professional bodies and associations.
(5) The veterinarian.
(6) Other staff of the practice (nurses, vets, reception).
(7) Law enforcement agencies.
(8) Government welfare bodies.

Solving the ethical dilemma

If we view this case as simply an animal welfare issue, then certainly a deontological approach would suggest this case must be reported to the authorities regardless of how little evidence there may be, as this ensures the animal is not put at any further risk of abuse and injury. In fact, when viewed simplistically, even a utilitarian approach would indicate that the best outcome for the greatest number of stakeholders would result if the veterinarian reported this incident. However, this case is not just about animal welfare.

This case highlights the complexities of the human–animal bond, and the interrelationship between animal and human welfare. Is it ethical to save an animal at the risk of causing harm to a woman or a child? Is it ethical to report an unconfirmed case of animal abuse at the risk of litigation or harm to the reporting veterinarian? If the evidence is not strong enough for authorities to charge the offender and seize the dog, then wouldn't reporting the case potentially result in more severe abuse of the dog, or indeed family members?

The complexity of this case makes it challenging to utilise even the broadest of ethical frameworks. If we take the example of principalism, an ethical framework commonly used in veterinary ethics, regardless of which action is taken in this case the preface that underpins principalism, "first do no harm", is difficult to adhere to. Any action aimed at protecting one stakeholder may result in harm to another.

The veterinarian should first and foremost treat the dog and manage the injuries that have been sustained. If the veterinarian wants to act on their suspicions they need to do so in a way that minimises potential harm and is most likely to have the intended outcome, that is, ensuring the animal abuse does not continue.

If the veterinarian is unsure how to proceed it may be useful to seek advice from those that have dealt with similar issues in the past. This could include the RSPCA, the NSW Veterinary Practitioners Board and the AVA.

Rather than accusing the owner of harming her dog, it may be better to engage in a conversation with the client and children about responsible pet ownership and the huge responsibility that comes with owning a pet. Animals can't speak up for themselves; this makes them vulnerable and we need to protect them. Perhaps identify some age-appropriate literature that you could recommend to the mother for her children.

The veterinarian could arrange a time to speak privately to Mrs V (without the children) about the concerns they have that this was not an accident but rather that the dog may have been harmed by a member of the family. It is important for the veterinarian to not put blame on the children or Mrs V but rather offer to listen to anything she may have to say and provide guidance or referral to others that may be able to support and help her if there is in fact a broader issue of family violence.

It is important that there is ongoing communication with the client so that the wellbeing of the dog can be monitored. This will enable further action to be taken, should there be repeat incidences that confirm the initial suspicions of abuse.

||||||||||||||||||||||||||||||||||||||||||||||||||||||||||||||||

ANIMAL abuse presents a number of challenges to the veterinarian, notably:

- Conflict between the obligation to respect client confidentiality and the interests of the animal;
- Distinguishing accidental from non-accidental injuries;
- Communicating to the client sensitively about animal abuse;
- Inconsistency in obligations around reporting suspected abuse;
- Concern about implications for reporting veterinarians.

In this case, our obligation to uphold confidentiality of discussions with clients that reveal a family member has deliberately injured an animal, conflicts with our obligations to look after that animal (by taking steps to prevent further abuse from occurring, such as reporting the abuse to relevant authorities). Much debate has centred around whether reporting of suspected animal abuse should be mandatory (Lachance 2016).

Proponents of mandatory reporting justify this position on the grounds that failure to report means that the animal, and other potential victims of abuse, remain vulnerable to further abuse. Breaking client confidentiality to report abuse is couched as the lesser of two evils. On the surface, mandatory reporting of suspected animal abuse seems to resolve the dilemma of the veterinarian to report or not to report by mandating the former.

One criticism of emphasising the link between animal abuse and interpersonal violence is that reporting perpetrators could be harmful. A false report of animal abuse is likely to irreparably damage the relationship between the client and the veterinarian. In the case of child offenders, the majority of offenders don't subsequently exhibit violence to humans, so there is a risk that authorities may falsely label and stigmatise children – which in turn could lead to greater deviance (Flynn 2011).

Differentiating deliberately induced or non-accidental injury (NAI) from accidental injury may be straightforward in some cases (for example, gunshot wounds, cigarette burns, asphyxia or microwave burns) but is quite challenging in others where the type of injury may be attributable to accidental or non-accidental causes – particularly high-impact trauma, for example, bruising, lacerations, fractures, head injuries or rupture of internal organs (Tong 2014). Veterinarians cannot simply extrapolate features of NAI in animals from those in the human literature, as there are significant differences in the nature of injuries (Tong 2014).

Given the sensitivities around and importance of identifying NAI, veterinarians may not be prepared to report suspected abuse without sufficient evidence, which may be difficult to gather. For example, in the case of investigating a non-accidental fracture, a combination of history, physical examination, radiographs, advanced imaging and specialist involvement may be required to characterise the injury. As tests are typically performed on a user-pays basis, the client may decline investigations. They may also be reluctant to discuss animal abuse. In a study of 26 women with experience of domestic violence, 92 per cent said they would be unwilling to discuss animal abuse with their veterinarian (Tiplady, et al. 2012).

Sensitive communication and investigation is critical. As mentioned in the scenario, the principle of non-maleficence may be challenging to maintain, especially if probing a client about suspected animal abuse is based on unfounded suspicion or if it exacerbates the abuse. Veterinarians may not feel comfortable in dealing with the myriad social issues associated with animal abuse, particularly where it occurs in the context of interpersonal violence. It may be possible to draw on the expertise of social workers with experience in domestic violence where animals and humans are involved (Holcombe, et al. 2016).

One area of confusion is inconsistency in requirements to report suspected or confirmed abuse, making the veterinarian's role in reporting animal abuse unclear (Holcombe, et al. 2016). In Australia, as mentioned by the author, reporting of abuse is not mandated. One justification of this, as stated by the Australian Veterinary Association, is that it "…may discourage owners from seeking essential treatment for their injured animals" (Australian Veterinary Association 2013).

But is there evidence to back up this assumption? A survey of community members found that 76 per cent incorrectly believed that Australian veterinarians are mandated to report suspected animal abuse (Acutt, et al. 2015). And where reporting is not mandated, veterinarians who report suspected animal abuse are without legislative immunity (Acutt, et al. 2015).

This is a case where further information is required to determine:

- Does reporting of animal abuse by veterinarians protect animal victims from further abuse?
- Does mandatory reporting result in improved welfare outcomes for animal and human victims?
- What resources are required to ensure that veterinarians minimise adverse impacts (further violence to animals or people, reprisals, legal liability) associated with reporting animal abuse?

There are an increasing variety of initiatives to support victims of domestic violence and their pets. For example, in the UK the RSPCA runs "PetRetreat", an advice and pet fostering service for families fleeing domestic violence situations: see www.rspca.org.uk/whatwedo/petretreat.

Similarly, the RSPCA in Australia runs the "Safe Beds for Pets" programme, designed to provide temporary housing for animals: see www.rspcansw.org.au/our-work/programs-community-services/safe-beds-for-pets.

In the USA, organisations such as Sheltering Animals of Abuse Victims provide a referral service (http://www.saavprogram.org/) and an increasing number of shelters for victims of domestic violence provide some accommodation for animals as well.

Another important issue to consider is what constitutes abuse. The literature on animal abuse focuses on illegal animal abuse committed by individuals or groups of individuals. But a number of authors have asked why the investigation of potential associations between animal abuse and human violence should be limited to incidents classified as illegal or socially unacceptable (Flynn 2011). If indeed a broader definition of animal abuse is considered, encompassing any act that causes pain or death, for example, then practices such as hunting, factory farming and animal experimentation would be subject to similar analysis.

ANIMAL ABUSE

What do you think?

ONE   **What specific features of interactions with animals characterise animal abuse?**

TWO   **What are the differences between animal use and animal abuse?**

△

13.21 Is there a difference between animal abuse and neglect?

PHOTO ISTOCK

Conclusion

One Health is a construct that helps us to understand how inextricably linked humans, animals and the wider environment are. Whether this is seen through an anthropocentric lens depends on our philosophical orientation. There are synergistic approaches to human and animal health and welfare that can be more than the sum of their parts. Veterinarians and associated animal health professionals need to be clear about their role in a One Health or One Welfare framework, and ensure they operate within their scope of practice.

References

Acutt D, Signal T, and Taylor N 2015 Mandated reporting of suspected animal harm by Australian veterinarians: community attitudes. *Anthrozoos* **28**: 437–447.

Akhtar S 2011 Animal pain and welfare: can pain sometimes be worse for them than for us? In: Beauchamp TL and Frey RG (eds) *The Oxford Handbook of Animal Ethics*, 495–511. Oxford University Press: Oxford.

Australian Veterinary Association 2013 Animal Abuse. AVA.

Baquero F, and Campos J 2003 The tragedy of the commons in antimicrobial chemotherapy. *Revista espanola de quimioterapia: publicacion oficial de la Sociedad Espanola de Quimioterapia* **16**: 11–13.

Boston S 2014 *Lucky Dog: How Being a Veterinarian Saved My Life*. House of Anansi Press Inc.: Toronto, Canada.

Clower TL, and Neaves TT 2015 The Health Care Cost Savings of Pet Ownership. Human Animal Bond Research Initiative Foundation.

Cousquer G, and Meredith A 2014 Pseudotuberculosis. https://www.vetstream.com/lapis/Content/Freeform/fre00324

Deviney E, Dickert J, and Lockwood R 1983 The care of pets within child abusing families. *International Journal for the Study of Animal Problems* **4:** 321–329.

Dvorak G, Roth JA, Gray GC, and Kaplan B 2013 *Zoonoses: Protecting People and Their Pets, 1st Edition.* Center for Food Security and Public Health, Iowa State University.

Febres J, Brasfield H, Shorey RC, Elmquist J, Ninnemann A, Schonbrun YC, Temple JR, Recupero PR, and Stuart GL 2014 Adulthood animal abuse among men arrested for domestic violence. *Violence against Women* **20:** 1059–1077.

Flynn CP 2011 Examining the links between animal abuse and human violence. *Crime Law and Social Change* **55:** 453–468.

Garcia-Alvarez L, Dawson S, Cookson B, and Hawkey P 2012 Working across the veterinary and human health sectors. *Journal of Antimicrobial Chemotherapy* **67:** i37–i49.

Gardiner A 2007 Everyday ethics. *In Practice* **29:** 553.

Gibbs EPJ 2014 The evolution of One Health: a decade of progress and challenges for the future. *Veterinary Record* **174:** 85–91.

Hadley J, and O'Sullivan S 2009 World poverty, animal minds and the ethics of veterinary expenditure. *Environmental Values* **18:** 361–378.

Hardin G 1968 The tragedy of the commons. *Science* **162:** 1243–1248.

Hardin G 1998 Extensions of "The tragedy of the commons". *Science* **280:** 682–683.

Holcombe TM, Strand EB, Nugent WR, and Ng ZY 2016 Veterinary social work: practice within veterinary settings. *Journal of Human Behavior in the Social Environment* **26:** 69–80.

Jani AA 2013 Pseudotuberculosis (Yersinia). http://emedicine.medscape.com/article/226871-overview

Joffe M, O'Shannessy D, Dhand NK, Westman M, and Fawcett A 2014 Characteristics of persons convicted for offences relating to animal hoarding in New South Wales. *Australian Veterinary Journal* **92:** 369–375.

Johnson RA 2010 Psychosocial and therapeutic aspects of human-animal interaction. In: Rabinowitz PM and Conti LA (eds) *Human-Animal Medicine: Clinical Approaches to Zoonoses, Toxicants, and Other Shared Health Risks,* 24–36. Saunders Elsevier: St Louis.

Lachance M 2016 Breaking the silence: the veterinarian's duty to report. *Animal Sentience* **1:** 1–16.

McDonald SE, Collins EA, Nicotera N, Hageman TO, Ascione FR, Williams JH, and Graham-Bermann SA 2015 Children's experiences of companion animal maltreatment in households characterized by intimate partner violence. *Child Abuse & Neglect* **50:** 116–127.

McEwen FS, Moffitt TE, and Arseneault L 2014 Is childhood cruelty to animals a marker for physical maltreatment in a prospective cohort study of children? *Child Abuse and Neglect* **38:** 533–543.

Mellanby RJ 2015 Our time is now – how companion animal veterinarians can transform biomedical science. *The Journal of Small Animal Practice* **56:** 689–692.

Meredith A, and Redrobe S 2002 *BSAVA Manual of Exotic Pets.* BSAVA: Gloucester.

New South Wales Government 2013 Veterinary Practitioners Code of Professional Conduct. Sydney, Australia.

One Health Initiative 2016 One Health Initiative. http://www.onehealthinitiative.com/

Page S, Prescott J, and Weese S 2014 Antimicrobial resistance: the 5Rs approach to antimicrobial stewardship. *Veterinary Record* **175:** 207–208.

Pinillos RG, Appleby MC, Scott-Park F, and Smith CW 2015 One Welfare. *Veterinary Record* **177:** 629–630.

Porco TC, Gao D, Scott JC, Shim E, Enanoria WT, Galvani AP, and Lietman TM 2012 When does overuse of antibiotics become a tragedy of the commons? *Plos One* **7.**

Public Health England 2015 Antibiotic Guardian. www.antibioticguardian.com

Quesenberry KE, and Boschert KR 2011 Disorders and diseases of guinea pigs. In: *The Merck Manual Pet Health Edition.* Merck Sharp & Dohme Corp.: Whitehouse Station.

Rabinowitz PM, and Conti LA 2010a Legal and ethical issues in human-animal medicine. In: Rabinowitz PM and Conti LA (eds) *Human-Animal Medicine: Clinical Approaches to Zoonoses, Toxicants and Other Shared Health Risks,* 7–11. Saunders Elsevier: St Louis.

Rabinowitz PM, and Conti LA 2010b Preface. In: Rabinowitz PM and Conti LA (eds) *Human-Animal Medicine: Clinical Approaches to Zoonoses, Toxicants and Other Shared Health Risks,* xiii–xiv. Saunders Elsevier: St Louis.

RCVS 2015 Code of Professional Conduct for Veterinary Surgeons. http://www.rcvs.org.uk/advice-and-guidance/code-of-professional-conduct-for-veterinary-surgeons/

Rollin BE 1999 *An Introduction to Veterinary Medical Ethics: Theory and Cases.* Iowa State University Press: Ames.

Rollin BE 2002 The use and abuse of Aesculapian authority in veterinary medicine. *JAVMA-Journal of the American Veterinary Medical Association* **220:** 1144–1149.

Shaw JR, Adams CL, Bonnett BN, Larson S, and Roter DL 2008 Veterinarian-client-patient communication during wellness appointments versus appointments related to a health problem in companion animal practice. *JAVMA-Journal of the American Veterinary Medical Association* **233:** 1576–1586.

Shaw JR, and Lagoni L 2007 End-of-Life communication in veterinary medicine: delivering bad news and euthanasia decision making. *Veterinary Clinics of North America Small Animal Practice* **37:** 95–108.

Speare R, Mendez D, Judd J, Reid S, Tzipori S, and Massey PD 2015 Willingness to consult a veterinarian on physician's advice for zoonotic diseases: a formal role for veterinarians in medicine? *Plos One* **10:** 8.

Spillman R 2015 Gold Coast vets refuse to treat injured and sick horses that are not vaccinated for Hendra virus. *Gold Coast Bulletin.*

Tiplady CM, Walsh DB, and Phillips CJC 2012 Intimate partner violence and companion animal welfare. *Australian Veterinary Journal* **90:** 48–53.

Tong LJ 2014 Fracture characteristics to distinguish between accidental injury and non-accidental injury in dogs. *Veterinary Journal* **199:** 392–398.

Villalobos A 2002 Ethics in practice – decision-making issues with euthanasia. *North American Veterinary Community Clinical Briefs* **May:** 23–24.

Weese JS, Giguere S, Guardabassi L, Morley PS, Papich M, Ricciuto DR, and Sykes JE 2015 ACVIM consensus statement on therapeutic antimicrobial use in animals and antimicrobial resistance. *Journal of Veterinary Internal Medicine* **29:** 487–498.

Yeates JW, and Main DCJ 2010 The ethics of influencing clients. *JAVMA-Journal of the American Veterinary Medical Association* **237:** 263–267.

CHAPTER 14
WILDLIFE

Introduction

The term "wildlife" is generally used to refer to animals that have not been domesticated and are living in the wild. While we may encounter wild species in captivity, these are – in the main – bred in captivity from stock originally wild-caught several generations previously, and have not been subject to the same degree of selection pressure for desirable traits as livestock and companion domestic species.

Wild animals may evoke awe or a sense of wonder, or perhaps fear – particularly when encountered in uncontrolled surroundings. There is no consistent regulatory approach to the treatment of wildlife in the law – in some circumstances restrictions on the use and treatment of wildlife are imposed by animal welfare legislation, in other cases environmental law or property law may be invoked (Cao 2015). In different contexts, wild animals may be perceived as a resource (for example, as a source of game or bush meat), pests (for example, rabbits, foxes, possums or badgers may be culled to protect human interests, including the lives and wellbeing of other domestic animals) or a source of entertainment (for example, animals kept in captivity or used to promote tourism).

Veterinarians and allied animal health professionals may encounter wildlife in a variety of contexts. This may include encounters in the wild or even urban environments, conservation activities, clinical scenarios where wild animals are presented for care, participation in wildlife culling, scientific research including observational and interventional studies, or as part of educational or training activities.

Working with wildlife presents a variety of ethical challenges, the first of which is whether to intervene at all. Veterinarians are trained to treat individual animals, but injured wildlife may be a vital source of prey for predators. Intervention may result in survival, reproduction and passing on of genes of wildlife that are injured because they are less fit or diseased. Thus there can be potential negative impacts on the wider population, other species and offspring.

In most cases there is no owner, although an animal may be presented by a member of the public or carer who is willing to take responsibility, whether financial, practical or both, for the animal's care. Because they are not domesticated, wildlife may suffer greater iatrogenic harms (for example, capture myopathy in some species) and it may be difficult to assess their interests. Whilst they make novel and interesting patients, it is important to ensure that we prioritise their interests. For example, the stress of a prolonged recovery, or the prospect of release in an uncertain environment,

◁

14.1

PHOTO JENNA MOSS-DAVIS

14.0
INTRODUCTION

△
14.2a–b Wild caught animals may be kept as pets, both legally and illegally.
PHOTO ANNE FAWCETT

△
14.3 Releasing wild animals such as this koala into areas where the species is over-abundant may be associated with poor outcomes where there is competition for resources and territorial fighting.
PHOTO ANNE FAWCETT

may yield much greater suffering to a wild animal than for domestic animals.

Other potential challenges include:

- Balancing the interests of the individual patient against those of the population, particularly with at-risk and endangered species. For example, there may be a possibility of releasing an animal with a resistant strain of bacteria or a novel virus into a naive wild population;
- Rehabilitating and releasing members of over-abundant species where there is not enough habitat;
- Rehabilitating and releasing members of a species for which there is no longer suitable natural habitat;
- Treating wildlife kept as pets (for example, wild-caught birds and reptiles);
- Treating wildlife that have been injured due to sport (for example, hunting);

△
14.4 Neonates and juvenile animals like this fox kit may be orphaned due to hunting.
PHOTO ANNE FAWCETT

- Performing surgical procedures on wildlife cases that render animals non-releasable (for example, wing amputations in birds);
- The role of concerned members of the public or carers in decision-making around case management (for example, objection or refusal to euthanase).

As with all other species, our efforts to help – no matter how well intended – can lead to harm.

14.1

To treat or euthanase?

For the reasons discussed in the introduction of this chapter, assessment of injuries or disease in wildlife may be significantly different from equivalent injuries or disease in domestic species. The following scenario explores decision-making in the case of injured wild animals.

SCENARIO
WILDLIFE CASUALTY

▶ You are called to an adult female roe deer that has been found sheltering under a hedge by a member of the public. The deer is very fearful, but despite this you capture her easily as she has severe lacerations to her hind legs, with one particularly bad one around one tarsal joint. You don't know how this has occurred but the injuries are consistent with being caught in barbed wire. The injuries will take several weeks to heal and require at least one general anaesthetic to clean and dress the wounds.

What should you do?

RESPONSE
PETE GODDARD

▶ In the case of this casualty roe deer, the main concern of the attending vet is the wellbeing of the injured animal. An *in situ* welfare assessment will need to include both initial concerns (is immediate euthanasia the only option or is recovery possible) and also medium- to long-term aspects once the animal is no longer under his or her care. This assessment will likely differ from those where a patient is passed back to the responsibility of the owner and where progress can be monitored, for wildlife generally have no "owner". However, in this case (or when a wildlife casualty is presented at a veterinary surgery), there will also be a need to consider

△
14.5 Roe deer are sometimes injured on fences or hit by cars.
PHOTO ISTOCK

carefully the expectations of the member of the public. There may also be some legal restrictions on what can be done (see later) and this will need to be explained. To avoid the difficulty of deciding these issues on-the-hoof without time for adequate consideration, it is a good idea to have a practice policy in place to make it easier to deal with these cases when they are presented.

The person reporting the animal or bringing in a casualty to a clinic will likely be well meaning and have the impression that the vet can restore the animal, a roe deer in this case, to normal function for it to later be released and to lead a natural life thereafter. Perhaps not for this roe deer but for many individual animals, life in the wild will be on a knife-edge and any impediment to functioning at full efficiency (predator avoidance, searching for food) will prejudice their survival. This will need to be factored into an initial assessment and also reviewed post-treatment, prior to release.

The initial assessment of the patient is a crucial step. If a full recovery is unlikely then an early decision about euthanasia should be made. Having clear protocols about this will help both in determining whether veterinary care is appropriate and in dealing with the public. If incomplete recovery is predicted then the only other option is for long-term (lifetime) confinement, though even if facilities exist for this it may not necessarily be in the animal's best interest in terms of a quality of life analysis. The vet may view wildlife casualties as a challenge and be motivated to try heroic procedures. It is, however, important to look at the long-term likely outcomes in terms of quality of life once the animal is passed into someone else's (or commonly, no one's) guardianship. Deer can generally be expected to make a good recovery from injuries such as those described in this case; even those with lower limb fractures usually do very well, surprisingly sometimes with minimal intervention.

In some cases, following treatment there may be a need for long-term rehabilitation at a wildlife centre. It is recognised that many wildlife rehabilitation centres have considerable experience in working with injured or orphaned wildlife. These may provide better medium- to long-term homes for casualties while they recover, away from the hustle and bustle of a busy veterinary practice. They may also allow for social animals to be housed together. During rehabilitation it may transpire that release may never be possible and the option of euthanasia should be reviewed again. Some animals (prey species in particular) will experience significant stress due to captivity. If captivity is prolonged there could be considerable suffering; were this considered "unnecessary suffering" this must be critically reviewed. We probably don't know enough about the use of Long Acting Neuroleptics (LANs) to reduce captivity stress in these situations but they have been used to mitigate boma stress in translocated ungulates in Africa; these may help.

Much of the success of any casualty treatment will be due to the effectiveness of encouraging the deer to feed properly and not to injure itself further in any attempt to escape. This latter concern can be a particular issue for deer and it is important to consider this before any treatment is commenced. Some specialist centres may be able to help. In the specific case of this roe deer, it is noteworthy that even bottle-reared roe deer are seldom suited temperamentally to captivity. In the case of an adult, adaptation would be even less likely unless the facilities are extensive and good cover is available to offer seclusion.

It is important to ensure that during the period of residence at the veterinary practice or a wildlife centre the casualty animal does not contract a novel (to that population) disease that could be introduced back into the recipient wildlife community (though admittedly many other routes for this to happen usually exist through contact with domestic animals and livestock).

If a medication is used for an animal, particularly a deer which might be released to the wild, only to be shot by a stalker and subsequently consumed, there could be a risk to the human food chain. This consideration could include anaesthetics, tranquillisers and antibiotics. The real risk might be slight but cannot be controlled by the vet after the animal leaves their care. In the case of this deer, choosing products with authorisation for use in food-producing animals (under the cascade system in the UK) and observing withdrawal times would be a good idea if possible. (In the past, the UK's Veterinary Deer Society has issued "Do Not Eat" ear tags to be applied to red deer darted with anaesthetics to enable management operations such as relocations or release of entrapped stags to proceed in an attempt to protect potential consumers). It is generally held that the sooner the patient can be released, the better the chance of survival.

If a veterinary practice is known to err on the side of euthanasia, even following expert and well-guided triage, they may be frequented less by members of the public who consider that all wildlife must be saved. Euthanasia is more acceptable if a clear and rational argument can be made and explained. The presenting individual may still take the animal away; the practice needs to have a clear policy about what to do if a moribund animal is presented but subsequently not left with them.

As a native species, the Wildlife and Countryside Act (1981) in the UK permits release of roe deer, but were the subject a non-native species such as a sika, muntjac or Chinese water deer [see Schedule 9 of the Wildlife and Countryside Act (1981) for a full list], restrictions would apply. So any treatment of these latter species must be accompanied by a consideration of the next steps if recovery can be achieved. Is a long-term captivity option available? Is this likely to be in the best welfare interest of the subject (i.e. is this

proposed option likely to offer a life worth living or even a good life or will it present a life not worth living)? These are difficult questions that the attending veterinarian has the main responsibility to answer.

More commonly, the vet will need to consider the realistic likelihood of self-sufficiency once the animal is released, especially as for many species there will not be an option for a "soft" release whereby food is provided for a period in the release area. In general, the animal should only be released if it is judged to have an equal chance of surviving to conspecifics living in the same habitat. Treating and subsequently releasing an individual from a species where some form of population control or hunting occurs or placing a prey species back into an area where the risk of predation is high raises a different type of concern. If release is back into the "home" environment then the animal will be aware of issues of terrain and food supply but if the origin is not known, release to an unfamiliar area may be an issue. Also for some very territorial animals, the vacuum left when an individual is removed will soon be occupied by a competitor. Will the period of treatment/rehabilitation have made the animal less able to fend for itself or possibly less fearful of hunters? For any animal held in temporary captivity but destined to be released, human contact should be kept to an absolute minimum. For the roe deer in question, once recovery is complete a view may need to be taken about possible release back into an area where it may be shot. Equally, what is the moral position if the deer is released back into an area of cultivation where deer are having a negative impact (e.g. feeding on agricultural crops)?

The declaration (oath) made by vets in most countries on admission to their regulatory bodies will generally include a requirement to ensure the welfare of animals under their care, but what would an injured wild animal choose to be in their

14.1
TO TREAT OR EUTHANASE?
RESPONSE
PETE GODDARD

best interests? Is this a more difficult question when the animal doesn't really have an owner (so the vet may have to assume this role directly) or the "finder" has taken on this role and feels a particularly strong motivation to do good as they see it by getting the animal back to normal function and so modifying the vet's own personal position?

Other, more generic issues are raised by this scenario. Is there more imperative to treat if the animal is from an endangered or otherwise threatened population? Many people believe that we have a moral obligation to conserve species or ecosystems. They believe that wild species have intrinsic value, that we have an obligation to preserve them and care for them but this may come at a cost to the individual.

Undertaking a harm:benefit analysis is potentially difficult here but the attempt will highlight important issues. In the case of the injured roe deer coming from a flourishing population which is potentially causing some environmental damage, the balance may be on the side of euthanasia. Another, more extreme, concern could arise if the vet is asked to treat an animal from a species where there is a programme of control, such as for the grey squirrel, which is regarded as a pest in the UK and where it is an offence to release one back into the wild or allow it to be released.

From the above it can be seen that the main challenge is not really one of veterinary expertise necessary to deal with wounds, but to balance the immediate and longer-term welfare of the roe deer, the impact on the finder, and sometimes the effect on the species at large or the wider ecosystem. To end on a prosaic note, not to be forgotten in the real world is the associated question of payment for treatment, especially if expensive procedures such as orthopaedics are necessary and the recovery at the practice is likely to be prolonged. Many members of the

public consider this pro bono activity a legitimate professional role, but if there are many cases or treatment costs are high, this can have a negative impact on other practice activities, though there may be a partial offset through the recognition of this goodwill.

||||||||||||||||||||||||||||||||||||||||||||||||||||||||||||||||||||||||

IN this scenario the author uses a cost:benefit or utilitarian approach, in which stakeholders include the individual deer, the wider population, the veterinarian, the member of the public/ animal carer and the environment.

One issue raised by this scenario is whether it is the role of the veterinary team to make decisions based on perceived duties to the environment or ecosystem, rather than the best interests of the individual animal in its current circumstances. A potential conflict can arise between the perceived interests of an animal and the perceived interests of the local ecosystem. This will be discussed in a later scenario on "pests".

A relevant question is how a wild animal has come to be injured in the first place – occasionally injuries are incurred due to underlying pathology (for example, metabolic or neurological disease leading to weakness or impaired reaction time, visual deficits and so on) which may ultimately impact on prognosis.

One of the challenges is predicting both short- and medium-to-long-term outcomes of decision-making. This can be particularly difficult as in most cases there is no central repository of post-release data, particularly when animals are discharged from veterinary facilities into the care of rescue organisations, sanctuaries and carers who provide rehabilitation and release. On the one hand, it could be argued that we have little information to demonstrate successful release following medium-to-long-term hospitalisation, and thus it is not worth putting the animal through certain stress and suffering for an uncertain outcome. Equally, it could be argued that while we don't know that the overall outcome will be negative, we are obliged to try.

It may be the case that a deer, once released, survives until the age of three. Whether this constitutes a good outcome or not may depend on the average age of survival for its conspecifics. If life in the wild really is nasty, brutish and short, we need to ensure our standards for what makes a life worth living in the wild are not impossibly high, or the threshold for euthanasia may be too low. Another consideration is quality of life, which in wild animals is very difficult to assess. It is possible for an animal's life to be extended post-treatment, but if much of that is spent in hospital, that will negatively impact overall quality of life.

Wildlife exhibit a strong "fight, freeze or flight" response which suggests an underlying interest in survival and potentially survival without intervention. Our threshold for intervention for wild animals may differ significantly from that for domestic animals for this reason.

TO TREAT OR EUTHANASE?

What do you think?

In the previous scenario, the roe deer was injured due to human activity (placement of barbed wire). How would your approach differ if you encountered a wild animal injured by itself (for example, in trying to catch prey) or another animal (for example, a fight over territory)? What ethical justification would you use for your position?

△

14.6 Would you treat a banded mongoose that had been injured through fighting?
PHOTO ISTOCK

TO TREAT OR EUTHANASE?

What would you do?

A raptor presents with a broken wing. What factors will you consider in deciding whether to treat this animal?

△
14.7 What factors will you consider when deciding to treat a raptor?
PHOTO ANNE FAWCETT

TO TREAT OR EUTHANASE?

What would you do?

An animal welfare charity was criticised after publishing photos on social media documenting dental treatment of a field vole. Critics argued that the treatment constituted a misuse of the charity's limited resources because the rodent was likely to end up being eaten by a predator shortly after being released into the wild. What would you have advised?

△
14.8 Field voles are one of the most common mammals in Europe and are the predominant prey of some populations of barn owls.
PHOTO ANNE FAWCETT

14.2

Euthanasia and killing of captive animals

It is reasonably uncontroversial that wild animals in captivity should be euthanased in their best interests, that is, to prevent suffering. For example, a captive Kodiak bear with debilitating osteoarthritis or an Asiatic lion with evidence of chronic kidney disease may be euthanased.

In zoos and sanctuaries, the decision to euthanase may be made by an individual or, increasingly, by a team incorporating veterinarians, curators, keepers and other staff. A structured quality of life assessment may or may not inform the decision-making process.

Some wild animals in captivity are killed because they are considered surplus – for example, if the holding capacity of the captive environment is exceeded or if the genes of the animal are considered surplus for regional breeding programmes. The practice of management euthanasia has been dubbed "zoothanasia" by critics (Bekoff & Ramp 2014). The following scenario explores a well-publicised case.

———
SCENARIO
KILLING ZOO ANIMALS SURPLUS
TO REQUIREMENTS

▶ Marius the giraffe was killed by Copenhagen Zoo in 2014. Marius was judged to be surplus to requirements based on the European Endangered Species Programme (EEP). After being killed, zoo keepers dissected Marius for the benefit of the viewing public and fed his remains to lions kept at the zoo. The killing of Marius caused considerable controversy and the story was extensively reported by the international media.

A number of zoos, conservation organisations and ecologists made statements either supporting or opposing the act.

How should zoo veterinarians advise zoos on policies for captive animals like Marius that are deemed surplus to requirements?

———
RESPONSE
STEVEN P MCCULLOCH AND MICHAEL J REISS

▶ The management of zoo animals considered surplus to requirements is a complex moral issue. The killing, public dissection and feeding of Marius the giraffe to lions at Copenhagen Zoo caused considerable international controversy (Eriksen & Kennedy 2014, Morell 2014, Rincon 2014). A number of questions can be asked to facilitate this brief ethical analysis and present a judgement as to veterinary advice on the scenario.

A starting point is to assess the impact of killing Marius on relevant individuals and groups. Individuals and groups impacted by the decision include Marius himself, the giraffe (sub-) species *Giraffa*

△
14.9 Zoo staff prepare to dissect Marius the giraffe in front of onlookers.
PHOTO COPENHAGEN ZOO

camelopardalis reticulata, lions at the zoo and relevant human publics. The first question to ask is whether killing Marius constituted a moral harm to Marius. Secondly, does the giraffe species to which Marius belongs have intrinsic moral value? Thirdly, did lions at the zoo, which were fed Marius' remains, benefit from the act? Fourthly, which human publics were impacted by the decision, and have they benefited or were they harmed?

Was killing Marius a harm *to* Marius? In the animal ethics literature, there are broadly two accounts of the harm of killing creatures such as Marius, which are considered to be sentient but not, or only minimally, self-conscious (a description generally restricted to adult mammals excluding the great apes).

The first account is the desire-based account. In the desire-based account, individuals must have a desire to continue living to be harmed by the act of killing (DeGrazia 1996, Frey 1987). Since individuals that are not self-conscious do not have a desire to continue living, the desire-based account of the harm of killing holds that killing is not a harm to individuals, such as Marius, who are sentient but lack self-consciousness. Indeed, a statement by the Scientific Director of Copenhagen Zoo is based on the desire-based account: "None of our animals have any expectations of how long they will or can live. But they sense the quality of their actual life, and that part must be as good as ever possible" (Holst 2014).

In opposition to the desire-based account is the opportunities-based account. The opportunities-based account claims that sentience is sufficient for death to be a harm (DeGrazia 1996). Being killed forecloses future pleasures and other goods. This holds whether or not individuals are self-conscious, have some future-oriented desires – such as the desire to live – and have an understanding of the concepts of life and death. Importantly for scenarios such as this, the opportunities-based account holds that killing Marius is

a harm if, and only if, he could be expected to continue to have a life worth living (LWL), i.e. a life of net positive value, from the time at which he would have been killed to the end of his life – whether that end is natural or not.

Most progressive animal ethicists defend some version of the opportunities-based account of the harm of killing (Cochrane 2012, DeGrazia 1996, Garner 2013). It seems unnecessary for an animal to have the more developed cognitive capacities of self-consciousness for death to constitute a harm. Killing sentient animals, which would otherwise have a LWL, deprives them of that life worth living. This claim holds whether or not they are self-conscious and have future-oriented desires involving life and death.

The second key question is whether the giraffe species *Giraffa camelopardalis reticulata* has intrinsic moral value. Does the giraffe species, as a group of animals, have moral value independent of the value of individual members of that species, as well as the extrinsic value it has for other species, most notably humans? Individual animals have intrinsic value due to their sentience. Individuals, such as Marius the giraffe, can experience pleasure, but are harmed if they are caused to suffer. Similarly, the extrinsic value of a species, to humans or other species, is clear. For instance, the conservation of species may lead to instrumental value in terms of advancing human knowledge, or aesthetic value through the appreciation of the existence of a beautiful species. Species also have extrinsic value by virtue of their places in an ecosystem; for example they provide energy and nutrients for their predators and parasites.

However, the idea of intrinsic value of a species seems more difficult to defend. This is an important point, because the rationale provided for killing Marius was based on prioritising the genetic diversity of the species above the intrinsic value of Marius as an individual sentient giraffe. Marius' genes were not sufficiently different from those of other males of his species, both in the Copenhagen Zoo and in Europe more generally, for it to be considered appropriate for him to participate in the captive breeding programme and, as a two-year-old male, he was close to reaching sexual maturity (Holst 2014).

The question of whether lions benefited from being fed Marius' remains seems the least controversial. The footage of lions consuming Marius suggests they enjoyed the act. Lions are a carnivorous species that hunt giraffes in African savannahs. Rollin argues that welfare is related to the actualisation of *telos*, the species-specific purpose of animals (Rollin 2006, 2007). Thus, the feeding of Marius to captive lions in Copenhagen Zoo mimics closely the natural predator–prey relationship, save putting a live giraffe in with the lions. The latter, of course, would cause substantial stress both to Marius and to many humans who observed or were otherwise aware of the act. Furthermore, live-feeding in a captive situation removes the possibility of the prey escaping.

The fourth question to ask is how various human publics are impacted by the decision. Broadly, there are three groups to consider. First, a subplot of the Marius controversy has been whether it was appropriate for children to watch the dissection. Secondly, killing Marius, and surplus zoo animals in general, may have a more general impact on zoo visitors who have enjoyed watching Marius and other animals. Thirdly, there is a very broad public whose views may be impacted, either positively or negatively, by their heightened awareness that zoos kill surplus animals.

Whether it was beneficial or otherwise for children to watch the public dissection of Marius is clearly influenced by the broader moral question of whether it was right or wrong to kill him in the first place. To explain, consider if it is found that Copenhagen Zoo was morally justified in killing Marius. In this case, the claim that the public, including children, should be educated about the reality of killing surplus animals, together with any beneficial consequences from public dissection, gains credence based on the widely held belief that knowledge is a human good. In contrast, consider if objective analysis finds the decision to kill Marius morally problematic. Here, the argument that children might become desensitised to the killing and disposal of zoo animals – in effect the objectification and instrumental use of animals – becomes stronger. We can also note that in the second case, when the killing of Marius is morally problematic, it might have been permissible for children to watch his dissection had he not been killed but died of natural causes.

Lock and Reiss (1996) argue that it can be ethically appropriate for children not only to watch animal dissections but to undertake them, provided a number of conditions are met. Chief among these are that such participation should be optional and that it should be educational. It is also preferable for animals not to be bred specifically for the purposes of dissection. All three criteria seem to have been met in the case of Marius. In schools the educational argument for dissection principally centres on the fact that students learn something of anatomy that even the best videos and simulations cannot provide. For a subset of school students, a particular argument in favour of such dissection is that they may thus be more or less likely to choose to study medicine or veterinary science after they leave school. It seems better to find out that one loves or abhors dissection before one finds oneself on a university course that requires it. These arguments hold in the case of dissecting Marius. In addition, in watching the dissection of Marius or, indeed, his being served to the lions, children might more meaningfully reflect on such moral questions as the purposes of zoos and the nature of human dominion over animals.

The discussion above about the relation between the morality of culling Marius and its impact on humans can, to some extent, be applied

14.2

EUTHANASIA AND KILLING OF CAPTIVE ANIMALS
RESPONSE
STEVEN P MCCULLOCH AND MICHAEL J REISS

to other publics. Firstly, there are those who visit or who have visited the zoo. Secondly, there are those who do not visit zoos, but who form views about the morality of the existence of zoos and their practices. For these publics, if killing Marius is objectively right or wrong, then divergent public opinion is, to a large extent, an information/educational issue. If there is no objective rightness or wrongness about the act of killing Marius, then divergent public opinion is the inevitable result of moral subjectivism. Furthermore, public opinion can be considered more as a pragmatic issue of publicity for Copenhagen Zoo, and the wider zoo industry, to consider seriously. If, for instance, making a public show out of killing surplus animals gives zoos a negative impression in the public eye, it would be unwise for zoos to continue such practices.

Based on the above considerations, the following preliminary conclusions can be made. Firstly, based on the opportunities-based account of the harm of killing, since giraffes are sentient, and Marius is a giraffe, killing Marius harms him. Secondly, whereas species are composed of individuals with intrinsic value, and the species (group) has extrinsic value to humans and others, it is more difficult to locate intrinsic value in the species itself. Thirdly, lions benefit from consuming giraffes as prey, both through hedonistic pleasure and, in a broader sense, based on the actualisation of their *telos*. Fourthly, various human publics are likely to be impacted in a range of ways, both negatively and positively, by the decision to kill Marius or other animals "surplus" to requirements. Despite this, the impact on these groups may be, to a significant extent, related to the intrinsic rightness or wrongness of killing Marius *simpliciter*. Furthermore, it could be argued that Copenhagen Zoo had a particular duty to Marius as an individual in a similar way that a pet owner has one to his or her pet (cf. Haidt, et al. 1993). Arguably, zoos have some form of duty of "guardianship"

(FAWC 2009) to the individual captive animals in their care.

Based on the above, and particularly the opportunities-based account of the harm of killing, it seems very possible that it was morally wrong to kill Marius. Therefore, on the *primum non nocere* principle (first do no harm), we should have good reasons to kill Marius for it to be morally justifiable. Additionally, of course, there should be no reasonable viable alternatives, such as sterilisation/neutering or rehoming. When empirical factors about the issue are examined, the moral grounds for killing Marius seem weak. Krakow Zoo and the Yorkshire Wildlife Park both offered to rehome Marius. Since Krakow Zoo and the Yorkshire Wildlife Park are both members of the European Association of Zoos and Aquaria (EAZA), there were no regulatory obstacles to rehoming Marius at these establishments. Krakow Zoo claimed that Copenhagen Zoo did not give a reason for turning down its offer. Copenhagen Zoo advised the Yorkshire Wildlife Park that, since it already housed Marius' biological brother, it should use the space for a more genetically valuable giraffe (Eriksen & Kennedy 2014).

In its decision to (1) kill Marius, and (2) refuse offers for him to be rehomed, Copenhagen Zoo seems to have adopted the premise that the good of the genetic pool for the species *Giraffa camelopardalis reticulata* should trump other considerations. However, this argument seems problematic. *Giraffa camelopardalis reticulata* is not an endangered species. Was Copenhagen Zoo right to refuse an offer to rehome Marius, based on its reason that Marius would be taking the place of a genetically superior individual? The strong weighting given to genetic arguments and the correlative elevation of species-based arguments above those protecting the individual seem difficult to justify.

CHAPTER 14 WILDLIFE

14.2
EUTHANASIA AND KILLING OF CAPTIVE ANIMALS
WHAT DO YOU THINK?

THIS scenario highlights the way the same framework can be used to justify different approaches, based on the assumptions underlying what constitutes harms and benefits. For example, if – unlike the authors – one takes a desire-based account – then humane killing of Marius does not constitute a harm.

This seems like a neat solution to an ethical dilemma, but is it? If we subscribe to the desire-based account, it is then perfectly acceptable to kill any healthy animal as long as it is killed humanely. Such a claim seems out of step with the predominant social ethic.

That aside, dissecting Marius publicly for educational purposes and feeding his remains to the lions is a way of maximising good from the situation. An argument may be advanced that inviting the public to view the dissection and being transparent about zoo operations (after all, surplus animals are killed in other zoos and wildlife institutions) respects the autonomy of the public. It could be argued that the approach to stakeholders is just, in that all animals including giraffes and lions are given the opportunity to enjoy what the zoo conceives as good quality of life (for example, the ability to reproduce). It may also be argued that Marius' death was the lesser of two evils. Many prey species breed prolifically, and females prefer to live in groups, with males essentially "surplus" to these herds.

EUTHANASIA AND KILLING OF CAPTIVE ANIMALS

What do you think?

You attend a heated panel discussion around the role of zoos and sanctuaries in wildlife conservation. One panellist, a conservationist who works in a consultancy role, argues that "if zoos are really about conservation they should humanely kill all animals that have passed their breeding age and can no longer contribute to the gene pool. This way zoos make room for young, healthy animals that can actively increase the population."

Another panellist, a zoo curator, vehemently disagrees, and argues that "as stewards of captive animals we need to ensure they enjoy long, healthy lives, because even after they have produced and raised offspring they remain ambassadors for their species."

ONE   How would you provide an ethical justification for one or both of these approaches?

TWO   Which do you support and why?

△
14.10 What obligations do zoos have to individual animals, and how does this align with conservation goals?
PHOTO ANNE FAWCETT

14.3

Conservation

Conservation efforts may involve interventions that harm individual animals or expose them to risk, with the overall aim to benefit a group or population of animals. The following scenario explores the decision-making around conservation interventions that may involve the veterinary team.

dramatic increase in the volume and sophistication of poaching, and conservation and protection measures are ever important. In 2014 over 1000 rhinos were killed by poachers in South Africa alone. At a weight of up to around 3000 kg, rhino capture, anaesthesia and translocation represent a significant challenge.

It is important to frame any ethical consideration about such activities by asking why this is being done using the particular species or

SCENARIO
DARTING RHINOCEROS

▶ A number of conservation initiatives involve trapping and translocation of animals, with some risk of morbidity and mortality to the affected animals (for example, anaesthetic death, or misadventure during anaesthetic recovery).

You are a veterinarian that has been asked to join a team that will anaesthetise, trap and translocate six adult male white rhinoceros. Surplus males from the breeding programme will be relocated to a game hunting reserve. This will involve darting the animals and performing field anaesthesia.

What should you do?

RESPONSE
PETE GODDARD

▶ With around 20,000 southern white rhinos in existence in Africa they represent a major success story (since there were estimated to have been as few as 20 in the late nineteenth century) compared to the northern white rhino; at the time of writing there are only 3 living individuals from this subspecies. Although poaching appeared to be on the decline, since 2007 there has been a

△

14.11a–b Conservation initiatives such as translocation carry some risk of morbidity and mortality, and may impact social interactions and wellbeing of animals such as these white rhinoceros.
PHOTO ANNE FAWCETT

individuals as this may affect the outcome of the ethical cost (harm):benefit analysis which should accompany any wildlife conservation intervention.

Looking at the necessity (potential benefit) we can identify four key types of benefit:

(1) Benefit to the subjects – for example, if the home range is experiencing some natural disaster or anthropogenic threat (such as extreme poaching pressure) and animals need to be moved in order for them to survive.

(2) Benefit to conspecifics – for example, the remaining rhinos will have a better chance of survival on existing resources if the population pressure becomes too great and the ecological carrying capacity (ECC) has been exceeded. Some people maintain that *not* moving animals in these situations is the wrong rather than safe option; this is equivalent to the stance noted in chapter 1 where performing an act of omission (neglecting to do something) may lead to a worse outcome than the act of commission. Unhealthy (including lame or poor body condition) individuals should not normally be considered as suitable relocation subjects. However, in some cases health can only be checked once the animal has been darted – a decision not to move the animal may be challenging at this point. There is a need to consider the risk of the translocated rhino introducing novel disease into a susceptible population. There is a subclass of this "species" benefit, sometimes called sustainable utilisation, which is more ethically challenging: moving prize specimens to sporting or game-hunting parks/reserves where individuals are shot for a fee (trophy hunting) generates a large amount of money, some of which has been claimed to be fed back and used to support local same-species populations or more general conservation

measures. Similarly, capturing an individual for a zoological collection can be part of a breeding programme and generate a huge amount of financial support to feed back into conservation measures; but again, there is no direct benefit to the individual.

(3) Dual benefits – these can accrue to animals and humans if the main function is species conservation, for example where animals are moved to new areas with sufficient resources and there is an interchange of genetic material with animals that are already present. Thus there is a wildlife population / species survival benefit and a new opportunity for wildlife watching / ecotourism. Unfortunately this short section does not provide the opportunity to consider, from a philosophical perspective, whether endangered species have an inherent concern about becoming extinct: briefly, do the three remaining northern white rhino "care" that they are the last in their line or are humans trying to salve their conscience by recognising that they have a moral responsibility to conserve the species – undoubtedly too late in this case.

(4) Human benefits – for example, moving wildlife from livestock areas to preserve vegetation, reduce disease transmission or other inter-specific conflicts. There are also simple commercial benefits if the animals are moved to game parks to be shot subsequently and there is no conservation feedback benefit in kind as under point 2. Interestingly it has been reported that some rhinos are kept on private land and the horn harvested periodically, in what may lead ultimately to a legal rhino horn market. 3D-printed rhino horn may soon appear – with an uncertain market influence.

Since the action of translocation represents a potential hazard to the subject, do these different benefits identified above imply a different level of risk is acceptable? One point of view is that where the benefits to individuals are greater, then a commensurately higher risk is acceptable. Others view the costs as independent from the ultimate benefit.

Looking at the costs then we can consider some as immediate (rounding up, darting/anaesthetic risks), and some as medium term (transport risks and release into a new location – either directly on arrival or after some local acclimatisation). It is also necessary to consider longer-term effects on the subject(s), the donor community and the recipient community. In many cases these costs cannot be completely known or well quantified but best estimates should always be attempted through, for example, gaining as much evidence as possible from previous translocations. Here are just some of the considerations relating to the longer term:

- White rhinos have a very complex social structure and tend to live in small and isolated populations. Within this structure, males are usually solitary and associate with females only for breeding. Removing one or a small number of individuals will not only impact on the subjects but also on those who remain. So, unless specific individuals are required in the new area, conservationists should consider the availability of younger males who would be more likely to adapt to new territory and conspecifics, while their loss will have less impact on the source community. Often a new group of around 25 per cent breeding males and 75 per cent females is created. The current scenario concerns adult males so the cost would be mainly in relation to the donor population if they were important individuals.
- It is important to consider the fact that animal translocations can play a role in the emergence of infectious disease in the recipient community. This is one reason for a veterinarian's involvement and advice, to ensure that

full health checks and disease implications are considered.
- The arrival of animals in a new territory will inevitably disrupt the existing ecological balance and if there are rhinos there already, the introduction of mature breeding males may result in territorial conflicts at the very least – such that it is important to consider whether the translocation can be conducted in the non-breeding season. The introduced animal will be at a significant disadvantage as the territory will be unfamiliar, both in terms of finding food and water and in escaping aggressive approaches.

The team veterinarian will more likely be involved with the immediate effects surrounding capture and removal and thereby reducing the short-term costs/risks to the individual animal, maximising the chances of success. Here are some issues to consider, the attention to which should reduce the chances of the subject sustaining an injury or dying:

- Does the veterinarian have sufficient experience? This will allow appropriate planning, choice of appropriate agents, and the ability to prepare for and manage complications, including the need for humane destruction.
- Be clear about who is directing the overall operation on the ground and from whom the team will take instructions (it could be the veterinarian of course but it is better for the vet to concentrate solely on the anaesthetic task in hand). This needs to include prior consideration of when operations might be suspended.
- What agents to use – tranquillisers/immobilising agents/sedatives and their speed of onset, dart volume and antidotes (and when these will be given). For example, the use of butorphanol as part of the anaesthetic regime may provide handling and physiologic advantages (Miller, et al. 2013).

- Who will do the darting – their experience, the effect of weather conditions, and ability to get close to the subject.
- Anaesthetic monitoring, including avoidance of hyperthermia and minimising the risk of capture myopathy.
- Health monitoring once the animal is anaesthetised (and what to do if certain conditions are found).
- Placing in transporter and oversight during transport.
- Release at site (initially usually into a restricted area or boma – this also allows any post-translocation injuries to be identified) and ensuring the best anaesthetic combination has been used during capture to minimise inappetence. Consider methods to encourage translocated animals to eat following release from the boma.

Finally, there is a potential risk of human injury which could result from physical consequences of handling such a large animal (including, for example, hoists breaking leading to crush injuries), incomplete anaesthesia allowing the animal to partially recover during a handling process, or chemical effects of exposure to potent anaesthetic agents. The veterinary team will be responsible for identifying and managing these hazards, and developing protocols for actions to be taken in the event of something going wrong. This consideration should include the availability and use of appropriate anaesthetic antidotes including the ability of someone to treat the veterinarian if s/he is accidentally exposed.

Since you have been invited to join the translocation team, in coming to a decision you need to consider whether you are sufficiently experienced in the first place. If that is the case, you need to make a personal judgement about your willingness to participate, given the specific purpose of the intended movement set in this scenario. Many people would consider it wrong to kill an animal for sport. Even if

you are dubious about the rationale, which may be primarily for financial gain, you may consider that you are better able than others to do a good job and therefore wish to take the assignment.

|||||||||||||||||||||||||||||||||||||||||||||||||||||||||||||||||||||||||

THE above scenario considers the costs and benefits of translocation of surplus rhinoceros to a game reserve. A cost:benefit analysis is undertaken, and the benefits to the wider population and to humans involved are emphasised. Consideration of moving animals in immediate danger of poaching to a safe, protected environment may result in a different analysis, with costs to individual animals easier to justify given those individuals may stand to benefit more directly from the translocation.

The greater chance that harms may be minimised and benefits maximised, the higher costs we are likely to accept – although in this and the previous case, the costs are borne largely by the individual rhinos that are translocated.

A deontological approach which respects the autonomy and intrinsic value of animals may not permit intervention to facilitate translocation in such circumstances, particularly if the animals will be hunted as trophies. The concern is that the veterinary team involved may become complicit in an activity that is argued by many to be against the interests of animals.

Is there an alternative? In the case of protecting rhinos from poaching, a number of strategies have been employed including removal of rhino horns or infusing the horns with a chemical that not only alters their colour but has a toxic effect on persons who subsequently consume that horn – undermining the myth that the horn, made entirely of keratin, has medicinal value (Platt 2012). Other approaches include promoting rhinoceros-based tourism, to convince locals that rhinoceros are worth more alive than dead, and introduction of artificial rhino horn to collapse the current market.

What do you think?

The belief that rhino horn has medicinal properties underlies consistent demand for this product. In order to deter would-be purchasers, some horns have been infused with drugs that cause nausea and convulsions in consumers. Others have proposed infusing rhinoceros horn with toxins such as cyanide that – while not harming the rhinoceros – are potentially fatal to consumers (Platt 2012).

One of your colleagues supports this approach. "Rhinos are so threatened by human behaviour, it's okay for a few humans to suffer if it shuts down the market and associated suffering," she says.

"Is it?" asks another colleague. "But then you're saying it's fair that a person who buys rhino to cure their cancer – and has already been duped out of their money by charlatans – should suffer illness or even death. Isn't that unethical?"

ONE   **What is your view?**

TWO   **What are the important ethical considerations to you?**

△ **14.12** Rhino horn may be sold as medicine in some countries.
PHOTO ISTOCK

14.4
Wildlife as pest species

The term "pest" is a relative one. A pest is an animal, insect or plant that is actually or potentially detrimental to human interests. Thus a crop-eating locust is a pest, as is a noxious weed that represents a threat to livestock, as is an elephant rampaging through a village.

Depending on the context, members of one species may be respected or revered, while in another place they are actively targeted in

△ **14.13a–b** Possums are considered a pest in some countries and not in others.
PHOTO ANNE FAWCETT

the name of pest control. Thus for example it is common in Australia for wildlife carers to seek veterinary attention for injured possums, and invest a good deal of time in rehabilitating and releasing these creatures, while in New Zealand possums are considered a pest.

Similarly, it is common to keep companion cats, rabbits and dogs but free-roaming cats and rabbits and dogs are considered pests. Yet our treatment of animals labelled as "pests" is very different.

According to a growing body of literature, both pests and pest control strategies can have desirable and undesirable impacts on people, animals and the environment – directly and indirectly, and in the short, medium and long term. For example, negative impacts on target and non-target animals include "pain and distress due to lethal and sub-lethal poisoning, capture by traps or chronic injury following escape from traps" (Littin, et al. 2004). Other negative impacts include rebound of pest numbers following control, or predation of threatened prey species by predators that would otherwise have consumed the pest.

The following scenario explores decision-making around whether culling of a "pest" species is ethical.

△
14.14
CARTOON CARTOONRALPH.CO.UK

grey squirrel population, but causes high morbidity and high mortality in the susceptible red squirrel population.

Is culling of one species in order to save another justifiable on ethical grounds?

SCENARIO

RED VS GREY SQUIRREL

▶ Red squirrels (*Sciurus vulgaris*) are native to the UK. Populations of red squirrels declined in Great Britain and Ireland, due to a combination of habitat loss and the deliberate introduction of the grey squirrel (*Sciurus carolinensis*) in 1876. The red squirrel population has reduced dramatically and local extinction is widespread. Grey squirrels, a larger species with a broader dietary repertoire, competitively exclude red squirrels and act as a reservoir for squirrelpox virus (SQPV) (Collins, et al. 2014). SQPV is endemic in the

---

RESPONSE

RICHARD GREEN

▶ The question of whether culling one species to save another is justifiable ethically presents a perfect opportunity to use an ethical matrix in order to evaluate the various ethical considerations of the dilemma. It also provides a perfect example of why ethical matrices can only provide an analytical tool to allow objective examination of the issues, and not an absolute answer to the

—
14.4
WILDLIFE AS PEST SPECIES
RESPONSE
RICHARD GREEN

dilemma, since the "answer" will depend on the ranking of the ethical impacts, which will remain a personal and subjective opinion.

In this instance, the primary stakeholders in the situation are the individual squirrels, both red and grey; the species – also both red and grey squirrels; and the environment/ecosystem, or biota, and they can be assessed for the impacts on wellbeing, autonomy and justice.

If one looks at the very simple matrix in Table 14.1 – and clearly one could expand greatly on each of the box entries – then it is apparent that the conclusions as to whether a cull is right or wrong will depend on the sympathies or allegiance of the individual who is ranking the impacts.

A strict utilitarian "welfare-ist" proponent (e.g. Singer) might oppose a cull on the basis that the welfare of those animals (the grey squirrels) currently in existence should be the primary concern, and that any alternative stance would be "species-ist".

A more animal rights-based view (e.g. Regan) might also oppose a cull on the grounds that the interests of those animals already "subjects-of-a-life" should outweigh the interests of those as yet non-existent animals (the reds who would replace the greys).

An alternative utilitarian view might counter-argue that, as long as the cull was humane (although it would be hard to argue that any cull outside of a laboratory could be completely humane), then since death has no welfare impact (also argued by some, since it represents loss of opportunity for those animals culled), it should not matter if one favours red or grey squirrels from an individual perspective (all other things being equal), but that when considerations for the benefit of a favoured species and the environment are taken into account, then the balance would tip in favour of a cull of the greys. This version of a utilitarian argument could also be used despite the objections that no cull would be completely humane,

and that loss of individual opportunity (for the greys) would be unavoidable, by weighting the interests of the environment more highly so as to negate these factors.

There might also be some argument that virtue ethics might favour a cull since this would be "doing the right thing" by nature, in attempting to correct a situation initially created by human introduction of a non-indigenous species.

Current social ethics (common morality) would seem to favour the indigenous species, and the environment, over non-indigenous or "unnatural" species, and one only has to look at the ongoing

△
14.15a–b What are the morally relevant differences between these red and grey squirrels?
PHOTO ISTOCK

argument about whether it is acceptable to cull badgers to protect dairy cows (and humans) to see how strongly some people feel about issues such as this.

The ethical matrix here has proven an excellent checklist, as well as a means of identifying assumptions that can be challenged. For example, the assumption that extinction is bad may be challenged. Extinction could be argued to be a component of evolution, where in this case one species is no longer fit to live in an ecosystem where another can thrive.

Ecosystems are complex, and it is difficult to predict outcomes when a single species is targeted. For example, if habitat division and destruction is a factor, simply ridding an ecosystem of one particular species (if that is indeed possible) will not solve the problem.

	WELLBEING	AUTONOMY	JUSTICE
Grey squirrel (individual)	Reduced, though mitigated by cull methods. Loss of opportunity	Compromised	No
Grey squirrel (species)	Reduced	Reduced	No
Red squirrel (individual)	Increased assuming environment suitable. Increased opportunity	Increased	Yes
Red squirrel (species)	Increased	Increased	Yes
The environment – principles from Mepham (2013)	Principle: Protection of wildlife from harm (e.g. by pollution), with remedial measures taken when harm has been caused. Increased if one accepts that indigenous species are more in balance with environment	Principle: Protection of biodiversity and preservation of threatened species. Increased as having indigenous species may be thought of as more "natural". Extinction is a bad thing	Principle: Ensuring sustainability of life-supporting systems (e.g. soil and water) by responsible use of non-renewable (e.g. fossil fuels) and renewable (e.g. wood) resources; cutting greenhouse gas emissions. Limited impact. Redresses past injustice

△

Table 14.1 Ethical matrix on the effects of a grey squirrel cull.

14.4
WILDLIFE AS PEST SPECIES
RESPONSE
RICHARD GREEN

As the author notes, the predominant social ethic tends to favour indigenous species and considers the introduction of the grey squirrel a "past injustice" with ongoing repercussions which can nonetheless be redressed. But is that the case? Again, it could be argued that ecosystems evolve and it is perhaps a romantic notion to believe that culling grey squirrels will restore the red squirrel population to exactly as it was before the grey squirrels were introduced.

The notion that "indigenous" or "native" species have a value over and above "foreign" or "alien" species has been challenged on two main grounds. Firstly, it is noted that if we only exchange the word "species" for "human races" or "types of people", we see that the argument becomes frighteningly reminiscent of abhorrent racist proposals. Whilst appearing to parallel xenophobic ideas is never going to be helpful to a cause, the other main problem with such ideas is the practical difficulty of defining geographical and time boundaries and the relevance of human intervention to identifying "native" species (Warren 2007). Where should we place the Scottish red squirrel in this framework? This species, by the eighteenth century, had become extinct in Scotland due to deforestation, was subsequently reintroduced to Scotland from English and Scandinavian populations at the beginning of the nineteenth century and then was the subject of mass slaughter, with 82,000 killed in 20 years at the start of the twentieth century due to their ringbarking damage to plantation conifers (Forestry Commission 2015).

The environmental damage (for example, to crops and trees) done by grey squirrels is also cited as a reason this species is considered a pest (Signorile & Evans 2007).

The above, necessarily abbreviated, use of the matrix raises the question of whether a group or species or whole ecosystem are worthy of moral consideration and if so, how much weight should these be given relative to the interests of individuals?

ENVIRONMENTAL ethics has developed through several phases, as outlined nicely in a review by McShane (2009). Firstly, and still perhaps the most common argument voiced in the media, are anthropocentric views. Here, the environment, species and individuals have purely instrumental value to humans and should be afforded protection when it suits us, for example to preserve beautiful views or the chance to discover new medicines. The extension of this argument is that no harm would be done by the last human destroying the world as their final action (Routley 1973). Biocentric views began to develop where individuals other than humans were seen to have moral worth, whether utilitarian and on the basis of sentience (Singer 2015), or deontological based on being "subject-of-a-life" (Regan 2004) or extended more widely to all living organisms (Taylor 1986). The problem for some with such views was that they didn't value species or ecosystems in their own right. Others really saw the natural world as being valuable in and of itself, and felt that we are all connected through our links with nature (Naess 1973). Environmental ethics has received increasing attention yet the tension remains between individualistic and holistic ideas, leading to public debates and difficult policy decisions.

When a decision has been made to kill individuals to protect species or ecosystems there can be refinements to the process. Littin, et al. (2004) propose six principles that should underpin an ethically sound vertebrate pest control framework:

(1) The aims or benefits and the harms of each control programme must be clear;

(2) Control must only be undertaken if the aims can be achieved;

(3) The methods that most effectively achieve the aims of the control programme must be used;

But are these six principles setting the bar too high? For example, principles (1)–(4) seem to require an extensive evidence base. One criticism of this is that during the time it takes to gather such evidence, the pests in question can go on killing and decimating the populations of their prey, which may in some instances be critically endangered.

On the other hand, a pest control programme which results in harm to non-target species, or one that has no impact, can have a similar or worse impact. Potential alternatives to a cull include population control, for example by immunocontraception, exclusion of the species from a certain area, relocation or the use of repellants.

Littin, et al. (2004) propose three strategies to maximise the humaneness of control methods:

(1) The relative humaneness of all current methods must be assessed in the practical circumstances of their use and the most humane methods that are usable in any given situation must be employed. This step, conscientiously taken, should lead to an immediate reduction in animal suffering;

(2) Active attempts must be made to improve the humaneness of all current methods, not excluded in Step 1, that cause significant suffering. This step should lead to welfare benefits in the medium term;

(3) An active research programme to develop new, more humane methods may be implemented. This step should achieve improvements in the long term (Littin, et al. 2004).

14.16 Some people view the natural world as having its own value, beyond its value to humans.
PHOTO ANNE FAWCETT

(4) The methods must be applied in the best possible way;

(5) Whether or not each control programme actually achieved its precise aim must be assessed;

(6) Once the desired aims or benefits have been achieved, steps must be taken to maintain the beneficial state (Littin, et al. 2004).

The ideal pest control method is effective, easy to use, affordable, safe (for human users and non-target species), specific and environmentally friendly (Littin, et al. 2004).

This approach is very similar to the 3Rs, and is based on a utilitarian cost:benefit analysis. Alternatively, an approach based on compassionate conservation would question the need for control in the first place. The population of "pests" may increase due to factors affecting predators, such as resource availability, which may not be addressed by control methods.

WILDLIFE AS PEST SPECIES

What do you think?

ONE   Some people have raised parallels between the language of trying to eradicate "foreign" or "alien" species and xenophobic human sentiments. How much sympathy do you have with this view?

TWO   Do species have any intrinsic moral worth in your view or is it only individuals that count?

THREE   Few would argue that the natural world has no value to us but in your view what are the reasons for protecting, at least some of, the natural world?

△

14.17 Efforts have been made to eradicate hedgehogs from the Outer Hebrides, islands off Scotland, to protect sea birds.
PHOTO ISTOCK

WILDLIFE AS PEST SPECIES

What would you do?

The Southern cassowary, *Casuarius casaurius*, is a large, flightless bird native to Northern Queensland. Southern cassowaries have been known to injure and in one case kill people, particularly if cornered or chased, as they are very territorial. Their population has been in rapid decline over the last 50 years and they are listed as "vulnerable" on the IUCN Red List of Threatened Species (IUCN 2015). The local council contacts your veterinary hospital to seek advice about translocation of an adult female cassowary following complaints by a local resident running a bed and breakfast establishment. The resident has complained that the animal is frightening his guests. However, you are aware through social media that some guests have used fruit to tempt the cassowary to come closer so they can photograph the bird. You are aware that anaesthesia of a wild cassowary is risky, both to the bird and to personnel involved. You are also aware that the bird is likely to find its way back to its territory. What would you recommend?

△

14.18 You are called to move an adult female cassowary, a risky procedure for all involved including the bird.
PHOTO ANNE FAWCETT

14.5

Cultural and religious rights versus animal welfare

There are many instances where the right to cultural practices or freedom of religious expression is invoked as a justification for infringing on or ignoring the welfare of animals. For example, use of certain traditional hunting methods or the practice of the sacrifice of live animals may be considered essential acts of religious or cultural expression by one group and acts of unnecessary cruelty by another.

The following scenario explores this issue in relation to traditional hunting of protected wildlife.

SCENARIO
CULTURAL RIGHTS VS ANIMAL WELFARE

▶ You have been invited onto lands of Indigenous people to participate in a conservation workshop. During a break you wander down to the water to cool off. There you notice a docked boat. It contains a live adult green sea turtle on its back in the full sun. Legislation prohibits hunting of this species but permits an exception for traditional hunting by Indigenous people of this and other species. The community permits traditional hunting by initiated adult men. You are concerned that the turtle is suffering.

What should you do?

RESPONSE
JAMES YEATES

▶ Traditional hunting may be considered a valuable part of a valuable culture. It is hard to define exactly *why* a culture might have a value, but nevertheless many people again consider that humans have a right to maintain their traditional cultures. In many contexts, traditional people are usually traditional owners of the land they hunt on.

Subsistence hunting may be a necessary source of food for those individuals. If we consider that humans have a legitimate reason to kill an animal for food (and most of us do), then this seems legitimate.

These defences only apply to hunting that is genuinely subsistence and/or traditional. If so-called subsistence hunting is for commercial exploitation, then that seems to require another defence.

△
14.19 Marine turtles like this green turtle are protected.
PHOTO ANNE FAWCETT

If so-called traditional hunting is not genuinely traditional (and there are some surprisingly recently established "traditional bullfights" in some countries) or if it is not genuinely part of the culture (e.g. it is just a few individuals), then this might also require some other defence. Indeed, one can argue that it should be an *indispensable* part of that culture (that is, not merely incidental – for example, dog-fighting was part of British culture, but it is fair to say there still is British culture without dog-fighting). Being a traditional activity does not make it part of a traditional culture. Equally, by definition, the use of modern weapons (such as exploding harpoons) or other equipment (such as ultrasonic radar) cannot be traditional.

In addition, the legitimacy of traditional subsistence hunting might be ethically limited in certain ways. For example it does not mean that humans have a right to cause suffering that is not necessary to obtaining food, that is, humane methods should be used. Indeed, traditional subsistence hunting methods may stress respect for animals, and prohibit cruelty. Conversely, it is possible to be indigenous and cruel!

This is incompletely reflected in the law in this example. Green turtles are listed as endangered and cannot be traded commercially, being listed under CITES Appendix I. Although the example given is not a trade issue, there will be rules or legislation in place to account for the indigenous hunting exemption to the blanket ban on hunting – these will include rules on who can do it (in this case initiated males), how animals can be killed and how they can be distributed (for example, only to the local community and not for trade or selling as tourist trinkets). It may be that the local legislation specifies the methods allowed and might not allow for live animals to be left as this turtle is. Or it may be that traditional or subsistence hunting are completely exempt from animal welfare laws, as well as trade legislation. In some cases, authorities might be unwilling

to enforce local rules (perhaps to avoid accusations of racism), but often rules are enforced by the tribal council who might genuinely thank you for bringing any enforcement issues to their attention.

What to do
You have the options of:

(1) doing nothing;
(2) helping the turtle;
(3) informing the relevant authorities.

Doing nothing is, in this case, ethically problematic. While there may be reasonable limits on what an individual can be expected to do morally – we do not have a duty to save every turtle – there is a reasonable belief that we have a duty of "easy rescue". When we see a turtle in distress who can be saved, then we should do so.

How we help the turtle would depend on our clinical judgement. If the turtle is healthy enough to be released, then we may be able to simply place it back in the water. If the turtle is more seriously harmed and would not survive in the wild, then we might best either rescue it to a sanctuary (if that is feasible in the circumstances) or euthanase it. Killing it humanely may still allow the huntsmen to eat the meat.

Informing the tribal council or CITES/MA legislative authority in the country also seems legitimate and appropriate. Not only are the individuals harming the animal, but they are potentially jeopardising the exemption.

Conclusions
All else being equal, save the turtle and inform the authorities.

Consistency
It is important to reflect further, aiming for ethical consistency in one's beliefs. Non-traditional

methods of obtaining food can also be inhumane, and the suffering involved may be comparable to the turtle's. If so, then the same logic (without the defence of being traditional subsistence hunting) could apply to what you eat (or what you feed to your pet). It is important to be welfare-focused, and not to be hypocritical.

THE use of animals for cultural or religious purposes, particularly those that are protected species, is an emotive and divisive topic. This is in part because freedom of expression of culture or religion is seen as a basic right, which conflicts with the right of animals "not to be brutalised and killed unnecessarily" (Beauchamp 2011). At the same time it is very easy to condemn the practices of another culture as cruel or barbaric while ignoring animal welfare concerns in our own culture.

The question of what constitutes "traditional" is also fraught – who has the authority to judge what is traditional or not? For example, if a person hunts in a dugout canoe, but speaks English and wears clothing, does this count as traditional? How old and widespread does a culture have to be before claiming a tradition? These are contentious issues.

In some countries, laws permit hunting of protected wildlife in certain sites by Indigenous people. There may be conditions, for example under the Queensland Nature Conservation Act

"At the same time it is very easy to condemn the practices of another culture as cruel or barbaric while ignoring animal welfare concerns in our own culture."

1992 Aboriginal people may hunt protected wildlife on private land if this is permitted by a species conservation plan and permission is granted by the landowner (Ross 1994). In other countries, such hunting may be prohibited by law but informally permitted. There may be restrictions, for example, hunting may be allowed for traditional landowners only.

When it comes to Indigenous people, there is the added tension between the rights of Indigenous people to autonomy and rights asserted by non-Indigenous lawmakers.

For example, in South Africa, Animal Rights Africa (ARA) sought to prevent the slaughter of a bull or any other animal at a traditional festival celebrated by the Zulu king and members of his community. The ARA made the application on the grounds that the animal(s) would be subject to terrible cruelty. The respondents argued that important constitutional rights of the Zulu nation to practise its religion and culture were at stake. The judge, in rejecting the application, remarked that the application evinced intolerance and lack of respect for African culture and religion and in fact could spark civil unrest if granted (Bilchitz 2016).

Common objections to traditional hunting include the risks to already endangered species, the argument that "traditional" hunting should not involve modern methods and the claim that such permission is discriminatory because it creates one rule for Indigenous people and one for non-Indigenous people (Ross 1994), which is perceived to be unjust – although this depends on our working definition of justice.

Some authors argue that anti-cruelty legislation should apply without exception:

"One wonders whether the law, instead of granting extensive exemptions and defences, should prohibit acts of cruelty against animals done in the name of cultural, religious or

CHAPTER 14 WILDLIFE

14.5

CULTURAL AND RELIGIOUS RIGHTS VERSUS ANIMAL WELFARE
WHAT DO YOU THINK?

traditional practices. After all, animals suffer no less irrespective of the pretext that acts of cruelty are perpetrated under."

(Cao 2015)

It is possible to approach this scenario without challenging the practice of hunting the turtle in the first place. The key issue – at least from the point of view of the animal – is that it is alive and suffering. It may be ethically defensible to hunt these animals, as long as they are killed quickly and humanely. If this is not possible or deemed appropriate, it may still be possible to reduce the animal's suffering by turning it over and moving it into shade – if you are able to do so.

CULTURAL AND RELIGIOUS RIGHTS VERSUS ANIMAL WELFARE

What do you think?

While the commercial harvest of walruses is banned in the USA, Alaskan Natives (Indians, Aleuts and Eskimos) living in certain regions are permitted to harvest walrus for subsistence purposes or the creation and sale of Native handicrafts or clothing *if the harvest is not wasteful.* The numbers of animals harvested by Alaskan Natives are not limited by Federal Law, but should be limited to what can be reasonably utilised. Killing walrus just for ivory is considered wasteful and is therefore illegal. Walrus tusks must be tagged by Fish and Wildlife Service representatives. Handicrafts can be sold to anyone but must be made without using mass-production technology (Fish and Wildlife Service 2007).

ONE   What are the salient features of this scenario for you? Do you think this system is ethical? How would you justify your position?

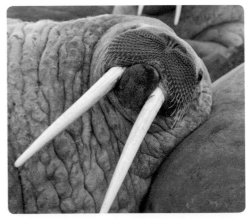

△

14.20 Walrus hunting is restricted and only permitted for certain people.
PHOTO ISTOCK

CHAPTER 14 WILDLIFE

14.5
CULTURAL AND RELIGIOUS RIGHTS VERSUS ANIMAL WELFARE
WHAT DO YOU THINK?

CULTURAL AND RELIGIOUS RIGHTS VERSUS ANIMAL WELFARE

What do you think?

Drive hunting of whales and dolphins has been practised in Japan for over 1000 years. Traditionally, non-powered boats were used to herd or drive dolphins into shallow water and kill them for food. Current drive hunts employ sophisticated technology to kill large numbers of animals and procure meat for commercial purposes, yet it has been claimed that "this killing method does not conform to the recognised requirement for 'immediate insensibility' and would not be tolerated or permitted in any regulated slaughterhouse process in the developed world" (Butterworth, et al. 2013).

ONE   What are the salient features of this scenario for you? Do you think this system is ethical? How would you justify your position?

△
14.21 Dolphin hunting has received much criticism in recent decades.
PHOTO ANNE FAWCETT

Conclusion

Our interactions with wild animals bring additional ethical considerations relative to other species. They may respond more adversely to veterinary care and require rehabilitation and release. The care they are afforded is often dependent on more than the welfare of the individual. Their abundance in the wild is often brought into play and so their usefulness to their species or an ecosystem can be a factor. To decide how to deal with wildlife we need to be clear on the value we place on individuals, species and ecosystems and whether that value stems from anthropocentric or other concerns.

References

Beauchamp TL 2011 Rights theory and animal rights. In: Beauchamp TL and Frey RG (eds) *The Oxford Handbook of Animal Ethics*, 198–227. Oxford University Press: Oxford.

Bekoff M, and Ramp D 2014 Compassion in conservation: don't be cruel to be kind. *New Scientist* **222**, 21 June: 26–28.

Bilchitz D 2016 Animal interests and South African law: the elephant in the room. In: Cao D and White S (eds) *Animal Law and Welfare – International Perspectives*, 131–155. Springer International: Cham, Switzerland.

Butterworth A, Brakes P, Vail CS, and Reiss D 2013 A veterinary and behavioral analysis of dolphin killing methods currently used in the "drive hunt" in Taiji, Japan. *Journal of Applied Animal Welfare Science* **16**: 184–204.

Cao D 2015 Regulation of wild animal welfare. In: *Animal Law in Australia Second Edition*. Thomson Reuters: Sydney.

Cochrane A 2012 *Animal Rights Without Liberation: Applied Ethics and Human Obligations*. Columbia University Press: New York.

Collins LM, Warnock ND, Tosh DG, McInnes C, Everest D, Montgomery WI, Scantlebury M, Marks N, Dick JTA, and Reid N 2014 Squirrelpox virus: assessing prevalence, transmission and environmental degradation. *PLoS ONE* **9:** e89521.

DeGrazia D 1996 *Taking Animals Seriously: Mental Life and Moral Status.* Cambridge University Press: Cambridge.

Eriksen L, and Kennedy M 2014 Marius the giraffe killed at Copenhagen zoo despite worldwide protests. *The Guardian*, 9 February.

FAWC 2009 Farm Animal Welfare in Great Britain: Past, Present and Future. Farm Animal Welfare Council: London.

Fish and Wildlife Service 2007 Hunting and use of walrus by Alaska Natives: fact sheet. US Department of the Interior.

Forestry Commission 2015 Red Squirrel Facts. http://www.forestry.gov.uk/fr/infd-8c8bhc

Frey RG 1987 Autonomy and the value of animal life. *The Monist* **70:** 50–63.

Garner R 2013 *A Theory of Justice for Animals: Animal Rights in a Nonideal World.* Oxford University Press: Oxford.

Haidt J, Koller SH, and Dias MG 1993 Affect, culture, and morality, or is it wrong to eat your dog? *Journal of Personality and Social Psychology* **65:** 613–628.

HMSO 1981 Wildlife and Countryside Act.

Holst B 2014 Euthanasia of a 2 year old male giraffe at Copenhagen Zoo. http://www.zoo.dk/files/2014_Giraffe_case-explanation_15_MAY.pdf

IUCN 2015 Casuarius casuarius. International Union for Conservation of Nature and Natural Resources.

Littin KE, Mellor DJ, Warburton B, and Eason CT 2004 Animal welfare and ethical issues relevant to the humane control of vertebrate pests. *New Zealand Veterinary Journal* **52:** 1–10.

Lock R, and Reiss M 1996 Moral and ethical issues. In: Reiss M (ed) *Living Biology in Schools,* 109–120. Institute of Biology: London.

McShane K 2009 Environmental ethics: an overview. *Philosophy Compass* **4:** 407–420.

Mepham B 2013 Ethical principles and the ethical matrix. In: Clark JP and Ritson C (eds) *Practical Ethics for Food Professionals: Ethics in Research, Education and the Workplace,* 39–56. Wiley-Blackwell: Oxford.

Miller M, Buss P, Joubert J, Mathebula N, Kruger M, Martin L, Hofmeyr M, and Olea-Popelka F 2013 Use of butorphanol during immobilization of free-ranging white rhinoceros (Ceratotherium simum). *J Zoo Wildl Med* **44:** 55–61.

Morell V 2014 Killing of Marius the giraffe exposes myths about zoos: for the Copenhagen zoo, it seems Marius was worth more dead than alive. *National Geographic,* 13 February.

Naess A 1973 The shallow and the deep, long-range ecology movement. A summary. *Inquiry* **16:** 95–100.

Platt JR 2012 Spiked. *Conversation Magazine.*

Regan T 2004 *The Case for Animal Rights.* University of California Press: Berkeley.

Rincon P 2014 Why did Copenhagen Zoo kill its giraffe? BBC News. http://www.bbc.co.uk/news/science-environment-26118748

Rollin BE 2006 *Animal Rights & Human Morality.* Prometheus Books: New York.

Rollin BE 2007 Cultural variation, animal welfare and telos. *Animal Welfare* **16:** 129–133.

Ross A 1994 Traditional Aboriginal hunting in Australia: a cultural heritage issue. *Cultural Survival* **18.**

Routley R 1973 Is there a need for a new, an environmental, ethic? *Proceedings of the XV World Conference of Philosophy*, 205–210.

Signorile AL, and Evans J 2007 Damage caused by the American grey squirrel (Sciurus carolinensis) to agricultural crops, poplar plantations and semi-natural woodland in Piedmont, Italy. *Forestry* **80:** 89–98.

Singer P 2015 *Animal Liberation.* Bodley Head: London.

Taylor P 1986 *Respect for Nature: A Theory of Environmental Ethics.* Princeton University Press: Princeton.

Warren CR 2007 Perspectives on the 'alien' versus 'native' species debate: a critique of concepts, language and practice. *Progress in Human Geography* **31:** 427–446.

CHAPTER 15
CHANGING AND CLONING ANIMALS

Introduction

Animals are changed or altered by humans on a number of levels. These can include physical changes (for example castration and dehorning of farm animals to improve handling and meat quality), genetic changes (through controlled breeding programmes or accelerated selection via molecular biology), behavioural modification (domestication, training, medication) – even changing the types of foods animals eat. These days, few farmed or domesticated animal species escape widespread physical alterations in at least some parts of the world. Such is the nature of animal science technology that scientists now ask whether indeed they should change the housing to better accommodate animals, or change animals to better accommodate the housing (Cheng 2007).

Animals may also be changed through breeding practices, for example selecting for fast growth rate of chickens or aesthetic features in chondrodystrophic breeds such as Dachshund dogs. Surgical and non-surgical techniques are used to change animals for functional and cosmetic purposes.

Another way that animals may be permanently changed is manipulation at the level of genes. This is particularly common in mice used in biomedical research, accounting for an estimated doubling in the numbers of experimental animals used worldwide since the advent of these techniques at the end of the twentieth century (Ormandy, et al. 2009). Ironically, at the same time as developing methods to genetically alter animals a parallel scientific endeavour has resulted in the ability to produce viable clones. Finally, animals may be altered by influencing their behaviour, through traditional selective breeding, for example producing the docile and often floppy Ragdoll cat, or by training – think of the extraordinarily unnatural circus tricks even undomesticated animal species can be made to perform.

Whether it's through concern for the welfare of the animals, being against unnatural interventions or wanting to preserve the fundamental integrity of the animal, all of these cases raise ethical considerations that veterinarians are very often at the centre of. In this chapter we will consider scenarios relating to mutilations, breed characteristics, cloning and training of animals.

◁

15.1

15.1

Permanent physical alterations and mutilations

Permanent physical alterations of animals may be referred to as mutilations within legislation. The use of such an emotive word is significant as a reflection of the societal concern about such procedures – the verb "mutilate" refers to the infliction of physical injury which may impair function. Yet the term is used to refer to common procedures, some of which may be undertaken in part for the benefit of the animal.

△

15.2 Ear cropping is a cosmetic procedure performed on dogs in some countries.
PHOTO ANNE FAWCETT

Further reflection of societal concern is shown in most jurisdictions where mutilations in principle are outlawed, and only specified exceptions are permitted. Table 15.1 outlines some mutilations and common justifications for them.

One of the main arguments for allowing mutilations is when they are expected to prevent future poor welfare, at least in some animals. This utilitarian approach is explored further for the case of docking working gun dogs in the next scenario.

SCENARIO

TAIL DOCKING OF DOGS

▶   You are asked to dock the tails of a litter of five three-day-old springer spaniel puppies by the breeder. The puppies are likely to go on to be working gun dogs where they will be searching in undergrowth for birds that have been shot down by hunters for sport. You are working in a country where it is neither legally nor professionally prohibited to dock dogs' tails.

How do you respond?

RESPONSE

DAVID MORTON

▶   There are several reasons for docking dogs' tails but why should we be concerned? In some breeds and in some countries it was, and still is, seen as the breed standard, and to show an animal it would have to be docked or otherwise it would be severely penalised for not being so. In some countries owners can do it themselves whereas in other countries it is seen as an act of veterinary surgery and can only be carried out

SPECIES	CHANGE/ALTERATION	TYPICAL JUSTIFICATION/S
Dog/cat	Ovariohysterectomy	To prevent oestrus behaviour and unwanted litters/ reduce health risks, e.g. mammary cancer, pyometra
Dog/cat	Castration	To reduce reproductive behaviour including marking or straying/ control population/ reduce health risks, e.g. perineal hernia, testicular cancer
Dog	Tail docking	To meet a breed standard/ to prevent tail injuries
Dog	Ear cropping	To give an appearance of being more alert
Cat	Declawing	To prevent damage to the home environment
Horse	Castration	To reduce risks of handling and riding
Cattle	Dehorning	To reduce injuries associated with horns/ risks of handling
Sheep/goats	Disbudding	To reduce injuries associated with horns/ risks of handling
Chickens	Beak trimming	To reduce injurious feather pecking and cannibalism
Pigs	Tail docking	To reduce tail biting
Sheep	Mulesing	To prevent blowfly strike

△
Table 15.1 Examples of procedures that permanently physically alter or change animals.

△
15.3 Spaniel puppies with docked tails.
PHOTO ISTOCK

△
15.4 Spaniel puppies with intact tails.
PHOTO ISTOCK

15.1

PERMANENT PHYSICAL ALTERATIONS AND MUTILATIONS
RESPONSE
DAVID MORTON

by a veterinarian. In general terms, veterinarians try to avoid harming animals if it is not in the best interests of those animals.

So the question is whether it is in any animal's best interests to have part or all of its tail amputated. The harms in the case of docking comprise the pain at the time of removing part of the tail, and also the subsequent post-amputation pain and distress. There may be further sequelae such as infection, neuroma at the end of the cut nerves, a failure to communicate naturally with other dogs and animals, short dock interference with the muscles of the anus and possibly defecation, and compromised gait when cornering as the tail acts as an important balancing organ. The downsides of docking therefore seem quite marked. In terms of feeling pain, some believe that very young animals do not feel pain, or at least soon forget it (arguments that have been advanced to support circumcision in human babies). However, at the time of docking, puppies show behavioural signs of feeling pain, wriggling and squealing, and post-amputation comfort behaviours, e.g. suckling. Research by Fitzgerald and others (Fitzgerald 1994, Fitzgerald & Koltzenburg 1996) has shown that very young animals of many species are likely to experience more pain in the first 2–3 weeks of life than when the nervous system has matured to an adult-like state.

Therapeutic docking is likely to be the most straightforward and legitimate type of docking (though not if the tail was damaged deliberately to encourage a therapeutic dock!), and is obviously being done in the best interests of the dog. This is sometimes necessary in boisterous dogs such as Labradors and Great Danes, in confined spaces where they damage their wagging tail on furniture, doorways and so on. It may also happen to dogs that go through rough terrain, e.g. when used in hunting and shooting, mainly spaniels, pointers and retrievers. It follows that if we can

prevent this happening by docking the dogs when they are very young, then we could prevent them from damaging their tails. Unfortunately, dogs can still damage docked tails. From a utilitarian perspective, one might discuss whether it is right to cause pain and suffering in 100 per cent of animals at the time of docking, to prevent later pain and suffering, in what is only a small percentage of affected animals requiring a therapeutic dock, at a later date (Cameron, et al. 2014, Lederer, et al. 2014). Moreover, if docked as an adult, animals would be given an anaesthetic as well as post-operative pain relief.

The arguments can become even more difficult when considering breed standards for dogs that are not used for hunting but only for show (Bennett & Perini 2003a, 2003b). Is it right to surgically reassign/adjust/mutilate/change/redesign animals to fit a human concept of beauty or aesthetic appreciation?

How far should a preventive argument be taken? Docking may reduce the level of damage to a tail (not completely) but if the incidence of fractures could be reduced by amputating limbs, or the incidence of ear disease by removing the pinna, how far should we take that argument? And why don't we preventatively dock Great Danes and Labradors which also damage their tails?

‖‖‖

IN his response David Morton questions just how far we can take the utilitarian argument. How much benefit is required to find universal mutilation acceptable? In the case of docking of dogs' tails it has been estimated that in the UK 232 working dogs (Cameron, et al. 2014) or 500 general dogs (Diesel, et al. 2010) would need to be prophylactically docked to prevent one single case of tail injury that would require therapeutic docking. Both of these studies quantifying the risks of injury and benefits of tail docking were funded by UK governments to provide an evidence base to guide policy. In the UK, exemptions exist to allow docking of working dogs. By any calculation, still far more animals need to be docked than are injured. Moreover, no studies were carried out on the number of injuries to docked working dogs for comparison (Lederer 2014).

Sometimes it is argued that the mutilations are only required in order to limit the problems associated with poor husbandry, for example in the cases of tail docking of pigs to prevent tail biting and beak trimming of chickens to prevent injurious pecking and cannibalism. On some farms with inadequate environments, the highly motivational foraging and exploratory behaviours are thwarted and the animals may investigate and injure others in the group (EFSA 2007, Rodenburg, et al. 2013). Legislation is in place across the EU prohibiting "routine" tail docking (European Union 2001) of the 148 million pigs (Eurostat 2015), requiring that farmers first address underlying husbandry problems; however, despite this at least 90 per cent of EU pigs are estimated to be docked (EFSA 2007). If certain negative behaviours such as tail biting or injurious pecking are seen only in particular husbandry systems, it is appropriate to determine whether the system itself is ethically justifiable.

One important concept related to changing animals is their bodily integrity. Generally speaking, the bodily integrity of humans is held in high regard. Whilst neutering of companion animals to control undesirable sexual behaviour is widespread in some countries, the neutering of people to control unwanted sexual behaviour would be met with almost total resistance, with concerns about infringement on personal autonomy looming large.

The ethicist Bernie Rollin has written extensively on the concept of *telos*, an Aristotelian term encapsulating the very essence of a being, describing it in the case of a spider as

"a nature, a function, a set of activities intrinsic to it, evolutionarily determined and genetically imprinted, that constitutes its 'living spiderness'"

(Rollin 2006)

Rollin proposes that *telos* is useful in framing all that is important to animals, that their welfare is bound up in it and that anything that detracts from this *telos* is an affront to the animal, including physical mutilations. But it is argued that respecting *telos* also involves offering opportunities for animals to live a good life, to "live flourishing lives in accordance with the kind of beings that they are" (Harfeld 2013).

15.1

PERMANENT PHYSICAL ALTERATIONS AND MUTILATIONS
WHAT WOULD YOU DO?

PERMANENT PHYSICAL ALTERATIONS AND MUTILATIONS

What would you do?

You are working in companion animal practice in a country/region where declawing cats (phalangec-tomy) is neither legally nor professionally prohib-ited. Mr E requests that you declaw Sukie, his six-month-old kitten, as it is a requirement of his landlord that any cat living on the property must be declawed. The American Association of Feline Practitioners (AAFP) have published a position statement on declawing stating that it should be undertaken only when certain conditions are met. These include:

*   Owners must be informed, with reference to anatomic details, that declawing entails amputation of the third phalanx (P3) and that this is a procedure that is associated with acute and chronic pain;
*   Owners must be informed of all risks and benefits to surgery;
*   Owners must be informed that scratching (which declawing is designed to prevent) is natural, normal feline behaviour which is an important means of visual and olfactory communication;
*   Owners must be informed of alternatives to declawing (described in the position statement);
*   Deep digital flexor tendonectomy is not recommended;
*   Multimodal analgesia should be employed (American Association of Feline Practition-ers 2015).

What do you do? How has the AAFP position statement influenced you?

△

15.5 Scratching, which declawing is designed to prevent, is a normal feline behaviour.
PHOTO ANNE FAWCETT

△

15.6 The provision of environmental enrichment may reduce unwanted scratching.
PHOTO ANNE FAWCETT

PERMANENT PHYSICAL ALTERATIONS AND MUTILATIONS

What do you think?

Some veterinary professional associations actively promote routine neutering of dogs and cats, while others recommend the decision be made on a case-by-case basis (Palmer, et al. 2012). Advocates of routine neutering argue that it carries health benefits and reduces the incidence of unwanted litters. Those against routine neutering stress that surgical neutering is an invasive procedure that is associated with risks, both short and long term.

ONE   How do you justify your position on
            neutering cats?

TWO   Does your position differ for the neu-
            tering of dogs?

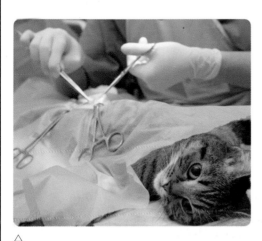

△

15.7 Cat neutering: if prevention of unwanted litters is the primary concern then neutering of only one sex is required.

PHOTO ISTOCK

Given widespread access to analgesia and increasing evidence that physical alterations and mutilations are painful, we are in the position to refine these procedures by minimising pain and mitigating harm where such procedures are carried out.

15.2

Breeding animals

In this next scenario we move from considering the implications of physically altering animals surgically to the effects of breeding on an animal's phenotype and subsequent welfare.

SCENARIO

BREEDING FOR WELFARE?

▶   As a specialist soft tissue surgeon, a high proportion of your caseload is operating on animals with heritable conditions that you consider ultimately preventable, from "face lifts" for Shar Peis to correcting elements of brachycephalic airway obstructing syndrome (BAOS) in bulldogs. This is your bread and butter income, but also saddens you and now you have the opportunity to act as part of two panels to advise breed societies on breeding to improve welfare.

Knowing how strongly breeders feel about the breed and their own dogs, what do you advise?

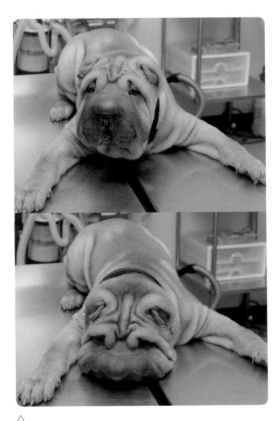

△ **15.8** The distinctive skinfolds in the Shar Pei are associated with entropion which may require corrective surgery.

PHOTO ANNE FAWCETT

△ **15.9** Freddie the Cavalier King Charles Spaniel had syringomyelia and died 24 hours postoperatively. George the pug suffers from brachycephalic airway obstructive syndrome.

THESE STILLS FROM "PEDIGREE DOGS EXPOSED" ARE REPRODUCED WITH PERMISSION FROM JEMIMA HARRISON

RESPONSE

IMKE TAMMEN

▶ The BBC Television documentary "Pedigree Dogs Exposed" (BBC 2008) created concerns about how dog breeders deal with heritable conditions in dog breeds, and as a very positive outcome prompted scientists and breed societies to investigate and implement changes (Nicholas 2011, Nicholas & Wade 2011). Some comments on inherited diseases caused by single genes are provided in a different scenario in this book and the focus here will be on what geneticists describe as multifactorial or complex disorders, i.e. medical problems that are not due to genetic variation in a single gene but are caused by the combined effects of multiple genes and environmental factors (such as diet, "lifestyle" and animal husbandry).

It is important to note that most medical problems in animals and humans have some genetic contribution – and are therefore multifactorial disorders. For some of these multifactorial disorders the genetic contribution is relatively large and thus, at least in animals, the incidences of such diseases in future generations can be reduced via careful breeding. These multifactorial disorders can be considered from a geneticist's point of view as "quantitative traits". Animal geneticists have developed very effective methods to influence quantitative traits relating to production and health in livestock. The effective application of the same methods in companion animals could lead to great improvements in animal health (Thomson, et al. 2010), and it has been encouraging to see recent initiatives in this area, e.g. Estimated Breeding Values (EBVs) for hip and elbow scores in the UK (The Kennel Club 2015) and the USA (Cornell University College of Veterinary Medicine 2014).

Key issues

The genetics of multifactorial disorders is more complex when compared to genetics of single gene disorders, and this is a rapidly evolving area of knowledge due to continuous improvements in molecular and quantitative genetics. It can be very difficult for individual animal breeders, small breed societies that largely rely on volunteer contributions, and veterinarians to stay up to date with these developments. It would therefore be advisable for the members of the panels who have been asked to advise the breed societies, to seek expert advice or "genetic counselling" from an animal geneticist, and to tackle these issues in a team of dog breeders, veterinarians and animal geneticists to achieve "good" outcomes for all involved.

When considering whether genetic intervention (changes to the breeding programme and/or breed standard) should be implemented to

reduce the incidence of multifactorial disorders in dog breeds, a range of scientific issues come to mind. Facts and data relating to the disorder and the breed in question need to be gathered. Some of the information might be readily available; other aspects might require research involving input from veterinarians and geneticists.

It is important to first develop an agreed definition of what the disorder in question entails. Clinical presentation of many multifactorial disorders can vary, for example in relation to severity and time of onset. Once such a definition has been agreed on, agreed ways on how the disorder is diagnosed or measured need to be developed, i.e. how the disorder or predisposition to the disorder can be measured as a quantitative trait. Ideally this would involve quantifiable measurements for more than one trait that correlate well with presence of disease and/or severity of disease, e.g. for hip dysplasia standardised procedures for scoring of X-rays.

Once these issues are clarified (i.e. the phenotype is defined), accurate information about the incidence of the disease in the population can be identified and the impact of the disorder on health and welfare can be more accurately assessed. Pedigree information, information about how individual animals in the population "measure up" in relation to those traits that have been identified to correlate well with the disease, and – if available – information about the environments that these dogs have been exposed to, can then be analysed to estimate the heritability of the disorder and identify important environmental factors. The heritability provides an estimate of the relative importance of genetic and environmental factors to the aetiology of the disorder. The analysis would identify key environmental factors that predispose animals to develop the disorder and this information can be used directly to implement changes to the environment (e.g. dietary changes or changes in animal husbandry).

15.2

BREEDING ANIMALS
RESPONSE
IMKE TAMMEN

It is important to note that the heritability of a trait is specific to the breed – so studies in other breeds can only be used as a guide. If the disorder has a reasonably high heritability, then it would be expected to respond well to genetic intervention. Selection based on the phenotypic values could be implemented immediately. However, the calculation of estimated breeding values (EBVs) would present a more advanced and accurate approach, and has been implemented for some multifactorial diseases in dogs already, e.g. hip dysplasia (Wilson & Nicholas 2015) and syringomelia (Lewis, et al. 2010). An even more advanced approach would be the development of so-called genomic EBVs (gEBVs) (Meuwissen, et al. 2001). These gEBVs have been implemented successfully in some livestock industries but would require substantially more research which would include genotyping of a large number of animals with phenotypic information with dense genetic markers (e.g. SNP chips) and more advanced statistical analysis.

Considering that most dog breeds represent relatively small numbers of animals (at least when compared with livestock breeds), but are not restricted to specific countries, any of these approaches benefit from an international approach to increase the number of animals involved and thus increase the accuracy of the results; to share the financial burden related to the research; and to have a wider impact on animal health (Fikse, et al. 2013, Hedhammar, et al. 2011, Wilson & Wade 2012). However, such international approaches between researchers and breed societies require high levels of collaboration and leadership.

When deciding on how to implement genetic interventions via artificial selection – based on phenotypic data, EBVs or gEBVs – it needs to be considered how such selection would impact on other traits as well as the effective population size and levels of inbreeding in the population, e.g. very strong selection against the disease may have inadvertently negative impacts on other

traits or other aspects of genetic health of the population.

Ethical considerations relating to breeding of dogs with multifactorial diseases can be found in the literature (Mullan 2010, Palmer 2012). The veterinarian would like to see the welfare for future generations of dogs improved, but appears concerned that her/his involvement in an advisory panel to the breed society could alienate some clients, and that a breeding scheme that reduces incidence of disease could have a negative impact on caseload. The ethical challenge to balance animal welfare issues with economic viability of a business is not an uncommon scenario for veterinarians; and depending on one's views on what the duties of a veterinarian are towards animals, clients, the public and the business they are working for, this can be a difficult balancing act.

In relation to management of client relationships, a team approach that acknowledges the expertise of the breeders and reaffirms their role in the decision-making process is recommended. The veterinarian and the animal geneticist should advise on what research should be conducted to gather necessary information and what options for genetic and non-genetic interventions exist, but the ultimate decision on what changes are implemented lies with the breed society. Breeders are well aware that breed standards are man-made constructs, and that adjustments to breed standards to improve animal welfare ultimately improve the breed. The responses to the abovementioned BBC documentary show that with an increasing availability and awareness of tools to address problems, these are taken up by breed societies – although at variable pace. Some breeders might be more resistant, oppose change, have strong emotional attachments to dogs that they fear might do less well in a changed breeding scheme and might hold the veterinarian's involvement in the process against them. However, if managed carefully, it would appear more likely that the veterinarian's involvement in the panel would appeal to new clients, who are attracted by the veterinarian's expertise in breed-specific health issues and commitment to animal welfare.

In relation to caseload, it needs to be considered that any impact of proposed changes (to breeding and/or the environment) on animal welfare will occur in future generations and the improvements to disease incidences and/or severity of clinical signs will only be incremental, thus any decrease in caseload would not be expected to be immediate or dramatic. Furthermore, breeding schemes against multifactorial disorders often require veterinarians to measure associated traits in all breeding animals (e.g. X-ray rating for hip dysplasia), thus potentially creating new opportunities of engagement with clients.

More importantly, advising clients to take actions to prevent disease in their animals is commonly understood as the duty or responsibility of a veterinarian. A veterinarian who would not provide dietary advice to the owner of an obese dog or not advise to vaccinate dogs to maintain a caseload of obese dogs or dogs with infectious diseases, respectively, would surely be considered unethical. Ignoring the evidence that genetics plays a major role in the aetiology of many diseases or not acting on this knowledge is unethical, and if in doubt veterinarians are expected to act "*in dubio pro animale*" (Kuhlmann-Eberhart & Blaha 2009).

Conclusion

If a medium to strong genetic contribution to a multifactorial disease is suspected, veterinarians ought to voice their concerns to breeders and the corresponding breed society. Jointly with animal geneticists they should encourage and assist breed societies with further research and propose the implementation of a management programme to improve animal welfare.

Acknowledgements

I would like to thank Frank W Nicholas for very useful feedback on a draft of this scenario.

COMPANION animal veterinarians will frequently encounter occasions when clients have bred animals with genetically influenced negative welfare traits. How each of these individual cases is dealt with is part of the overall solution. It may be that breed societies have policies to guide the practising vet such as requiring reporting, registration or neutering of animals that have had surgical corrections of inherited undesirable traits. In the UK, the Kennel Club has teamed up with the veterinary profession to provide a centralised

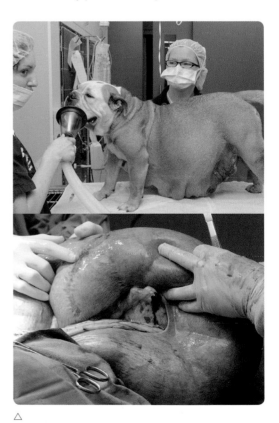

△
5.10a–b Veterinarians recommended ovariohysterectomy for this bulldog which required a caesarean section.
PHOTO ANNE FAWCETT

point to report "(1) any **caesarean operation, and/or (2) any operative procedure carried out by a veterinary surgeon** on a dog which alters its natural conformation" (emphases in original) so they can:

"1. Provide information regarding breeding lines appearing to perpetuate certain defects.
2. Help to deter breeders from breeding from animals displaying evidence of hereditary problems, or whose natural conformation has been altered, or where breeding routinely relies on caesarean operations (vide Regulation B22c (5)).
3. Assist dog show judges to be sure that they are not being called upon to assess dogs with hereditary defects which have been so well corrected by surgical intervention that these cannot be detected.
4. Collect and utilise clinical data over time, to assist in inferring prevalence and monitor changes and trends to support health and welfare."

(The Kennel Club 2016)

The focus here is on purebred dogs and showing but veterinarians may need to advise on non-purebred animals, for example on the problems with breeding siblings or from animals that are the product of close familial matings.

BREEDING ANIMALS

What would you do?

Ms F brings you Sophie, a lovely six-month-old white domestic cat that she is thinking of breeding from, to get her microchipped. She hasn't noticed anything wrong with Sophie but you discover the cat is deaf. Before you have a chance to discuss anything about breeding from Sophie, Ms F pipes up cheerily, "Oh well, deaf people can have children, can't they?" How do you respond?

△
15.11 Should an owner breed from a deaf kitten?
PHOTO ANNE FAWCETT

Selection for phenotypes deleterious to welfare also occurs in farm animals. One notable example is Belgian Blue cattle, a relatively recent breed characterised by extreme conformation known as "double-muscling". Double-muscling spontaneously occurs in other breeds, including Piedmontese and Charolais (Fiems 2012).

Double-muscling is caused by a genetic mutation leading to inactivation of the myostatin gene, which causes skeletal muscle hypertrophy and hyperplasia in affected animals (Fiems 2012). While this trait is desirable because affected animals have an extremely high carcass yield, myostatin mutations are pleiotropic, affecting a number of different body systems, mostly negatively. Double-muscled animals have a reduced organ mass and have increased susceptibility to respiratory disease, urolithiasis, lameness, nutritional stress, heat stress, infertility and dystocia (Bassett 2009, Fiems 2012). While double-muscling in cattle has been reported since the 1800s, it was only

△
15.12 The Belgian Blue has been selectively bred for the genetic mutation that causes double-muscling.
PHOTO ISTOCK

systematically exploited after the Second World War because the availability of anaesthesia, antibiotics and new surgical methods facilitated caesarean sections (Fiems 2012), required because of the large calf size and reduced pelvic cavity of cows.

Caesarean sections are routinely performed in purebred Belgian Blue herds, with one review citing 80 to 90 per cent of births requiring a caesarean (Bassett 2009). Caesareans themselves may be associated with scar tissue formation, uterine adhesions and subsequent infertility and risk of abortion (Fiems 2012). The number of caesareans that can be performed in a single cow is physiologically limited and routine caesareans may substantially reduce the lifespan of cows (Bassett 2009).

Interestingly, while double-muscling is a desirable feature in the Belgian Blue despite the problems it causes, the British Charolais society discourages breeding from "the double-muscled heavy shouldered bull" *because* it is associated with infertility and calving problems (British Charolais Cattle Society 2016).

While Belgian Blue bulls are often used as a source of semen to produce heavily muscled calves from dairy cows (usually Holsteins) for beef production, the ethical question for producers is whether it is acceptable to use semen from a breeding programme that "is known to be an abuse to welfare" (Webster 2013).

The current approach of breeders is to mitigate harms by taking measures to accommodate the extreme phenotype: providing adequate, appropriate nutrition to avoid nutritional stress, providing suitable husbandry to reduce the risk of respiratory disease and heat stress and minimise the effects of locomotive disorders, and using elective caesarean to reduce calf loss due to dystocia (Fiems 2012).

However, given the welfare problems documented, others have argued that the breeding and importation of live animals and semen from breeds with undesirable genes – such as the Belgian Blue – should be banned altogether (Bassett 2009).

In all of the above cases, the role of the veterinary team is important in addressing proximate or immediate welfare issues (for example, performing a caesarean to avoid or relieve dystocia). However, without addressing longer-term or ultimate welfare issues there is a risk of the profession becoming complicit in "animal welfare abuse". According to the BVA's Animal Welfare Strategy:

> "Veterinary surgeons must, of course, treat breed-related health problems as they arise. But if we assist an animal to give birth, when the animal is otherwise incapable due to selective breeding, and we accept money for this without taking measures to address the underlying problem (e.g. reporting Caesarean sections undertaken in pedigree dogs to the Kennel Club), then it has been suggested that as a profession we are enabling poor animal welfare to persist and we are simply facilitating the status quo."
>
> (British Veterinary Association 2016)

BREEDING ANIMALS

What would you do?

You are part of a veterinary team providing services to a number of clients, including a Belgian Blue breeder. All calves on the property are delivered by elective caesarean section, with a high rate of survival for cows and calves. However, there are also high rates of infertility and you are concerned about the welfare of the animals. Your client is happy with his carcass yield at sales. How would you discuss the welfare of the cattle with the farmer?

△

15.13 How would you approach a discussion with a Belgian Blue breeder about improving the welfare of the cattle?
PHOTO ISTOCK

BREEDING ANIMALS

What would you do?

Scientists are developing gene-target technology as a means of creating new strains of animals with favourable traits. Preliminary work has shown some success in knocking out the myostatin gene in dogs, leading to double-muscling (Zou, et al. 2015), a trait which the authors suggest may be particularly helpful in dogs that herd livestock, hunt, guard homes, and perform police, rescue and assistance work. You consult for the regional police dog unit. Their commander reads about gene-target technology and seeks your advice about sourcing double-muscled dogs. What would you advise?

△

15.14 It has been suggested that police dogs could benefit from gene manipulation to produce double muscling.
PHOTO ISTOCK

BREEDING ANIMALS

What would you do?

You work with a pig-farming family whose husbandry is, in your view, barely acceptable. You audit the farm annually and each time they take some, but not all, advice, leading to a reasonably high rate of infectious disease. They've been offered breeding stock that, they are told, have proven lower rates of infectious disease, and want your advice about purchasing these animals. What factors would you take into consideration in advising your clients?

△

15.15 Infectious disease may be able to be reduced through selective breeding.
PHOTO ISTOCK

15.3

Artificial breeding techniques

Artificial breeding techniques are widely used in farm animals, in particular artificial insemination (AI), resulting in rapid changes in population phenotypes. By the turn of the twenty-first century 110 million female cattle (20 per cent of the global population) and 40 million female pigs (50 per cent) were bred via artificial insemination. This occurs less frequently in sheep and goats (Thibier, et al. 2004). The aim has been to improve productivity through such breeding, a benefit to humans, but there is the potential to improve welfare if certain welfare-protective traits are selected for, such as resistance to lameness in dairy cattle (D'Eath, et al. 2010). With artificial breeding technologies this could happen rapidly, within a few generations. But, an extension of either natural or artificial breeding may be the ability to produce animals that are better able to cope with their inadequate environments by being in some ways less reactive, zombie-like shadows of their former generations. D'eath, et al. (2010) caution against this; Sandøe, et al. (2014) explore the different philosophical origins of problems associated with breeding blind hens to improve welfare through reduced injurious pecking. While "disenhancing" hens may indeed violate their *telos*, Sandøe, et al. (2014) raise the question of whether this should matter if it improves their welfare.

What do you think?

One of the challenges for the dairy industry is the problem of unwanted "bobby" calves, which are often killed shortly after birth or farmed intensively for veal. The use of sexed semen is one potential solution, as it will ensure that only female calves are born. But there are some risks associated with the technology, including infertility, and it is currently expensive.

ONE   How would you advise a client seeking your opinion about the use of sexed semen in his herd?

△

15.16 Many male dairy calves are killed shortly after birth.
PHOTO ISTOCK

The manipulation of mouse genes isolated from stem cells of a blastocyst in the laboratory has led to the development of thousands of types of genetically modified mice, sometimes termed "knockout mice", used to determine the effects of particular genes on physiology and disease. Some people, and regulatory frameworks, use a utilitarian cost:benefit analysis to determine the acceptability of such procedures, weighing up the expected benefits, to humans mostly, and the harms to the mice suffered through housing, breeding, the results of the genetic disruption and any additional experimentation. Alternative concerns may stem from the unnaturalness of such a procedure rather than the negative welfare effects, as in one study people reported they were less willing in principle to support research using both genetically modified pigs and corn than non-modified equivalents (Schuppli & Weary 2010).

In our next two scenarios we will look specifically at the ethical issues surrounding a different form of artificial breeding technology, namely cloning.

SCENARIO
EQUINE CLONING

▶ You have been working in equine fertility practice for some years and have developed a good reputation in this area. Your clients aim to breed high-performing sport horses. You have helped them achieve their dreams of producing offspring that have competed at the highest levels, including the Olympic Games. You already offer a range of techniques including artificial insemination and embryo transfer and a company has approached you to see if you would be willing to partner with them to be the first centre offering cloning for horses in your country.

What should you do?

RESPONSE
MADELEINE CAMPBELL

▶ This scenario raises two major ethical questions. Firstly, there is a question of professional ethics. What exactly does the company mean by "partnering"? Secondly, there is the moral question of whether cloning of horses is ethical.

In relation to professional ethics, as an equine reproduction specialist I would want to know what exactly my involvement is expected to be. Would my role be simply to take skin biopsies which are to be used as a source of somatic cells by others

△
15.17 Knockout mice are used to determine the effects of particular genes on physiology and disease.

PHOTO ANNE FAWCETT

△
15.18 Successful showjumpers have already been cloned.
PHOTO ISTOCK

undertaking the "cloning" process in a specialised laboratory elsewhere, or am I expected to provide the laboratory services? If the latter, will the company provide me with the training necessary to satisfy me that I am suitably qualified and experienced to undertake somatic cell nuclear transfer (SCNT, commonly known as "cloning") as part of the service which I am offering to my clients? How would my clinical autonomy and my relationship with my existing clients be affected? Are there any potential conflicts of interest which might arise in the course of my more general equine reproduction practice as the result of "partnering" with the cloning company? How would the process of referral and reporting back between referring practices, myself and the cloning company operate? Additionally, I would want to review whether I considered the request of a client to clone their horse as ethically reasonable. Veterinarians have a primary duty to the welfare of their patients, and should not agree to undertake a procedure simply because the client wishes them to do so (the example of thermocautery of equine tendons springs to mind).

This leads on to the question of whether the cloning of horses is something which I consider ethical, and with which I would therefore be happy to associate myself. One starting point to answering this question is to consider whether equine cloning is inherently any more or less ethical than the other equine assisted reproductive techniques (ARTs) which I already provide for my clients, such as artificial insemination (AI) and embryo transfer (ET).

All ARTs differ significantly from every other veterinary technique except blood donation, in the sense that the procedure which is being undertaken does not confer any obvious benefit on the animal on which it is being performed. Even simple veterinary procedures such as vaccination involve a cost to the animal in terms of stress and mild discomfort. This is usually outweighed, in a cost:benefit analysis, by the perceived benefit to the animal (or group of animals) of the treatment. However, in the case of ARTs there is no perceived benefit to that animal (unless you believe that mares benefit psychologically from the process of giving birth and raising a foal; or you argue that the fact that she is carrying a valuable foal is likely to guarantee that the mare is treated better by her owners than she would otherwise be). Thus, every time I undertake any ART I am imposing a cost upon the mare(s) involved (Campbell & Sandøe 2015) with no anticipated benefit to those animals. This is no more true of cloning than it is of AI or ET. (The nature of the "cost" will vary with procedure. There is currently no evidence base for whether AI, ET or any other equine ARTs cause stress and pain. One assumes that more invasive procedures, for example oocyte retrieval, are likely to cause greater stress and pain than a simple transcervical AI.)

However, there is something fundamentally

different about cloning, since it aims to reproduce an existing animal, whereas all other ARTs aim to produce a novel animal. (In fact, because clones carry the mitochondrial DNA of the donor oocyte, they are not genetically identical to the "cloned" animal from which the somatic cells were taken.) Some consider that this difference is in itself sufficient to make cloning unethical – the idea of reproducing an individual seems to them to be beyond the limits of how far humans ought to interfere with nature, and morally repugnant (EFSA 2008). Such arguments are hard to refute, as they are based primarily in moral conscience.

Even if one does not see anything inherently morally wrong with the idea of recreating an individual, there may be other ethical reasons to refuse to become involved in equine cloning. One is the high rate of embryo wastage which the process of equine cloning involves (Galli, et al. 2008, Hinrichs 2005, Vanderwall, et al. 2006). Other equine ARTs (for example ET) inevitably involve some aspect of embryo loss when the process is not successful, but the rate of embryo loss during cloning is significantly greater. Should this be a matter of ethical concern? From an animal welfare point of view, the answer must currently be "no", since there is no evidence to suggest that early equine embryos are capable of suffering (Campbell, et al. 2014). If one believed that equine embryos had an inherent moral worth and a consequent right to life, then the high rate of embryonic loss would be of moral concern and a sufficient basis to stop cloning.

The need to respect the moral worth of non-sentient equine embryos is not an ethical argument which would itself dissuade me from becoming involved in equine cloning. However, what does seem to me to be deeply ethically relevant is the effect of the cloning process on the offspring which result from it. In other species such as cattle and sheep, it is known that offspring resulting from SCNT suffer a wide range of health (and thus of welfare) problems, ranging from embryonic abnormalities to foetal abnormalities (often associated with placental abnormalities), neonatal abnormalities and abnormalities in later life (Gjerris, et al. 2006, Houdebine, et al. 2008). These welfare issues are, in my view, enough to tip the balance of the cost:benefit ethical analysis in favour of not allowing cloning in the species for which such abnormalities have been demonstrated. (It is, however, worth noting that many of the abnormalities seem to relate to technique, and that the rate of abnormalities has decreased as techniques improve).

In horses, evidence about such abnormalities is sparse, and comes mainly from work published by one group relating to very small numbers of animals (Hinrichs & Choi 2015, Johnson, et al. 2010). Generally, foetal abnormalities seem to be fewer than in farm animal species. However, cloned foals do seem to require more neonatal intensive care than their non-cloned contemporaries. There have not, to date, been any long-term studies of the health of horses created using SCNT.

The combination of known abnormalities in other species and unknown abnormality rates in the short and long term for horses makes me wary of involving myself in equine cloning. Owners' reasons for requesting cloning are either sentimental (wanting to "recreate" a much-loved horse) or economic (wanting to benefit financially either in terms of stud fees or, less commonly, of competition earnings). The uncertainty about equine welfare costs in terms of abnormalities and poor health in foals created using SCNT in my view currently outweighs such emotional and financial benefits to people. I would not be happy to involve myself in equine cloning until such time as evidence becomes available to demonstrate that short-, medium- and long-term health problems are not significant in horses created using SCNT. Such evidence may be dependent upon improvements in technique, or it may simply require well-executed studies on adequate numbers of animals.

MANY of the problems associated with cloning are likely related to current techniques, which may diminish as techniques improve. The author has argued elsewhere that, paradoxically, techniques are likely to improve only if cloning continues (Campbell 2016). Accordingly, the onus is on providers of equine cloning technologists to provide a stronger evidence base, based on data about the short-, medium- and long-term health and welfare of equine clones (Campbell 2016). This process should involve not just those providing the cloning service, but data provided by veterinarians and owners of cloned horses.

Others may object to the cloning of animals on the grounds that it violates the *telos* of an animal, or violates the dignity of an animal. According to the author, our current use of animals renders these weak ethical arguments (Campbell 2016).

One area that may be explored further is client expectation: why are they seeking to clone this animal in the first place? It is argued that while cloning can produce a *genetically* identical animal, it can never produce a *phenotypically* identical animal due to the influence of variable environmental factors. Clients looking to replicate sporting success by cloning an equine athlete may be disappointed, as there is no evidence that cloned horses have a competitive advantage (Campbell 2016). Veterinarians have a professional obligation to inform clients and manage their expectations.

The following scenario explores whether cloning could be useful for those seeking to replace a beloved companion.

SCENARIO
CLONING DOGS

▶ Mr and Mrs P belong to the class of Google-assisted clients. When advised that their beloved dog Marmaduke is in the terminal stages of kidney failure, they announce that they have discovered a company in Korea who operate a successful cloning service that could recreate Marmaduke, in form and in character. All you, their vet, have to do is take a few biopsies and arrange shipment to Korea. They clearly expect you to proceed.

Should you?

△

15.19 Mr and Mrs P seek your assistance in cloning their beloved dog Marmaduke.
PHOTO ANNE FAWCETT

RESPONSE

JOHN WEBSTER

▶ Although this scenario is hypothetical, it is based on fact. A company, Sooam, for a mere US$ 100,000 will clone dogs from biopsy material (Taylor 2016), even offering a competition with a prize of free cloning (Sooam 2016)! The evidence to date is sparse but indicates that cloned puppies carried to term in surrogate mothers are healthy and do not suffer from developmental problems similar to those reported from bovine calves cloned by nuclear transfer.

Marmaduke may therefore be recreated as a cloned puppy without apparently causing any sentient (post-natal) dog to suffer. This creates for GPs a new question of practical ethics: "Should I assist my clients in their aim to clone their beloved dog or, if not, why not?"

> "It is necessary, first, to acknowledge that an act does not become unethical simply because you, or I, or the majority, deem it to be tasteless, self-indulgent or just plumb crazy."

It is necessary, first, to acknowledge that an act does not become unethical simply because you, or I, or the majority, deem it to be tasteless, self-indulgent or just plumb crazy. It is valid to question the motives of someone who wishes to spend over US$ 100,000 just to obtain a dog that's like one they had before, when they could pick up an equally charming individual from a rescue centre. It is likely that for many such the motivation would be to create the canine equivalent of a Fabergé egg: a shining example of their capacity for conspicuous consumption. On the other hand, they *could* be motivated by pure love, which would present a stronger case than, for example, that of an owner who sought to clone his highly successful (but gelded) eventing horse to produce a stallion with a high breeding value.

Two of the central pillars of practical ethics are "Do no harm" and "Respect the rights of the individual". Neither principle can be addressed in isolation. In this case, the owner can claim the right to clone their pet, provided it does not cause harm, or compromise the right to life of any other animal. To date, Sooam claim to have produced 700 cloned puppies, and there is no evidence that the puppies are abnormally prone to disease or developmental abnormalities. While I am of the firm belief that kidney transplants in cats are unethical because they are likely to add to the sum of suffering in geriatric pets, I could only marshal an ethical case against cloning pets if I could be persuaded that some pets were coming to harm. Would I do it myself? No, because I balk at pandering to the self-indulgence of clients with more money than sense. But that is one reason why I am not a pet vet.

||||||||||||||||||||||||||||||||||||||||||||||||||||||||||||

BOTH responses to the preceding scenarios consider it important to weigh up the costs and the benefits of cloning. John Webster uses a principalist approach, in particular focusing on non-maleficence and autonomy. It could be argued that any discomfort or pain suffered by the surrogate, as well as embryo loss, violates the principle of non-maleficence.

Whilst it's not clear how members of the public frame their reasons for opinions on animal cloning, in the USA the proportion of polling respondents who disapprove of the practice remained consistent at around 60 per cent in various polls between 1997 and 2013. Women and religious people tended to be more likely to be against cloning and support was lower for pet cloning, with 80 per cent of people disapproving, but cloning was

15.3

ARTIFICIAL BREEDING TECHNIQUES
RESPONSE
JOHN WEBSTER

better accepted in some polls for research animals (Center for Genetics and Society 2014).

In the above case, is there a risk that the grieving owners are being exploited? After all, no procedure can ever bring back *their* Marmaduke. While cloning can produce a *genetically* identical animal, it can never produce a *phenotypically* identical animal due to the influence of variable environmental factors. Clients looking to replicate their beloved pet may be surprised when it looks and behaves very differently. The first cloned cat, Copy Cat, was tabby and white, whilst her genetic "donor", Rainbow, was a tortoiseshell/calico.

Likewise those looking for sporting success by cloning an equine athlete may be disappointed, as there is no evidence that cloned horses have a competitive advantage (Campbell 2016).

If, in the process of cloning, the company creates 2 or 3 or 10 "Marmadukes", what happens to these animals? How might the owners feel if these animals are sold to other families or even companies? Do Mr and Mrs P have an interest in owning Marmaduke's genotype?

Veterinarians have a professional obligation to inform clients and manage their expectations accordingly.

△

5.20a–b The first cloned cat, "Copy Cat", was tabby and white; her "donor", "Rainbow", was calico.
PHOTO ISTOCK

ARTIFICIAL BREEDING TECHNIQUES

What would you do?

Mr D, a beef farmer, approaches you to ask your advice about cloning one of his best bulls. In the USA, where you work, approximately 1000 beef bulls have already been cloned. Although it's not currently likely to be profitable Mr D has always been an "early adopter" of new farming techniques and has heard this could be the future of farming. Mr D would like you to work closely with the cloning company to achieve his first clone. What do you do?

△

15.21 Mr D is considering cloning one of his best Hereford bulls.
PHOTO ISTOCK

15.4

Changing animal behaviour

Animal behaviour is the product of genes ("nature"), environmental factors ("nurture") and previous experience or learning. For sentient animals, behaviour can lead directly or indirectly to pleasant or unpleasant experiences. Humans have attempted to influence animal behaviour, for example through domestication of animals. In addition, there are many attempts to influence the behaviour of wild animals, both directly and indirectly, for example trying to prevent elephants from foraging in crops by using beehive "fences", sounds of tiger calls or chilli rope (Dasgupta 2014).

Animal behaviour, whether influenced by humans or not, is shaped by internal and external drivers and the feelings they induce. Animal

△
15.22 Animals such as these elephants are trained to provide entertainment, often employing techniques that are considered inhumane.
PHOTO ANNE FAWCETT

training utilises an animal's motivation to avoid negative states, such as fear or pain, but there is evidence that methods based on tapping into their motivation for positive experiences are more effective and reduce behavioural "problems" for owners (Hiby, et al. 2004, Hockenhull & Creighton 2013). In this next scenario we consider some of the ethical issues surrounding shaping animal behaviour using positive, reward-based methods.

SCENARIO
TRAINING ANIMALS

▶ You have always been interested in behaviour problems and training and since you've been doing this work professionally you feel you've had some great responses and really helped owners and their animals improve their relationship and welfare. You've also had clients that have gone on to use your reward-based clicker training methods and taken them further than you ever have. You can see some of their videos on YouTube: dogs that bring tissues to sneezing owners, horses that will take a bow and even a parrot that will ride in a toy electric car!

Are you right to feel uneasy about the power of this type of training?

RESPONSE
SUE HORSEMAN

▶ Clicker training, as a form of reward-based training, is a relatively well-established training method within the dog world, but has also grown in popularity as a training method for other species, including parrots and horses. I will discuss some of the ethical considerations associated more generally with clicker training and where

15.4
CHANGING ANIMAL BEHAVIOUR
RESPONSE
SUE HORSEMAN

△
15.23 Horse trained to take a bow.
PHOTO ISTOCK

appropriate link this to particular considerations for its use in equids, as this is where I have personal experience of using clicker training.

"Traditional" horse training methods are based on pressure and pressure release, where the pressure applied encourages an action by the horse, which when performed results in the pressure being relieved, for example, applying tension to a lead rope to encourage a static horse to walk forward and allowing the rope to go slack once a step is taken. More recently clicker training has been promoted as a "non-harmful" approach to training horses. It is argued that, in contrast to pressure and pressure release training systems, when using clicker training there is no negative "cost" to the horse when the desired behaviour is not performed. During clicker training, a horse not performing the desired behaviour is simply ignored, i.e. they are simply not rewarded.

Whilst on the surface this "non-reward" may appear a "neutral" experience for a horse, a horse may get frustrated during the training session if, for example, they cannot work out which behaviour is needed to gain the reward or if the reward is poorly timed (Hart 2008). As such, there is the potential for negative emotional states to be elicited during clicker training and when done incorrectly, clicker training offers just as much opportunity to cause mental harm to a horse as other more "traditional" training methods. Avoiding negative emotional states during training involves good timing on the part of the trainer and setting up training scenarios where it is easy for the horse to succeed.

There is some evidence that reward-based training can be a positive experience for animals. For example, Melfi (2013) reported that training of zoo animals can be enriching for the animals if the consequence of the training is considered by the animal to be enriching. In the context of clicker training this suggests that it can be a positive experience if the reward is valued by the animal. Based on anecdotal evidence and personal experience I believe that horses can "enjoy" clicker training sessions when these are carried out well (i.e. when the horse does not become frustrated) and particularly where the reward is valued by the horse. There may also be wider benefits to the horse whose owner trains them using clicker training, beyond any immediate enjoyment gained. Through the process of using reward-based training the owner may build a more positive relationship with their horse, may learn what motivates their horse and can learn how to use the method "well" in terms of shaping behaviours and correct timing. They will then be able to apply these skills to many training contexts and due to the better relationship with their horse, they will have more enjoyable experiences with them. This I would argue not only benefits the owner but also the horse.

Reward-based training of horses offers a unique opportunity to promote the autonomy of the horse when compared to training methods based on pressure. For example, the horse may have more choice about whether or not to perform the given behaviour and can even choose whether or not to engage with training at all. Consider the scenario where the owner has a clear signal for the horse that a training session is possible. For example, the owner may be wearing a bum bag that the horse associates with training. If the owner enters the field, the horse sees the bum bag and then chooses to approach the owner, then we could say that the horse was "choosing" to participate in the training session. Of course, the horse's autonomy is only preserved if, when the horse decides not to come over, the owner does not then catch the horse using more conventional pressure-driven means. We also need to consider the context in which these decisions are made. If, for example, the horse is hungry, they may choose to come and train for a food reward, but there is a degree of unfairness in the choice.

So far we have considered the use of clicker training in general terms without considering what is being trained. In horses, clicker training can be used to train "necessary" processes such as picking the horse's feet out, or to teach the horse to perform "tricks" such as bowing. It is possible to argue that the benefits to the animal are greater when clicker training is used to train tasks that "need" to be done. For example, when used to teach horses to pick their feet up, a task that they would "have" to learn how to do, the horse avoids being trained with methods involving pressure or punishment and possibly has the added benefit of a "reward" for performing the correct behaviour. When trained to perform tricks, which the horse possibly wouldn't be trained to do if the owner wasn't using positive training methods, the benefits are less clear as there is no avoidance of other training methods. Here, the "enjoyment" and wider benefits to the horse become key. If the horse does not find the "training" itself rewarding, and perhaps if given a fair choice would choose not to participate, and there are no wider benefits to the horse, then there is little justification for training "unnecessary" behaviours.

However, some wider benefits, specifically associated with training horses to do tricks, should be highlighted. It is possible that some owners may be reluctant to use clicker training to teach "necessary" behaviours as other approaches are more established and seen as more "acceptable" by the wider equine community. As such, the training of "tricks" offers an opportunity for owners to try out the approach without the same type of scrutiny and judgement as might occur if they were training other behaviours. Therefore training the horse in "tricks" offers owners a more comfortable way of learning the new method which they can then apply to other contexts. It also provides a good forum for promoting the approach and engaging others in exploring the technique. The training of "tricks" by one horse owner may instigate the use of positive training methods by other horse owners. Again, this is only a solid justification for training "tricks" if frustration is avoided and if the horse values the training "reward".

I believe that incorrect usage of clicker training has just as much chance of causing "harm" to horses as other training methods. If the training causes frustration, the horse is trained to do something it finds physically difficult, the horse does not "enjoy" the training session and/or there are no wider benefits to the horse, then the training of tricks becomes hard to justify. However, when used correctly clicker training offers more opportunity for training to be a positive experience for the horse. I have seen first-hand how training horses to do "tricks" can have wider benefits to the horse and the horse–human relationship. As such I do not think that, providing the horse finds the process rewarding and negative emotional states are avoided, there is anything wrong with training horses to do tricks.

||||||||||||||||||||||||||||||||||||||||||||||||||||||||||||||||||

IN the above scenario Sue Horseman's approach aligns most with principalism, in that there is an emphasis on minimising harm done by training, on promoting good and on respecting the autonomy of the animal. In such a case it is important that the animal is trained in such a way that its interests (basic healthcare or play/enjoyment) can be met. Deontological objections are not raised, yet some people feel that training animals to do crass circus tricks is just wrong, appealing to the notion of it being unnatural and disrespectful of *telos* or intrinsic value. These arguments do not consider the consequences of the training, whether the animal enjoys it or whether people derive any benefit from doing or watching it.

CHANGING ANIMAL BEHAVIOUR

What would you do?

A client of your practice owns a Jack Russell terrier that incessantly chases and attacks snakes, some of which are venomous. The dog has been treated for snake bite envenomation on two occasions. On the second occasion the dog had an anaphylactic reaction to anti-venom and required cardiopulmonary resuscitation. The owner lives in an area where snakes are common, but she is keen to do everything she can to protect her dog from another bite. Recently she has engaged a self-taught trainer who uses aversion techniques, including electric-shock collars and exposure to "de-fanged" snakes, to train her dog to avoid snakes. You are concerned about the welfare of both the dog and the snakes, yet the client feels this is the only way to save her dog from a life-threatening snake bite. What do you do?

△
15.24 How would you go about modifying the behaviour of a dog that incessantly chases and attacks venomous snakes?
PHOTO ANNE FAWCETT

Conclusion

The issue of changing animals is of particular societal concern, with legislation governing mutilations and reproductive techniques reflecting societal views. For this subject in particular there appear to be relatively popular deontological arguments against the unnatural and disrespectful elements of permanently changing animals. For others a utilitarian analysis suffices to derive their views on these matters. Given the general interest in the topic it is right that the views of society are taken into account to determine the correct level and type of regulation for procedures, and that such a process is transparent to all.

References

American Association of Feline Practitioners 2015 Feline focus. *Journal of Feline Medicine and Surgery* **17:** 829–830.

Bassett A 2009 Animal Welfare Approved Technical Advice Fact Sheet No. 1: Welfare and Belgian Blue Cattle. Animal Welfare Approved: Marion, USA.

BBC 2008 Pedigree Dogs Exposed, 19 August.

Bennett P, and Perini E 2003a Tail docking in dogs: a review of the issues. *Australian Veterinary Journal* **81:** 208–218.

Bennett P, and Perini E 2003b Tail docking in dogs: can attitude change be achieved? *Australian Veterinary Journal* **81:** 277–282.

British Charolais Cattle Society 2016 Breed description. http://www.charolais.co.uk/society/breed-description/

British Veterinary Association 2016 Vets speaking up for animal welfare: BVA animal welfare strategy. BVA.

Cameron N, Lederer R, Bennett D, and Parkin T 2014 The prevalence of tail injuries in working and non-working breed dogs visiting veterinary practices in Scotland. *Veterinary Record* **174:** 450.

Campbell MLH 2016 Is cloning horses ethical? *Equine Veterinary Education.*

Campbell MLH, Mellor DJ, and Sandøe P 2014 How should the welfare of fetal and neurologically immature postnatal animals be protected? *Animal Welfare* **23:** 369–379.

Campbell MLH, and Sandøe P 2015 Welfare in horse breeding. *The Veterinary Record* **176:** 436–440.

Center for Genetics and Society 2014 Animal and Pet Cloning Opinion Polls. Center for Genetics and Society.

Cheng H 2007 Animal welfare: should we change housing to better accommodate the animal or change the animal to accommodate the housing? *CAB Reviews: Perspectives in Agriculture, Veterinary Science, Nutrition and Natural Resource* **2:** 14 pp.

Cornell University College of Veterinary Medicine 2014 Breeding a better dog. http://www.vet.cornell.edu/research/bvhip/

D'Eath RB, Conington J, Lawrence AB, Olsson IAS, and Sandøe P 2010 Breeding for behavioural change in farm animals: practical, economic and ethical considerations. *Animal Welfare* **19:** 17–27.

Dasgupta S 2014 How to scare off the biggest pest in the world. BBC, 7 December.

Diesel G, Pfeiffer D, Crispin S, and Brodbelt D 2010 Risk factors for tail injuries in dogs in Great Britain. *Veterinary Record* **166:** 812–817.

EFSA 2007 The risks associated with tail biting in pigs and possible means to reduce the need for tail docking considering the different housing and husbandry systems. *The EFSA Journal* **611:** 1–13.

EFSA 2008 Outcome of public consultation on the EFSA Draft Animal Cloning Opinion. http://www.efsa.europa.eu/en/supporting/pub/834

European Union 2001 Council Directive 2001/88/EC amending Directive 91/630/EEC laying down minimum standards for the protection of pigs.

Eurostat 2015 Agricultural production – animals. http://ec.europa.eu/eurostat/statistics-explained/index.php/Agricultural_production_-_animals

Fiems LO 2012 Double muscling in cattle: genes, husbandry, carcasses and meat. *Animals (Basel)* **2:** 472–506.

Fikse WF, Malm S, and Lewis TW 2013 Opportunities for international collaboration in dog breeding from the sharing of pedigree and health data. *The Veterinary Journal* **197**: 873–875.

Fitzgerald M 1994 The neurobiology of fetal pain. In: Wall PD and Melzack R (eds) *The Textbook of Pain 3rd Edition*. Churchill-Livingstone: Edinburgh.

Fitzgerald M, and Koltzenburg M 1996 The functional development of descending inhibitory pathways in the dorsolateral funiculus of the newborn rat spinal cord. *Developmental Brain Research* **24**: 261–270.

Galli C, Lagutina I, Duchi R, Colleoni S, and Lazzari G 2008 Somatic cell nuclear transfer in horses. *Reproduction in Domestic Animals* **43**: 331–337.

Gjerris M, Lassen J, Meyer G, and Tveit G 2006 Ethical aspects of farm animal cloning. A synthesis report, 1–16. http://curis.ku.dk/portallife/files/8026132/Cloning_in_Public-report

Harfeld JL 2013 Telos and the ethics of animal farming. *Journal of Agricultural & Environmental Ethics* **26**: 691–709.

Hart B 2008 *The Art and Science of Clicker Training for Horses. A Positive Approach to Training Equines and Understanding Them*. Souvenir Press: London.

Hedhammar AA, Malm S, and Bonnett B 2011 International and collaborative strategies to enhance genetic health in purebred dogs. *The Veterinary Journal* **189**: 189–196.

Hiby EF, Rooney NJ, and Bradshaw JWS 2004 Dog training methods: their use, effectiveness and interaction with behaviour and welfare. *Animal Welfare* **13**: 63–69.

Hinrichs K 2005 Update on equine ICSI and cloning. *Theriogenology* **64**: 535–541.

Hinrichs K, and Choi HY 2015 Health of horses produced by ART. In: *AAF IETS Equine Reproduction Symposium*, 42–43. IETS: Paris.

Hockenhull J, and Creighton E 2013 Training horses: positive reinforcement, positive punishment, and ridden behavior problems. *Journal of Veterinary Behavior-Clinical Applications and Research* **8**: 245–252.

Houdebine L-M, Dinnyés A, Bánáti D, Kleiner J, and Carlander D 2008 Animal cloning for food: epigenetics, health, welfare and food safety aspects. *Trends in Food Science & Technology* **19**: S88–S95.

Johnson AK, Clark-Price SC, Choi YH, Hartman DL, and Hinrichs K 2010 Physical and clinicopathologic findings in foals derived by use of somatic cell nuclear transfer: 14 cases (2004–2008). *JAVMA-Journal of the American Veterinary Medical Association* **236**: 983–990.

Kuhlmann-Eberhart I, and Blaha T 2009 Codex Veterinarius der Tierärztlichen Vereinigung für Tierschutz e. V. (TVT). Ethische Leitsätze für tierärztliches Handeln zum Wohl und Schutz der Tiere.

Lederer R, Bennett D, and Parkin T 2014 Survey of tail injuries sustained by working gundogs and terriers in Scotland. *Veterinary Record* **174**: 451.

Lewis T, Rusbridge C, Knowler P, Blott S, and Woolliams JA 2010 Heritability of syringomyelia in Cavalier King Charles spaniels. *The Veterinary Journal* **183**: 345–347.

Melfi V 2013 Is training zoo animals enriching? *Applied Animal Behaviour Science* **147**: 299–305.

Meuwissen THE, Hayes BJ, and Goddard ME 2001 Prediction of total genetic value using genome-wide dense marker maps. *Genetics* **157**: 1819–1829.

Morton DB 1992 Docking of dogs: practical and ethical aspects. *Veterinary Record* **131**: 301–306.

Mullan S 2010 The ethics of the management of diseases present at birth: individuals versus populations. *Scientific Meeting of the Association of Veterinary Soft Tissue Surgeons*, 1–2 October 2010.

Nicholas FW 2011 Response to the documentary Pedigree Dogs Exposed: three reports and their recommendations. *The Veterinary Journal* **189**: 126–128.

Nicholas FW, and Wade CM 2011 Canine genetics: a very special issue. *The Veterinary Journal* **189**: 123–125.

Ormandy EH, Schuppli CA, and Weary DM 2009 Worldwide trends in the use of animals in research: the contribution of genetically-modified animal models. *ATLA-Alternatives to Laboratory Animals* **37**: 63–68.

Palmer C 2012 Does breeding a bulldog harm it? Breeding, ethics and harm to animals. *Animal Welfare* **21**: 157–166.

Palmer C, Corr S, and Sandøe P 2012 Inconvenient desires: should we routinely neuter companion animals? *Anthrozoos* **25**: S153–S172.

Rodenburg TB, van Krimpen MM, de Jong IC, de Haas E, Kops MS, Riedstra BJ, Nordquist RE, Wagenaar JP, Bestman M, and Nicol CJ 2013 The prevention and control of feather pecking in laying hens: identifying the underlying principles. *World's Poultry Science Journal* **69**: 361–373.

Rollin BE 2006 *Animal Rights & Human Morality.* Prometheus Books: New York.

Sandøe P, Hocking PM, Forkman B, Haldane K, Kristensen HH, and Palmer C 2014 The blind hens' challenge: does it undermine the view that only welfare matters in our dealings with animals? *Environmental Values* **23**: 727–742.

Schuppli CA, and Weary DM 2010 Attitudes towards the use of genetically modified animals in research. *Public Understanding of Science* **19**: 686–697.

Sooam 2016 The 1st Dog Cloning Competition UK. http://en.sooam.com/dogcn/sub06.html

Taylor D 2016 British couple celebrate after birth of first cloned puppy of its kind. *The Guardian*, 26 December.

The Kennel Club 2015 Mate select – estimated breeding value. http://www.thekennelclub.org.uk/services/public/mateselect/ebv/Default.aspx

The Kennel Club 2016 Caesarean operations and procedures which alter the natural conformation of a dog.

Thibier M, Humblot P, and Guerin B 2004 Role of reproductive biotechnologies: global perspective, current methods and success rates. In: Simm G, Villanueva B, Sinclair KD, and Townsend S (eds) *Farm Animal Genetic Resources, 25-27 November, 2002.* BSAS: Edinburgh.

Thomson PC, Wilson BJ, Wade CM, Shariflou MR, James JW, Tammen I, Raadsma HW, and Nicholas FW 2010 The utility of estimated breeding values for inherited disorders of dogs. *The Veterinary Journal* **183**: 243–244.

Vanderwall DK, Woods GL, Roser JF, Schlafer DH, Sellon DC, Tester DF, and White KL 2006 Equine cloning: applications and outcomes. *Reproduction, Fertility and Development* **18**: 91–98.

Webster AJF 2013 *Animal Husbandry Regained: The Place of Farm Animals in Sustainable Agriculture.* Earthscan, Routledge: London.

Wilson BJ, and Nicholas FW 2015 Canine hip dysplasia – towards more effective selection. *New Zealand Veterinary Journal* **63**: 67–68.

Wilson BJ, and Wade CM 2012 Empowering international canine inherited disorder management. *Mammalian Genome* **23**: 195–202.

Zou Q, Wang X, Liu Y, Ouyang Z, Long H, Wei S, Xin J, Zhao B, Lai S, Shen J, Ni Q, Yang H, Zhong H, Li L, Hu M, Zhang Q, Zhou Z, He J, Yan Q, Fan N, Zhao Y, Liu Z, Guo L, Huang J, Zhang G, Ying J, Lai L, and Gao X 2015 Generation of gene-target dogs using CRISPR/Cas9 system. *Journal of Molecular Cell Biology* **7**: 580–583.

FINAL WORD

"The worst sin towards our fellow creatures is not to hate them, but to be indifferent to them, that's the essence of inhumanity."
George Bernard Shaw, Irish playwright and essayist

When we asked our contributors to respond to scenarios, we gave them free rein to answer as they liked. Their voices have emerged loud and clear. Whilst many would agree with each other on some issues they are certainly not all singing from the same song sheet. We deliberately offer a diverse set of views and approaches to ethical problems. But, that doesn't mean that there is no right answer. Each author has written what is right for them. It's up to us to agree or disagree and try to offer valid reasons as to why we do.

In response to the question "What makes an ethical decision different from any other?" the commonest response we experience during undergraduate and postgraduate veterinary ethics teaching is, "There's no right or wrong answer." This isn't the place to discuss the various merits of relativism, where truths or moral principles are always framed within a context, versus realism, where truths exist independently of the characteristics of the observer and other philosophical paradigms. What is more important for practical veterinary ethics is to recognise that there *is* a right, or preferred, answer for each individual person in each case. There may also be more than one acceptable answer or action. Reflecting on what is important to us, our values, can help us identify the right thing to do.

Value conflicts are a source of stress for people in all sorts of jobs and it's not surprising that those involved in providing veterinary services also find ethical conflicts stressful. Using frameworks can be particularly helpful in understanding other stakeholders' positions and identifying areas of agreement and disagreement on which to focus a resolution process. Respecting that other people can have different values and then challenging our own values in the light of counterarguments ensures that our values remain robust.

Our contributors have used a range of frameworks in addressing the scenarios, with most emphasising a utilitarian approach, weighing up the costs and benefits of various options to find the best outcome. Protection of individual rights or interests was sometimes achieved through combination with a deontological "bottom line", consistent with the common morality approach. Utilitarian approaches are commonly employed by the wider population but there may be gender, cultural and other differences, depending on the nature of the scenario, which affect their appeal.

Few contributors invoked an entirely deontological argument but where it was employed it tended to be in response to scenarios that encapsulated a fundamental principle about animal use, care or treatment. Virtue ethics was prominent in some responses, perhaps unsurprisingly given its revival

◁
End of day.
CARTOON AILEEN DEVINE

in medical ethics. Some of the virtues proposed included responsibility, discernment, trustworthiness, wisdom, integrity, fairness, kindness and compassion. There were appeals to care ethics in some responses, where close relationships are valued highly, for example between a human and their companion animal.

Several contributors used a framework to try to encapsulate the interests of a range of stakeholders. Some directly applied the four principles of medical ethics: non-maleficence, beneficence, autonomy and justice, whilst others used the adapted form of these in an ethical matrix considering effects on wellbeing, choice and fairness for all. But no framework yields the answer by itself. At some point, one's values have to be overlaid to give appropriate weights to each element and enable the preferred option to be identified. We may not be aware of our values until they are challenged.

Our contributors have provided detailed analyses, often over and above the type of analysis one may be able to carry out on the spot in a busy work setting. We hope, however, that these examples will challenge you to examine your own decision-making. Whatever sphere veterinarians and associated practitioners work in we can have a huge impact on the wellbeing of others in a way that is rare in many jobs. With that power comes the responsibility to exercise it ethically, with animal welfare our central concern.

CARTOON FRANKO, GRRINNINBEAR.COM.AU

INDEX